Management System Dynamics

Management System Dynamics

R. G. Coyle
System Dynamics Research Group
University of Bradford

A Wiley–Interscience Publication

JOHN WILEY & SONS
London · New York · Sydney · Toronto

Library of Congress Cataloging in Publication Data:
Coyle, R. G.
Management system dynamics.

A Wiley–Interscience publication.
Bibliography: p.
1. System analysis. 2. Industrial management–Mathematical models. I. Title.
HD20.5.C67 658.4'032 76-40144

ISBN 0 471 99444 8 (Cloth)
ISBN 0 471 99451 0 (Paper)

Typeset in Great Britain by Preface Ltd. Salisbury, Wilts. and printed by The Pitman Press, Bath

For

Julie, Jonathan, and Robert

Preface

All organizations – business firms, cities or economic systems – are liable to be diverted from their chosen paths of growth or stable operation by the shocks which fall on them from the outside world, or by pressures generated within the organization. The 'shock' may be an opportunity, such as a firm attempting to react to an enlarged market, or it may be unpleasant, as when changes in international market prices affect the balance of trade. The shocks may be unexpected, foreseen, or self-inflicted, but the essential problem for managers is that of controlling the organization so as to take advantage of favourable opportunities while defending it against unpleasant upsets.

This book covers the principles and techniques for applying control concepts to business and economic problems. The subject is called 'System Dynamics' or 'Industrial Dynamics'. It owes a good deal to Control Theory, and other disciplines, but has important differences from them. System Dynamics is an extension of the existing area of Management Science. It addresses problems of controllability and integrated policy design, but it is not an optimizing approach, nor does it cope with the stochastic processes of queues and decision theory.

The book is intended for practising management scientists, corporate planners and economists who wish to know what System Dynamics is about, and if and how to use it in practice. It should also be useful to students of management, management science, economics, and other disciplines who require training in dealing with problems of controllability and policy design. Carefully used, i.e. without Chapter 4 to 8, it should benefit those students of management and administration who require the broader view of policy-making and its effects.

The economics of book publishing have prevented the inclusion of some non-essential material which it would have been useful to have covered. In particular the number of case studies has had to be restricted and those which are included are drawn from the business firm. It is hoped that readers who are interested in other areas will be able to visualize the applications, perhaps with the aid of the bibliography in Appendix C. The techniques which have been treated should, however, be sufficient to carry the practitioner through most of the problems he is likely to meet in the real world.

The material has been arranged so that the reader can study only those parts of the text which relate to his needs without losing the thread of the argument. New ideas are brought in carefully as they are needed, and the reader who invests the necessary effort should be able to follow it all.

The principal requirements for successful work in System Dynamics is understanding of the way in which the organization actually works, rather than advanced mathematical skill. For this reason the techniques described here are mathematically rather simple, so they should be easily learned by readers who have not had training in the mathematics of control systems.

The student who lacks practical experience may be led to think that, because the techniques are comparatively simple, so are the problems. This is not the case. The experienced management scientist will recognize that we are dealing with problems which come near to the very heart of the managerial task.

The application of System Dynamics is not limited to the type of problem described in the case studies. A careful reading of Chapters 1, 9 and 13 will suggest how to identify problems suitable for dynamic analysis.

Professor J. W. Forrester of the Massachusetts Institute of Technology was the originator of System Dynamics. I am grateful to him for initiating me into System Dynamics and for sparking my own interest by his powerful intellect. Dr. Carl Swanson, lately of MIT, influenced my early work in System Dynamics by his own enthusiasm and patient help.

My colleagues Dr. John Sharp and Ajita Ratnatunga have been an enormous source of help and criticism both in the writing of this book and in building up the Bradford System Dynamics Group. I am deeply grateful to them, and to the students who have tested the material in teaching. Mrs. Margaret Swift, Christine Packer and Brenda Akers typed the manuscript from my horrible handwriting and, in particular, Mrs. Akers bore the brunt of completing it under pressure, and the tedium of its revision.

Contents

Hints on Using the Book

The book has been written so as to be flexible in use for teaching or private reading. Experience of using it as a teaching vehicle suggests several alternatives:

1. For a quick general view of the nature and applicability of system dynamics read chapters 1, 9 and 13, glancing at Chapter 10 to 12.
2. A first course in the techniques can be based on Chapters 1 to 8 supplemented by careful use of parts of Appendix B for more examples of equation formulation.
3. The practical application can be studied via Chapters 9 to 13 simply reading the case studies but not running the models in Appendix B.
4. To improve the student's facility in dealing with complex models Chapters 10 and 11, and Appendix B.10 and B.11 can be used to set up and run models for student experimentation.
5. A course in corporate modelling, to link with a Business Policy course, can be based on Chapters 9 and 12. The model of Chapter 12 can be used to allow the student to experiment with policies with either the instructor or the student carrying out the model runs as appropriate.

The programs in Appendix B are given in full to provide either many examples of equation formulation (which is the greatest hurdle encountered by the student of mathematical modelling), or vehicles for model handling and experimentation depending on available time and achieved skill. There is a lot of material in the book and the reader should not attempt to digest it all within a term. Ideally it should become a work to which he refers from time to time as the need arises.

Suitable problems with solutions, and further case material will be found in R. G. Coyle and J. A. Sharp, *Problems and Cases in System Dynamics* which has been prepared as a companion to this book.

How to Read the Computer Graphs

The book contains a large number of graphs illustrating system behaviour. The graphs were produced directly on the CALCOMP Plotter of Bradford's ICL

Computer and I am grateful to Mr. A. K. Ratnatunga for developing the software to do this. It is, however, necessary to understand how the graphs are interpreted and the reader should make sure he is clear about this before trying to make sense of the system performance shown by the graphs.

Each graph contains the dynamics of one or more system variables on a horizontal scale of time. Time is measured in weeks or months as appropriate to the system, and this time unit is printed under the horizontal scale.

Each variable has its own line pattern, for example, a solid line or one of several variations on chain dotted or dashed lines. The same pattern is used to draw the vertical scale for that variable and is continued below the vertical scale and then to the right until it points to the variable name and definition which appears under the time axis. For example, in Fig. 2.19 Production Ordering Rate is shown by a solid line which is the first scale encountered as one reads to the left. The solid line is then carried down in an 'L' shape until it points to the variable name ORATE and then the text definition of this is the Production Ordering Rate, measured in units per week. Similarly, Inventory has the variable name INV and is plotted as a chain dotted line which appears as the lowest of the curves, and at the extreme left hand of the scales. The first three variables in Fig. 2.19 are plotted to a common scale so that the three patterns representing them are drawn close together with common numbering on the vertical axis.

This CALCOMP Package is a standard feature of the DYSMAP language described in Chapter 4 and has flexible options for common or individual scaling as appropriate to the problem.

Chapter 1

Introduction to System Dynamics

1.1 The Basic Problem

All business firms, economies and social organizations show DYNAMIC BEHAVIOUR. That is, as time passes, the variables by which we measure their condition (such as sales, profits, stocks, balance of payment, unemployment, and many others), fluctuate noticeably, sometimes alarmingly as when cash reserves fall, and sometimes, of course, gratifyingly such as when profits rise. Inhibiting the undesirable variations, and/or promoting the beneficial ones, are objectives for the managers concerned, and this book describes the analysis of such problems. First we show, by an example, that there is a problem to analyse.

A firm holds an inventory of finished goods, which is depleted by sales and replenished by production. Such inventories rise and fall, frequently by serious amounts. It might be argued that these variations are simply due to external fluctuations in sales. The subject of System Dynamics exists because this supposition is not necessarily correct.

Let INV.START and INV.END be the inventory at the beginning and end of a time period, a week perhaps. Let PROD denote the amount manufactured in the period and SALES the quantity sold, then

$$\text{INV.END} = \text{INV.START} + \text{PROD} - \text{SALES}$$

Now:

1. The variation in SALES from one period to the next certainly are at least a partial cause of the observed variations in inventory.
2. PROD is subject to random variations due to machine breakdown, material variations, etc., but these are usually comparatively minor and do not help very much in explaining the behaviour of inventory.

   These points are fairly obvious, the next is the important factor

3. PROD usually varies because management choose to alter it. For example, if inventory falls too low management may decide to increase production to compensate.

Clearly, MANAGERIAL POLICIES AFFECT DYNAMIC BEHAVIOUR and dynamic behaviour results from the way in which external variations, such as in SALES, are responded to by management. The policies, or rules by which these responses are selected, may diminish the effect which the external shock has on the firm or they may, of course, make matters worse. It often happens that a policy in one part of the firm conflicts with one in another part in such a way that neither policy is effective.

The essential aim of a System Dynamics study is to find policies which will control the firm effectively in the face of the shocks which will fall upon it from the outside world, which will not make things worse than they need to be, and which will be properly attuned to one another.

We can therefore say that System Dynamics is: *A method of analysing problems in which time is an important factor, and which involve the study of how a system can be defended against, or made to benefit from, the shocks which fall upon it from the outside world*. Alternatively: *System Dynamics is that branch of control theory which deals with socio-economic systems, and that branch of Management Science which deals with problems of controlability*.

In a practical situation there as so many complexities that the common sense approach is insufficient, and likely to be misleading, and a formal discipline is needed. It is with such a discipline, and its techniques, that this book is concerned.

This example was about inventory. It could equally well have been labour force, production capacity, cash reserves, balance of payments, unemployment, housing stock, or many other variables, and the organization could have been the economy, a city or region, an ecology, rather than a business firm. The reader should try to envisage these, and any others he can think of. In each case he should list the factors which affect the variable and think about why these factors vary and, in particular, the extent to which the variation is due to chance, or on the other hand, to 'policy'. This exercise should illustrate the widespread nature of this kind of policy-influenced variation.

In some cases, such as an ecology, there is little or no conscious choice as there was for the inventory case. For these 'unmanaged' situations variables such as the birth rate of a species of fish change because of response to, say, food supply. There is still variation, however, and we may simply wish to understand it better by explaining how it comes about, or we may be interested in how man might effectively interfere by, say, a fishing control policy which would give good yields without annihilating the species.

In most cases the reader will notice there is a kind of structure. For example, a company which sells from stock usually takes the stock level into account when planning future production. If inventory is too high, production will usually be cut back to some extent and vice-versa. This creates a loop, in which inventory affects production which affects inventory which, in turn, influences production again, and so on. System Dynamics is concerned with detecting these loops and, by analysing them, improving performance. Usually there are many loops, some of them very complicated, and we need special tools for the analysis.

1.2 Fundamental System Concepts

There are many definitions of 'system', a convenient one being *a system is a collection of parts organized for a purpose*.

The parts may be something recognizable, such as a cash balance, a stock of product, or a factory – the size depending a great deal on the point of view and the level of detail required. However, the parts also include the decision rules or policies by which the system is controlled, and the information linkages within the system. It is easy to recognize the functional parts and it has to be stressed that policies and information paths are also parts.

The way in which the parts are connected together, or *organized*, is an important aspect of a system, and the information link parts are the major means by which organization is achieved. The minor features of organization are the physical flows of material and orders from one part to another. These usually arise from physical necessity in that the output from the production department 'part' has to go somewhere, such as into the inventory 'part'. The important effects on system performance will derive from the way in which the production department arrives at how much to produce. This will depend on whether the production department uses, say, sales forecasts *and* the state of inventory. If it does, then the department is organized with the inventory part and with the forecasting part. Furthermore, another part enters into the problem, namely the decision rule by which information from the parts with which production is organized is transformed into actions within the production function. Clearly, the overall behaviour of the system will be strongly affected by the parts themselves *and* by the manner of their organization.

All practical systems have a purpose, which is usually a strong influence on their operations. The system organization, and the collection of parts it possesses, arise from its purpose, though often in a very haphazard and vague way. It is a truism that parts eventually acquire purposes of their own, survival being one, and purpose conflicts are common and damaging.

Although definitions and their elaboration are useful, a far better understanding of the system concept, and System Dynamics as a subject, can be gained by studying systems in operation. That is what this book is about, but we shall consider two simple examples at this stage.

1.3 Illustrations of System Ideas

Consider two simple subsystems: a car suspension mechanism and a production–inventory control rule.

The first has the following obvious parts: a wheel, a spring, a shock absorber, a car and its occupants. The spring and the shock absorber physically connect, or organize, the wheel and the car. There are two other parts, the stiffness of the spring and the damping of the shock absorber, which together control the way in which the suspension will respond when the wheel is shocked by passing over a bump in the road surface. In a very real sense the stiffness and the damping 'decide' how the car body is to respond to movements of the wheel, and we can regard them as decision rules or policies for controlling system response in the face of outside shocks. These decision rules are built into the car at the choice of its designer and can only be changed with some labour and within limits.

The purpose of the system is to insulate the passengers from the road surface, and to keep the car in one piece.

As the car is driven over a country road the wheel is subjected to shocks from the road surface but, if the suspension is well designed and in good order, the passengers have a smooth ride. If, however, the designer has made a poor choice of the decision rules, the suspension will amplify the shocks and the passengers will feel them as being worse than they really are. Clearly, poor rules can make behaviour far worse than it needs to be.

The suspension has to be able to deal with minor road variations, and it also has to cope with major changes such as from the level to an uphill slope. It has no way of knowing whether a particular shock from the road is a minor bump, which it should try to smooth out as much as possible, or the start of an uphill grade which it must not ignore. Its design must be a compromise, between reacting too readily to the first of these and too sluggishly to the second.

Even a well-designed suspension will give poor performance if the driver does not operate it properly and, by driving too fast or too slowly for the road conditions, tries to make the system give performance which it is inherently incapable of providing.

Notice that we do not expect the suspension to adjust its behaviour now to allow for the road conditions ahead. Such 'forecasting' would be impossible, so the suspension is designed to give good performance whatever the shocks it meets. Naturally, the driver 'forecasts' and, by controlling his speed, attempts to regulate the shocks, but this is a very uncertain process which is much affected by visibility, the driver's experience and his desire for speed. The suspension thus has to be designed to cope with his mistakes and the designer does not rely on the driver being able to make perfect forecasts. The success of this design process can be judged by the generally excellent performance of popular cars in a wide variety of road conditions and an even greater variation of driver skills.

The contrasting production–inventory system has, as parts, an inventory, a manufacturing delay dictated by technical factors, a production start rate and a production completion rate. Its decision rule part determines the production start rate. There are many possible information stream parts which could feed the decision rule and the designer's problem is to choose which ones to use, and how to transform their content into a production choice.

The purpose of the system is usually to maintain inventory at a satisfactory level whilst keeping production fairly stable in the face of shocks from the outside world. These shocks are variations in the rate at which inventory is depleted by customers' orders. As before, however, the design has to be a compromise between the need to smooth out noise and to respond rapidly and smoothly to major changes.

The similarity between this and the suspension system is very marked and they can be described by the same differential equation.

The designer of the production system has many more options open to him than the designer of the suspension system, but his problem is very similar, namely how to make his system controllable in the face of external shocks. Attempts may be made to forecast consumption, though there will be error, as in the case of the driver attempting to predict road conditions around the next bend. The production system, therefore, has to be designed to cope with shocks and errors and to give a good, controlled response under a wide range of conditions.

The wide range of options available to the industrial system designer complicates his problem. The car designer only has springs available to him and simply has to choose the right stiffness. The industrial system designer has many alternatives which can be used in combination and, for each one, many more factors to play with than just the spring stiffness. Furthermore he usually has to look much more widely and consider the interactions of finance, marketing, research and development (R and D), capacity acquisition etc.

To reiterate, System Dynamics is that part of management science which deals with the controllability of managed systems over time, usually in the face of external shocks.

1.4 Models of Systems

The subject of what a model is, and what it can be used for, is a large one in management science, economics, engineering, and other disciplines. Like 'system', the word 'model' is much overused.

Basically, a model is simply a means by which we attempt to represent some aspect of the external world, in order to be able to influence, control or understand it more effectively. There are, however, various types of model, the distinction between which is by no means as clear-cut as we shall make it.

For the scale models used to plan machine layout, the nature of the modelling process is apparent. The scaled-down version looks like the real thing and the nature, purpose, and consequences of the model are readily understood. This is because making even a rough scale model of a piece of machinery can be thought of as using a 'language' which gives a good, clear, description of the real problem and which is easily understood by some person other than the modeller.

Alternatively, a model might be an intuitive mental understanding of human behaviour, as we have experienced it, which we employ in order to persuade some group of people to follow a particular course of action. A teacher unconsciously employs such a 'model' while he is lecturing, so that he may so choose his words and illustrations as to enhance his chances of conveying his meaning to his students.

This case is vastly different from the first, scale-model, example. There is no really clear 'language' by which the second type of model can be described, partly because concepts such as rapport, comprehension etc. are practically impossible to define and analyse, let alone convey to someone else, such as a trainee treacher.

We shall, in this book, reserve the term model, for a sort of half-way-house between these two extremes. In our usage we shall mean by 'model': *any formal description, in words, diagrams and/or mathematical equations, of the structure and workings of a system, together with unambiguous, acceptable, definitions of its parts.*

The three alternatives of verbal, diagrammatic and mathematical form *must* be exactly equivalent to one another, in order to be able to build an adequate model of a system. Any one form should merely serve as an aid to understanding for someone who is not fluent in the other languages.

This definition of a model implies that it may well be downright misleading to seek to analyse such intuitive processes as the dynamics of interpersonal relationships, using the methods described later in this book for analysing comparatively simple socio-economic processes, such as the dynamics of a firm, an industry, or an economy. To do so would be to attempt to impress formalism and structure on something which, by its very nature, is fluid and evolving. It must, in short, be very firmly pointed out that it is simply not the case that a formal model is always better than intuitive understanding because many of our mental processes cannot be described in words without so distorting and simplifying them as to remove most of the meaning from them.

1.5 Examples of Dynamic Behaviour

Having said that dynamic behaviour is virtually universal in systems, we now examine some cases of such behaviour. The examples are from four levels: the economy as a whole, industries within the economy, and behaviour of and within a firm. The purpose of the examples is to illustrate the following points about the dynamic behaviour of systems.

1. Dynamic systems occur very widely indeed, and at all levels of complexity and scale.
2. Systems in totally different areas often have surprisingly similar behaviour patterns.
3. In order to improve a system's behaviour (if it is bad, and not all systems are) we usually start by trying to understand why it behaves as it does. This involves trying to discover its working processes and seeing whether the knowledge, or ideas, we have about how the system operates can be supported by examining its dynamic behaviour. In this way we link together statistical data on the system's dynamics, with our comprehension of the inner working of the system. This is particularly illustrated in Section 1.7.
4. The dynamics of the system as a whole raise a wide range of problems, for its parts may well start to operate in conflict.

The examples have been chosen to illustrate points of general applicability, and the reader who is interested in business problems is urged not to omit the other sections.

1.6 Economic Dynamics

The first example is that of unemployment in the UK economy. Fig. 1.1 shows the percentage of the UK labour force unemployed at two year intervals over 90 years; the economic data have been chosen to reflect interesting system behaviour, rather than topical economic problems. The pattern of behaviour from 1880 to 1945, apart from the artificially low unemployment during the World Wars, was of oscillations of considerable amplitude and a generally high level. This behaviour was most extreme between 1920 and 1940, during the Great Depression. Although the data are not reliable, there is pretty convincing evidence that the same situation had existed for very many years before 1880.

After 1945 there is a very striking switch to a lower level of unemployment, with a smaller amplitude of fluctuation.

It is fairly evident that after 1945 the behaviour of unemployment underwent a radical alteration. This is usually attributed to the maintenance of full employment having become a principal concern of all governments during the post-war years. There have been plenty of external shocks to the UK economy in the years since 1945, but it is fairly clear, from Fig. 1.1 that it *is* possible to change the internal policies of a system in such a way as to improve its behaviour, in spite of upsets administered to it from outside.

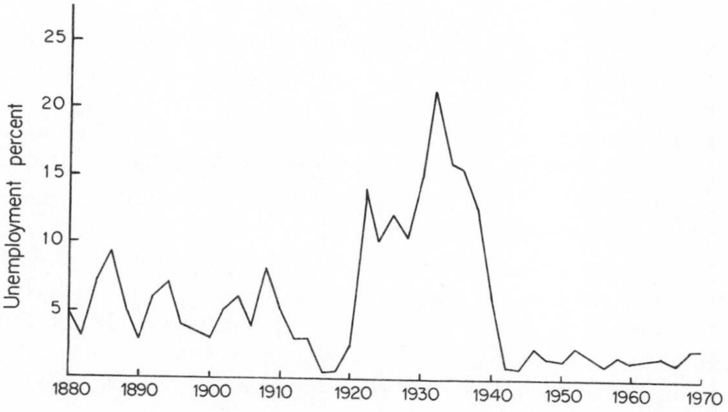

Fig. 1.1 A 90-year history of unemployment in the UK

Figure 1.1 illustrates a number of additional points which usually apply to all systems, at least to some extent.

Policies and Modes of Behaviour

In considering whether or not to introduce a particular policy it is often more useful to look at modes, or general types, of behaviour than at specific numerical values.

In the case of Fig. 1.1 the important thing to any government would be that unemployment had been generally lower and more stable since 1945 rather than the precise percentage in, say, 1963. This is, of course to focus on unemployment and it is certainly true that the UK Balance of Payments since 1945 has been far less satisfactory. This highlights the need to look at a system fairly broadly in order to design policies for its operation.

It is also true that system behaviour changes one's standards for judging it. The level of unemployment in 1970–1972 would have delighted the politicians of the 1880s but is appalling to those of today, simply because they know that the system can do better.

'Data' should be Treated with Suspicion

The numbers on which Fig. 1.1 is based are of no great accuracy and come from several sources. For example, the early years are based on the number of trade union members unemployed, since there are no equivalent Government figures available. Trade Union membership was low in the late 19th Century so the figures in the diagram are hardly representative of the country as a whole.

Although the number unemployed in the war years is low, one could only say that the men in the Forces were 'employed' in a very limited sense of the word. It is also rather hard to find a representative period in the data, though the 'unrepresentative' period from 1920 to 1940 is a long time.

The data contain errors and omissions at which we can only guess. One tends to assume that Government statistics will be accurate, but there have been recent exceptions to this.

It is essential to remember in interpreting data that the agency which collected them did not do it for fun. They had a purpose in data collection which may not be the same as that which one has in studying the data. The purpose for which the data were collected will affect the way in which they are defined and presented, and may make it very hard to interpret them in the light of some other purpose. This is particularly so when the purpose for which the data were collected meant that certain aspects of the system could be ignored, but the purpose for which they are now being analysed is so different as to create a need for the missing data. How much attention should one pay to data which covers only one part of the question?

As a final point we should note that data can *only* be interpreted with the aid of a model of the system. For example, Fig. 1.1 makes a great deal of sense if one is told about the change in economic policy and attitudes which took effect after 1945. Someone without that information about the system, but with knowledge that the low point in 1916–1918 was connected with a war, might feel that the behaviour of the system from 1942 onwards was due to a very long war or its equivalent. (Perhaps it was, and the 'economic policy' explanation is superficial.)

In general one has to be very careful about what the definitions of data actually mean. In particular, it is important to be sure that the dynamics are real and not due to changes in the basis on which the data was collected.

The second example of large-scale dynamic behaviour is that of the UK Visible

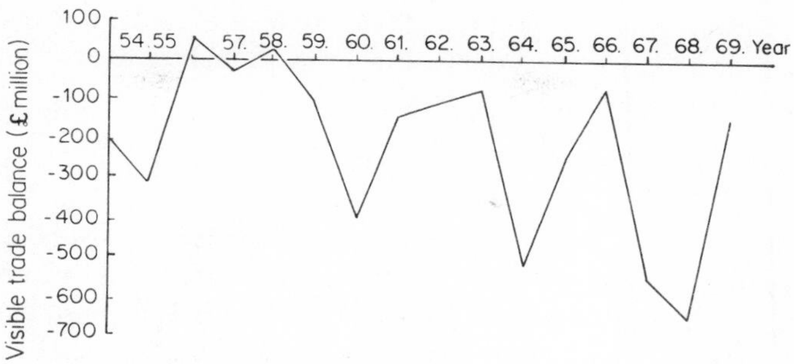

Fig. 1.2 Visible trade balance in the UK economy 1954–1969

Balance of Trade, shown in Fig. 1.2. This shows, very clearly, a system in which an adverse situation leads to corrective action in the form of exhortation, credit squeezes, import surcharges and, ultimately, devaluation. In each case the corrective action has been partially successful (though often only after a time-lag which was much longer than had been anticipated). One is left, however, with the impression of some underlying force or trend which, sooner or later, overcomes the effect of the corrective action. Apart from the strictly economic problems this graph suggests, there are also some interesting, and very difficult, system questions. Some of these are:

1. How large do the dynamics have to become, and in which direction, before one should take action?
2. How powerful a corrective action should be applied, and how long will it be before it takes effect? (If it will be two years, there is not much point in basing decision-making on a forecast for six months.)
3. Are any other significant variables showing undesirable behaviour at the same time, and are their dynamics synchronized or not with those of the first variable?
4. Are the fluctuations in the variable in which we are interested becoming more severe? If they are, there is some instability in the system, and the attempts to correct the undesirable behaviour may be making matters worse.
5. The answers to the first four question groups suggest others such as whether the corrective action is a mere palliative, which is ineffective in the face of underlying forces in the system. The aim is to find out if one has really identified the salient features of the system, or whether the boundary will have to be drawn more widely in order adequately to reflect the behaviour of, and the possibilities for effective control over, the variables in which management, whether it be that of the firm or of the economy, are interested. One also asks if they are interested in variables which really reflect the underlying condition of the system, rather than mere surface symptoms.

These question groups have been discussed in the context of an economic example, but they apply to all systems. The reader should try to keep them in mind in considering the examples which follow.

1.7 Industry Dynamics

Having discussed two examples of economic dynamics we now come down the scale of magnitude to examine a case of industrywide dynamic behaviour.

Many industries exhibit cyclical fluctuation, one particularly severe case being the Chemical Plant Construction Cycle—the plant cycle, for short. Data on the plant cycle for some recent years are shown in Fig. 1.3, which shows the salient features of typical cyclical behaviour. Only one cycle is shown as, to show more, we should have to go back so far that it would be rather hard to find a representative period. The difference in the lengths of the cycles of the three variables illustrates the effects of data-collection difficulties in this industry.

The plant industry is 'driven' by orders placed by the chemical industry, and these vary quite sharply from one year to another. This, naturally, leads to a backlog which fluctuates. The construction completed by the industry follows the rises and falls in orders, but about a year later, and to a lesser degree. Clearly, the plant industry smooths out the load variations placed on it by the chemical industry, and this is a common system characteristic.

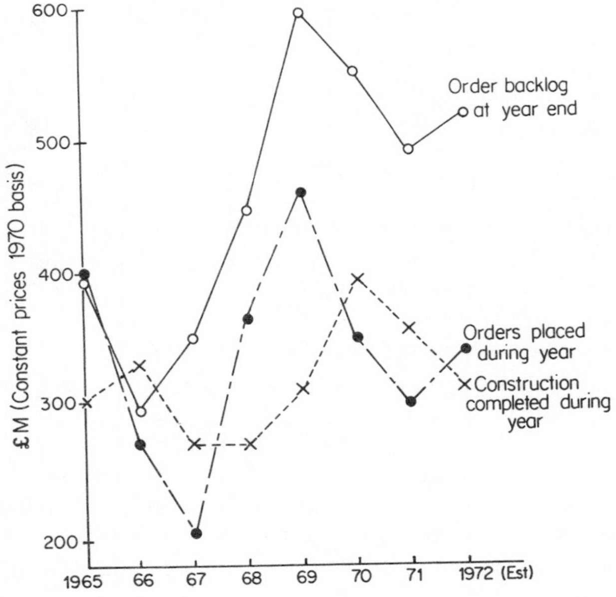

Fig. 1.3 UK chemical plant orders construction and backlog 1965–1972. (Note suppressed zero scale on the vertical axis)

The delivery delay which a chemical company will experience, before its plant is installed and operational, is obviously important to it. One could well understand a chemical company bringing forward its orders so as to get a good position in the queue when it anticipates a moderate increase in delivery delay. Conversely, the company might hold back orders when the delay becomes very large, on the grounds that it might as well wait until the anticipated demand for chemicals is clearer, since it won't be able to get the plant in any case. Alternatively, the order may be placed with an overseas plant builder. In either case, the effect is the same in that the chemical firm orders less than it intended to from UK manufacturers.

This description of how part of the system works is widely believed by people involved in the industry (though there are other views) and it therefore has to be taken into account along with the data shown in the diagram. It would be unreasonable to ignore either of these pieces of knowledge, imperfect though we *know both* of them to be. The question is whether they support one another and whether the two pieces of information can be used to add to our comprehension of the system and its problems.

The data in Fig. 1.3 does not refute the foregoing view of the plant industry and the chemical industry driving each other, rather than the plant industry simply responding to whatever the chemical industry demands. As a first approximation, dividing the Order Backlog at the end of each year by the Construction Completed during the year (which is a rough measure of the plant industry's capacity) gives an estimate of the delivery delay which an order placed in the following year might expect. If we plot this delay against the orders placed during the following year, we get Fig. 1.4, in which the numbers by the data points refer to the year in which the orders were placed.

The curve drawn on Fig. 1.4 has no pretension to accuracy, but it does show the way in which the data support the qualitative analysis which implied that the chemical and plant industries were linked dynamically.

The curve also shows two other features. There is, presumably, some minimum

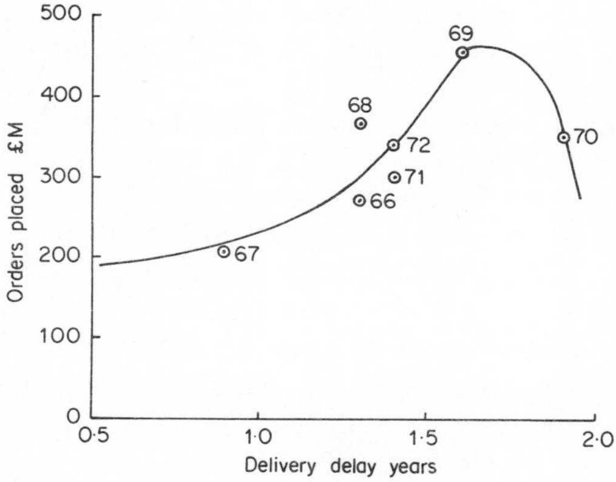

Fig. 1.4 Orders and delivery delay for the UK chemical plant industry

value of orders which represent plant replacement, and it also appears that orders drop off sharply when the delivery delay gets up to the horizon beyond which chemical firms do not like to forecast.

It is easy to see how a dynamic link between the chemical industry and the plant industry, governed by a curve such as Fig. 1.4, could lead to amplification of dynamic shocks. Demand for chemicals appears, to the chemical industry, to rise, or some other cause leads one or two companies to place orders on the plant industry. The ensuing moderate rise in delivery delay causes other firms to bring forward tentative construction plans and to place firm orders so as to ensure a place in the queue. This drives up backlog and the plant industry increases its capacity, but, in the meantime, the increased delivery delay has led to a fall off in orders. The increased plant construction capacity now reduces the backlog, and hence the delivery delay, to the point where the plant industry's relationships with the chemical industry can again amplify the variations in demand for plant which would always occur in any case.

We need not pursue the plant industry in any more detail, but it does serve to illustrate some important points about industry-wide systems.

1. Industries, as well as economies, have dynamic characteristics which derive from a whole range of technological, economic, organizational, and historical factors. These characteristics may include factors which make dynamic behaviour worse than it needs to be. Chapter 10 is a case study of this in a firm.
2. The dynamics of the industry pose several problems for the firms which have to operate in it. In the case of this industry we give three examples, of the hundreds of possible problems. How should a firm in the plant industry adjust its production capacity in response to orders, backlog etc.? How should a firm in the chemical industry plan its plant ordering, in the light of its own needs and resources, and the behavioural characteristics of the rest of the industry, as reflected in Fig. 1.4? Thirdly, one might consider the industry's problems from the point of view of the Government. If the plant cycle is bad for the plant industry and for the chemical industry, what can be done about it by the judicious use of investment grants, tax credits or any other Governmental action?
3. Given the existence of the cycle and the importance of the problems which it generates, one could build a model of the interactions and study the way in which the dynamic behaviour of Fig. 1.3 could be improved, or might best be lived with. The form of the model will depend on the purpose for which it is constructed, but the example shows how one looks for *both* qualitative and quantitative information to help in defining the mechanisms which operate in the system, and to elucidate the way in which they work. The example also illustrates the dilemma of whether we should ignore an important part of the system simply because the quantitative data are poor, and, if so, what we can say about the quality of the resulting model.

Later in this book we shall examine the problem of sensitivity analysis in models, for the present we remark that the precise numerical values for a curve such as Fig. 1.4 rarely make much difference to the form of the system's dynamics, and what really matters is the shape of the curve. In using a model containing a curve such as Fig. 1.4 to design policies for a system to follow, the precise numerical values again rarely make much difference to the type of policy which emerges as being beneficial to the system. What is important, both in explaining dynamics and in designing policies, is to have the correct shape for the curve, and this often emerges from qualitative considerations rather than from a statistical test. Further details of this industry are given in a reference by G. W. Hill in Appendix C.

1.8 Corporate Dynamics

Dynamic behaviour occurs at all levels in the firm, from its overall performance indicators to the detailed departmental level. We call these extremes macro- and micro-dynamic respectively, though these are convenient labels rather than accurate technical terms.

Our example of macro-dynamic behaviour is taken from the mining industry and concerns the Rio-Tinto-Zinc Corporation Ltd.

The recent history of RTZ is shown in Fig. 1.5. Overall, there has been a very striking growth, which we may conveniently divide into two phases 1962–1970 and 1971–1973.

The first phase was generally one of smooth and rapid growth, the second starts with a drop-off in the growth followed by a rapid surge in 1973, though whether this surge will be sustained remains to be seen.

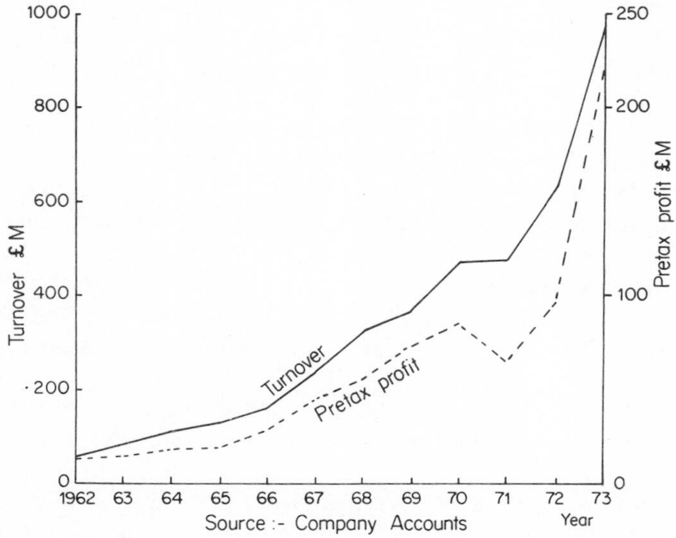

Fig. 1.5 The macro-dynamics of RTZ (Source: Company Accounts)

Basically RTZ uses profits from mining operations to finance new mining developments throughout the world. It is much affected by price movements in the metal markets, by the great uncertainties in mineral exploration, and by the long time delays before decision about new investments are turned into profits from them. We cannot say that these factors were constant throughout this 13-year period but they were certainly present during all of it. The problem for RTZ is striking a balance on the management of their resources of cash flow and engineering expertise to achieving growth in the face of outside shocks. We do not know from the data what that balance was, nor how it was arrived at but overall it was obviously enormously successful in producing growth which far outstrips that of other, general, indices for GNP etc.

The study of corporate policy-balances, and their effects on the dynamics of the system concerned, is the kind of problem for which the techniques of system dynamics were developed. Obviously not all dynamic problems are as complicated and difficult as this one, and in the next section we shall look at some lower-level examples. For the corporate growth example, the complexity and difficulty are inherent in the problem, not in the techniques. The importance of the problem is the justification for the considerable effort needed to solve it.

Before leaving this example, we can use it to illustrate a couple of points concerning the different types of system structures which we shall study in later chapters.

Up to 1970, Fig. 1.5 shows that there was a steadily-increasing growth in both the macro-variables. The growth in both variables is very like the growth curve of a capital sum at compound interest in which growth provides the fuel for yet further growth. This is a dynamic mechanism called positive feedback or self-reinforcing growth and Fig. 1.5 suggests that, up to 1970, RTZ was being driven by just such a process of investment providing the source of further investment, leading to smooth growth despite all the severe fluctuations in metal prices which took place during these years.

There is a marked change in the behaviour pattern from 1971 onwards and it does seem possible that the policies which RTZ followed from 1962 to 1970, and which were very successful in producing growth, do not work as well at higher levels of turnover, or could not withstand the changed market conditions from 1971 onwards.

It is also possible that some other policy balance might have been even more successful in the period covered. An interesting feature is the very long time-delays in this industry. The profit drop in 1971 was affected by decisions taken 8–10 years earlier.

Fig. 1.5 is a strong contrast with the ill-controlled goal-seeking dynamics shown in Fig. 1.2, which is an example of the operation of a process known as negative feedback.

We shall be studying these two types of feedback in much more detail in later chapters. For the moment our interest in the RTZ example lies in noting that the Company's policies have clearly been comparatively well adjusted to produce positive feedback. This implies that, in the study of corporate growth, one would look very closely at the way in which the policy-parts are organized, in order to locate the positive feedback structures which are capable of producing growth. One would certainly also find negative feedback processes and at least some of these would act as stabilizing, growth-inhibiting mechanisms. The purpose of the study would then be to try to improve the 'tuning', or balancing, of the company's policies so as to enhance the positive, growth-producing, processes and to minimize the effect of the negative, growth-limiting, forces. We shall have more to say about this later, when we have dealt with the theory of positive and negative feedback in appropriate detail.

1.9 Corporate Micro-dynamics

The first example of this is Total Company Net Inventory (i.e. physical stocks less any orders outstanding) and data from a domestic appliance company are plotted in Fig. 1.6. This shows a steady rise in inventory and then a decline, in response to strenuous efforts to bring the inventory down. In fact there was a reduction of about £1.5M between February and September 1971 and this involved considerable changes to production schedules and so on. Fig. 1.6 shows dynamic behaviour which starts to reach worrying proportions and where major efforts are needed to bring the situation under control.

The questions which this raises are: why do they get excessively large inventories, how big should inventory be allowed to become, how vigorously should it be controlled and, most importantly, does the attempt to control inventory cause problems elsewhere in the system and how can these be avoided? It is, of course, quite possible, and quite common, that a perfectly good inventory control policy

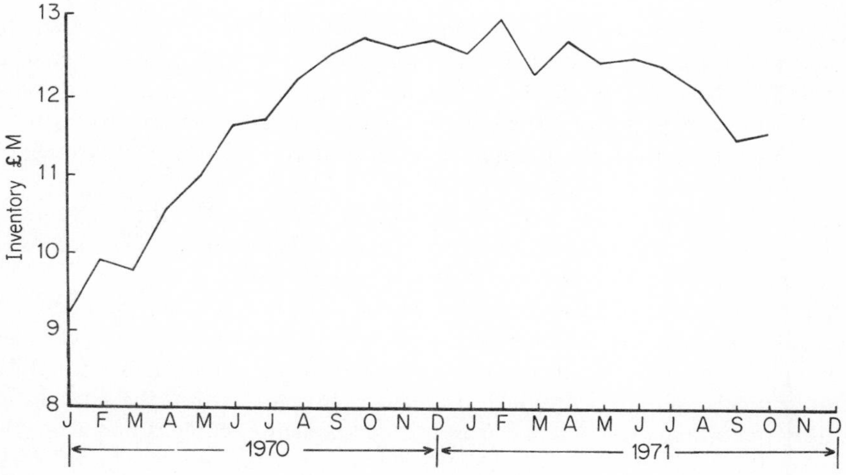

Fig. 1.6 Total company net inventory (Note suppressed zero scale on the vertical axis)

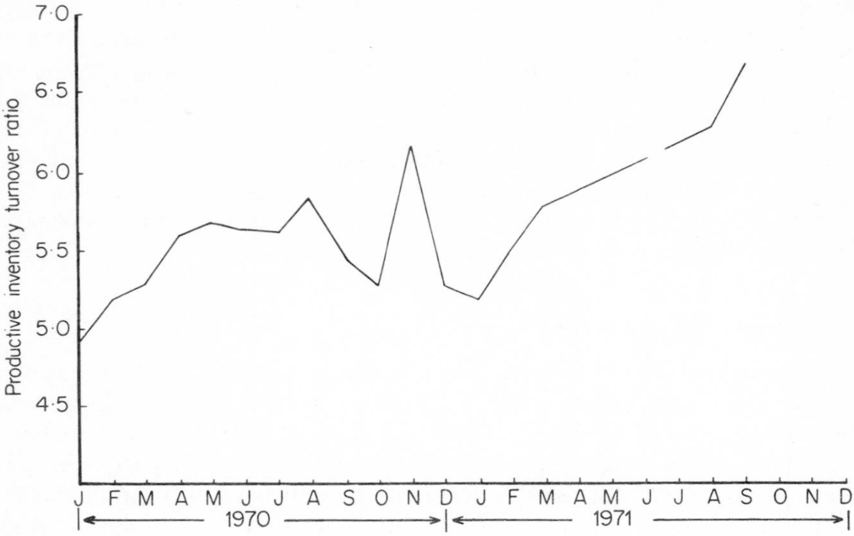

Fig. 1.7 Productive inventory turnover ratio (Note suppressed zero scale on the vertical axis)

could be completely vitiated by coming into conflict with another policy (e.g. cash control). The competing policies are usually perfectly good, when considered in isolation, but both are rendered ineffective by not having been designed to operate in conjunction with each other, and other policies, in the firm's information and control system.

Similar data from the same company are shown in Fig. 1.7. This shows the Productive Inventory Turnover Ratio during the inventory crisis depicted in Fig. 1.6. The Ratio relates volume of business to the size of inventory and is low when stocks are excessive. The graph shows a steady increase with some fluctuations and portrays the attempts to reach a value of 7.8 which was chosen by the company's Board.

This raises questions. Why 7.8? Should control be exercised at the last moment, or more gradually? Should the Productive Inventory Turnover Ratio be monitored and controlled continuously, or only in special circumstances?

Dynamic behaviour in two related variables may not be consistent without additional evidence. For example, the peak in the turnover ratio for November 1970 implies the larger throughput in anticipation of the Christmas boom in sales of this particular product. Although productive inventory is not the same as net inventory one would expect to see an associated fluctuation in Fig. 1.6 at about the same time but there is hardly a ripple. This kind of inconsistency has to be looked for, and explaining it will often provide a good deal of insight into how the system works. In this case the explanation lies in the difference between Net Stock and Actual Stock, caused by dealers who place their orders well before the busy period.

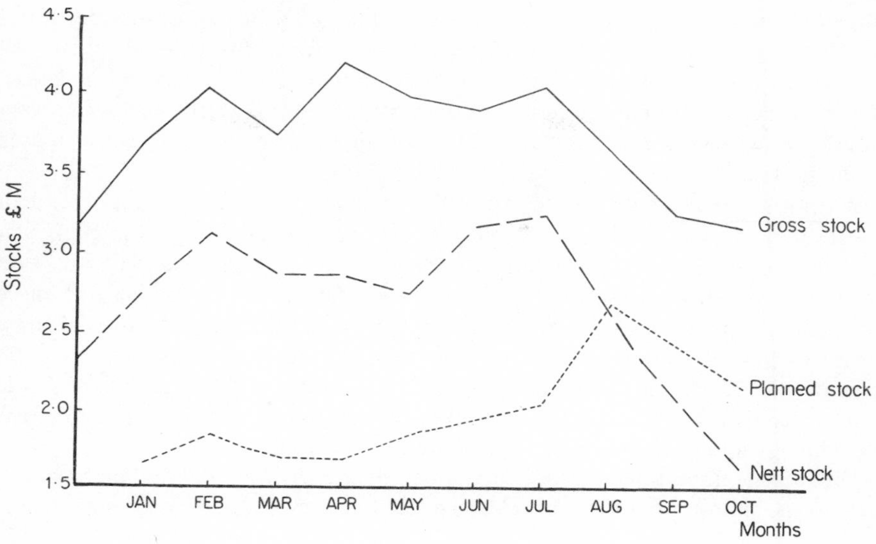

Fig. 1.8 Gross net and planned stock in 1971 (Note suppressed zero scale on the vertical axis)

The implication is that the dealers in a very real sense 'drive' the company. This may or may not be bad, that is quite another question, at the moment it is simply a statement about the workings of the system inferred from a study of its dynamics.

Fig. 1.8 shows an example, again from stock control, of an attempt to bring Gross Stock in a company division into line with its planned value. Planned stocks intended to be a control variable and the difference between planned and actual influenced decisions about the production target to be adopted. The graph shows the very long time needed to bring the actual and planned values together.

The differences between Gross and Net Stocks is roughly constant, which implies that production and sales are more or less in harmony over the long run However, there is little progress towards bringing Planned Stock and Gross Stock into line, which leads one to suspect that the control system is incapable of doing what it is supposed to do. In this case 'control system' means the decision rules by which the company arrives at the manufacturing schedule in the light of information about stocks and demand forecasts.

Fig. 1.9 shows the production schedules planned for two products – the plans being drawn up in the latter part of 1970 in the light of demand forecasts for the following year. In each case, factory management aimed to have the schedule as level as possible and proposed to allow stocks to vary in order to cope with expected demand changes. In the case of product A, this involved increasing the schedule from a little over 11000 units per week to 12000 which takes time, hence the steps in the early part of the year. Product B was expected to show only a small change from the present level. In the event, marketing needs changed and the schedules actually used are shown by the dotted lines. In the case of product A this involved large alterations to the planned schedule and this naturally raises the question of how the company should respond to indications of a need for change. In particular, how rapidly should the company attempt to close gaps, which gaps should be watched, e.g. should it be the gap between planned and required production, planned and actual stock etc?

A simple view would be to say that they simply need to improve their forecasting. In practice this is not quite as easy as it seems, and in this particular market it is practically impossible to forecast with any pretensions to accuracy (in Chapter 9 we shall investigate some of the subtleties of 'forecasting'). A more realistic assessment is that any forecasting in this market is liable to be overtaken by events so that an indication of a need to abandon the original forecast is itself a forecast and, as such, is a liable to error as the first one was. This leads one to try to create a control system which would, in some way, correct for errors in forecasts and which would lead to different dynamics from those in Fig. 1.9.

One might, in fact, wonder about the advantages of forecasting at all. Since the forecast is almost bound to be changed by the passage of events, as shown in

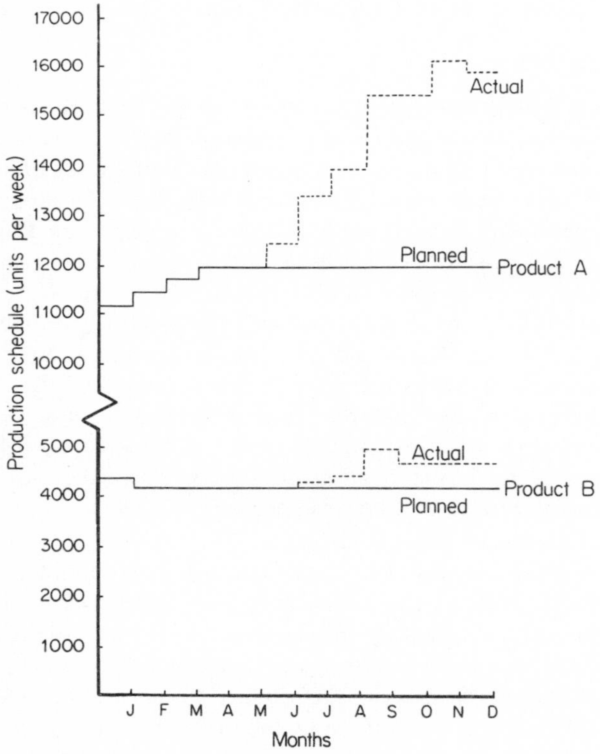

Fig. 1.9 Planned and actual schedules in 1971

Fig. 1.9, is there much point in trying to draw up the plan shown in Fig. 1.9? It might, in fact, be worth attempting to devise a control system which would not depend on forecasts at all but which would simply respond to current events in a smooth and efficient manner.

1.10 The Object of a System Dynamics Analysis

So far we have discussed what a dynamic system is, and we have considered several examples of dynamic phenomena arising in economic, industrial and corporate systems. The examples show the widespread occurrence of dynamic phenomena in economic and managerial systems. The same type of phenomena arise in biological and ecological systems, but we have no space to consider those. The examples also show, and this is their most valuable feature, the way in which a study of a system's dynamics can stimulate questions about the way in which the system should work. Attempts to propose solutions to the questions lead one to the investigation and analysis of the system's properties, and we now turn to considering the kind of objectives which such an investigation could have.

In general a System Dynamics study has a two-fold objective:

1. Explaining the system's behaviour in terms of its structure and policies.
2. Suggesting changes to structure, policies, or both, which will lead to an improvement in the behaviour.

A very common situation is that of the firm within an industry which it cannot dominate, let alone control, or operating in competition with other firms which it cannot control. The analyst then has two options. He can pursue objective 2 and make generalized recommendations for the whole industry in the knowledge that there is only a remote chance of his being listened to. Alternatively, he can rephase objective 2 to make it:

3. Suggesting changes to structure and policy in a small system which will enable it to survive, or better still, take advantage of, what the larger system does to it.

The obvious example, in this chapter, would be the individual chemical firm or chemical plant manufacturer, who has to make the best of living with the plant cycle depicted in Fig. 1.3.

While these are general objectives the details of the individual model will depend on the point of view adopted by the analyst. For instance, an analyst modelling the chemical plant construction cycle, from the standpoint of a chemical company, would produce a different model from that which he would build if he worked for a plant company. This is in no way surprising, nor would it be worth mentioning were it not so often forgotten. There is no such thing as *the* absolutely correct model of a system. All that can ever be done is to build the right model of that part of the system which is relevant to the purpose at hand. This last sentence reinforces the point that practically all system dynamics investigations have the same, two-fold, objective, but the details which the model includes when the study has been completed will depend on the purpose for which the work was done.

The remaining chapters fall, roughly, into three parts, designed to develop the techniques necessary to achieve each of the two objectives, together with illustrative material. The next four chapters are devoted to an explanation of system structure, and the way in which it can be modelled, Chapters 6 to 8 examine the properties of systems and, hence, the way in which system performance can be improved. Finally, the case studies in Chapters 10 to 12 attempt to draw some of this material together, and show how the ideas have been applied in specific practical contexts. Chapters 10 to 12 are linked to 1 to 8 by Chapter 9 which discusses the theory of the firm from a dynamic viewpoint. Finally, Chapter 13 summarizes the practical application of system dynamics in the light of the earlier material.

An extensive bibliography on System Dynamics is given in Appendix C.

Chapter 2

System Structure and Behaviour

2.1 Introduction

The definition of a system as 'a collection of parts organized for a purpose' is so general that any phenomenon, or organization becomes part of some system or other. Often the same part belongs to many systems simultaneously, both 'up and down' and 'across'. For instance, a Bank is part of the systems by which people save money, obtain credit, and are employed. Its computer is part of all these systems and so on *ad infinitum.*

Clearly, mere repetition of 'system' adds nothing to understanding, and we need a language for describing systems in order to use the concept effectively in real problems. The dynamics of firms, economies, etc., are real problems, and they cause much concern to the decision-maker who has to deal with, say, fluctuating profits in a firm, or rising unemployment in a social and economic context. It is, however, only helpful to treat such problems in system terms, if we can describe the forces which cause the dynamics, and can then employ a theory of systems to study how these forces might be controlled, removed, or reinforced, in order to improve behaviour.

As we have seen, a common dynamic phenomenon in firms is fluctuating inventory. At times the inventory is so low as to involve at least the risk of lost sales and at other times so high as to lead to heavy costs and tying up of capital. The obvious influences on inventory are sales and production. Suppose, for simplicity, that sales are beyond the control of the company, and exhibit sharp day-to-day random fluctuations as well as longer-term variations which are also uncontrollable. Production is, however, the result of *conscious choice* by management. They can, for example, keep production constant, vary it to keep inventory constant, or allow both inventory and production to vary in proportions which can be chosen. They can try to forecast the general level of sales and feed this into the production choice and so on. In fact there are very many possibilities for reacting to external forces and creating the internal actions by which the firm evolves and changes. In short the company's policies are themselves forces which act to change the state of affairs in the firm.

In this particular example one could SIMULATE the consequences of various production choices. Suppose a firm has 160 units in inventory and has been asked to supply 110 next week. The firm's policy is to have two weeks sales in stock, so by the end of the forthcoming week they need 220 (twice the current sales rate of 110 per week) in stock. During the coming week they will need to produce 220 – 160 + 110 units, which gives a production rate of 170 units/week or 34 per day for next week. This calculation could easily be extended. For example, for a give pattern of weekly sales over a year we could see how this particular inventory policy would generate dynamics in production.

Now the '2 weeks cover' policy we have described is only one possible inventory policy and rather a simple one at that. One could easily simulate the dynamics of inventory and production for each of the possible production policies. This, together with a study of the characteristics of the system, would assist in the selection of a suitable policy for that firm. Generally, the equations needed to represent the decision rules in a real system are too numerous to be solved mathematically, and simulation by computer is standard practice in system dynamics. For reasons outlined in Chapter 4 the DYSMAP programming language is often used, and the equations in this book are expressed in a form suitable for DYSMAP. (See the bibliography in Appendix C.)

The production example shows that, since the forces can be described unambiguously, it is useful to use system concepts because it opens up an illuminating viewpoint for problem-solving. However, to speak of the 'system' of interrelationships between parent and child does not help in the choices which have to be made in rearing a child, precisely because we have no effective and unambiguous language by which we can describe systems as complex as those.

We now turn to the problem of classifying systems in such a way as to develop a language by which they may be described and their behaviour analysed.

2.2 System Classifications

There are two classes of system: OPEN-LOOP, and CLOSED-LOOP or FEEDBACK.

In the open-loop system, the output arises from the input but has no effect on the input. For example, in a simple ordering system orders are sent out to suppliers and, at some later time, goods are received. If these receipts have no effect on the amount which is ordered, i.e. the output has no influence on the input, the system consists only of a FORWARD PATH, and is OPEN-LOOP.

This system could easily be made into a closed-loop system by observing the rate of receipt of goods. If this does not match up with some desired rate of receipt (e.g. the rate at which the goods are being resold), the ordering rate can be adjusted accordingly. In this new system the output goods received) is being used to adjust the input(goods ordered). Diagrammatically, the difference between open-loop and closed-loop systems is shown in Fig. 2.1.

The distinguishing feature of the closed system is a FEEDBACK PATH of information, choice and action, connecting the output to the input. This creates a

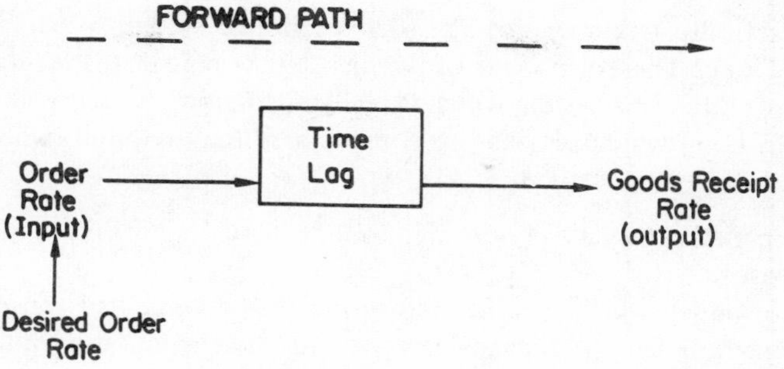

(a) An open-loop system

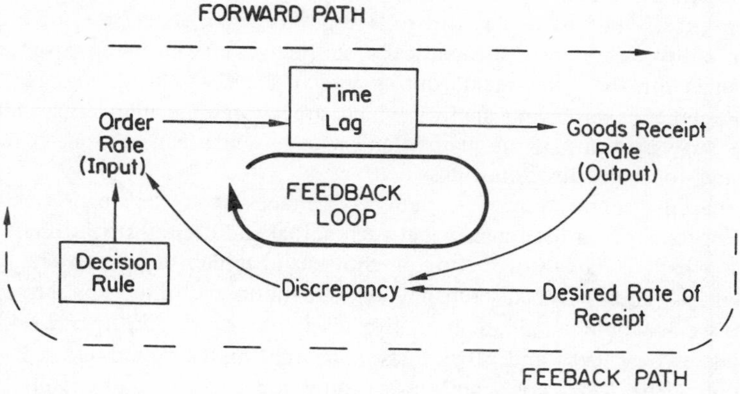

(b) A closed-loop system

Fig. 2.1 Open-loop and closed-loop systems (Note: The dotted lies and the broad arrows are merely intended to highlight system features and are not themselves parts of the systems)

closed chain of cause-and-effect, or FEEDBACK LOOP, consisting of the physical flow of goods and the associated feedback path of information.

The important thing about feedback loops is that they are the cause of dynamic behaviour. An open loop system *can* have dynamics arising from its response to external changes. The dynamics of a closed system on the other hand, are created by its attempts to control itself in the face of external variations.

The structure of a closed-loop system includes the way in which its flows of materials and/or information are connected *and* the way in which these flows are modified or transformed by delays and decision rules embedded in the loop. Obviously, the key to improving the dynamics of a closed system will lie in changing its structure, either by altering the connectedness of its flows and/or changing its decision rules.

It is practically unknown for any socio-economic system to be anything other than closed-loop. There is no way of proving this statement; the reader is, however, challenged to think of a system which is really open-loop.

A Closed Loop which is trying to regulate a system, requires two factors for its operation:

1. A discrepancy between a desired value and an actual value. These may be current or forecast values.
2. A rule or policy which specifies the actions to be employed for a given size of discrepancy. If the discrepancy is zero and all is going according to plan the decision rule may specify that no changes are to be made and everything is to carry on as before. This 'do-nothing' rule is perfectly valid.

The decision-rules in engineering systems are, in effect, inherent in the equipment itself. For example, the rolls used to make sheet steel of a desired thickness adjust themselves automatically to changes in the incoming steel. The time needed for the adjustment, during which sheet of the wrong thickness is produced, depends on many factors which are chosen during design. Short of replacing large portions of the plant, there is little which can be done if it has to process steel for which it was not designed.

In managed systems, however, there is a much larger degree of freedom of choice. For example, a discrepancy between actual and desired stock can be closed at a rate which is largely a matter of choice and which can be varied to meet circumstances. Thus if the discrepancy is 5000 units, a production rate of 1000 units per week would eliminate it in five weeks, a rate of 2500 per week would clear it in two weeks, and a purchase of 5000 units would close it almost immediately. The usual constraints are production capacity and material availability, but the possibility of purchasing goods or raw materials in many cases makes even these constraints less severe than they seem.

There are, of course, elements in a managerial system which are fixed, in the short to medium term, by technical considerations, or by the equally potent forces of custom and practice; production delays and marketing arrangements are, respectively, examples. Other factors, which are not literally fixed by technology or custom, affect the system as though they were; for example market growth is affected by population growth.

In general, however, many more of the parameters of a 'managerial' system can be chosen than is the case with an engineering system.

2.3 Internal System Structure

In practical system dynamics work, and as a conceptual framework, it is useful to look at systems in the light of how much certain knowledge about their workings it is possible to acquire, at any rate in principle, and how far it is possible to exert actual and direct control over what goes on. We can discern three distinct sectors in the structure, for example, of a firm.

The first is its own decision-making processes. For this sector it is possible, at least in principle, by sufficiently close enquiry, to find out exactly where people obtain their information, by what rules or operations they transform it into decision choices and to whom those decisions are communicated. In practice this may prove to be a very difficult task, but this is a common feature of all management science work and the problem of finding out what goes on is solvable, at least in theory. (Anyone who has tried this in practice will know that 'Mr. Rational Man' hardly inhabits the modern corporation.)

This sector is also characterized, in principle, by being capable of being changed if there is some advantage in doing so. For instance, suppose information about sales forecasts and target stocks is used in deciding next month's production. If production and stocks would both be better controlled by using the previous month's production level and the rate of change of stocks, and the new decision rule is adopted, the structure of the sector has been altered.

The first sector is, therfore, in principle, knowable and changeable, and we call it THE CONTROLLER. A socio-economic system may contain many competing controllers.

The second sector, THE ENVIRONMENT, is that with which the controller interacts but of which the workings are not unequivocally discernible, even in principle, and which cannot be directly changed, only influenced. This sector will usually be the market. The firm, i.e. the controller, hopes to influence the market by setting prices, launching new products, and in many other ways. The market responds by demanding the firm's products, demanding those of a competitor, or not demanding this product-type at all, depending on what the firm and its competitors have done.

The important distinction is that, while we can say with certainty how the controller reacts to the environment, we cannot do so for the environment's response to the controller. Thus, the characteristic of the environment is that it has mutual interactions with the controller, but it is impossible to know exactly how it works or to change its operations with any guarantee of getting the effect which is sought.

This raises the question of how we are to know how to describe the environment with any degree of confidence. Basically, there are six approaches, as discussed in Chapter 3.

The third sector is THE COMPLEMENT, and contains those parts of the system which belong neither to the controller, nor to the environment. If the firm is the controller, and its market the environment, the complement would be the national economy, the Government and, indeed, anything capable of affecting the controller or the environment, but incapable of being affected by either. This ability to affect without being influenced is the key to whether a particular system part belongs to the environment or the complement.

The purpose of the model comes into the question so that if, for example, the object was to model competitive strategies, the analyst would have to consider whether the competitors interacted with the firm, or whether they were independent of it and simply imparted shocks to it from time to time. In the former case, the competitors would have to be modelled as part of the

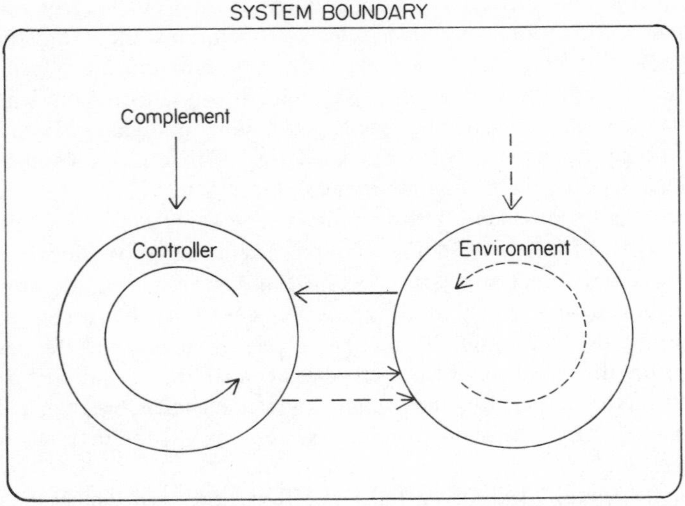

Fig. 2.2 A system's controller, environment and complement

environment, which would be difficult, but essential if a significant part of the system was not to be ignored. In the latter case, the competitors would be parts of the complement. Putting parts of the system into the complement when they really belong in the environment is appealing because it is easier, but it may be unrealistic and seriously misleading.

This classification is shown in Fig. 2.2, in which the solid lines with arrows denote causal influences which can be determined with certainty, and the dotted lines represent those which are based on assumptions, beliefs or statistical inference, rather than direct observation The lines within the controller and the environment simply indicate that there are internal influences within the sectors, as well as those which act between them. Naturally, the actual structure of a sector may be very complicated.

In practical cases, complications can arise. The controller may send 'signals' to the environment in the hope that they will influence it, but without the certainty of this being the case. For example, the firm *believes* that price affects demand by the environment, but suppose that delivery delay also affects demand. The firm has a fairly conscious policy of price adjustment, but delivery delay simply arises and is recognized by the market. If the latter outweighs the former the market will respond to something which the firm is not trying to control, and largely ignore the firm's price signals. This can only be found by analysing the feedback between controller and environment and either by changing the firm's policies to make price a more effective controller of demand, or devising new policies for the firm which will control demand via delivery delay, which is what the market takes notice of.

For the controller, the problem is to arrive at production, pricing, and other policies which will enable the firm satisfactorily to follow movements in demand. For the analyst the problem is to elucidate the structure of the controller by

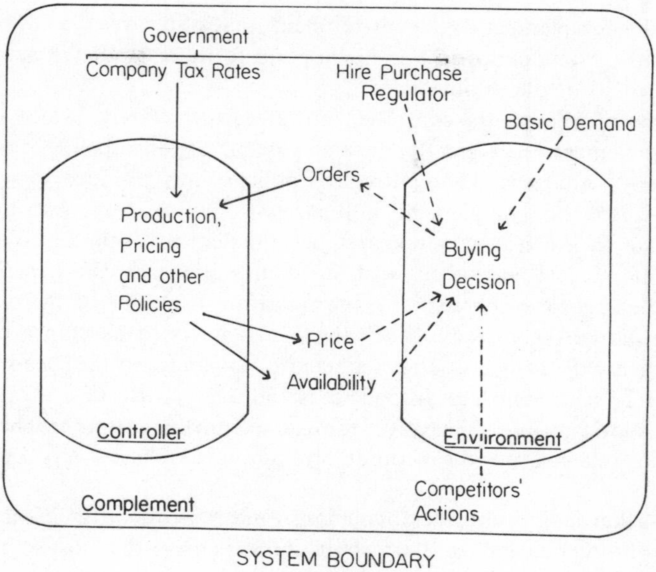

Fig. 2.3 An illustration of system substructure

observation, and of the environment by forming plausible hypotheses, backed up, where possible, by statistical analysis, in sufficient detail to help in this policy-design problem.

We can make this rather more vivid by considering a simple example, shown diagrammatically in Fig. 2.3. The diagram is a highly simplified version of a consumer goods firm. The controller has to decide on price, and on production to create available stocks. In doing so, it takes into account the level of orders, and profitability as affected by company taxation.

Orders are the result of buying decisions by the public in the environment. These are influenced by price, availability, competitive actions, the general level of demand and by hire purchase regulations.

2.4 Driving Forces and System Operation

The dynamics of a system arise from three sources:

1. Shocks imposed by the complement on the controller and/or the environment.
2. Operating policies within the controller.
3. Policies and reactions within the environment.

A study of the dynamics of the system is, by definition, done from the viewpoint of improving the operation of the controller. However, the controller cannot control what the complement does, either to itself or to the environment nor, in practice, can the contoller do very well at predicting the future moves of either of the other two sectors. Even if the controller can forecast perfectly the future actions of the complement or the environment, that still leaves the controller with a large amount of freedom as to how to respond to those predicted actions so as to achieve its, the controller's objectives.

Regardless of whether the controller can forecast accurately, or not, or indeed at all, the complement has a great variety of possible actions open to it which can be used in many sequences. The controller's policies may produce very satisfactory dynamics when the complement influences it in one way, but may be very inappropriate to a different action by the complement. The controller's policies must, therefore, deal adequately with different actions by the complement, and must also be capable of coping with the transition from one of the complement's actions to another. The problem for the controller is to develop a collection of policies which will always produce satisfactory dynamics in the face of any action by the complement. Such a set of policies is said to be ROBUST.

Ideally, robustness would involve, for example, making sure that the firm got its share of an enlarged market without attempting to chase every little whim of demand.

On the other hand, robustness would also mean an effective defence in the face of a down turn in demand. In short, robust policies make the most of opportunities and the best of catastrophies.

2.5 Decision Structures in Managed Systems

The discussion of system classification shows that:

1. All systems of any practical interest are closed-loop.
2. All systems contain a controller and it is from the point of view of the controller that we try to model the system, with the aim of designing a controller which will function well in that system.
3. One needs to model only those parts of the controller, complement and environment which are relevant to the purpose of designing robust policies. This provides a key to modelling techniques which we shall use in Chapter 3.

We now develop a language of system description which will be necessary and sufficient for the construction of models capable of fulfilling these requirements.

The FIRST STEP is to notice that the fundamental processes in socio-economic systems are FLOW and ACCUMULATION. For example, production flows accumulate into stocks, cash flows build into cash balances, labour recruitment creates a workforce, foreign exchange flows are summed into a balance of payments, plant construction creates productive capacity, net birth rates have their effect as population totals and so on. We shall often use INTEGRATION as a synonym for accumulation.

The SECOND STEP is that, although flow and integration are both inherent in systems, only an integration can be observed. Flows are instantaneous and can only be measured as averages over a period, but Section 2.7 shows that time-averaging of a flow is mathematically equivalent to integration. Therefore, the integrations are the variables which can be measured, they form the practical basis for control in the system, *and it is the results of accumulation which lead to changes in the process of accumulation*. This is the fundamental point for understanding and modelling system structure and it is best established by some examples.

In a very simple case of the choice of production rate when manufacturing for stock, the accumulated stock, as compared with some target level, affects the future rate of production. In many cases, of course, consumption, averaged over some recent period, also affects production rate, but averaging over time is mathematically equivalent to integration.

As another example, the size of the labour force, compared to some target, leads to a decision to recruit workers or to allow wastage to take place. For instance if we have 100 men but would like to have 150, we may choose to take on 10 men/week until the gap is made up. The choice of 10, rather than any other number, is the result of managerial policy acting within the various constraints, such as the availability of labour in the area, the ability of the firm to absorb and train them and so forth.

A third example, from economics, leads to the same ideas of system structure. A balance of payments is made up of the accumulation of previous flows of exports and imports. If there is a serious deficit steps may be taken to increase the flow of exports and/or to decrease that of imports so as to move the balance in the right direction within an acceptable period of time. The policy problem is to decide what constitutes a 'serious' deficit, and to select the measures to be adopted and the severity of their application so as to control balance of payments without causing harmful effects elsewhere. Other aspects of the economy, also expressed as integrations, such as unemployment, average real income, and so on, are also being controlled at the same time, so the problem is to select an harmonious collection of economic and social policies which will be robust for the whole economy in the face of external shocks from the economic controller's complement and environment.

The reader should now think of six other illustrations before proceeding any further. In each case he should identify the integration which acts as the indicator, the flows which are controlled as a result, and some policy options for control.

In general, Fig. 2.4 summarizes the way in which control is exercised.

The key variable in this diagram is the integration or SYSTEM STATE. This latter term is literally accurate; it is by looking at the present state of affairs that we decide what, if anything, to do.

The system state may be the current value, an average over some period of time, a forecast value, or an old value. Many 'current' states are actually out of date to some degree, and the system state is almost certainly inaccurately known, either because it is old information or because of errors in counting, transmission, forecasting or *interpretation*. It is often more appropriate to refer to apparent or perceived system state.

The system state is compared to a required state, which may also be a forecast value, and this comparison leads to a positive, negative, or zero discrepancy. The sign of the discrepancy is *usually* found by taking (Desired Value–Actual Value).

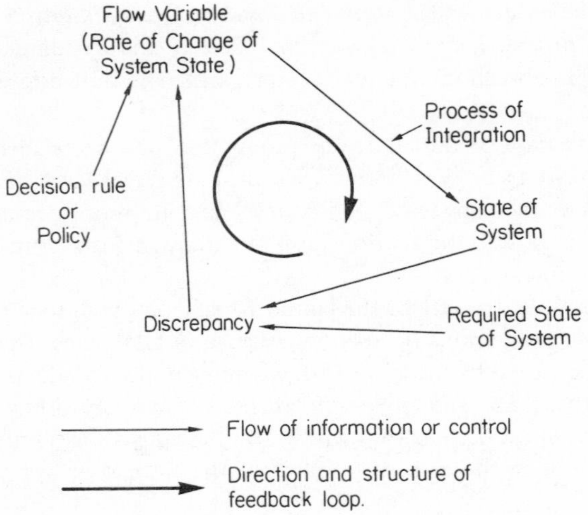

Fig. 2.4 The general structure of decision processes

A decision rule or policy specifies the action to be taken for a discrepancy of a given sign and magnitude. This gives the magnitude (and possibly the sign) of the flow variable which will lead to a change in the system state. In the inventory example, the difference between required and actual inventory dictates the choice of production rate.

Thus, our definition of a POLICY is: *a rule for regulating a flow in the hope of achieving a target level*.

The reader should now express the other examples in this section, and those he worked out for himself earlier, in terms of Fig. 2.4, specifying the nature of the flow variable involved, and deciding how the decision rule controls the flow variable for given values of discrepancy.

Note that the discrepancy may be such that no action should be taken. This may mean either that the flow should be zero or that it should not be changed from its previous value and the reader should think out which of these is to apply in each case. The corollary of this is that, at some point, the discrepancy may become large enough to warrant action being taken, which means that the policy will have to specify how large is large enough, and how rapidly the flow variable should be changed if the discrepancy is larger than this critical value.

The feedback loop mentioned in Section 2.2 is clearly present in Fig. 2.4. In this system the input is the flow variable and the output is the value of the system state. The feedback path is modified by the discrepancy and the decision rule, so that it incorporates a transformation in which information about the output is altered before being used to control the input.

In an actual controller there would usually be many variables characterizing the system state which would interact in a complex way to determine the flow variables. Such a system would, therefore, contain many feedback loops and, in system dynamics, *the feedback loop is the basic unit of system structure and of analysis*.

2.6 Microstructure of Feedback Loops

Having considered system structure in terms of feedback loops, we now develop a mathematical language for the description of the microstructure of these loops.

The essential process in managed systems is integration or accumulation and this is the key to microstructure. The inventory problem furnishes an example to illustrate the concepts. First we define the notation to be used.

The convention is to use abbreviations of the variable names, so that INV denotes the amount of inventory, PROD the rate of production over a period of time, and CONS the rate of consumption. This is more cumbersome than normal mathematical practice, but is easier to understand, and very close to computer simulation programming. The reader can use mathematical notation if it suits his problem, and it is, in fact, employed in various chapters of this text, and extensively in Appendix A.

The units of measurement of these variables are important. INV is measured in the same way as the product is normally counted – tons, gallons, or whatever. We shall stick to 'units'. CONS and PROD, however, are variables which deplete and increase INV, that is they are flow variables. Both should, therefore, be measured in, say, units per week. Thus if PROD were 1000 units/wk and sales had stopped, then after one week INV would increase by 1000 units, by 10000 after ten weeks, by 100 after 1/10th of a week and so on. The effect of CONS would be analogous but in the opposite direction, and the joint effect of PROD and CONS could be determined for any length of time during which CONS and PROD were known. In general, in a period of time which is DT weeks long, and during which neither PROD nor CONS varies, the effective change in inventory will be

$$\text{DT} \times (\text{PROD} - \text{CONS})$$

We extend the notation to take account of the time factor, by writing after each variable name the time to which we are referring, separating the two by a dot. Thus INV.9 would be the inventory at time = 9.

To be more general, consider three succesive points in time, denoted by J, K and L. The durations of the intervals between J and K and between K and L are both DT weeks. The successive values of inventory are INV.J, INV.K and INV.L. Since production and sales both take place *during* these intervals we write PROD.JK for the production rate during the interval from J to K. We have to assume that DT is such that PROD and CONS do not alter during the time intervals. We then have

$$\text{INV.K} = \text{INV.J} + \text{DT} \times (\text{PROD.JK} - \text{CONS.JK})$$

and this is a CONSERVATION EQUATION There are a few points to note

1. The conservation is, mathematically, an integration of PROD and CONS to give

INV, which is a system state, PROD and CONS being flow variables which jointly affect INV.
2. The accumulation variables are called LEVELS (in control theory they are called STATE VARIABLES, but 'level' has become established in system dynamics). The flow variables, which are the *components* of the derivatives of the levels, are called RATES.
3. If there is one rate into a level, the value of the rate is the derivative of the level, but a level may be affected by several rates.
4. It is conventional to regard K as the current moment of time. Levels are then expressed as in the previous equation, which shows how the current value of the level was determined by the old value and the rates which *flowed* during the *last* time interval JK.
5. Rate variables have a two letter postscript either JK or KL. It is conventional, when writing the equations which describe rates, to put RATE.KL on the left-hand side of the equation to show the rate which *will* exist during the *coming* time interval KL

We now turn to major principles. The first cardinal point about feedback loop microstructure is, as we have remarked, that the state of the system at any given time is described by the levels.

The second point was expressed in Fig. 2.4, but we may now restate it more exactly: *levels, and only levels, create rates and the rates in turn create new values of the levels, thereby moving the system state with the passage of time*, recalling that variables such as 'Average Sales Rate' are, mathematically, levels.

The third cardinal point is that there are necessary relationships between levels and rates which mean that the type of a particular variable depends on its position in the feedback loop relative to other variables and not merely on its units of measurement. Thus, the fact that a variable is measured in, say tons/week *does not necessarily mean that it has to be a rate*, a level may very well have such dimensions.

Levels and rates are necessary and sufficient to provide the microstructure of feedback loops and thus of the system as a whole. In practice it is convienient to allow for the presence of a third variable type, the AUXILIARY, because systems often involve complicated processes by which the decision rules give the rates from the values of the levels, and intermediate variables may be useful. For instance, in Fig. 2.4, we had the intermediate variable 'discrepancy'. Such intermediate, or auxiliary, variables are not strictly necessary, because they are subsidiary parts of the rate equation, and could be substituted into them. However, this can make the rate equations very cumbersome and the auxiliaries are a considerable practical simplification.

An auxiliary has a single time postscript e.g. AUX.K to show that it is a subsidiary value, existing only at K, and used in determing rates for period KL.

Three other types of quantity are also found in system dynamics models. CONSTANTS specify parameter values, e.g. the number of weeks of average sales which form target inventory, SUPPLEMENTARY VARIABLES indicate, to the analyst, the performance of the system, e.g. Cumulative Lost Sales, but do not form part of the system itself. In order to start the simulation process it is necessary to have INITIAL CONDITIONS for all levels and for some rates, as described in Chapter 5.

The structure of decision processes in Fig. 2.4, and the ideas of flow and accumulation mean that a system consists of a collection of feedback loops, each of which includes variables of different types. There are necessary relationships about the way in which variable types are connected by the loops, and these must be understood and used in order to get a correct and useable formulation of the structures of a system.

The relationships are:

1. Since a level is an integration of a flow, *a level in a loop can only be preceded by a rate.*
2. Levels are the means by which the system acts to control itself through its decision rules, which produce the flows or rates. The decision rule may be so involved as to call for the use of auxillary variables, therefore, *a level may be followed by an auxiliary or a rate.*
3. An auxiliary is a step in the determination of a rate, but it may not be the only one and, therefore, *an auxiliary may be followed by another auxiliary, or by a rate.*
4. A rate has to be accumulated into a level and *a rate must be followed by a level* (subject to an important class of exceptions which are examined in Section 2.7).
5. *A level may not directly affect another level.* The connection between two levels can only be through an intervening structure of at least one rate and, possibly, one or more auxiliaries.

The axioms of system structure mean that a feedback loop MUST contain at least one level *and* at least one rate. A closed chain of auxiliaries would not be a feedback loop and would have no dynamic properties.

At this point we must comment on the important similarities and differences between rates and auxiliaries. Auxiliaries are components of the processes by which the rates of flow are to be determined by the decision system. They can be substituted into the rate equations if we wish to do so and they are, therefore, not necessary to the system in the way that levels and rates are. Thus, mathematically, there is no real difference between rates and auxiliaries and the only advantage of the latter is that they break cumbersome rate equations into simpler components and make the model easier to understand.

In terms of the way in which systems work and can be modelled, auxiliaries are much more important and are significant components of the system structure. They shows how the stages in decision-making stem from the values of the levels; they are observable and their values can be recovered from the decision process and used in other ways if we wish. At the end of the decision-making chain they become rates, but rates cannot be observed, either in theory or in practice. The auxiliaries thus trace the fine structure of the system and we, therefore, treat them as a category of variables in their own right even though we do not really need to in strict mathematical logic.

In Chapter 3 we discuss a method for determining exactly what type a particular variable should be and thereby detecting errors in model formulation. Later in the book when we have developed more skill in system modelling we shall be able sometimes to simplify model programming by relying on the mathematical identity between rates and auxiliaries and relaxing some of the formal rules of the next few chapters. For the moment we summarize them as follows:

1. *Levels* are usually easily recognized as they are the results of accumulation e.g. stocks, or of averaging. They are the variables whose values would not drop to zero if all the flows in the system were stopped.
2. *Rates* are the flows in the system. They cannot be observed and their values used except by accumulation or averaging – both of which create levels. Rates can only flow into and out of levels, except when they feed a DELAY which, as we shall see, involves hidden internal levels.
3. *Auxiliaries* give the fine structure of the system in the way in which the levels govern future rates. They are observable, because they stem from levels, and they can be used to create new system stucture if we wish to do so.

2.7 Delays

A common element in a system are delayed flows which are usually very important in determining the system's behaviour.

Examples of delays abound in industry, for example production is ordered at some time, but completed at a later time, workers are recruited to the labour force, but become fully trained later, production capacity ordering and commissioning are separated by a time lag, and so on.

These examples show that a delay involves the presence of a quantity or accumulation of the material being delayed. A delay is, therefore, another occurrence of the level in a system, and a simple delay consists of an inflow rate, a level in which the material is held during the delay duration, DEL, and an outflow rate.

The equations for the level and the inflow are:

$$\text{LEV.K} = \text{LEV.J} + \text{DT}*(\text{IN.JK} - \text{OUT.JK})$$

IN.KL – *exogenous to the delay*

Now, in the steady state, the inflow is constant, and the average duration of delay is DEL. The amount in the level must be, dropping the time postscript,

$$\text{LEV} = \text{IN} * \text{DEL}$$

The outflow rate in the steady state is equal to the inflow and must therefore be

$$\text{OUT.KL} = \frac{\text{LEV.K}}{\text{DEL}}$$

This is an EXPONENTIAL delay and it is said to be of FIRST ORDER because it contains one intermediate level. More complicated delays can be constructed by considering the outflow rate of the first delay to be the input rate to another first order delay. The outflow late of the second delay is the final outflow. The two delays are said to be CASCADED and, jointly, comprise a SECOND-ORDER delay. If DEL is the average total time that an entity spends in a second order delay, each of the two first-order delays is given a delay duration of DEL/2.

Any number of first-order delays can be cascaded to produce a higher-order delay. In practice, third-order delays are very common and will be denoted by DELAY3. For short-hand we write

OUT.KL = DELAY3(IN.JK,DEL)

in which the time postscript on IN and OUT should be noted, DEL is the *total* delay and each of the three first-order delays has a time-lag of DEL/3. (This is all provided automatically in DYSMAP.)

Very high-order delays can be approximated by cascading the required number of third-order delays (see the Annexe to this chapter). If n third-order delays are cascaded each has a delay duration of DEL/n and the individual delays making up the standard DELAY3 will have a duration DEL/3n.

The limit to increasing the order of a delay by cascading more and more delays of ever shorter duration is the hypothetical infinite order delay. This is also called a perfect, pipeline, or discrete delay, and *exactly* reproduces the time-form of the input, DEL time periods delayed.

So far we have dealt with delayed flows of material, but rather similar equations can be used to represent the delaying of information by averaging it. The well known exponential smoothing equation is:

$$A_t = \alpha\, C_t + (1 - \alpha)\, A_{t-1}$$

where A is the average value of some variable C, t is time and α is a damping constant.

Rewriting,

$$A_t = A_{t-1} + \alpha(C_t - A_{t-1})$$

and if α = DT/ST we have, in the notation of the conservation equation

A.K. = A.J + (DT/ST)(C.JK – A.J)

where ST is called the smoothing time.

This first-order information delay has substantially the same properties as the ordinary first-order delay in that A, the average of C, follows the variations in C but later, and less sharply. Clearly, the averaging process is mathematically equivalent to integration of the term (C.JK – A.J)/ST which is the rate of change of the average.

The information delay is treated more fully in Chapter 5, but it is worth comparing it in more detail with the equations for the ordinary delay, by violating some of the conventions for equation forms which we shall later impose. The ordinary delay has the equations:

LEV.K=LEV.J+(DT) (IN.JK–OUT.JK)

and

$$OUT.KL = \frac{LEV.K}{DEL}$$

Rewriting, with appropriate changes in the time postscripts leads to:

$$DEL \times OUT.KL = DEL \times OUT.JK + (DT)\,(IN.JK - OUT.JK)$$

or

$$OUT.KL = OUT.JK + (DT/DEL)\,(IN.JK - OUT.JK)$$

The similarity between this equation and the information delay is very clear and justifies our use of the word 'delay' in the information context.

2.8 Behaviour of Delays

In its steady state, a delay has no effect on the dynamics of the system because the outflow and the inflow rates are equal. If, however, the inflow rate varies, the outflow rate adjusts to the new inflow by following a pattern of TRANSIENT BEHAVIOUR. Since the inflow rate is under its own controls, the transients in the output modify the control effectiveness and *this accounts for the importance of delays in the dynamics of a system*.

The behaviour of the output from a delay of various orders when the input steps up from 100 to 150 units per week and when the input has an impulse and then reverts to its original value of 100 are shown in Fig. 2.5 a and b respectively. (Note suppressed-zero vertical scales). In all cases the delay has a total duration of 10 weeks.

The behaviour of the delays is mostly easily understood by considering Fig. 2.5b. For the first-order delay the equations are:

$$IN.KL = \textit{Impulse}$$

$$LEV.K = LEV.J + DT \times (IN.JK - OUT.JK)$$

$$OUT.KL = LEV.K/DEL$$

The impulse in IN causes LEV to rise sharply by the amount injected i.e. by IN. OUT immediately rises to LEV/DEL but this is much less than IN. The level is, therefore, steadily depleted and this causes the progressive decline of OUT.

The third-order delay consists, as we have said, of three levels in a row. The first is fed by IN and the last is depleted by the final output, OUT. Between the first and second levels and between the second and third there are two hidden flows D1 and D2 which deplete one level and feed the next. (Write the equations.)

The impulse in IN stokes up the first level and D1 behaves exactly as the first-order output shown in the diagram. This feeds the second level, but less dramatically. D2 now starts to increase, feeding the third level which in turn allows the final output, OUT, to start to rise. In the meantime, LEV2 has been built up by

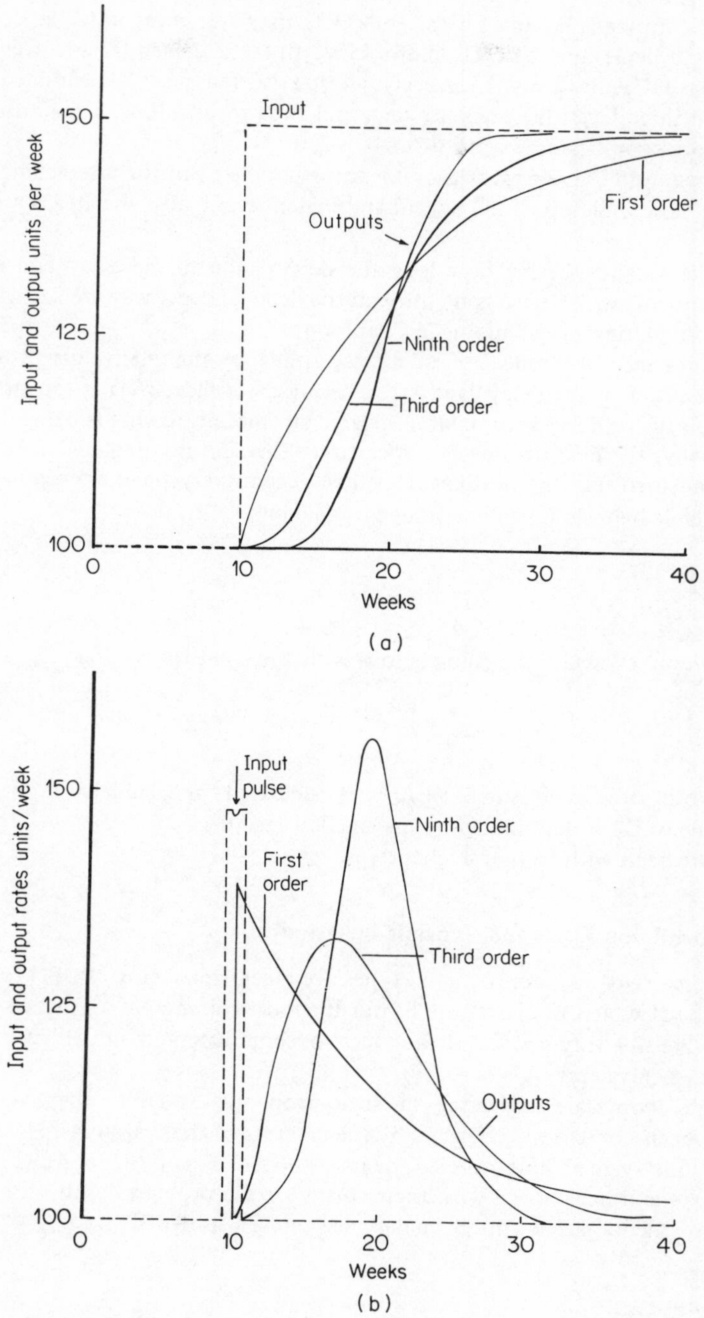

Fig. 2.5(a) Response of delays of various orders to a step input (b) Response of delays of various orders to an impulse

D1 which in turn feeds D2 and LEV3. Thus OUT rises even more until the decay in D1, due to the depletion of LEV1, allows LEV2 to fall and then D2 and eventually LEV 3 falls as OUT dies away. Ultimately, all that was fed in by IN is drained away by OUT, the three levels die away to zero, and the process stops. The ninth-order delay is a more complicated version of this.

The behaviour in Fig. 2.5a works on the same principle but the intervening levels produce the slow build-up of the output and match, eventually, the steady value of the input.

The reader should observe how long the delays take to die away and should study the amounts which have gone through the delay at successive weeks, when he has studied simulation programming in Chapter 4.

The performance of a model of a delay is affected by the relative magnitudes of DEL, n (the order of the delay) and DT. The reason is that, if DT is too large, the internal levels of the delay accumulate too large an amount of what is being delayed and, eventually, the outflow rate becomes large. This, in turn, depletes the internal levels too much so that the outflow rates then become very small or even negative.

This unstable behaviour can be avoided by making

$$\mathrm{DT} < \frac{\mathrm{DEL}}{2n}$$

It is, however, shown in the Annexe to this chapter that for accuracy,

$$\mathrm{DT} < \frac{\mathrm{DEL}}{10n}$$

where, in both these expressions, DEL/kn is calculated for each delay in the model and the value of DT is determined by the smallest result.

Delays are dealt with further in chapter 5.

2.9 Feedback Loop Types and Dynamic Behaviour

There are two and only two types of feedback loop, POSITIVE and NEGATIVE, of which the latter are by far the most common in managed systems. The names are not very good, but they have become accepted, in the face of more cumbersome equivalents.

Although loops are universal in socio-economic systems, they have not necessarily been consciously designed by the creators of those systems. Most of our socio-economic systems have evolved, or just 'happened', and the loops which they inevitably contain have arisen with them. Many of the problems with our systems exist *because* the loops have not been designed or even studied.

Positive Loops

A positive feedback process is one which acts to reinforce a change in a system level and moves the level even further in the same direction as the initial change.

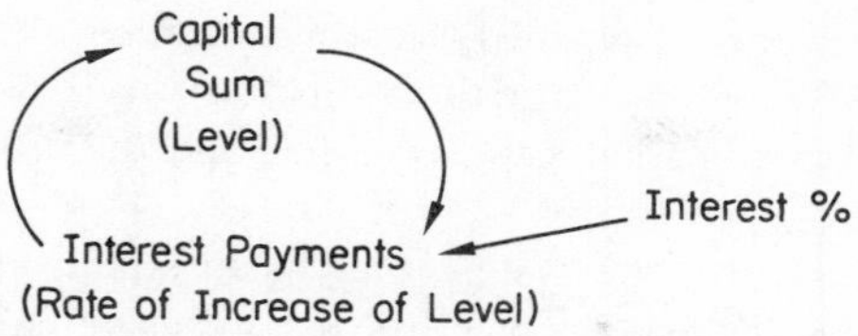

Fig. 2.6 Growth of a sum of money with reinvested interest

The classical example of this is the growth of a sum of money with reinvested interest as shown in Fig. 2.6.

The result is an ever-increasing growth in the capital amount. For example, £100 invested at 6% compound interest amounts to £180 after 10 years, to £33930 after 100 years, and to £2.023 x 10^{27} after 1000 years. This last amount is so large that the interest payments in the 1000th year would be about 10^{15} times GNP of the USA. Clearly, positive feedback cannot go on indefinitely, and it must come up against some kind of limit.

Before its limit is reached, the positive loop is capable of producing beneficial results for the system. For example a company's R and D expenditure produces profitable products which provide more R and D expenditure for more products and so on. Commonsense indicates that the shorter the delay between the R and D spending and the marketing of the product, or the higher the proportion of

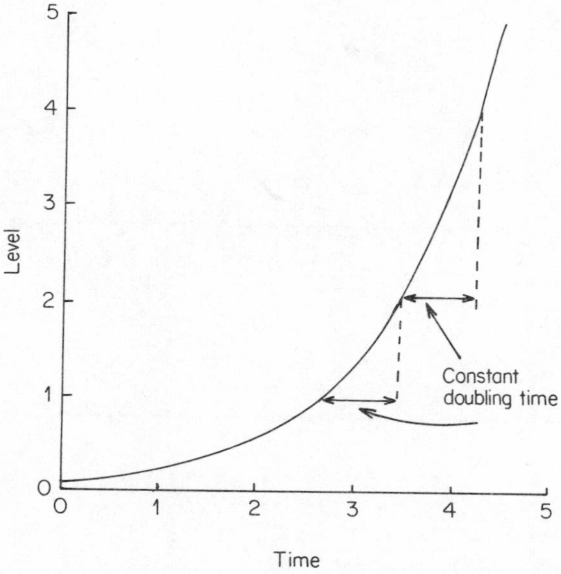

Fig. 2.7 Idealized behaviour modes of a positive feedback loop

successful products per unit of R and D expenditure, then the steeper the growth curve. However, the steeper the growth the harder the blow when the ultimate limit, perhaps one of total market size, is reached and the more suddenly will the limit be encountered.

A positive loop can drive the system ever more rapidly downw ards, as when losses are met by price increases, which lose sales causing larger losses and so on. The situation is made worse by the interest charged on the loans raised to meet the losses. This has typically been the experience on certain industries such as railways. In this case the complement may intervene by declaring that, for example, social considerations make the concept of loss irrelevant and this effectively removes the positive loop from the system.

In short, positive loops are GROWTH PRODUCERS, (though the 'growth' may be a crash) and the typical behaviour mode is shown in Fig. 2.7. Note the characteristic pattern of a period of time during which the level doubles. The DOUBLING TIME is the same whether the level is going from 1 to 2 or from 100 to 200.

Negative Loops

A negative loop is goal-seeking, that is it tries to move a level towards some desired target by creating action in the opposite direction from the discrepancy between the actual and the target values for the level.

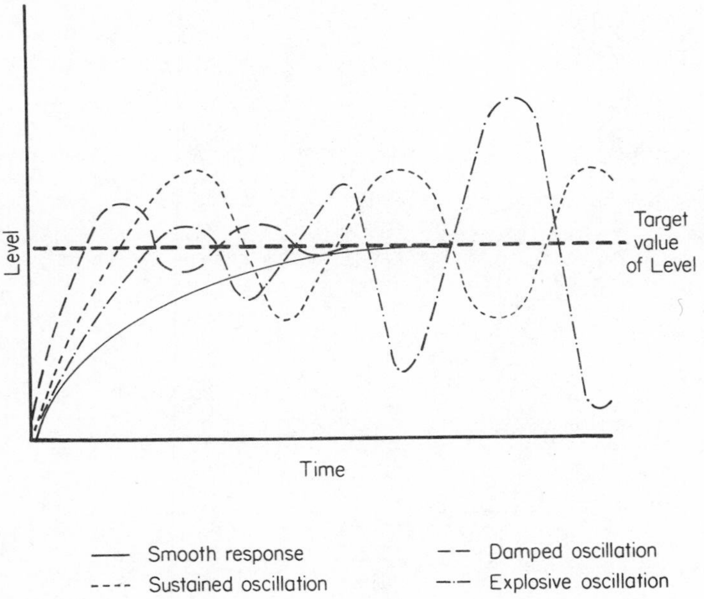

Fig. 2.8 The behaviour modes of a negative feedback loop

For example, many production systems are intended to adjust the rate of production in order to make actual inventory move to some desired level. Such systems act by increasing production when inventory is too low and cutting it back when stocks are excessive. They are, therefore, negative feedback systems because they are acting to eliminate errors in the system. Their subtleties lie in how large an error has to be before an action is generated, and how severe that action is.

Delays in a negative feedback loop may defeat the loop's attempts to achieve its purpose of bringing the actual value to the target value. There are, in fact, four possible modes of behaviour for a negative loop, which are best visualized by considering the loop's attempts to close some gap between the actual and target values. The four cases are:

1. Smooth response
2. Damped oscillation
3. Sustained oscillation
4. Explosive oscillation

These are shown diagrammatically in Fig. 2.8.

Although the last of these modes leads to ever larger changes it is NOT a positive feedback process.

2.10 The Principal Types of Dynamic Behaviour

Dynamic behaviour can be more complex than the basic behaviour modes, in particular when both types of loop are present in a system. We now turn to a consideration of the six principal types of dynamics. At this stages we simply exemplify the principal types giving, without proof, explanations of the behaviour in terms of the presence, absence, or conjunction, of the two types of feedback loop.

S-curve dynamics (Fig, 2.9)

This has an initial period of low growth, followed by quite rapid growth, with final settling to a steady value. This suggests a system disturbed by a shock from its

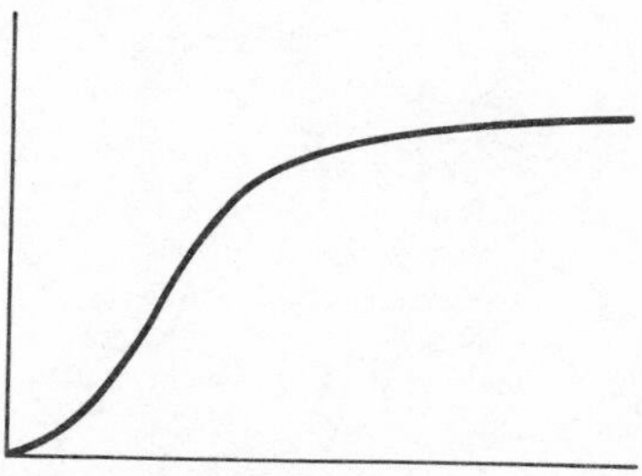

Fig. 2.9 Dynamic behaviour: S-curve dynamics

environment and adjusting to the new conditions. Obviously, there could be adjustment to a lower, rather than a higher value.

Such behaviour can be desirable, in a system which is largely dominated by its environment, and which simply responds to whatever the environment demands. For such cases, one might wish to reduce the time needed to make the transition to the new value, and to make the settling as smooth as possible. Alternatively, for a system which is subject to very frequent shocks, which are often of opposite sign, one might wish to increase the time needed to move to the new value. This would prevent the system from chasing a constantly-moving target, and would, therefore, insulate it from the environmental shocks. This is what stocks are supposed to do, whether they be of product, money, or any other resource.

This dynamic mode illustrates a classical dilemma. If the environmental shocks are frequent, and often of opposite sign, for example short-term market movements, we would like the system to be as insensitive as possible, thus smoothing out the external effects. On the other hand, a system should adjust as rapidly and smoothly as possible to permanent changes, for example to meet a genuine change in the market. In advance, even with the aid of forecasting methods, it is in practice very hard to know what kind a particular shock is. The designer of the system must, therefore, compromise between the two alternatives.

Oscillation (Fig. 2.10)

In this case, the variable moves up and down, either about a steady average value, or with respect to an upward or downward trend. The behaviour is often very irregular. The oscillations may be stable, explosive or damped. It is very common to have oscillations about the steady final value of the s-curve.

The oscillations shown in Fig. 2.8 arise from a single feedback loop and should not be confused with those referred to above, which may be the consequence of a *system* of many loops.

Oscillations may or may not be undesirable, depending on the variable concerned. For example, Fig. 1.4 shows the oscillatory behaviour of the balance of payments, and this is certainly not desirable. On the other hand oscillations of inventory would be perfectly acceptable, providing they indicated that the

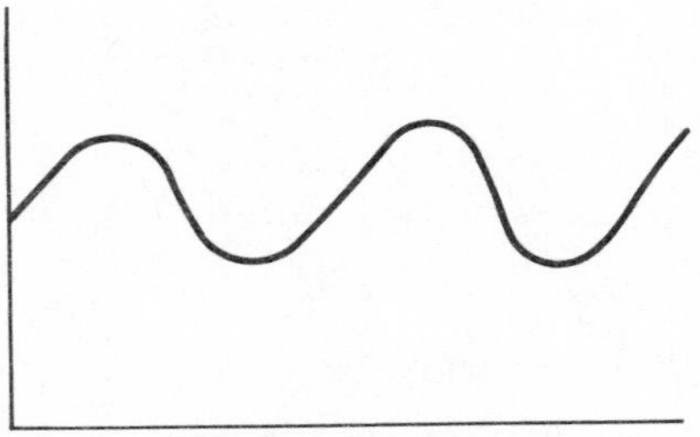

Fig. 2.10 Dynamic behaviour: oscillation

inventory was fulfilling its function of insulating production rates from short-term market variations.

For undesirable oscillation one would aim to reduce them, or to make sure that they were heavily damped and died away rapidly. In all circumstances explosive oscillation would be undesirable.

Oscillatory behaviour suggests an attempt to attain some kind of goal or stable value. Damped oscillation implies success, the heavier the damping the greater the success. Continued oscillation implies a system which either cannot attain its goal or which is continually knocked away from it by outside shocks, which, because the oscillations don't become any larger, it is fairly succesful at overcoming. If the oscillations are explosive, the system is obviously completely unsatisfactory.

Oscillatory behaviour is often attributed to seasonality in some external factor, when the duration of the oscillation and the location of its peaks coincide with annual occasions. This is a very tempting suggestion, particularly since it is manifestly the correct one in many instances. It is not, however, always the correct explanation because, and the apparent 'seasonality' may simply be the system's response to a single shock from its environment.

Peaked Oscillation (Fig. 2.11)

This is very common for commodity prices, for example, the spot price of copper shows this pattern, though the peaks may last for a year or more. The behaviour arises in a system which seeks to adapt itself to movements in external demand by bringing forth an additional supply. Because the time needed to expand production in a copper mine is in the order of a couple of years, the peaks tend to last that long. The new supply then reduces the peak, and the oscillations between peaks are the result of the system's attempts to adjust to small variations in demand. They may also be the result of inherent slight instability in the system.

Ideally one would also like to see the peaks eliminated, but this is very hard to achieve, given the very small price elasticities of supply and demand in a commodity market. For such a situation, a system dynamics study might have as its purpose the design of a policy for commodity stock-piling to stabilize the price. This would take account of the funds available for operating the stockpile, the points at which buying and selling should take place, and the policies for arriving at the amounts to be bought and sold.

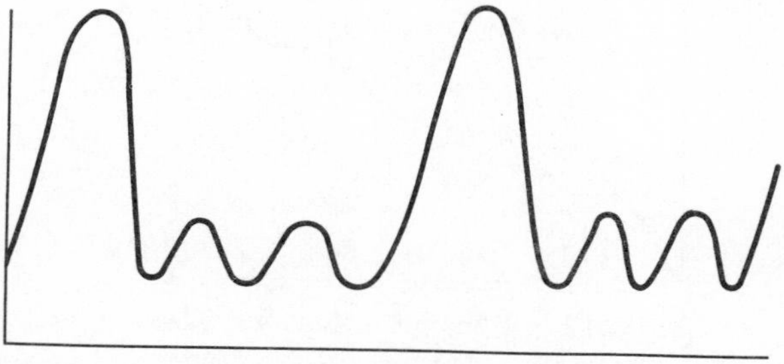

Fig. 2.11 Dynamic behaviour: peaked oscillation

It should now be clear that the first three types of behaviour are the result of the operation of negative feedback loops. Systems exhibiting these modes may contain positive loops, but the negative loops are dominating the system in order to produce these behaviour patterns. We shall now examine the converse situation.

Exponential Growth (Fig. 2.7)

Fig. 2.7 shows the ideal behaviour of a single positive feedback loop. In a finite world, it is impossible for such growth to proceed indefinitely, but growth does occur in managed systems, as was shown in Fig. 1.5 and System Dynamics investigations often focus on growth patterns.

Where the growth is desirable one would seek to strengthen the positive feedback. If it was undesirable, one might try to weaken the positive feedback, so as to slow the growth. This is not a real solution, as the growth does go on, and sooner or later will reach the original rate.

An alternative approach might be to try to introduce a negative feedback, or to strengthen existing ones. If this is possible and successful the situation will stabilize, though oscillations may set in. The stabilization may have unforeseen results, in that the positive feedback may reverse itself and go ever steeply downwards.

It is possible to have a double positive feedback, with one variable getting ever larger while another, related, variable gets progressively smaller. Some types of public transport have experienced this, where ever-mounting losses have been met by fare increases, which have led to an ever-increasing fall in the number of passengers and further losses. Such a process cannot continue, of course, as sooner or later the number of passengers reaches zero, or some captive lower limit. This is an example of a positive feedback process eventually being limited by some natural stop.

It must be pointed out that, even though the exponential 'growth' curve may be pointing downwards, it is still the result of *positive*, or self-reinforcing feedback.

Stair-step Growth (Fig. 2.12)

Stair-step growth is due to a mixture of positive and negative loops. During the growth phases the positive loops are dominant, while the negative feedbacks govern

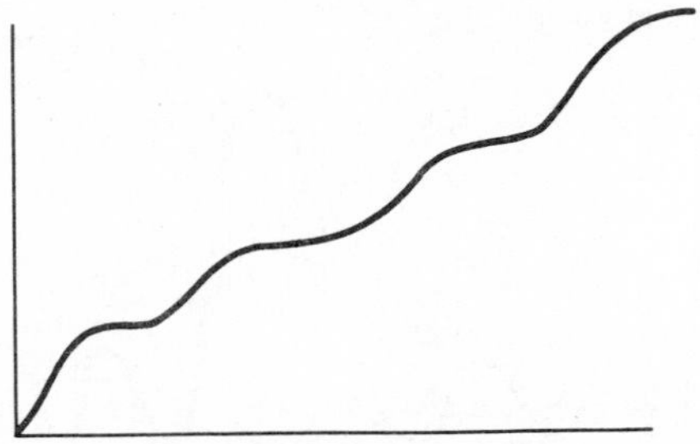

Fig. 2.12 Dynamic behaviour: stair-step growth

the behaviour during the static periods. Clearly there is some mechanism operating in the system which is making the dominance move back and forth. The purpose of analysing the system would be to find the dominance-shifting mechanism and to control it. This control could be aimed at keeping the positive feedbacks dominant, in order to enhance growth, or at reinforcing the stabilizing effects of the negative feedback if the system was declining.

Stair-step growth is very common in industrial firms. Periods of rapid expansion are followed by periods of quiescence. The latter may be due to the company having outrun the market, or to a need for time to absorb the earlier growth. Both of these situations are negative feedback processes.

Overshoot and Collapse (Fig. 2.13)

The system is attempting to grow towards some limit, and there must be such a limit in a finite world. Although the limit exists it is not necessarily impassable and, in the overshoot mode, the system does so. This releases forces which attempt to restore the system to the limit, and the collapse occurs because so much damage has been done to the system that it cannot sustain its existence at the limit.

The classical example of this mode is the firm which achieves a rate of growth, or an ultimate size, beyond what its capitalization can sustain and collapses into bankruptcy, or, as shown in the diagram, into a much lower level of activity than could have been achieved if a proper transition to the limit had been achieved.

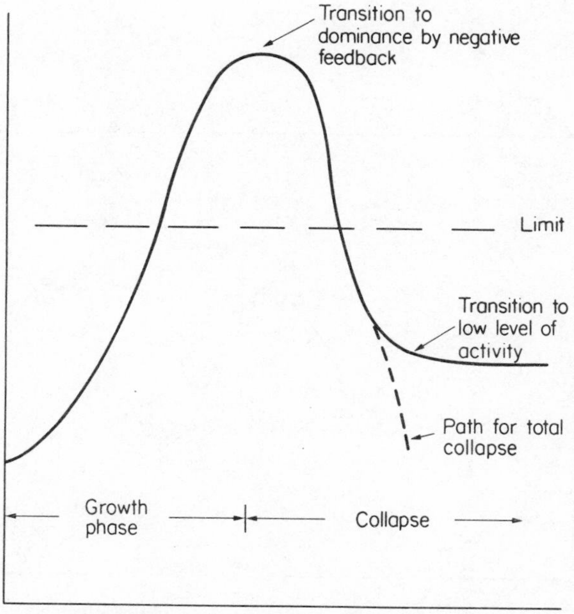

Fig. 2.13 Dynamic behaviour: overshoot and collapse

The overshoot and collapse mode is characteristic of a system containing positive feedback to produce the growth and negative feedback which attempts, unsuccessfully, to adapt the growth to the system's limit.

So far we have examined types of behaviour, without considering the time-scale, or the simultaneous, but possibly different, behaviour of two or more variables.

2.11 Time Factors in Dynamic Behaviour

Nearly all dynamic behaviour of practical interest contains a mixture of two or more of the principal behaviour types. There may be transition from one type to another, or short-term movements super-imposed on a long-term trend of the same, or another, type. This latter situation is shown in Fig. 2.14a for oscillations imposed on s-curve dynamics, and in 2.14b for oscillations about other oscillations. The diagrams relate to company sales over a three-year period, and production rate over a three month period, respectively. These variables are chosen for illustration only, and the two examples are not related in any way.

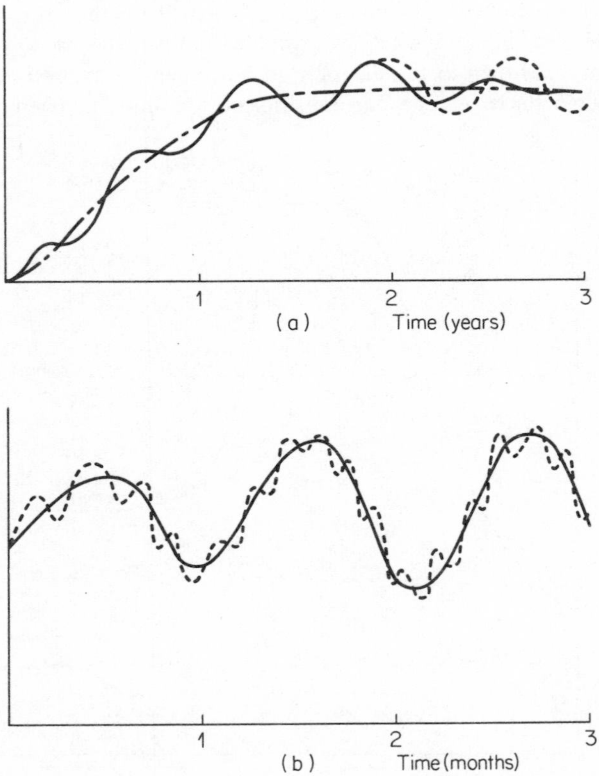

Fig. 2.14 Superposition of behaviour types

In Fig. 2.14a, the solid line shows the actual values and the chain-dotted line picks out the general trend of the s-curve. The dotted line showing fluctuations during the third year illustrates the possibility that they do not die away, as do those of the solid line.

Clearly one could be interested in two quite different questions for this data.

Looking at the s-curve (or sigmoid) one might analyse how the growth could be improved, perhaps by inhibiting the forces which led to the flattening, so that it takes place at a higher level, or by inducing the growth forces to come into play again, thus leading to something like stair-step growth. This kind of study would involve a model with a time-scale of years and many of the purely short-term factors in the system could be ignored.

On the other hand, examination of the same data reveals the fluctuations, and one could conduct an entirely different analysis of the same system designed to reduce them. This would have a time-scale of months, and many of the slower-reacting features of the system could be omitted.

The result is that, since the data can be viewed from two different perspectives, it is possible, and, in fact, reasonable, to build two quite dissimilar models of the same system, designed to answer different sets of questions. Thus, the form of a model is determined by the purpose for which it is being constructed, and it is only relevant to judge a model by how well it enables one to answer questions. If one model happens to fit the data better than another, it does not follow that the first is better than the second, unless both are equally useful for answering the questions about the system which we wish to ask.

The second example in Fig. 2.14, shows oscillations of actual daily production rate superimposed on oscillations in monthly average daily production rate. Again the different time scales pose different questions.

The monthly average figures may be the result of conscious managerial policy acting on the system through monthly production-planning meetings. One could study this system with a view to designing decision rules which would reduce the fluctuations; always assuming, of course, that they are undesirable, and that better control of production will not make matters worse elsewhere in the system.

The short-term variations are a different matter entirely as they may be due to attempts to adjust, on the shop floor, to the changing production planning targets set at the monthly meetings. Because the dynamics are relatively small, compared to the monthly average values, it may not be worth trying to tackle this problem by system dynamics even though it is a dynamic problem.

If the daily fluctuations arise from random variations due to machine-breakdown, machine interference, etc., the problem is not a system dynamics one at all, and other approaches would be appropriate.

Phasing in Dynamic Behaviour

Examining the behaviour of two or more variables over a period of time to see if there are any apparent connections or PHASING between them, usually improves understanding of system behaviour.

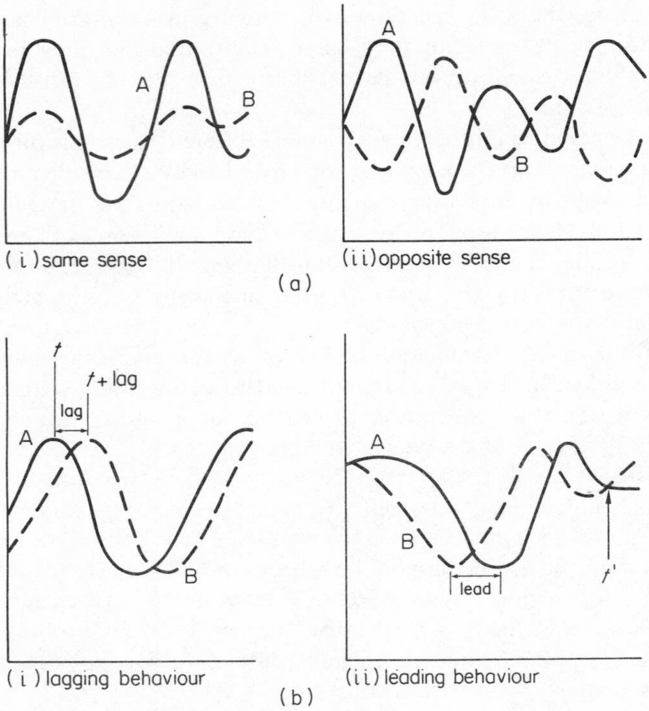

Fig. 2.15 Phasing in dynamic behaviour

Fig, 2,15a shows related dynamics for two variables A and B. The behaviour is IN-PHASE in that A and B reach their peaks and troughs simultaneously. There are two possibilities, either B peaks when A does (SAME SENSE), or it troughs when A peaks (OPPOSITE SENSE).

The production-inventory system furnishes examples of these behaviour types. If A is the production completion rate then, Fig. 2.15a (i) could show the relationship between production completion and inventory when consumption is fairly constant. There are no vertical scales in Fig. 2.15 so the variations in B are not necessarily to be compared directly with those of A.

Similarly, in Fig. 2.15a(ii), if A is the consumption rate, B could still be inventory, if production is fairly uniform.

The opposite of in-phase dynamics is, of course, out-of-phase behaviour, two examples of which are shown in Fig. 2.15b.

The first possibility is that B LAGS A, but has essentially the same behaviour pattern. In Fig. 2.15b(i), B is lagging behind A and not, despite appearances, leading it, becuase B reaches its peak (or any other point on its pattern), a period of time after A does. Thus, if A peaks at t, then B will do so at 't + lag' i.e. 'lag' time units *later* than A.

In leading behaviour, B reaches its trough before A does, as in Fig. 2.15b(ii).

We have loaded the argument considerably by stating, for example, 'B lags on A'. How can we tell from case b(i), say, whether B is lagging behind A, or A is lagging B? From the diagram alone, we cannot tell which is which, and we cannot necessarily infer anything other than a purely coincidental relationship between the two variables, *simply* by looking at their dynamic behaviour. We need more information about A and B to say whether there is any connection between them and, more importantly, to say how the connection operates, i.e. does A affect B, or does B affect A? This kind of causal inference is the key to building dynamic models and we shall study it in more detail in Chapter 3.

Certainly the dynamics of the system may throw some light on how the system works. For example, the dynamics of Fig. 2.15b(i) would suggest a connection between two variables in the same system which showed that behaviour, but this would not be sufficient evidence to sustain the case that changes in A led directly, but later, to changes in B. Practical dynamics are not as neat as those of Fig. 2.15 which makes it more difficult to infer causal relationships from dynamic evidence alone.

In the absence of causal evidence it is hard to say what 'leading' and 'lagging' mean. In Fig. 2.15b(i) for example, we say that B is lagging behind A but in fact it could be leading it by quite a large amount. In fact it would be difficult to say much about the two cases Fig. 2.15a. In a(ii) it might even be that what we have called opposite sense is really a fairly large degree of lead or lag. (The technically-orientated reader should review his knowledge of phase relationships in periodic functions.)

One helpful key to using dynamic behaviour as an aid to understanding the system it to take account of the vertical scale in the figure, which we have deliberately excluded from Fig. 2.15 and to consider whether the variables involved are likely to be levels or rates by thinking about their units of measure.

For example if, in Fig. 2.15b both variables are measured in terms of units per period of time and drawn to the same vertical scale, we should have some grounds for supposing that the one is leading or lagging the other, and that 'lag' or 'lead' is a measure of the time difference between them. A very obvious instance would be if, in Fig. 2.15b(i) A was called 'production order rate' and B was 'production completion rate', then we should be surprised *not* to see that kind of dynamics and we should be perturbed if the value of 'lag' did not tie in with other information about the magnitude of the delay in the production process.

Similar considerations apply to Fig. 2.15a when one variable is considered as a rate and the other as a level. We have already exemplified this by cases of production and consumption rates and associated inventory variations, and the reader should, perhaps, pause to think of examples from, say, biology and economics.

These are simple cases involving only two variables and the reader should review Section 1.6 for an example involving three variables.

In practice, simple comparison between A and B is clouded by the effects of other variables in the controller, the environment, or the complement and very complicated behaviour patterns will arise. It is, however, useful to look at the historical dynamics of the system, where the data can be regarded as reliable, in order to improve understanding of how the system behaves. This is a process which lends itself better to practice than to description. The reader should, therefore, study actual situations and attempt to use the dynamic evidence, as a supplement to other available information, to infer the workings of the system. He should also study closely the cases in later chapters.

2.12 An Example of System Behaviour

Having discussed dynamic processes and the basic components of system structure, we now apply these concepts to a simple example. The computer program for the model appears in Appendix B, Section 2.12 and a mathematical analysis in Appendix A, Section 8.

The system is a simplification of that used by many manufacturers of consumer goods, which are sold from stock. The stock is depleted by consumption which varies in two ways: firstly it moves fairly abruptly from one level to another; secondly, even at a given level it still varies quite sharply from one week to another, as a result of all the myriad factors which affect purchasing decisions. We assume that it is not possible to tell in advance, or at the time, whether a movement in consumption is random noise or a step change. The only way is to wait and see, and the consumption pattern cannot be forecast in any satisfactory way.

The firm must decide the rate of production for stock, which involves a delay. The firm has to construct a policy, or decision-rule, stipulating which information will be used in selecting the ordering rate, and how rapidly any gaps between actual and desired states are to be closed. The system structure is shown in Fig. 2.16.

The Order Rate can be based on any number of factors. The first is simply to

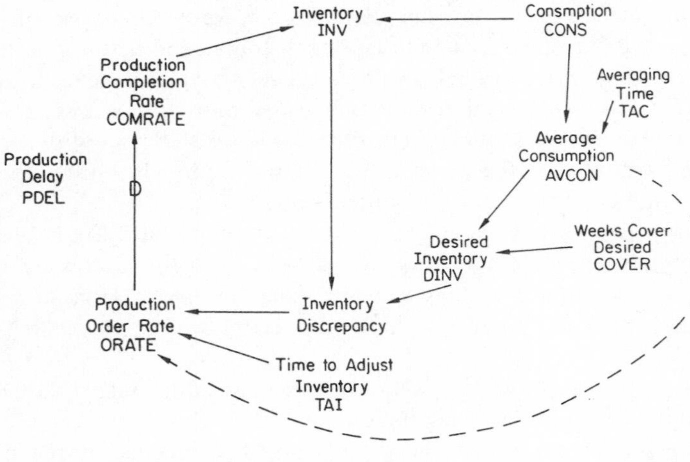

Fig. 2.16

consider the inventory gap and to have:

$$\text{ORATE} = \frac{\text{DINV}-\text{INV}}{\text{TAI}} \text{ and DINV} = \text{COVER}\times\text{AVCON}$$

A second option is to take account of the average consumption, AVCON, as indicated by the dotted line, and write:

$$\text{ORATE} = \frac{\text{DINV}-\text{INV}}{\text{TAI}} + \text{AVCON}$$

We shall refer to these two options as Inventory-based (IB), and Inventory and Order-based (IOB), respectively.

We wish to determine how each of these rules would stand up to the shocks imposed by the external environment. To do this we use two test inputs: a step change in consumption from one steady value to another, and a sine-wave in consumption with various frequencies. This tests the system's response to step changes and to noise. The reader should also use a RAMP input, in which CONS increases steadily, to see if there is any final error.

Step Response of the System

The response to a step-change in consumption with the IB order rule is shown in Fig. 2.17, with the behaviour of inventory shown for three different values of TAI.

With TAI = 1 the inventory oscillates explosively, which is one of the behaviour modes of a negative feedback loop, (The reader should identify such a loop in the system diagram). As TAI increases, the oscillations in the inventory become less acute until, at TAI = 6, the overshoot is quite small. When TAI = 10, (not shown) the oscillations practically vanish, and inventory exhibits the smooth response which is another possible behaviour mode of a negative loop. The dynamics in this case are not quite the same as the smooth response because inventory first falls below its initial level before starting its smooth rise to the new higher level. Why?

The reader should, when he has read Chapter 4, conduct simulation experiments of his own with this ordering policy, using various values of TAI.

The behaviour for TAI = 1 is woefully unstable. Inventory level goes from one extreme of massive positive stock to the other of huge negative inventories, with the extremes getting progressively further apart. Obviously, the system cannot do this in practice because the order rate has to do the same and this would make our implicit assumption of no constraint on the order rate increasingly unrealistic. In this case the model is predicting not the future course of inventory, but that the IB order rule, with that value of TAI, is not a practical proposition. The model predicts a crash but it does not, and cannot, say what form that crash will take.

The behaviour modes for the other values of TAI call for two comments, one on the level to which inventory settles, and the other one on the length of time it takes to get there.

We should expect this system, eventually, to settle to a stable level, even without the aid of the concept of negative feedback, and commonsense dictates that there is

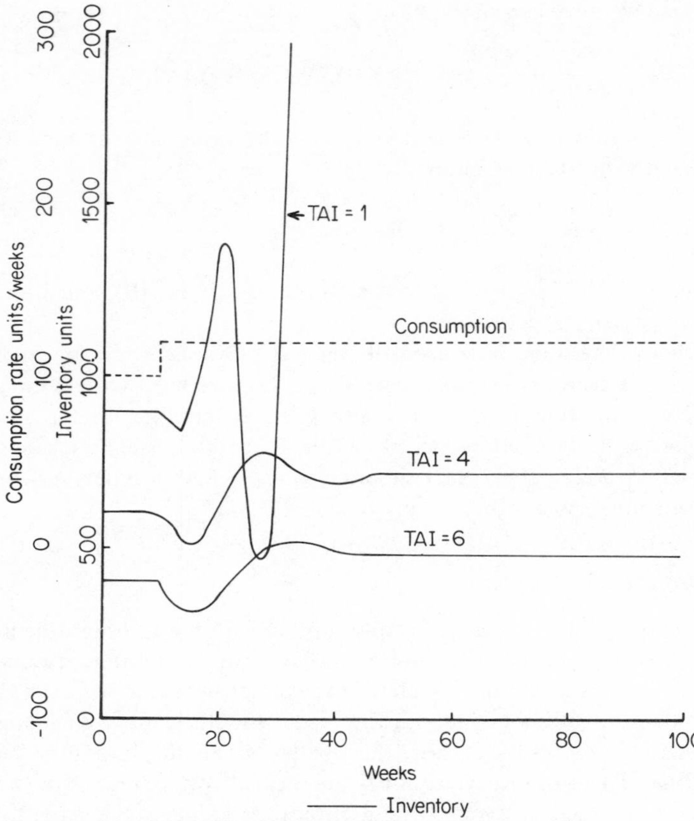

Fig. 2.17 Dynamics of the example with inventory-based control

something wrong with a system which does not settle to a steady value when a steady input is applied. (Stability is a very difficult concept, which will be explored fully in Chapter 6.)

In this case we should expect that the steady level should bear at least some relation to the Desired Inventory, DINV. This, in turn, depends on AVCON and AVCON does *not* depend on TAI. Why, then, does Fig. 2.17 appear to show that the steady level of inventory does depend on TAI?

The equation for ORATE shows that when INV = DINV, ORATE = zero. However, CONS is not zero so there will be depletion of INV and ORATE will start to rise. Indeed ORATE can, and does, rise so far that INV overshoots DINV because of the amount of production already ordered and in the production delay. When that happens, ORATE has to be negative, corresponding to cancelled orders. If TAI = 1, there is a very extreme case of this phenomenon. Clearly, for a steady state to arise, ORATE has to be equal to CONS so that the IB equation could be rewritten as

INV = DINV -- TAI x CONS

Since DINV depends on AVCON and, eventually, AVCON will be very close to CONS, we may put

INV = (COVER – TAI)×CONS

Thus, the steady state of this system is to fail to meet its target stocks by TAI weeks of consumption. In the model COVER = 10, so when TAI = 10, the steady-state is one of zero stocks and, if TAI > 10, the steady-state is to be out of stock.

The other dynamic feature is the SETTLING TIME, which is the time taken to return to the original stable state *or* to adjust to some new stable state after a finite shock. For this system, Fig. 2.17 shows this to be about 40 weeks when TAI = 4 and about 35 when TAI = 6. (The settling time is undefined for TAI = 1.) The model has a delay in the production process of 4 weeks, so settling time is surprisingly large.

The settling time can be calculated from the system equations, but it is very tedious to do so. The general point is that systems nearly always have very long settling times and the student should experiment with different values of PDEL *and* TAC to see what effect these have on settling time.

The first of these illustrates how one can use a system dynamics model to assess the effect on the system's *controllability*, as opposed to its economics, of a change in hardware. The second, and the changes in TAI, show how policy and controllability interact. We may, in fact, classify the system variables as POLICY or HARDWARE VARIABLES.

The long settling time suggests that the system may not have time to adjust to one shock before the next comes along. The reader should simulate this system with shocks of ± 20 units per week in consumption, spaced every 15 weeks, for TAI = 6. These are fairly small shocks, spaced well apart by real-life standards. The graph of Order Rate (which represents the factory load) will show the common phenomena of production panics followed by slack times. This happens to a greater extent than the actual consumption changes would indicate, and the system is, in short, making matters far worse than they need to be. How can matters be improved?

The performance, with the IB control rule, can be summed up as follows:

1. It can produce any one of the four behaviour modes, depending on the value of TAI.
2. It is slow in settling down after a shock.
3. The control rule is fundamentally incapable of doing what it is supposed to do, i.e. keeping inventory at its desired level.

Note that this system is not all that oversimplified. Control rules which are very like this simple IB rule are widely used in industry, perhaps the most widespread

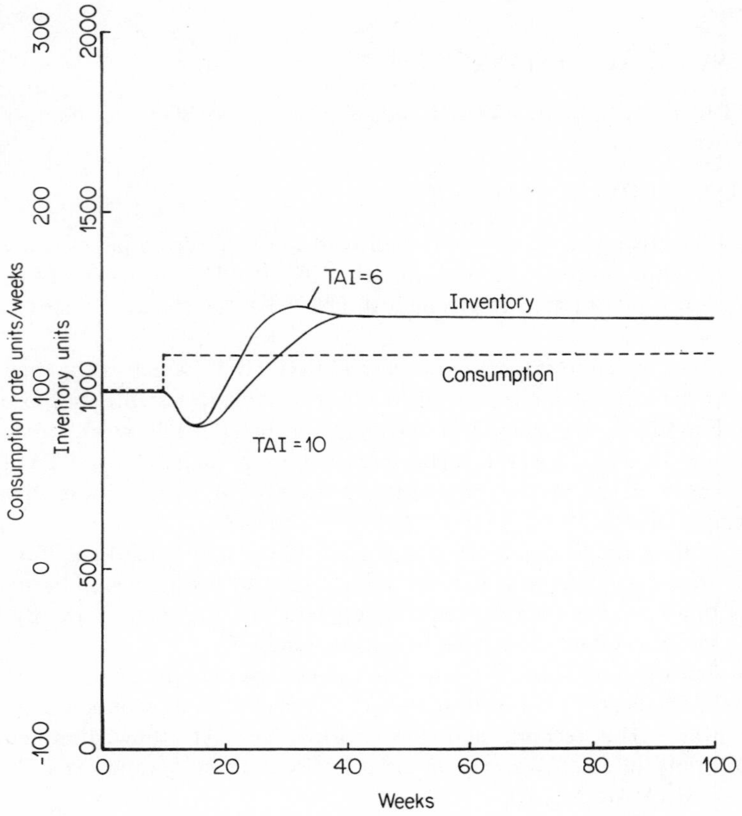

Fig. 2.18 Dynamics of the example with inventory and order-based control

example being the re-order point rules which are the basis of much of modern inventory theory.

Since the IB rule behaves fairly badly, we may study the step reponse of the system to the IOB rule for various values of TAI. The result appears in Fig. 2.18 which the reader should compare with the earlier figure. The reader should also conduct further tests on the model with both rules to study the effects of varying other parameters.

Frequency Response of the System

The response of a system to noise, for a given setting of its parameters, depends on the frequency of the noise. If the system is unduly sensitive to noise of the frequency which it encounters we shall seek to alter its parameters to tune out that noise.

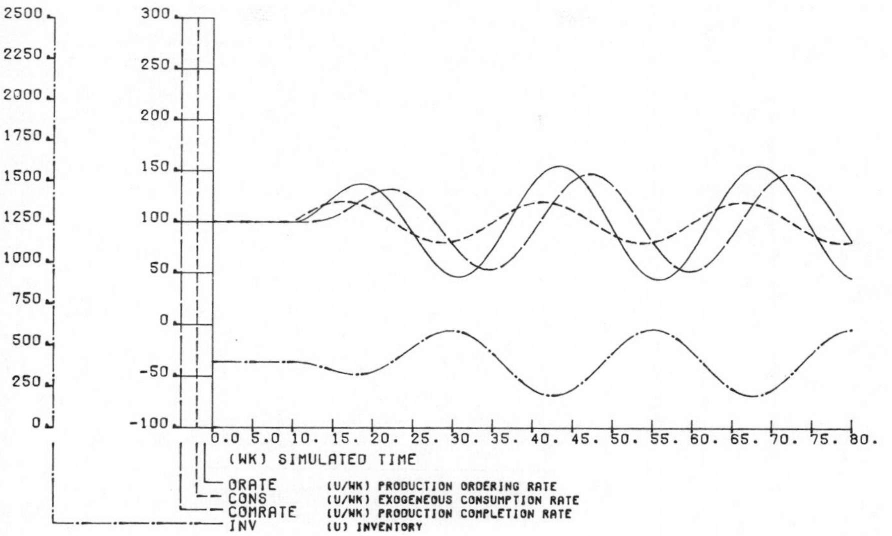

Fig. 2.19 Response to a sine input (IB control)

We shall illustrate for the IB case and the reader should himself study the other option.

The system behaviour in response to a sine wave of 25 weeks period with TAI = 6 and TAC = 4 is shown in Fig. 2.19. It settles very quickly into smoothly sinusoidal variation, that of inventory being about a level of 400 units (why?). The variation of order rate is about three times as large as that of consumption so, for this input frequency at least, the system amplifies the input frequency.

The amplitude ratio of order rate to consumption rate is plotted against the periodicity of the input in Fig. 2.20 for TAI = 6 and TAC = 4. The diagram shows that, with a monthly average of consumption, which affects DINV, the system amplifies considerably any noise periodicities greater than about 10 weeks. The amplification is serious for periodicities above about 15 weeks. The periodicity of 23 weeks, at which the amplification is at its maximum, is called the CRITICAL or NATURAL PERIODICITY. In short, this system would not behave at all well if it was subject to input noise with, say, quarterly, half-yearly, and annual, components, such as might arise in an industry dependent on movements in consumer buying power. On the other hand, it would very effectively damp down the short-period noise such as is caused by day-to-day or month-to-month variations in consumer habits.

It would be desirable to improve the response of the system by, in effect, shifting the curve of Fig. 2.20 to the right so as to reduce the amplification of the system for noise components of, at least, quarterly periodicity.

Consider TAC, the consumption averaging time. If TAC is short, AVCON will

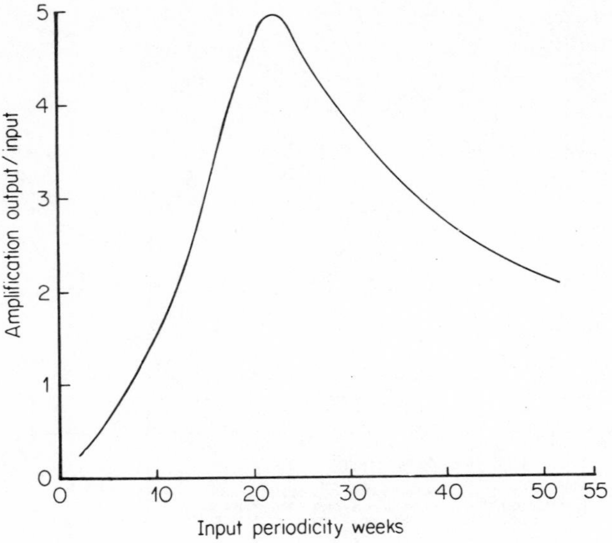

Fig. 2.20 Amplification in the system with TAI–6 and TAC=4

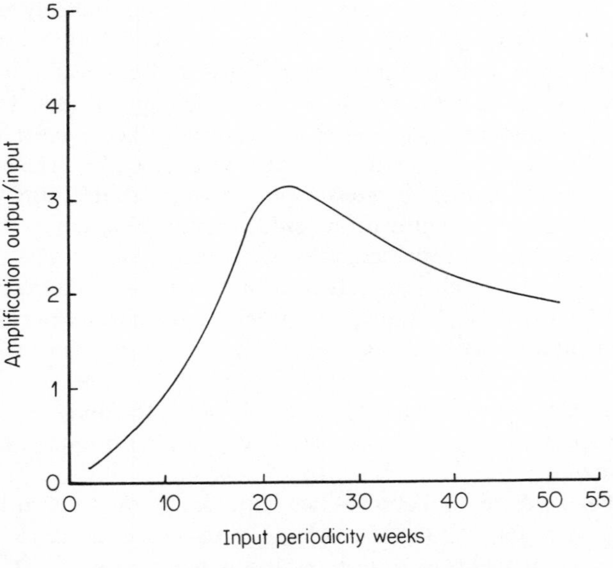

Fig. 2.21 Amplification with TAC=13 weeks

follow the sine wave in consumption fairly closely, more and more accurately as the input periodicity increases. This will make DINV follow CONS and will inevitably amplify ORATE as the system will be trying to build stock when consumption is rising and vice-versa when it is falling.

The effects of changing TAC are shown in Fig. 2.21, the change being quite dramatic.

The reader should conduct further experiments on this system. It will be found that, if the performance can be improved in one respect by a parameter change, it will almost certainly be degraded in some other respect by the same change. In short, *system design always involves compromises of one sort or another*. While this is strictly true, it is worth remarking that some systems are so bad that there is no trade off between improving one variable's behaviour at the expense of another, and overall improvement can be achieved.

2.13 Policies and Linearity

In Section 2.6 it was stated that, even though there will be many complicating factors and constraints in an actual case, the essential feature of decision processes was an action in response to a discrepancy (or ERROR SIGNAL) between the required and actual state of the system. The state may be as measured at present (though in fact that means as it was in the recent past) or it may be what it is expected to become in the time which must elapse before the remedial action which is contemplated can take effect. This viewpoint allows one to take actions intended to overcome the present discrepancy, together with any further deterioration which may take place and, perhaps allowing for the effects of any actions ordered in the past which have not yet had their expected effect. Even with these qualifications, the basic decision process is:

discrepancy → action

A given discrepancy gives rise to an action by means of the decision rule or POLICY. A policy can be represented by a graph and two production control policies, to give the amount of extra production (units per week) to be scheduled over and above the normal production level in response to a given shortfall of inventory from its planned value, are shown in Fig. 2.22.

Policies A and B both provide for production to be cut back if actual inventory exceeds the target level, but they are quite different policies and would lead to different dynamic performance if the system was attempting to follow a fluctuating exogenous consumption pattern.

There are infinitely many curves which could be drawn on Fig. 2.22 and the aim of System Dynamics is the design of the policy curves which will give the best system performance. The reader should contrast this concept of policy with the usual use of the word, and should try to draw curves such as Fig. 2.22 for policies in some system other than a simple production-inventory controller.

A linear system is one in which all the policy curves are straight and a non-linear one is one in which at least one policy curve is not perfectly straight. All systems of

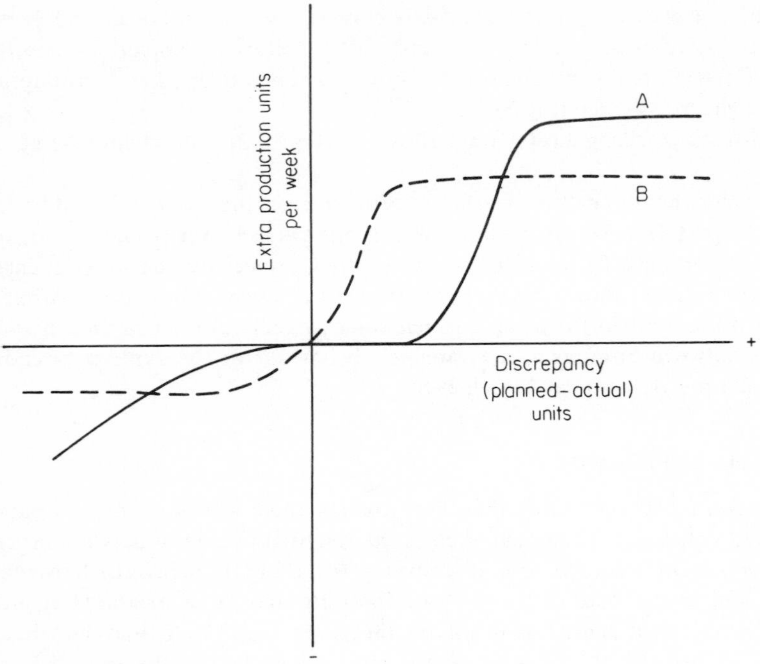

Fig. 2.22 Some production control policies

practical interest are non-linear, but the mathematics of control theory is essentially linear. This leads to the use of computer simulation, as exemplified in Section 2.1, to permit us to deal with systems which are very complicated and which contain many non-linearities.

The system examined in Section 2.12 is linear and, for such a system, the settling time, dynamics, and frequency response are not affected by the magnitude of the input nor by the state of the system before the input is applied. For a non-linear system neither of these propositions is necessarily true, which makes the analysis of such a system all the more difficult.

Annexe: Measuring the Order of a Delay

This section deals with a method, suggested in Holmes, D . S., *A probabilistic interpretation of the delays of Industrial Dynamics*, Union College, Schenectady, New York, 1970. By which the order of a delay can be estimated.

Holmes' method regards the level in a delay as containing a large number of discrete objects. This is a convenient approximation to the concept of a continuous system which is inherent in system dynamics, and accords with the manner in which the data needed to estimate the order of the delay would be collected.

A first-order delay has the equations:

$$\text{OUT.KL} = \text{LEV.K}/\text{DEL}$$

and

$$\text{LEV.K} = \text{LEV.J} + (\text{DT})(\text{IN.JK} - \text{OUT.JK})$$

and the derivation which follows is independent of the time form of the inflow.

The approach is to assume that the inflow has fallen from a steady value to zero. This is in no way a restrictive assumption.

If IN.JK = 0

$$\text{LEV.K} = \text{LEV.J} - \text{DT.}\frac{\text{LEV.J}}{\text{DEL}}$$

or

$$\text{LEV.K} = \text{LEV.J}\left(1 - \frac{\text{DT}}{\text{DEL}}\right)$$

The *fraction* of the contents of the level which is withdrawn during DT, is DT/DEL and the probability that any one item in the level will leave in the time period DT is also DT/DEL.

The probability that an item will leave the level in the second time period from now is therefore:

$$\left(1 - \frac{\text{DT}}{\text{DEL}}\right)\left(\frac{\text{DT}}{\text{DEL}}\right)$$

Let P(k) denote the probability that an item remains in the level for k time periods i.e. it leaves at some unspecified instant during period (k + 1), and

$$\text{P}(k) = \left(1 - \frac{\text{DT}}{\text{DEL}}\right)^k\left(\frac{\text{DT}}{\text{DEL}}\right)$$

The average number of DTs which an object will wait in the delay, E, is the Expected Value of P(k).

$$\text{E} = \sum_{k=0}^{\infty} k\text{P}(k) = \left(\frac{\text{DT}}{\text{DEL}}\right)\sum_{k=0}^{\infty} k\left(1 - \frac{\text{DT}}{\text{DEL}}\right)^k$$

Putting

$$1 - \frac{\text{DT}}{\text{DEL}} = x$$

to simplify the notation

$$\text{E} = (1-x)\sum_{k=0}^{\infty} kx^k$$

Since

$$\sum_{k=0}^{\infty} kx^k = x(1-x)^{-2}$$

$$E = \frac{x}{1-x} = \frac{1 - \frac{DT}{DEL}}{\frac{DT}{DEL}}$$

or

$$E = \frac{DEL}{DT} - 1$$

Each object spends, on average E DTs in the delay, another ½ entering and another ½ leaving, so the average number of DTs in the system, E′, is given by

$$E' = \frac{DEL}{DT}$$

The average time spent delayed in the system, T, is thus found from

$$T = E'.DT = DEL, \text{ as required.}$$

In calculating the variance of the time spent in the system we must distinguish carefully between this and the variance of the number of DTs spent.

The equation

$$P(k) = \left(1 - \frac{DT}{DEL}\right)^k \left(\frac{DT}{DEL}\right)$$

refers to the probability of spending k DTs waiting and, in applying the usual definition of variance, we have to incorporate DT. Thus, if V_T is the variance of the waiting time,

$$V_T = \sum_{k=0}^{\infty} (DT.k - DT.E)^2 P(k)$$

Using the simpler notation in x this reduces to

$$\frac{V_T}{DT^2} = (1-x) \sum_{k=0}^{\infty} k^2 x^k - E^2$$

or

$$\frac{V_T}{DT^2} = \frac{x}{(1-x)^2}$$

To apply this analysis to a nth order delay, we keep the overall delay at DEL and consider a series of n first order processes each with a delay of DEL/n. Since the variance of the sum of n independently distributed random variables, denoted by V_n, is the sum of their individual variances and since the individual variances and since the individual variances will be equal by virtue of their having a common delay we have

$$\frac{V_n}{DT^2} = n\frac{x}{(1-x)^2}$$

where now $x = 1 - \dfrac{nDT}{DEL}$

This leads to

$$V_n = \frac{DEL^2}{n} - DEL.DT$$

If DT <0.1 x DEL/n, V_n will be within 10% of its true value and $n = DEL^2/V_n$ will be an approximation to the order of the delay. Forrester suggests that if DT < 0.5 DEL/n there will be no serious numerical instability in the model, but this is a value rather in excess of that argued above. The penalty for using

$$DT < 0.1\frac{DEL}{n}$$

will be increased computing time but this is not usually an important factor in the overall benefit/cost pattern of a system dynamics study.

The application of this analysis of delays may be illustrated as follows, using data derived from a real example.

Goods are shipped from Britain to depots in Germany and the journey usually takes 10 working days. The distribution manager asserts that 'practically all – about 95% – of the shipments take between 6 and 14 days'. Company records are vague, for the perfectly good reason that there has never been any need to record journey durations, so that we must proceed with this 'data' or not at all. In order to estimate the standard deviation of the delivery time we have to assume that the information that 95% of the despatches arrive within an 8 day variation can be treated in the usual way by equating 95% to 4 standard deviations. This implies that the standard deviation, σ_n, is approximately given by

$$\sigma_n = \frac{8}{4} = 2.0$$

and

$$V_n = 2.0^2 = 4$$

With DEL = 10 we have

$$4 = \frac{100}{n}$$

and therefore $n = 25$ approximately. The nearest way of achieving this would be 8 cascaded DELAY3 functions.

In some cases this approach leads to very large delay orders, which are cumbersome to program, and very small values of DT which are wasteful of computer time. This will however only arise when exceptional numerical accuracy is being aimed for and if this occurs we must trade off the accuracy achieved against the effort and time involved.

For example to model an ageing population in which most deaths occur between the ages of 50 and 80 we find

$$\sigma_n = \frac{30}{8} \quad \text{and} \quad \text{DEL} = 65$$

so

$$n = \frac{65^2}{3.75^2} = 300$$

and

$$\text{DT} < 0.1 \times \frac{65}{300} = .0216$$

This would involve 100 cascaded DELAY3s or 200 lines of program as each delay needs an initial condition (see Chapters 4 and 5). A Pipeline delay would be easier to program but less accurate as it would involve assuming that everybody died at exactly age 65.

In practice such great accuracy is usually superfluous as we are often more interested in the effects of policy on the *pattern* of behaviour rather than the *exact* numerical behaviour. One can usually (but not always) achieve perfectly adequate and usable results by cascading 2 or 3 DELAY3s when many more are strictly called for.

Chapter 3

Influence Diagrams

3.1 Dynamic Modelling – The Influence Diagram

The basic tool for developing a dynamic model is the INFLUENCE DIAGRAM. This is not as simple as it looks, and has a far-reaching effect on the construction and analysis of a model.

The influence diagram records the way in which the system works. This is done by writing the names of the variables concerned and connecting them by an arrow, or influence line (or link). *The direction of the arrow shows the direction of causation*. The links are often obvious, e.g. production and consumption both affect inventory because there are physical flows. This is illustrated in Fig. 3.1. The signs at the heads (points) of the arrows show the sign of the effects. These are deduced by considering the effects on the variable at the head of the arrow of a *change* in the magnitude of the variable at its tail. The rule is: *if the head variable changes in the same direction as the tail variable, use a + sign, but if it changes in the opposite direction use a – sign; if the result is sometimes in the same direction and sometimes in the opposite direction use an asterisk*.

Thus an increase in production leads, other things being equal, to an increase in inventory and a decrease in production leads to a decrease in inventory. The change in inventory is always in the same direction as the change in production, so we use a + sign. An increase in consumption leads to a decrease in inventory and vice-versa, so we use the – sign.

Fig. 1.4 provides an example of a situation in which the sign of the effect may vary. According to Fig. 1.4, if delivery delay rises from, say 1.0 to 1.2, the rate of placing orders will increase, so that there will be a positive link, but for a rise from 1.8 to 2.0 the ordering rate falls off and there is a negative link.

We also need to represent delays and control actions. In the example we should, for realism, replace 'production' by 'production order rate' and 'production completion rate', with, of course, a delay between the two, shown by a D on the link.

In the early stages of a model, control actions are shown by dotted lines, to distinguish them from physical flows. The control links connect the variables which are chosen for use in the control action, with the variable being controlled. The

Fig. 3.1 Influence lines

signs are worked out by considering whether the control policy would convert an increase in the variable being used for the control into an increase or decrease in the variable being controlled. The resulting sign is shown in brackets on the influence line to denote that it is conditional upon a particular decision rule which cannot itself be shown on the influence diagram.

For example, if, in the production–inventory control example, production order rate is based on a comparison between Inventory and a fixed Desired Inventory, with the gap between the two to be closed over a fixed Production Adjustment Time, we get the influence diagram of Fig. 3.2.

The link from Desired Inventory to Production Order Rate has no sign because Desired Inventory was stated to be a fixed parameter, and the same applies to the link from Production Adjustment Time.

The solid lines represent those flows which are, in some sense, fixed and physical. The dotted lines, on the other hand, are flows of information which management *chooses* to use in its decision-making. They are nothing more or less than that, and they can be changed if there is some virtue in doing so. They may, of course, be eminently sensible and represent perfectly good ways of controlling the system. They are, however, optional and should be questioned and investigated so that they can stand or fall on their own merits. The dotted line notation, with signs in brackets, serves to remind us of the need for questioning. As the model evolves

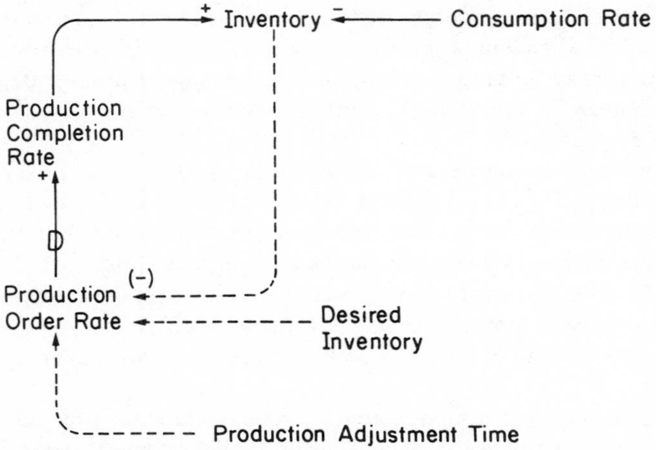

Fig. 3.2 Example on an influence diagram

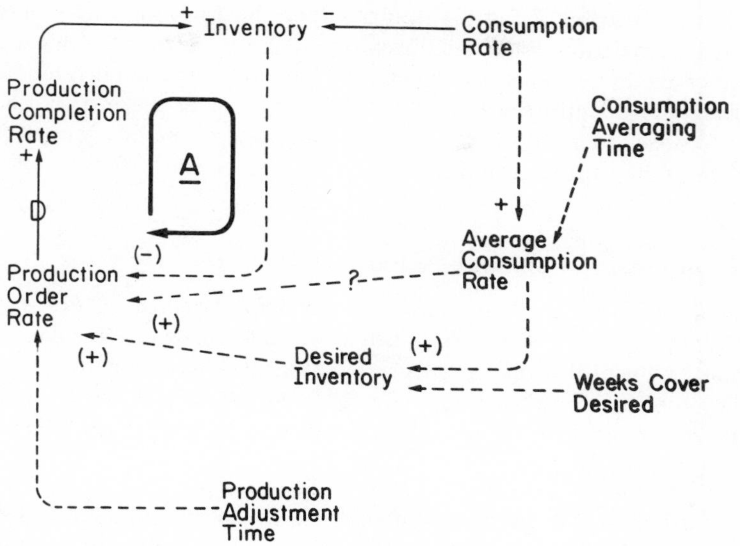

Fig. 3.3 Influence diagram for the production control case

the dotted-line-and-brackets notation can be dropped because, as we shall see, the influence diagram notation will become very cumbersome if it is not.

We shall, however, use the full notation once more, in Fig. 3.3, to elaborate the influence diagram for the production control case, to show the inclusion of more complicated control rules which take account of Average Consumption.

The diagram now shows a fairly small, but complete, influence diagram. There are a number of detailed points.

The + sign on the link from Consumption Rate to Average Consumption is not bracketed because it is an arithmetical averaging process. The fact that the link is dotted shows that the calculation is optional and not inherent in the system, in the way that the flows shown by the solid lines are, but the unbracketed + indicates that the sign of the effect *is* inherent in the process of averaging.

The + signs on the links from Average Consumption Rate to Desired Inventory and from Desired Inventory to Production Order Rate, and the – sign on that from Inventory to Production Order Rate, show that the signs of these effects are a matter of choice, depending on the decision rules being used. Different rules, even with these same, optional, links could be conceived of which would lead to the other sign on a given link and, furthermore, such other decision rules could be used, in the model and in actual practice, if there was any advantage in doing so.

The question mark, and the bracketed plus, on the link from Average Consumption Rate to Production Order Rate, shows that this is an information stream which could be used directly in determining Production Order Rate, as well as, or instead of, its indirect use in converting Desired Inventory from a parameter into a variable.

The heavy black line round the feedback loop through Production Order Rate, Production Completion Rate and Inventory simply identifies and highlights the existence of the loop, and is not structurally part of it. The letter A in the middle of the heavy line identifies the loop, and the – sign under the A shows that it is a negative feedback loop. In Chapter 6 we show that the type of a loop depends on the pattern of the signs on its links.

3.2 The Justification of Influence Diagram Links

Building up an influence diagram involves making statements about the way in which the system works. Thus a link such as A → B is a diagrammatic condensation of a verbal statement that the factor A acts as a direct cause of B and that variations in A will, possibly after a delay, be reflected as variations in B. Naturally in many cases we need to record that, not only A but another variable C, are joint causes of B, or $\begin{matrix} A \searrow \\ C \nearrow \end{matrix}$ B.

The situation may be even more complicated than that, but these are simply elaborations on the A-causes-B theme. This raises a very important question: how do we know that it is A which 'causes' B? In short, how are we to justify incorporating links into an influence diagram?

There are six basic methods of link justification:

1. Conservation considerations
2. Direct observation
3. Instructions to that effect
4. Accepted theory
5. Hypothesis or assumption
6. Statistical evidence

Conservation Considerations

These are derived from commonsense or 'what comes in must go somewhere'. If there are tangible flows of physical entities, such as raw materials, product, people, capital plant, or of information about orders, cash payments and receipts, production instructions, etc., they do not simply vanish from the system. At any rate they do not disappear until the analyst declares that they are of no further relevance to the system he is studying and allows them to flow over its boundary and be dropped from further consideration.

Thus we can readily justify links which show orders flowing into backlogs which are then depleted by production start orders. These last then start the flow of actual production which flows into and out of inventories, and so forth. Similarly, in a social or economic system we have birth rates augumenting populations, school leavers affecting work forces, cash flows accumulating into a balance of payments.

In building influence diagrams it is good practice to be alert for stocks and other levels as these must have flows which feed and deplete them. Detecting a stock therefore automatically locates one or more flows which must, in turn, have control actions. Careful identification of the stocks thus automatically leads one to a good system model.

Direct Observation

The direct observation method is only practicable in the system's controller. It consists in saying, for example, 'I have found by observation and questioning that the production manager takes account of the anticipated level of inventory three months in the future in deciding on the production start rate'.

The great advantage of this method is that it is open to verification. In conducting a practical study it is excellent practice to write down what has apparently been discovered so that people who have supplied the information can check and correct it. The disadvantage is that the method is affected by the prejudices, assumptions, and interpretations of analyst and manager and it requires more skill than may at first appear.

Instructions

We can justify links by making the system operate in accordance with them. This is, or course, the last stage in a study directed towards policy design rather than system elucidation, but the method may be adopted to overcome ambiguities about how the system does operate by, in effect, giving orders as to how it shall be operated.

In practice, persuasion rather than command has to be the order of the day. However, the effect is the same in that the system is made to operate in a stipulated way, rather than described as functioning in an observed way.

Accepted Theory

The part of a system most likely to present problems of modelling and, by its definition, only partially amenable to the three previous methods, is its Environment. One possible approach to this is the use of accepted theory. In modelling a hardware system, the control engineer can draw on the theories of physics, thermodynamics etc. A similar approach can be used, with reservations, in business or economic systems in that, in describing the functioning of the Environment, economic theory, or accepted theory from some other discipline such as marketing or behavioural science, may be useful. In an ecological system the relevant areas of biology, chemistry, etc. are available. There are, however, snags in this approach.

1. The theory of socio-economic systems, in particular, is far less clear-cut than, say, chemical theory. The use of concepts, such as supply and demand, is not the same as the employment of a stable, tested, experimentally-verifiable, theory. The analyst has to guard against using theoretical concepts, which are designed for problem visualization, as an aid in practical problem-solving.

2. The analyst needs to be sure that he really understands the theory he is using, rather than simply being generally aware of it. He has to verify that the assumptions underlying the theory *really* apply in the particular case. As a general rule, accepted theory is a very useful approach, but it should never be used simply to make a model seem respectable.

In short, accepted theory may well be useful but it needs to be handled with care and with a clear understanding of the purpose for which the model is being constructed.

Hypothesis, Assumption or Belief

In some cases there is little or no real evidence or theory about whether a particular link exists or how it operates. One way of proceeding is simply to make assumptions, or to state beliefs, about the link in question. The advantages are:

1. It enables the modeller to tailor the model precisely to his purpose in building it,
2. Any assumptions can be challenged or replaced by alternatives, providing the analyst is at pains to make clear that his assumptions are such, and not facts.
3. The modeller can draw on the accumulated experience and qualitative information which is available to the people who actually operate the system. (Qualitative information can be expressed in numerical form but is not open to theoretical or empirical verification because the methods and/or the data for doing so do not exist or are unsatisfactory. Quantitative information is data forming an historical record of some aspect of the system, or measurements collected as a result of experiments performed on it. It is sometimes possible to conduct a quantitative experiment to test qualitative information, but the practical and theoretical problems of experimenting in socio-economic systems, and consideration of time and expense, make this uncommon.)

Some limitations of data on socio-economic systems were referred to in Chapter 1, so, where they apply, we are faced with two alternatives – either we do not model the system's environment at all, or we use other information about the way in which it operates. If the environment has no appreciable feedback to the controller the first course is acceptable, but if the feedback is significant then either we must abandon the problem altogether, or we must rely on qualitative information, supplementing it by quantitative data where it is relevant to do so.

The arguments in support of the use of qualitative information are several:

1. Restricting oneself to quantitative data and no other involves the huge assumption that qualitative information is *not* relevant.
2. Managers always have qualitative beliefs about how the system of which they form part of the controller operates. These beliefs influence their decision-making and should, therefore, be included in any model which purports to represent decision-making. Thus if the manager, as part of the controller, believes that a price rise of 5% will reduce the demand by 3% then this qualitative statement will affect his decisions about pricing, even though it is

almost certainly in no way verifiable statistically. Since a model of pricing policy which excluded this kind of information would be irrelevant, we are led to making use of the information embodied in the statement. We reiterate, however, that a model incorporating subjective information is not an objectively verifiable model of the system, but it is a perfectly valid model of a controller, in which the controller's beliefs about, and experience of, its external world have been used to model the environment.

3. The precise form of the qualitative information will not usually make a vast amount of difference to the MODE of behaviour of the system. The presence or absence of the relationship may markedly alter the behaviour mode but if the 3% demand fall used in the last subsection is replaced by 5%, the system's dynamics will usually not be altered very much. Clearly this is a matter for sensitivity testing, which is dealt with in Chapter 7.

The disadvantages of using qualitative information are:

1. It seems unscientific, and may lead to unproductive controversy.
2. Information on a particular topic may vary from person to person,
3. The results depend on the information, and the policies designed by the model are only robust with respect to the hypotheses.
4. People sometimes attach too much weight to computer models, in fact more weight than the model can stand or the analyst intended to imply. Once again we are back to the behavioural relationship between analyst and client!

These factors have to be assessed for each case, and the reader should study the various case studies in the book. Note that the problems do not arise in the modelling techniques, but from the nature of the analysis of systems involving environments and complements.

Statistical Evidence

In certain cases it is possible to use statistical methods of path analysis to infer causal structures from statistical data (see, for example, the reference by Hamilton in Appendix C). This goes rather beyond the observation of statistical correlation, but it is a specialist topic which is beyond the scope of this text. It is however, important to distinguish between the use of path analysis for inferring causal mechanisms and the use of a variety of statistical methods for deducing parameters for links, the *existence* of which is established by other means.

To summarize, each link in an influence diagram has to be justified. Further, in order to be able to translate the link into equation form, we have to be able to state how it works. This is done by the methods used to demonstrate the link's existence and is subject to the same degree of certainty or uncertainty, depending on the source of the evidence.

3.3 Closure and Model Boundary

An influence diagram must pass a fundamental test in order to constitute a dynamic model, in that it must possess the property of CLOSURE. This means that it must contain at least one feedback loop, and that all its variables lie on a loop, have been defined as exogeneous inputs to a loop, or provide supplementary output from a loop.

The essential concept of a dynamic model is that its dynamics are produced by the operation of feedback loops in closed systems (review Section 2.2). Any model which does not contain a feedback loop is, therefore a static model. Paradoxically, a static model can produce dynamic behaviour when it is driven by an exogenous time-series, but this is simply due to the system state changing in response to the exogeneous input and not under the influence of control policies based on the system state. This was shown in Fig. 2.1a. As soon as a control law is introduced it must be based on some aspect of the system state, the system is CLOSED, a feedback loop is immediately created and a dynamic model ensues.

The test for closure is very simple: *starting from any point in the influence diagram it must be possible to return to that point by following the influence lines, in the direction of causation, in such a way as not to cross one's track.*

This test applies to all points in the diagram, with certain exceptions which are noted below.

The test can be simplified somewhat be exploiting the property itself. This means that, having chosen an arbitrary starting point and traced a path which returns to that point then a number of intermediate points will have been passed and these, of course, lie on the feedback loop just traced out. Since they lie on this feedback loop they lie on *a* loop and, therefore, satisfy the closure test and can be dropped from further consideration. Having taken one path in the diagram and proved that it is a feedback loop, one must still apply the closure test to any remaining paths in the diagram to see whether they pass it, whether their variables are covered by one of the exceptions or whether, indeed, the system is not totally closed.

There are three situations in which it is permissible not to be able to return to the starting point without violating the rule of closure: parameter values; input variables; and supplementary variables.

Parameter Values and Input Variables

If the chosen starting point is a parameter or an input then it is not possible, by following the influence lines, to return to the starting point. If, however, after one or more steps, one reaches a variable, X from which a closed path is possible, i.e. a path which leads back to X, then the model closes.

Supplementary Variables

A model may contain variables which act simply as indicators of system performance and which play no other part in the system or its control policies. Such variables are called SUPPLEMENTARY VARIABLES.

For example, in our production-inventory model we might wish to calculate the average value of stock and the cumulative orders received as indicators of the

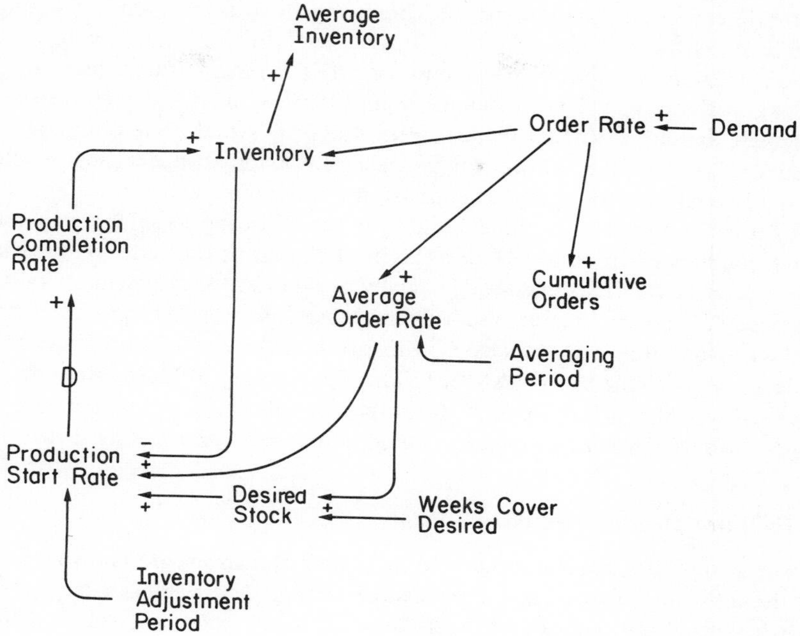

Fig. 3.4 Influence diagram including supplementary variables

efficacy of different production control policies which might be used. This would lead to the influence diagram in Fig. 3.4.

Average Inventory and Cumulative Orders are supplementary variables and terminate the influence lines on which they lie. If, in testing for closure, one arrives at a terminal variable it is essential to review whether or not this was *intended* to be a supplementary variable. If it was, all well and good, but if not, further thought will be needed. The list extension technique presented in the next section will be helpful in such a case.

In the foregoing diagram there is only one feedback loop, there are three parameters and two supplementaries; all the other variables being stages in the input process.

The concept of closure is very useful for another reason: the light that it sheds on the boundaries of the model and the derived notion of a hierarchy of increasingly complex models of a system.

The boundaries of the system itself are very far flung and there is practically no end to the complexity which can be built into a model. There is, however, the very practical consideration that a halt must be called somewhere, otherwise the model will become unmanageable. Obviously, then, out of the range of models which could be built of a particular system there is one which is sufficiently good to serve its purpose, but is not needlessly complicated, when the need for complication, or refinement, is judged against the purpose for which the model is being built. It is, therefore, necessary to select the model boundary or to decide what to put in the model and what to leave out.

These ideas imply that there is only one *TRUE* model: that which includes everthing in the system. The true model is impossible to build and we therefore aim to produce simply a *VALID* model – one which is 'sound, defensible, and well grounded'. For the moment we simply state that a sound model must be closed and, as a general rule, the simpler the model, the easier life will be.

This leads us to approach model building in the following way. We start with the simplest model we can build, M1, and this must contain at least one feedback loop. If it is not valid we develop into M2, usually by adding at least one more loop, but occasionally simply by creating more dynamics such as when Desired Inventory is modelled as depending on Average Consumption in the example in Chapter 2. At each stage we examine the model for validity, proceeding by the addition of more loops until we have a valid model. Questions of validation will be deferred until Chapter 7, for the present we concentrate on influence diagram construction.

3.4 The List Extension Method

Two of the difficulties encountered in System Dynamics, are knowing how to start the influence diagram, and knowing how to stop. The result is often a diagram which includes every conceivable variable. This is not good practice; it reflects a poor understanding of how the system operates and an even poorer one of what the study is supposed to be for.

The closure test shows when an influence diagram has become a model and discriminates between one model and a more complex one. We now discuss a technique which helps to get started, facilitates stopping with the aid of the closure test, and focuses attention on the purpose of the model. This is the LIST EXTENSION METHOD.

An influence diagram is a list of the variables in the model. To simplify drawing the connection patterns between the variables, the list is written out over the page rather than in a column, but it nonetheless remains a list. Adding new variables to the model extends the list, hence the name for the influence diagram building approach.

The list extension method starts from a series of columns on a piece of paper, about 6 or 8 being a good number to start with. From right to left, the columns are labelled the Supplementary List, the Model List, the First Extension, Second Extension, and so on.

The Model List contains the names of the variables the behaviour of which it is the purpose of the model to explain, or the control of which is aimed at. There should not be more than three to five such variables, and one or two is better at the very commencement of a new study. Even for a completed model, it is good practice to limit the number of variables in the model list, in order to have reasonably clear and coherent purpose.

For each model variable one writes, in the first extension column, the names of the variables which most immediately affect it, drawing the influence lines.

Remember that variables in the model list may affect other variables in the same list, as may be the case for any of the lists, and variables in an earlier list may affect those in a later list. The lists must, therefore, be scanned for these connections as they are built up and the influence lines drawn in. Even at this early stage an attempt should be made to add signs to the links.

After all the variables in the model list have been examined and the appropriate entries made in the first extension list, the resulting influence diagram is subjected to the closure test to see if an influence diagram which is also a model has been produced. If it is, the next question is whether it is a good model, a matter dealt with in Chapter 7.

If the influence diagram is not closed, attention moves to the second extension list. This contains, for each variable in the first extension which is not part of a feedback loop, which is not one of the exceptions to the closure rule, or which is not deemed to be modelled adequately, the names of the variables which most immediately affect it. Necessary links are drawn between the variables in the second extension and the variables in the first extension explaining, together with links denoting interconnections between variables in the second list, and between variables already entered in the first extension and the model list, and those being written into the second extension.

When the second extension has been completed the closure test is again applied, and the process either terminates or continues.

Simple Example of List Extension

The list extension technique really comes into its own in complicated systems. When the system consists of a controller and its complement, the practical need to satisfy management that a good model has been built usually forces one to include more detail than is really needed for an adequate model. However, it is a good plan to have a procedure for developing the early versions of the model, and the list extension method provides this.

We use, as our first example of list extension, the familiar case of the production–inventory system, shown in Fig. 3.5 and Fig. 3.6.

The purpose of the model is to explain the variations in inventory and production order rate and to enable the analyst to study the problem of devising improved control strategies. The first part of the list extension is shown in Fig. 3.5.

To the right of the model list, the Supplementary List contains Average Inventory, which is not part of the system, but is used as an indicator of system performance in order to discriminate between one control policy and another. It is enclosed in a lozenge to show that it is a supplementary variable.

The first extension includes the consumption rate, enclosed in a rectangle to show that it is treated as an exogeneous driving force, and the production completion rate. These links are justified on the basis of conservation considerations (Section 3.2). The same considerations show that the production completion rate is affected by the production order rate already entered in the model list. This link is therefore inserted, with D to show the delay.

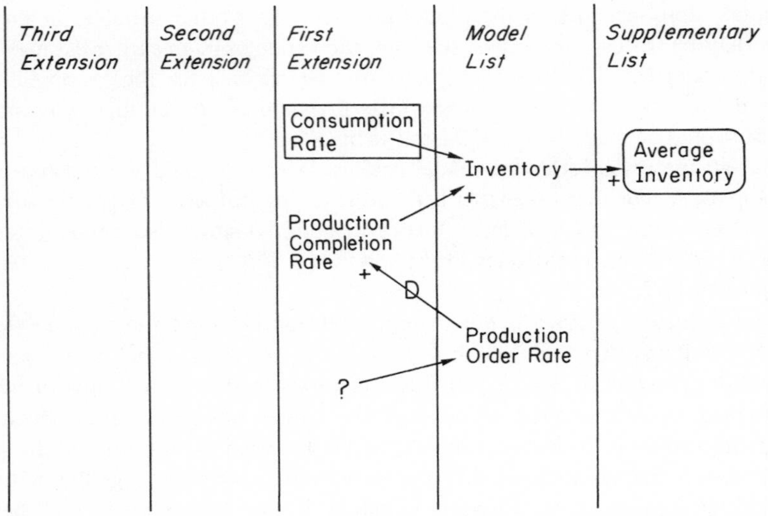

Fig. 3.5 First part of list extension for production–inventory example

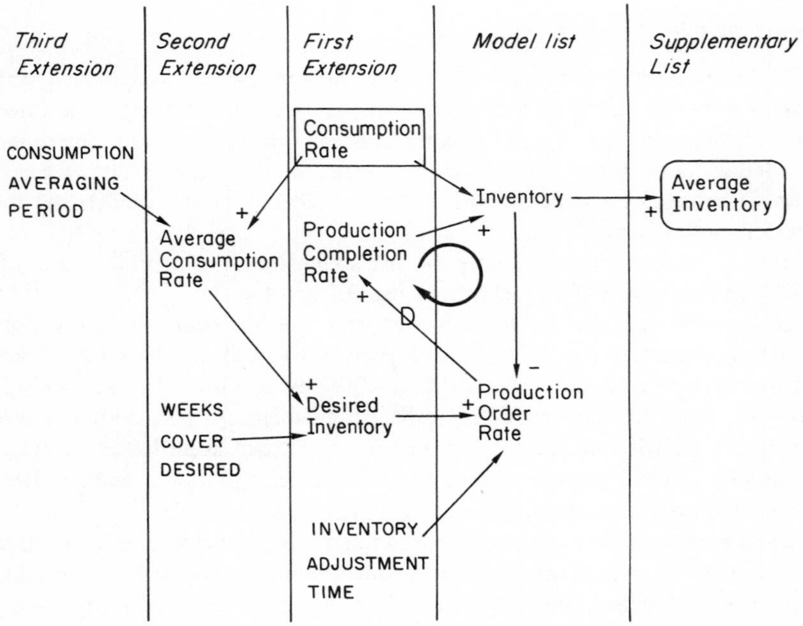

Fig. 3.6 Second part of list extension for production–inventory example

At this stage, the analyst does not know what affects the production order rate, and he therefore inserts a link with a question mark to remind him to find out. The production order rate could be determined in a host of different ways, but the analyst decides to find out what is done now. By the observation method, he discovers, say, that production order rate is determined by a comparison between actual inventory and its desired value and the latter is a fixed number of weeks of average consumption. This number of weeks, the consumption averaging time and the inventory adjustment time are all policy variables rather than technological variables, and they are therefore identified in capitals on the diagram. The results of this step are depicted in Fig. 3.6.

The diagram shows the use of a link denoting the interdependence of the two model variables. It also identifies the single feedback loop by a broad line.

The influence diagram passes the test of closure, but a study of it will raise many questions in the analyst's mind about how the system might be changed from its present form to a better one. For example, should the model parameters be fixed, possibly at values different from those in current use, or should they be made into true variables dependent on other parts of the system, perhaps including quantities not yet in the diagram? Maybe the average consumption rate should be used directly in determining production order rate as well as indirectly. Should desired inventory have a non-linear dependence on average consumption rather than the present linear relationship?

These questions relate to the possibility of other control strategies but there are matters to do with model adequacy which are suggested by the diagram *as it evolves*. For example, it seems rather unlikely that there are no constraints on production order rate, yet none are shown on the diagram. In addition, the diagram shows that the delay in the production process is assumed to be constant, yet it is by no means uncommon for the delay to rise as the level of activity increases. Is this the case in this system? These questions can only be resolved by further interrogation of the system. They are, however, suggested by a careful application of the list extension process.

In order to transform Fig. 3.6 into a final influence diagram, the column headings and the vertical rulings are dispensed with.

In this case the layout of the diagram is reasonable, but usually it needs redrawing to reduce the tangles in the influence lines.

A More Complex Example of List Extension

We now consider a case where the structure is far less clear-cut, where the analyst would probably have to rely fairly heavily on hypotheses and assumptions in developing his model, and where sensitivity testing would be especially important.

The purpose is to improve profitability in a mining enterprise and to control production.

The essential features are that the mine produces a metal, the market price of which is markedly unstable. Traditional policy has been to produce at a constant rate, but it seems plausible to management that gearing production to price might

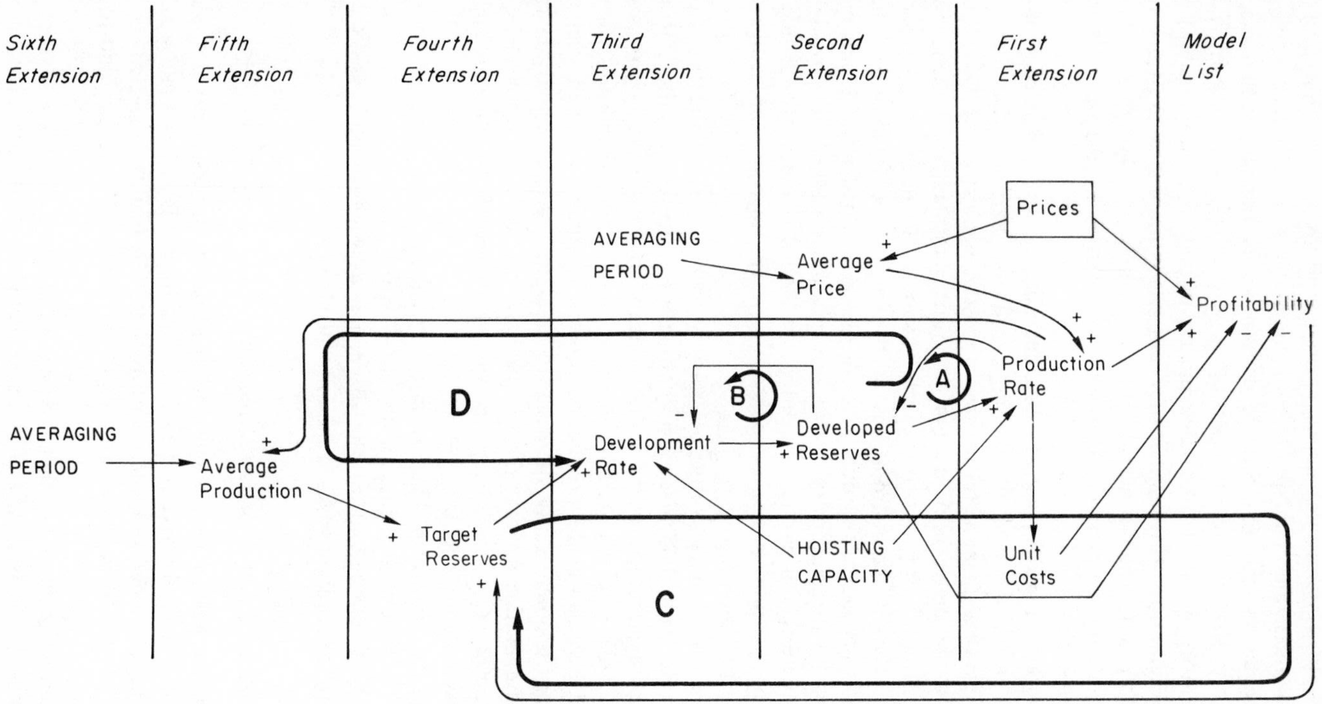

Fig. 3.7 List extension for mining company example

improve profitability. Since the price is so unstable, forecasting seems impossible, so the production is to be tied in to average price. World production is large compared with this mine's output so that this is unlikely to affect the price. We aim to design a production policy which will enable the company to do as well as possible in the face of fluctuating prices.

In a mine, preparatory tunnelling, or development, precedes production. Since only limited amounts of production machinery can be deployed in a given developed area the size of the developed reserves affects production. Investment in developed reserves affects profitability, and profitability affects the level of reserves which can be supported financially by affecting target reserves. Development produces waste rock which competes for shaft hoisting capacity with the ore from mining.

The reader should try to sketch the influence diagram, using the list extension technique. The solution is shown in Fig. 3.7.

The model contains six feedback loops. The first, A, appears at the second extension but the diagram is not closed by its emergence because profitability has not been declared to be a supplementary variable. A second loop, B, appears at the third extension but does not close the model.

The third, fourth and fifth loops, C and its two unlettered parallel branches, are found at the fourth extension. The diagram is, however, unclosed until the detection of loop D at the fifth and sixth extensions, at which point it becomes the simplest model which can be built of the system. Whether it is the adequate model is entirely another matter. It is unlikely that it is, but the modelling process has now started and the model itself, and its output, will guide its own elaboration.

From the building of influence diagrams we turn to two procedures which abstract important information from them and check that they are free from certain types of error. These two procedures – Type Assignment and Coherence Testing – will first be described from a theoretical standpoint. At the end of the chapter we consider their use in practical modelling situations.

3.5 The Order of a Model

The ORDER of a model is the number of levels it contains plus the total order of its delays. Thus a model with four levels and two third-order delays would be 10th order. The order is important because the dynamics of the system partly depend on it. The number of levels is not something which one dreams up, but is derived from the model structure.

Determining system order is fairly straight forward for a mathematical model, though it may be very tedious. The system equations derive from the 'basic physics' of the process. These equations reduce to a differential equation linking some chosen pair of variables. Such a process of mathematical manipulation is illustrated for several different cases in Appendix A. This transformation process is subject to the rules of mathematical operations, so there will be a 'correct' answer, in that a given set of system equations can only be transformed to a differential equation of a particular order. Any other order will be wrong, and the error will arise because the mathematical rules have not been properly applied. As a result, the dynamics of the system may be misinterpreted if they are studied via an equation of the wrong order.

In a system dynamics model, the system structure leads to a set of difference equations which are sufficiently close to the 'true' differential equation in numerical output, and which have the same order. The system equations are expressed in terms of the *names* of the variables involved, and we find the difference equations by deciding on the *types* of the variables. This means that an equivalent to the mathematical transformations illustrated in Appendix A is the process of deciding, for each variable in the model, whether it is a Level, an Auxiliary, or a Rate. By analogy with the differential equations, and given the delay orders, there is one correct collection of choices of variable types, and thus *a* correct order, or number of Levels, for the model. If we assign the variable types so as to get the wrong order we shall risk getting the wrong dynamics for the system.

Just as the ordinary rules of mathematical manipulation applied to the system equations will lead to the correct model order, we use formal rules for finding the model order from the influence diagram. These rules are explained and illustrated in the next section.

We must keep a sense of proportion about the 'correct' model order. If a system is, say, third order, then a second-order model may give serious errors in the dynamics. When the model is of a very high order, as is usually the case, the importance of getting *precisely* the right order diminishes a little, for a number of reasons.

1. Complexity. The behaviour of a system is usually sensitive only to particular parts of the model and the rest has little or no effect. Although the insensitive parts have to be modelled in order to find out that they are insensitive it is not likely to make much difference if a 48th order system is actually modelled as 47th order.
2. Delays. The order of a delay is part of the overall model order but, as was shown in Chapter 2, the order of a delay can only be estimated, so that some errors in determining model order will creep in from that source.
3. Simulation. The flexibility of computer simulation methods makes them fairly tolerant of details such as whether a particular variable is a level or an auxiliary. A given variable's type can usually be altered without making any detectable difference to the output from the model. This is a weakness in these languages rather than an asset, because it enables the programmer to get away with short-cuts in model formulation, which may be a cover for slipshod analysis.

For most of this section we shall denote variables by letters rather than by names, so as to show the technique of type-assignment unencumbered by preconceptions about a variable's type which might be generated by its name. In practice, a diagram with names should be used.

Consider a simple model involving only three variables, X, Y and Z (see Fig. 3.8a). In order to start the type-assignment process we have to make an initial choice for one variable; let us, therefore, suppose that we have some reason for

Table 3.1. *Relationships between Variable Types*

Types for Preceeding Variable	Variable Type	Types for Succeeding Variables
R	L	A or R
L or A	A	A or R
1. If a delay in the influence link R 2. If no delay in the link L or A	R	1. If a delay in the influence link R 2. If no delay in the link L

choosing variable X to be a level, perhaps because it is a stock of completed product. Write the variable type near the name, and enclose it in a box to show that X is the *chosen* starting point (see Fig. 3.8b).

We can now work either forwards to Y or backwards to Z. ('Forwards' and 'backwards' mean in the direction of the influence arrows and against them, respectively.) When a variable is a level, work backwards because Table 3.1 shows

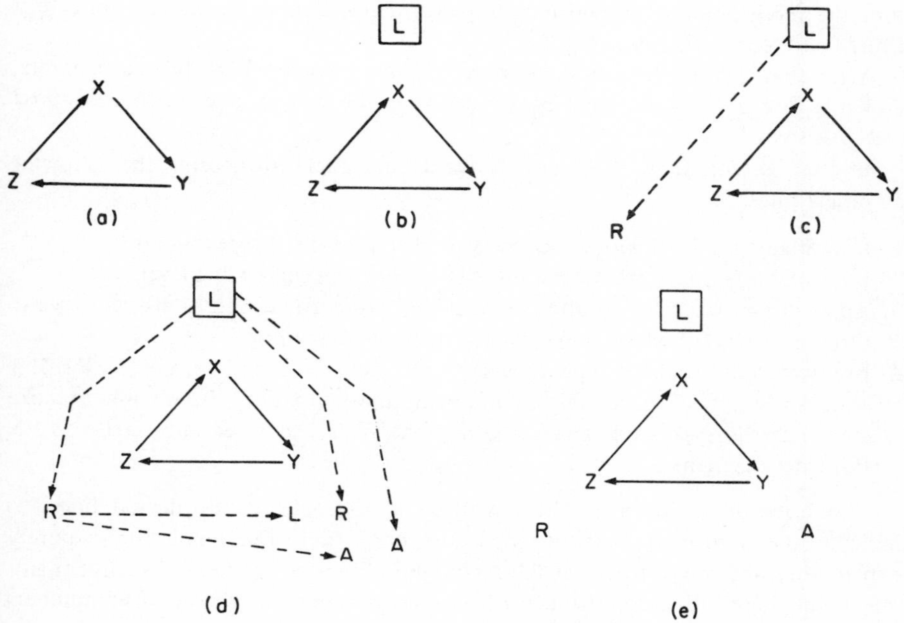

Fig. 3.8

that a level can only be preceded by a rate. Thus Z must be a rate. This is written on the diagram with a dotted line having an arrow head to show that the fact that Z is a rate derives from X being a level (see Fig. 3.8c). Dotted lines and their directing arrows indicate the sequence of derivation of the variable types and have nothing to do with the direction of causation in the feedback loops.

To type Y we first work forwards from X and then backwards from Z, noting that neither the Y–Z link nor the X–Y link contains a delay, so that the conditions 1 for a rate in Table 3.1 do not apply.

Considering the X–Y link, and referring to Table 3.1, Y can be either a rate or an auxiliary, and, from the Z–Y link and the left-hand column of Table 3.1, Y can be a level or an auxiliary.

We write these conclusions onto the diagram, leaving a space between them to show that we have yet to make the final choice of type for Y (see Fig. 3.8d).

We now appear to have reached the position that Y can be a level, a rate, or an auxiliary, but it is fairly obvious that only if it is an auxiliary will the forward and backward derivations of its type be consistent. The type-assignment for this model must, therefore, be as shown in Fig. 3.8e.

We drop the dotted lines, but retain the box for the statement 'X is a level' to show that this was our starting assumption, and that the variable types shown here depend on it.

In this case, the assumption about X led to consistent conclusions about Z and Y. An influence diagram for which such conclusions are possible is said to be COHERENT. The absence of coherence indicates that something is seriously amiss with the model, but its presence only means that the model is 'correct' in a very limited and technical sense.

After that simple, first-order, case we examine a more complicated example, shown in Fig. 3.9. Again, there are no delays in the system, and we shall suppose that W is to be level.

In the light of Table 3.1, work forwards and backwards, using the following empirical rules:

1. If a variable is a level, work backwards first to find the preceeding rate,
2. If a variable is a rate, work forwards first to find the succeeding level,
3. Apply these two steps as often as possible before attempting to assign types to those variables for which there is more than one possibility,
4. Even where a variable's type appears to be unequivocally determined by rule 1 or 2, always check if possible by following another path to the variable to make sure that the type assignment is consistent. This will reveal any errors in the influence diagram.

The notation includes numbers, written at the side of the dotted lines, to indicate the approximate sequence of the derivation. The numbers are purely explanatory and are not part of the technique of type-assignment. Usually, at any one time, there are alternative steps in the type-assignment process. The numbers do not, therefore, mean that step n *must* precede step $n + 1$ but, generally, the

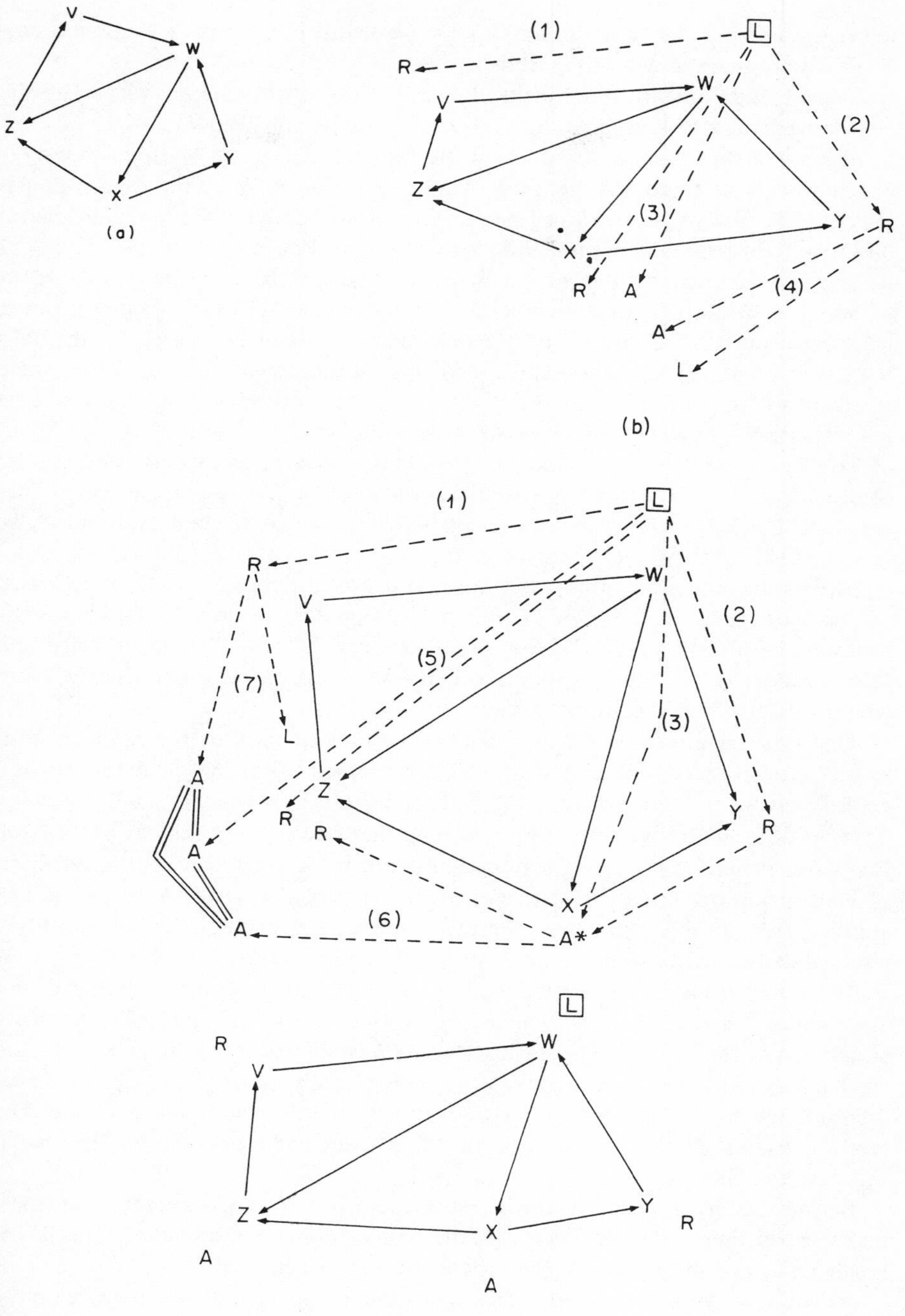
V
W
Z
Y
X
(a)
(1)
L
R
W
V
(2)
Z
(3)
Y
R
X
R
A
(4)
A
L
(b)
(1)
L
R
W
V
(2)
(5)
(7)
L
(3)
A
Z
R
Y
R
R
A
X
(6)
A
A*
L
R
W
V
Z
Y
R
X
A
A

Fig. 3.9

lower the number the earlier in the type-assignment process that step should occur. The early steps are shown in Fig. 3.9b.

Steps 1 and 2 are forced by the choice of W as a level. Steps 3 and 4 are the alternatives which are available for X, given the choices for W and Y.

Because of the interconnectedness of the system, the type of W affects that of Z, which is also affected by the type of V. Furthermore, Z's type affects, and is affected by, that of X. This can lead to a very complicated and illegible diagram if one tries to incorporate all these influences before making a final choice of type for X. We, therefore, choose X to be an auxiliary, which is the consistent outcome for X, based on W and Y, but this choice is provisional until it is seen whether it ties in with the deductions about X's type which can be made after considering the links V–Z, W–Z, and Z–X. If these links show that X should *not* be an auxiliary, there is obviously something wrong somewhere and the influence diagram would be INCOHERENT. In later sections we shall deal fully with incoherence.

For the moment we provisionally type as an auxiliary, writing A* see Fig. 3.9c on the diagram to denote the provisional assignment, and leaving single dotted lines for steps 3 and 4, to show the sources on which the tentative choice was based. We than proceed to use W and X to type Z.

Steps 5 and 6 indicate that there are two consistent choices for Z's type, based on the types assigned to W and X, and this can only be resolved by looking at the backward path from V, step 7. This gives a coherent choice of Z as an auxiliary, and this is shown by the triple = notation. If the V–Z link had included a delay, there would still have been a coherent assignment, with Z as a rate.

This type-assignment for Z is based on the provisional type for X which, as it leads to overall coherence, may now be confirmed. The final, and unequivocal choice of types is, therefore, as in Fig. 3.9d and this is a first-order model.

In both examples we started by assuming that one of the variables was a level. For a real system, the starting assumption is not made arbitrarily, but on the basis of what is known about the character of the particular variable. A variable which appears, from its name, to have the characteristic of accumulation is probably a level, while one which seems to be a flow may well be a rate.

The starting point may either be a level or rate, whichever is the more clear cut and obvious, though a level is to be preferred. One MUST NOT, however, start with an auxiliary. The type-assignment becomes impossibly laborious (the reader should try it), and, more importantly, auxiliaries are essentially dummy variables which are incorporated in the model to show the stages in the determination of a rate. The levels and rates are the true components of structure and must therefore be used as the starting points.

Before dealing with more complicated examples of type-assignment we must make a further point concerning the connection between the concept of incoherence and the natures of the variables in the model.

As we have defined coherence, it is a structural concept which derives from the pattern of connections in the diagram and, as such, either exists or does not. There is, however, a DEFINITIONAL COHERENCE which arises from the nature of the variables whose types have been inferred from the starting assumption. Thus, in our first example, we assumed that our knowledge of the character of X indicated that it could be a level, and from that we inferred that variable Z had to be a rate.

Now, the statement 'X is a level' derived from our knowledge of X as a system component, and the statement 'Z is a rate' was inferred from the structural relationships in the influence diagram. Clearly, this second statement must also marry with our knowledge of the nature of Z as a system part. If, for instance, Z has all the appearance of an integration then it is very unlikely that it can be a rate, and the influence diagram, although *structurally* coherent would be DEFINITIONALLY INCOHERENT.

The test for definitional coherence must be applied to those variables which have been inferred to be levels or rates from the starting assumption, once the test of structural coherence has been passed. If any of the inferred levels or rates fails the definitional coherence test then a mistake has been made in drawing up the influence diagram from the verbal description of the system or a definitional mistake was made in the starting assumption. In practice it is almost certain that something has been missed out, probably in the list extension process, and the only solution is to check the diagram against the system in the hope of discovering the error.

In the remainder of this chapter we shall assume definitional coherence.

3.6 More Complex Examples of Type-assignment

We now treat a more complicated example, incorporating a delay, as shown in Fig. 3.10.

For this influence diagram it is not necessary to make a starting assumption about the type of one of the variables. The delay between Y and N, and Table 3.1, show that both of these variables must be rates. Since these assignments stem from the structure of the influence diagram, rather than from an explicit assumption, we write them in circles rather than boxes. This distinction is important in practice when it comes to applying the definitional coherence test because it helps one to remember where one started.

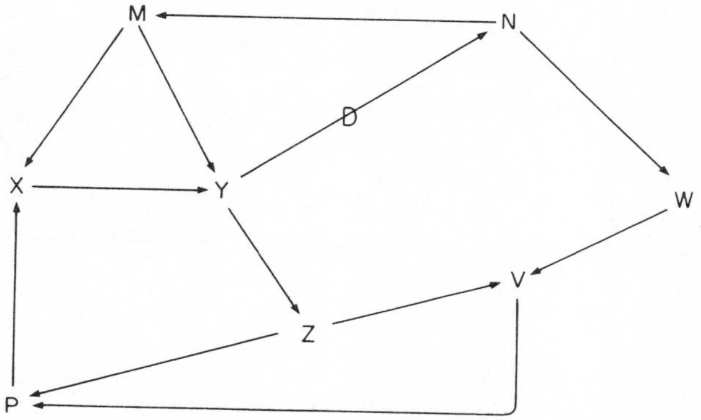

Fig. 3.10

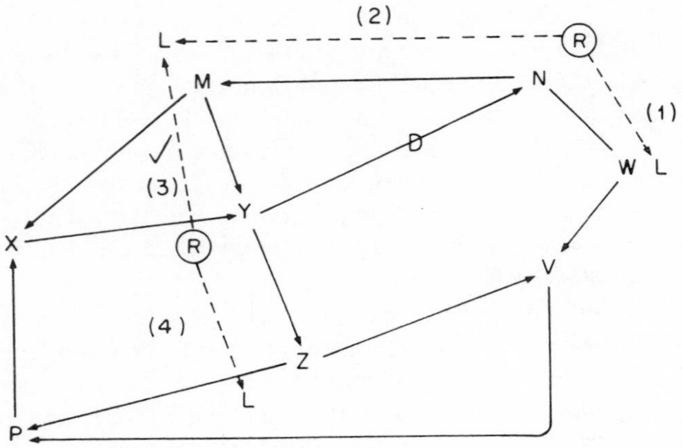

Fig. 3.11

The first four steps in the assignment process are shown in Fig. 3.11.

Step 3 closes the loop through Y, N, and M, and checks with the type-assignment via N–M. Link Y–N is not given a step number because it was the starting assumption.

Steps 5 and 6 for links M–X and Y–X, lead to a provisional assignment that X is an auxiliary.

We now have the option of typing P or V by working from X and Z, or from Z and W, respectively. Whichever we do first will affect the second result, and the second result will have to be coherent with the first. There is no guide in selecting either P or V to be typed first, so we arbitrarily choose V (see Fig. 3.12).

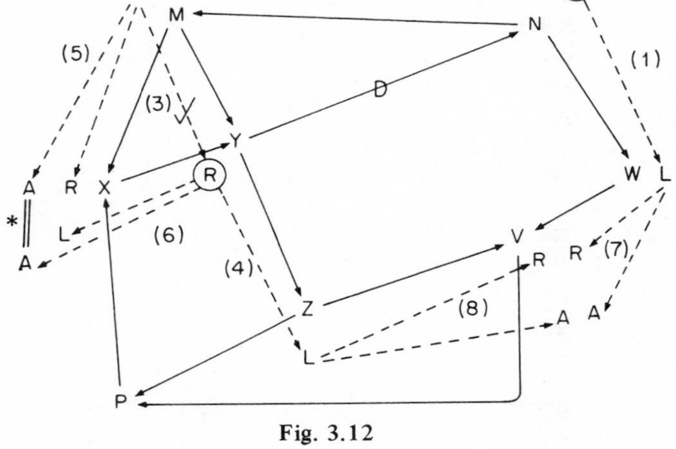

Fig. 3.12

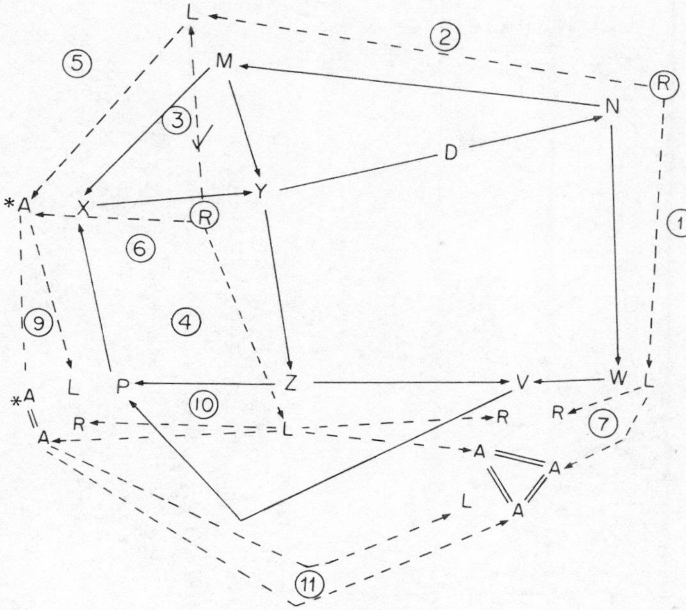

Fig. 3.13

Steps 7 and 8 show that there appear to be two consistent possibilities for V – either Auxiliary or Rate. There is no way of resolving this except by using the V link and to do that we must first type P (see Fig. 3.13).

Step 11 leads to a triply consistent assignment of V as an auxiliary which completes the process. The provisional assignments for X, P and V form a consistent set so the influence diagram is coherent and the final type-assignment is:

M	Level
N	Rate
P	Auxiliary
V	Auxiliary
W	Level
X	Auxiliary
Y	Rate
Z	Level

The order of the model is the order of the delay between Y and N, plus three for the levels M, W and Z.

We shall conclude by working a fairly complex example. The working and the results are given in Fig. 3.14, with a few explanatory remarks.

Despite the complexity of this influence diagram this is a simple assignment problem, because six variables – C, E, J, K, V and T – are fixed by the three

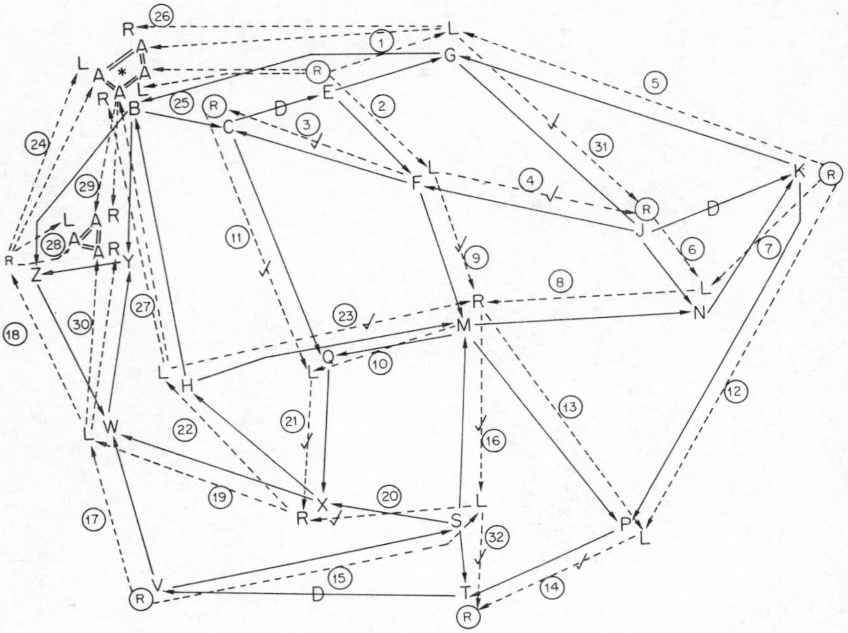

Fig. 3.14 A complex type-assignment example

delays. There are successor levels to these rates, G, L. N, P, Q and W so that 12 variables are typed automatically. The large number of ticks on the diagram shows the extent to which the system solves itself. The six levels immediately give the other rates which precede them, for example, step 17 gives to W as a level, which automatically makes Z a rate, step 18. The high degree of self-determination of this system makes it almost too easy to assign the types, and this is not uncommon for apparently large and difficult influence diagrams.

It is important to make sure that all links are checked to ensure system coherence – even if this is done at the end of the assignment calculation as in steps 31 and 32 in this case.

3.7 Structural Incoherence in Influence Diagrams

The presence of incoherence means that a mistake has been made, either in the type-assignment process or in the system description as it appears in the influence diagram. We shall devote the remainder of this chapter to a treatment of the causes of, and remedies for, incoherence arising from the second of thes causes.

The first example is the system shown in Fig. 3.15a. This is closed and appears to be an acceptable influence diagram. If V is assumed to be a level, we get Fig. 3.15b.

Step 3 indicates that Y *must* be a level, and step 4 shows that one of the possibilities for Y is that it *may* be a level. As far as links Y–Z and X–Y are

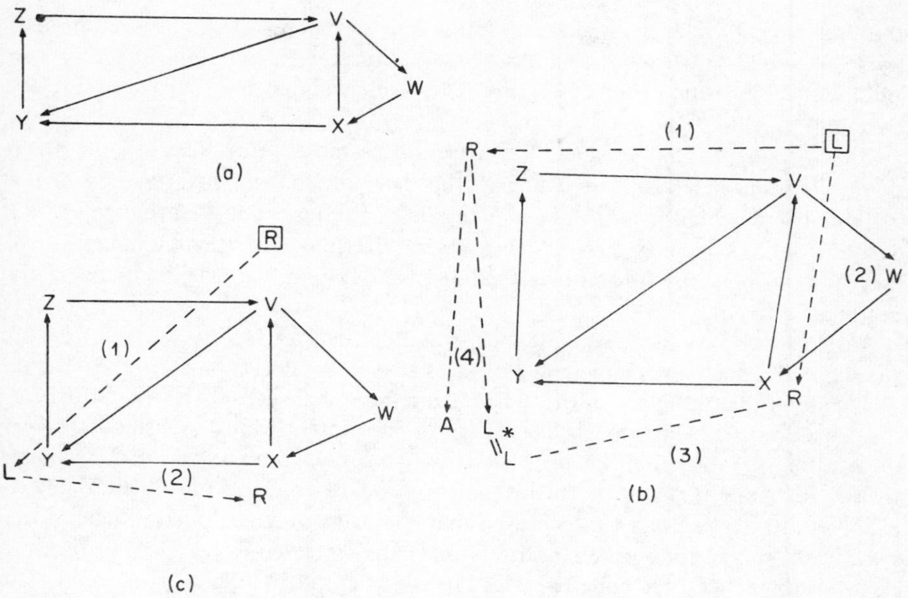

Fig. 3.15

concerned, Y is a level. However link V–Y shows that Y cannot be a level, and the diagram is incoherent.

The real importance of the concept of incoherence lies in what it tells us about the model as it has been developed. DYSMAP sometimes allows one to bend the strict rules of system structure far enough to permit the writing of a computable program from an incoherent diagram, so that incoherence does not always prevent *apparent* progress. It does, however, prevent real progress because the model contains errors which should have been cleared up.

There are two possible causes of incoherence; either the starting assumption was invalid, or there is some fundamental fault in the modelling. This may be that a link has been put in which does not exist in the real system, or that the influence diagram contains impermissible components.

Taking the first of these possibilities, we examine the other option for V, namely that it could be a rate. We obtain Fig. 3.15c.

Clearly X and V cannot both be rates, as there is no delay recorded for the link V–X and the diagram is still structurally incoherent. The incoherence is more than a matter of the starting assumption and whether there is reason to regard V as a level or, for a different system but the same diagram, to treat it as a rate, the diagram is incoherent. In neither case is the reason hard to find and, in both examples, it stems from a misunderstanding of the system structure.

For the first case, when V was a level we have X as a rate feeding two levels, V and Y, the character of which was inferred through the links V–Z and Z–Y. Although Z is a perfectly good rate, controlled by one level and feeding another, the link from V to Y is incorrect because a level cannot directly affect another level. The solution to the incoherence lies in a further examination of the system and the precise nature of the correction to be made will depend on the specific case. Certainly, the link from V to Y is suspect.

The second possibility, when V was a rate, is more easily disposed of. Again there has been an error in drawing up the diagram from the investigation of the system. The detected incoherence actually helps by suggesting that there may be a delay in the link from X to V which has been overlooked. If there is, and this can only be determined by further investigation of the system, the diagram immediately becomes coherent.

We have chosen this example deliberately because, if V is taken to be an auxiliary, the diagram is coherent with Z as a level and W as a rate (check it!). This is because of the simplicity of the diagram and, for a real system with this structure, there would have been enough knowledge of the system to justify using Z or W as the starting point in the first place, rather than V. The whole idea of the starting point is that one is saying 'with this system we know enough about this variable to use it as the starting point in checking that the diagram is a coherent model of the system'. In this chapter we use arbitrary diagrams unrelated to real systems but it is essential to realize that coherence testing is a method of relating a diagram to a system, not an abstract exercise performed on a theoretical diagram.

We have used the same influence diagram to represent two different systems each of which is incoherent because an error has been made in drawing up the diagram from the system description. In both cases the solution of the incoherence is a matter of 'back to the drawing board', that is, further investigation of the system itself is called for. We refer to such cases as SYSTEM-RESOLVABLE.

3.8 Structurally-resolvable Incoherence

It is possible to dispose of incoherence by arguments deriving from the fundamental concepts of system structure which were presented in Chapter 2. This

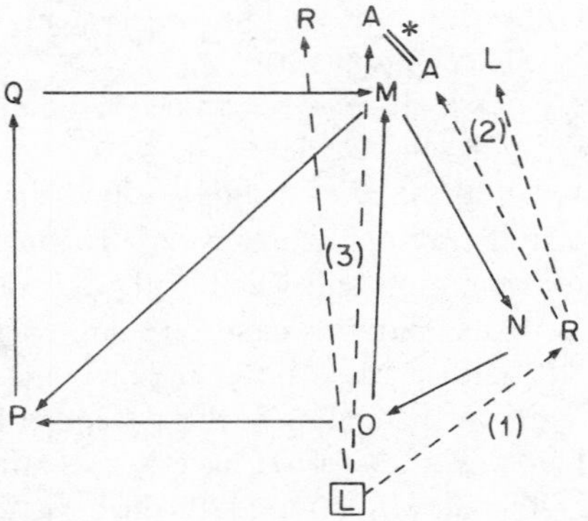

Fig. 3.16

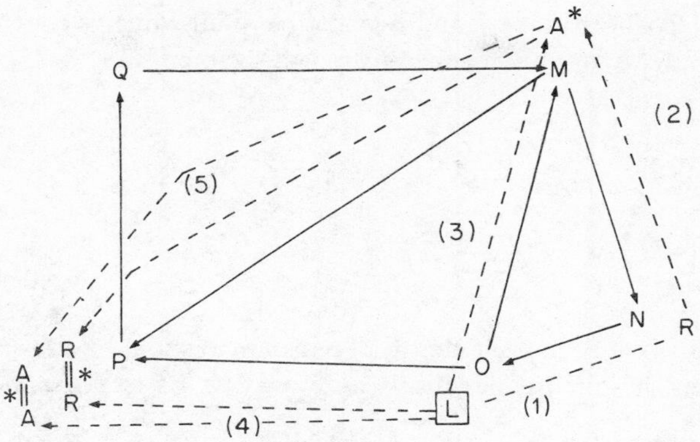

Fig. 3.17

is called STRUCTURALLY-RESOLVABLE INCOHERENCE, and we now consider an example of such a situation.

The first stages in the calculations are shown in Fig. 3.16, from the starting assumption that O is a level.

In the usual manner we make a provisional assignment of M as an auxiliary and proceed with steps 4 and 5 on links M–P and O–P (see Fig. 3.17).

There are two possible provisional assignments for P; as a rate or as an auxiliary. Definitional considerations will sometimes, but by no means always, distinguish between a level rate, but not between a rate and an auxiliary, nor between a level and an auxiliary. In order, therefore, to resolve this apparent case of what may best be called semi-coherence we must attempt to type Q and, since there are apparently two options for P, we have to use two steps for the link P–Q (see Fig. 3.18).

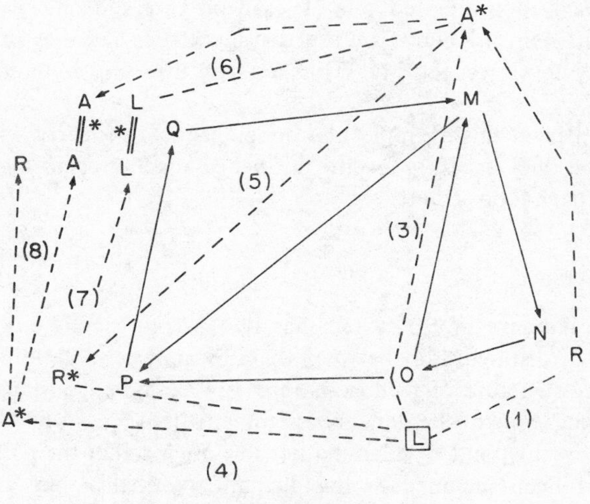

Fig. 3.18

Unfortunately, both steps 7 and 8 lead to a result which is coherent with step 6 and we can, it seems make two coherent type-assignments.

Variable	Type	
M	A	
N	R	
O	L	
P	R	A
Q	L	A

This appears to defy the rule that type-assignment should lead to a unique model order. The key lies in the feedback loops as the basic components of system structure.

This system contains three loops, the possible structures of which are:

Loop	Variable Names	Variable Types		
	M	A		
(1)	N	R		
	O	L		
	M	A		A
	N	R		R
(2)	O	L	or	L
	P	R		A
	Q	L		A
	M	A		A
(3)	P	R	or	A
	Q	L		A

Now, a feedback loop must contain at least one rate and one level. Loop 3 will violate this requirement if P and Q are typed as auxiliaries and they must, therefore, be a rate and a level respectively. This reduces the semi-coherence to unique coherence.

It will be seen from this example that coherence is a property which relates to and derives from the structure of the influence diagram *and* the fundamental concepts of feedback-loop structure.

3.9 Double Coherence

It sometimes happens that a system has two or more coherent solutions and there is no way of distinguishing between them by appeals to definitional points or by the use of the structure of feedback loops. For example, consider the system in Fig. 3.19, dropping some of the detailed assignment steps.

The solution to this kind of dilemma lies in noticing that the parameters of the system have not been included in the diagram. Normally they are not needed

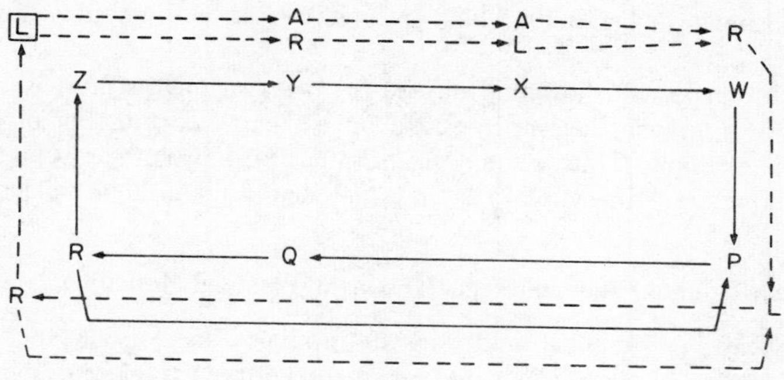

Fig. 3.19

because, as we have seen, quite complicated systems can be assigned without reference to the parameters.

For the disputed top line in this case, consider two possibilities shown in Fig. 3.20a, where D, E, F and G may be parameters or input variables (though not DT because that is a parameter of the simulation, not the system).

In case 1 the implied equations are:

$$Y = f(D, Z)$$

$$X = f(Y)$$

$$W = f(G, X)$$

and this means that X must be a level as only a level is parameter-free (recall the conservation equation from Section 2.6, and the ensuing discussion). A first-order information delay has a parameter, of course, but its correct influence diagram

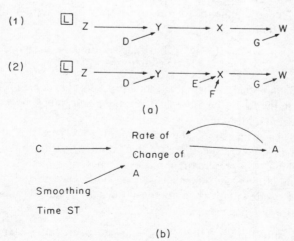

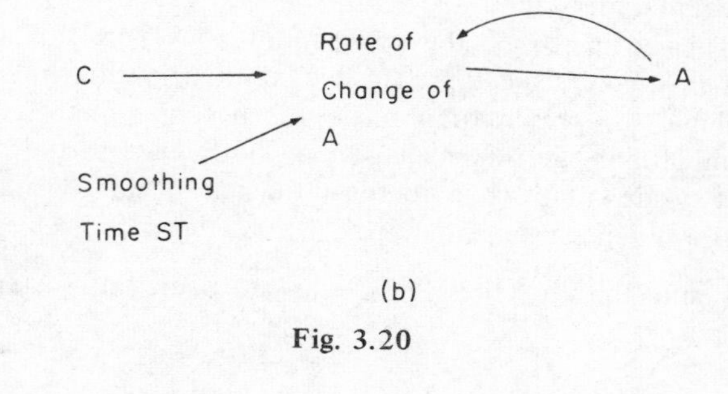

Fig. 3.20

structure is shown in Fig. 3.20. which is not the same as the above case of double coherence.

A little thought will show that in case (b) the assignments must be auxiliaries, and that this approach does not depend on the particular number of parameters used in this example. Thus we may restate the position that the type assignment *must* be unique, providing enough detail is included in the influence diagram.

3.10 Type-assignment and Coherence Testing in Practical Modelling

In many cases of practical modelling the influence diagram becomes very large, even to the point where it is hard to find a sufficiently large piece of paper to draw it on. There are two principal reasons for this:

1. Many practical problems really are very complicated with large numbers of interacting variables. Even with the aid of the list extension method it is not easy to see which variables are essential and which are, relatively, unimportant.
2. In order to have a chance of recommendations being implemented, and that is always the real object of the exercise in actual business situations, it is essential for the managers concerned to have a high degree of confidence that the relevant factors have been modelled.

   The advantage of the influence diagram is that it makes the model structure very clear and this is an aid to elucidating the system structure from management, who are in the best position to know what that structure is. However, the diagram also shows what has not been included, and this gives managers the opportunity of being the arbiters of relevance.

   Influence diagramming, and the simulation techniques to be discussed in Chapter 4, make it easier to include a factor to satisfy a manager who feels that it is relevant to system performance, than to convince him that it is not and may be omitted from the model. In any case, the analyst will usually find it hard to adduce convincing arguments for the irrelevance of a factor until it has been included and tested.

In practice then, we are likely to have large influence diagrams, and this has three drawbacks:

1. The type-assignment and coherence-testing procedures become rather tedious.
2. The larger the model is, the more difficult, time-consuming and expensive it will be to analyse and understand.
3. As the model size increases, the problem of conveying it and its results to management in a concise, comprehensible, fashion in the short time which is usually available for presentation of results becomes almost insolvable. The virtue of simulation – getting management confidence in the model – almost proves to be its downfall when to the influence diagram is added a welter of computer printout and system analysis.

The second and third of these problems are dealt with later, especially in

Chapters 8 and 13 respectively. For the moment we concentrate on the first, and ask if there are short-cuts which can be used.

In most models a lot of the levels will stand out very clearly. In particular, average values and stocks of product, cash, labour, and capital equipment have to be levels. If type-assignment showed them not to be such, the test of definitional coherence discussed at the end of Section 3.5 would not be passed. Since the remedy for this type of incoherence lies in altering the number and type of variables connecting the starting variable and the variable which has been found to be definitionally incoherent, in order to make the type of the suspect variable conform to what is known about its nature, we may just as well start out by typing *all* the model variables whose type is obvious, including delayed rates. This begs the definitional coherence test but it also exploits the concept on which that test is based in order to simplify model construction.

This simplified approach bases the type-assignment procedure on several starting points and, as was shown in the example of Fig. 3.14, this makes the procedure very easy. It also focusses the attention of type-assignment and coherence testing where it will do the most good, i.e. on the 'fine structure' connecting the system levels.

Note that the procedures should always be applied because they will reveal structurally-resolvable and system-resolvable incoherence in the fine structure, which is where they occur.

Chapter 4

Simulation Programming

4.1 Introduction

Chapter 2 described the idea of simulating a dynamic system. This chapter covers the principles and techniques of the necessary computer programming. Basic familiarity with computers is assumed, but no specific knowledge of programming is required.

The essential ideas in simulating a dynamic system is that of calculating the values for its levels at each of a sequence of points covering the required length of time. This requires:

1. The values of the levels at the starting point – the 'initial conditions'
2. The rates for the intervals spearating the points on the time scale at which the values of the levels are to be calculated,
3. A means of calculating the integral of the rate to give the next values of the levels.
4. Rates should, in general, depend only on levels or auxiliaries or, in the case of the output from a delay, on earlier values of a rate, as expressed in the values of the internal levels of the delay. Auxiliaries are stages in the calculation of a rate and they depend on levels and, possibly, other auxiliaries.
5. Thus, in general, we shall have to establish a computable sequence for the auxiliaries so that, i f,

   AUX2.K = AUX1.K+LEV.K/DEL

   AUX1 would be calculated before AUX2.
   We might also have

   RATE1.KL=AUX1.K+AUX2.K

   and RATE2.KL=AUX2.K
   This could equally well be written

   RATE1.KL= AUX1.K+RATE2.KL

and there is really no mathematical difference between rates and auxiliaries. We

must therefore have a method, automatic if possible, for sorting the rates and auxiliaries *together* into a feasible computable sequence.
We must NOT have a set of equations such as

RATE1.KL=AUX1.K+RATE2.KL

RATE2.KL=2*RATE1.JK

because this is a loop with no level and would produce dynamics in RATE1 which were completely spurious and dependent on the numerical value of DT. This example is manifestly absurd, but the situation could be rather hard to detect when the equations are scattered among hundreds of others. We therefore need to be able to detect and reject infeasible sequences of rates and/or auxiliaries, identifying those which form simultaneous equations, i.e. loops with no levels.

The gaps between successive time points for the levels are called the SOLUTION INTERVAL, and are DT time units long. The period of simulated time since the

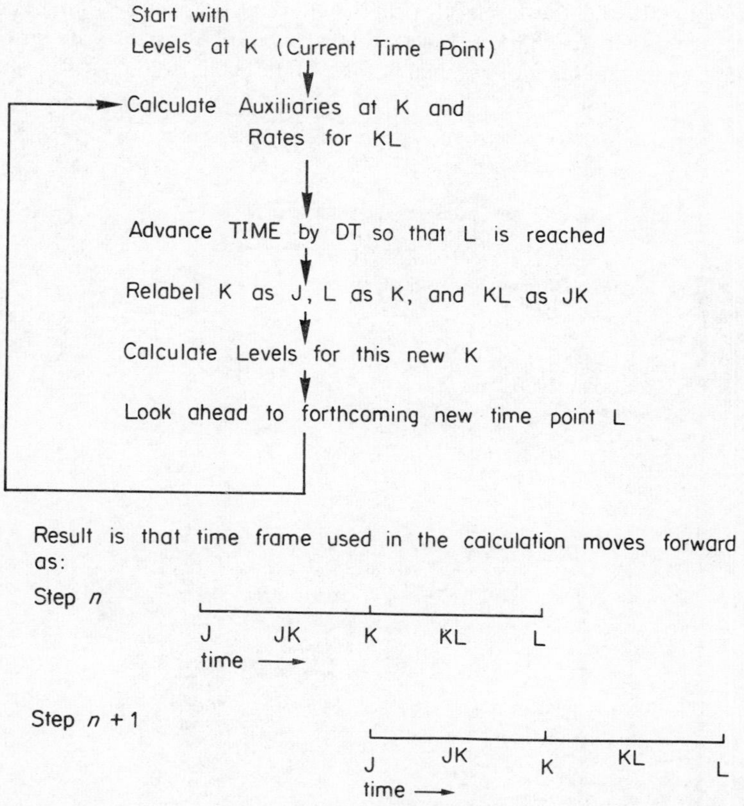

Fig. 4.1 Time-shift and relabelling in simulation

start of the calculation is denoted by TIME, and the total duration to be simulated is LENGTH.

Using this notation, the basic pattern of calculation for the simulation of a dynamic system is arrived at by fairly obvious considerations:

1. Since levels depend only on rates, and not on other levels, their current values can be calculated in an arbitrary order.
2. Rates and auxiliaries can be calculated from the prior values of the levels, providing a feasible computable sequence can be established.
3. A simulation may involve many hundreds of time points but only three need be considered in order to do the calculation. These are labelled J, K and L as shown in Fig. 4.1, and the periods separating them are denoted by JK and KL and are DT time units long.

   Point K is treated as the 'current' value and, knowing the levels at K, the auxiliaries at K and the rates for KL can be calculated. The simulated time now moves forward by DT so that what had been L, the next point in time, becomes K, the new current point. Values valid for old K, and KL, are relabelled to make them apply to J and JK, the time point and interval which have just passed. With the rates for JK and the levels at J the new levels at K can be found and the process is repeated. As TIME advances the values of the variables are stored so that they can be tabulated or graphed against TIME at the end of the simulation.

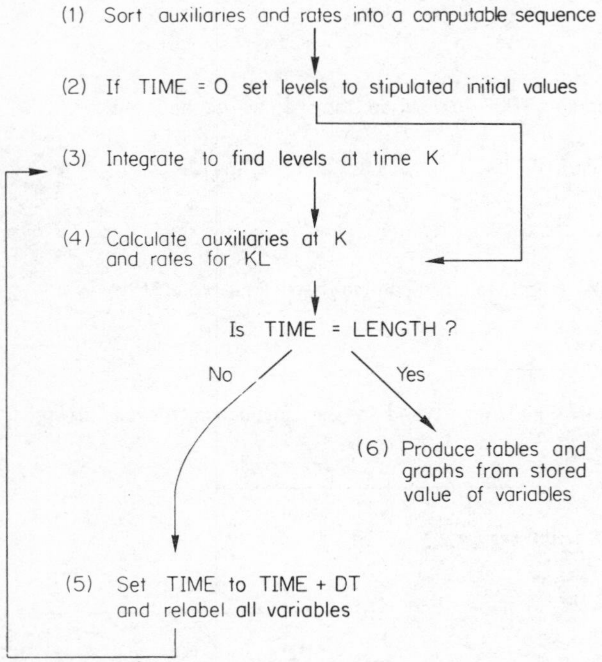
(1) Sort auxiliaries and rates into a computable sequence
(2) If TIME = 0 set levels to stipulated initial values
(3) Integrate to find levels at time K
(4) Calculate auxiliaries at K
and rates for KL
Is TIME = LENGTH ?
No
Yes
(6) Produce tables and
graphs from stored
value of variables
(5) Set TIME to TIME + DT
and relabel all variables

Fig. 4.2 Basic calculation steps

This basic pattern of calculation is shown in Fig. 4.2 in which step 1 is performed only once and step 2 is a special case of step 3 performed only once to get the calculation started. The sequence of calculation is, therefore:

1. Levels
2. Auxiliaries
3. Rates calculated from levels and auxiliaries or as delayed earlier rates.

This sequence should be firmly grasped by the reader.

4.2 Choice of Simulation Language

In teaching system dynamics on introductory courses it is very useful to carry out simple experiments in which the students act as the parts of, say, a production and distribution system. The instructor feeds in a demand pattern and the students make their own choices of decision rules for stock-holding, production, and so forth and trace the performance of the system by hand or on electronic calculators. (E.g. Jarmain, *Problems in Industrial Dynamics*.) For practical studies, and more sophisticated problems, this is simply not a feasible proposition and recourse must be made to the computer.

For digital simulation one may employ either the general high-level languages such as FORTRAN, PL1 etc, or special simulation languages such as DYSMAP, DYNAMO, CSMP etc.

Space does not permit a full treatment of the selection of a programming language, a choice which really hinges on:

1. Availability
2. The user's previous experience and skill
3. The need for special facilities in the calculation
4. The numerical accuracy required
5. The complexity of the problem and, therefore, the likely evolution of the computer model.

We shall simply summarize some of the main factors and come to a conclusion which is relevant for the purposes of this text but not necessarily, as will be argued, for all system dynamics work.

The advantage of the general languages, e.g. FORTRAN, is their wide availability and the greater flexibility they provide for modelling complicated decision rules. In theory, anything which can be programmed in FORTRAN can also be done in DYSMAP but, in practice, it may be rather more difficult and tedious and require more equations. The best example is arrays, which are easily handled in FORTRAN but which have to be written element-by-element in earlier versions of DYNAMO and DYSMAP.

On the other hand, the output from a FORTRAN program simulation model is usually much more tedious to arrange than from DYSMAP and CSMP, which have excellent output facilities for system dynamics simulation.

The main disadvantage of FORTRAN lies in the key to system dynamics modelling which is, essentially, to start with a simple model and to expand it until it is adequate for the purpose in hand. Since FORTRAN statements have to be carefully arranged in the required computation sequence there is a large possibility of creating errors in the program as the model expands. This problem disappears with the special simulation languages in which the equation cards can be placed in any arbitrary order. The compiler automatically checks and sorts them into a computable sequence, or identifies the error if no such sequence can be found. The equations in the model can thus be grouped according to the part of the system they described rather than in computation sequence, and the work of expanding, or contracting, a model is made very easy.

There are many special simulation languages of which we have space to mention only DYSMAP and CSMP. Both are easy to learn and have excellent output facilities.

DYSMAP (Dynamic System Modelling and Analysis Package) was developed by A.K. Ratnatunga, J. A. Sharp and the writer at the University of Bradford. It is syntactically equivalent to the earlier language, DYNAMO, developed by Pugh at MIT. Both these languages were specially designed for management system work, whereas CSMP was originally evolved for control engineering type studies.

The main advantage of DYSMAP is its simplicity of learning and use and the fact that its format relates very closely to the description of managed systems. Its numerical integration technique is extremely simple but is usually accurate enough for most purposes. CSMP, on the other hand, is slightly more difficult to learn for people who have not had an advanced scientific training, but its integration procedures allow the selection of sophisticated methods. However, it lacks many of the features of DYSMAP and experience of teaching, and practicing, system dynamics shows that, on balance, DYSMAP is to be preferred for student use and for many industrial applications. However, the advanced student and the analyst of very complicated problems will often find it helpful to go to the more sophisticated CSMP.

Both DYNAMO and DYSMAP are available for different computers, and they are equivalent for most practical purposes, but DYSMAP has more rigorous checking for level-less loops and has additional features and it is, therefore, the standard used in this book. The programs in Appendix B are in DYSMAP but can be run under DYNAMO by simply leaving out the documentation statements which define the variables. For the other differences between the two see the respective users manuals listed in Appendix C. Another advantage of DYSMAP over DYNAMO is that it produces a sorted equivalent FORTRAN program which can be preserved and used with the standard output subroutines.

Analogue simulation has considerable attractions, especially as a training device for managers and introductory students. The advantage is the interactive facility by which policies can be designed 'hands-on'. The disadvantage for practical use is the comparatively small model size to which one is sometimes restricted, the labour of setting-up and debugging the model, and the tedium of resetting the model to experiment with different system structures.

The choice of a particular simulation language is far less important than intelligent use of the language available. In general, time-sharing systems become very time-consuming once the model development stage is past, though time-sharing has enormous advantages in model development and debugging. Once the model is working, however, it is far better to use a batch system which will very rapidly produce a large amount of output which the user can analyse slowly and carefully without being driven by the demands of a time-sharing device for more input, or being held up by its slow output. Computer output is no substitute for thinking about the system.

In general, then, DYSMAP is the preferred language for the purposes of this introductory text and all the ensuing material is expressed in terms of it. The reader who does not have access to DYSMAP should have no real difficulty in expressing the material in terms of whatever language is available to him. DYNAMO is usually available on IBM equipment and DYSMAP is designed to be compatible with any computer having a FORTRAN facility.

4.3 Programming in DYSMAP

This section gives only an outline and we shall concentrate, in subsequent sections, on advanced use of the language. In Chapter 5 we shall deal with equation formulation and with initial conditions.

DYSMAP recognizes six equation types: levels, rates, auxiliaries, supplementaries, initial values and constants. The equation is punched on a card, using the appropriate time postscripts J, K, JK and KL and with the letters L, R, A, S, N or C, respectively in column 1 of the card, the equation proper usually starting in column 3 and continued on further cards if necessary. Continuation cards have X in column 1. Constants, and variables in initial condition equations, have no time postscript.

Constants may appear in one of two forms. For a simple numerical value, a C card is used, e.g.

```
C   CONS = 10.7
```

or

```
C   CONS = 10E06
```

for a large number in exponent form.

It is, however, often convenient to express a constant in terms of other constants. This helps to keep the model properly scaled and its conditions balanced when parameters are being changed. For example, if the number of weeks of production to be held in stock at factory A is to be some multiple of the stock cover required at factory B, which supplies components to A, and it is required to experiment with these parameters, then we might have

Table 4.1. *Subscript Conventions in DYSMAP*

	Quantity Type on Left of Equation	Subscript on Left	Subscripts on Quantities on Right if Quantity is					
			L	A	R	S	C	N
L	Level	K	J	J	JK	–	none	none
A	Auxiliary	K	K	K	KL	–	none	none
R	Rate	KL	K	K	KL*	–	none	none
S	Supplementary	K	K	K	KL	–	none	none
C	Constant	none	–	–	–	–	–	–
N	Initial value or computed constant	none	none	none	none	none	none	none

– means that the dependence is meaningless.

*Where there is an intervening delay, the subscript on the right must be JK e.g. OUT.KL=DELAY3(IN.JK,DEL). This is a convention used to remind one that there are levels in the delay whose values have been derived by integration of earlier rates and does not mean that the numerical value of the input rate at JK is actually used directly in calculating the output rate at KL.

N COVERA = RATIO * COVERB

C RATIO = 1.5

C COVERB = 6

where COVERA Weeks cover required at factory A
COVERB Weeks cover required at factory B
RATIO A's cover as a multiple of B's.
COVERA is called a COMPUTED CONSTANT and appears on *an N card*

The proper time postscript for equations are given in Table 4.1, which is derived from Pugh. We use the terms postscript and subscript interchangeably.

Notice that DYSMAP permits the use of auxiliaries in a level equation. This is in violation of the strict rules in Chapters 2 and 3, but it is a convenient simplification. The novice should, however, avoid it until he is quite sure what he is doing, as an apparently convenient device may cover up sloppy thinking about the way the system works.

More significantly, DYSMAP allows the use of RATE.KL in auxiliary and supplementary equations. This is a consequence of the ability to sort rates and auxiliaries together. This feature tidies up a loose end in DYSMAP but should *never* be used in modelling for the reasons discussed in Chapter 3. Although it is mathematically permissible it violates the rules of coherence and therefore indicates muddled thinking about the system and is symptomatic of poor modelling. We should avoid using computer conventions to fiddle our way round difficult modelling problems – solve the problem, don't fudge it.

By contrast, DYNAMO accepts RATE.JK in auxiliaries, supplementaries, and even in rates which do not have an intervening delay. This is done because rates and auxiliaries are sorted separately in DYNAMO.

With our convention for the subscripts in a rate equation, DYSMAP would reject the level-less loop in Section 4.1, item 5. It would assume that the second equation was

RATE2.KL=2*RATE1.KL

and would then refuse to proceed because the two equations could not be sorted into a computable sequence. The solution to this is explained in Section 5.7. Note that DYNAMO would tolerate the equations in Section 4.1 item 5 and could therefore produce false dynamics.

The duration to be simulated is denoted by LENGTH e.g. C LENGTH=120. DYSMAP does not care whether this is weeks, years, or centuries, so the user has to be sure he knows which it is. The time within the simulation is denoted by TIME which must not, therefore, be used as a name for a variable in the *system* as it is an attribute of the *simulation*. TIME can, of course, be used as an argument in a trigonometric function, or to calculate how long has elapsed since some event took place in the system.

4.4 Standard Functions

There are a number of standard functions, some of which must be mentioned briefly as they are used later in this book. The reader should carefully study the programs in Appendix B for examples of their use.

STEP

This generates sudden changes in a variable e.g.

V.K=100+STEP(STH,STM)

V= 100 if TIME<STM and from STM onwards it is 100+STH. An upward step followed later by a downward one will generate a square pattern and the step heights can be as large or small as required. A step only takes place once in a run as opposed to PULSE (see later) which can be used to create a similar, but regularly repeated, effect.

CLIP

This provides a conditional choice of two values for a variable, e.g.

V.K=CLIP(P,Q,R,S)

specifies that

V.K = P is R $\geqslant$ S

but

V.K = Q if R $<$ S

(Note the weak and strict inequalities)

PULSE

it is often useful to be able to make sharp changes in a level or to inject sudden shocks into the system. This is done by the PULSE function which injects a pulse of a given *height*, lasting for DT and repeated at intervals. Thus is

V.K=PULSE(HEIGHT,FIRST,INTERVAL)

the variable V will have the value HEIGHT *throughout the DT* starting when TIME = FIRST and repeated every INTERVAL time units, where HEIGHT can be a variable with the proper postscript.

To move a *total quantity* into or out of a level during 1 DT, e.g. to re-zero the level of cumulative sales at the end of each year, write

PULSE(HEIGHT/DT,FIRST,INTERVAL)

since the pulse will last for DT (see Section 4.8).

Only rate variables should be generated by PULSE functions.

SAMPLE

This function is very useful in calculating old values, such as last year's sales, and it also enables the modeller to incorporate discontinuous decision-making into a continuous system.

The format is

```
V.K=SAMPLE(X.K,PERD,INIT)
```

so V.K, every PERD time units, changes its value to that of X at that time, and then remains at that value throughout the ensuing PERD. This makes V have a stepwise behaviour. Until the first sample takes place, V=INIT, which may or may not be zero.

TABLE *functions*

Probably the single most useful features of DYSMAP are the two forms of the table function. These enable any functional relationship between two variables to be approximated by a piece-wise linear interpolation between equally-spaced points on the curve.

For example, the curve in Fig. 1.4 is a true curve but it could reasonably be approximated by a series of points connected by straight lines, such as:

Delivery Delay (years)	Order Placing Rate (£M per year)
0.6	190
0.8	200
1.0	220
1.2	280
1.4	330
1.6	450
1.8	440
2.0	280

This would be written in DYSMAP as

```
R OPR.KL=TABLE(TOPR,DD.K,0.6,2.0,0.2)
T TOPR=190/200/220/280/330/450/440/280
```

where

where	OPR	Order Placing Rate
	TOPR	Name of table holding data for OPR
	DD	Delivery Delay
	T	A special case of a constant.

(The table holds data and is really a special case of the Constant. Note the use of T in column 1 to denote this.)

Thus, OPR is to be found from a table called TOPR, by taking the value of DD and interpolating between the two nearest table values. The first value in the table corresponds to DD = 0.6 and the last to DD = 2.0, with intervening table values matching DD in steps of 0.2, i.e. 330 corresponds to DD = 1.4. Thus if DD = 1.285, the table function will automatically calculate:

$$\begin{aligned} \text{OPR.KL} &= 280 + (330 - 280) \times \frac{1.285 - 1.2}{1.4 - 1.2} \\ &= 301.25 \end{aligned}$$

The table function automatically checks that the proper number of entries has been included on the T card.

Table functions require one entry for each point in their range, including the last. Thus if

$$\text{V=TABLE(TV}, n_1, n_2, n_3)$$

where n_1, n_2 and n_3 define the range of the table as shown in the last example, the number of entries in TV must be

$$\frac{n_2 - n_1}{n_3} + 1$$

and $(n_2 - n_1)/n_3$ MUST BE AN INTEGER. In the example, TOPR needs

$$\frac{2.0 - 0.6}{0.2} + 1 = 8 \text{ entries.}$$

If the argument of the table can exceed its range, TABLE is replaced by TABHL. This puts a horizontal tail onto the curve at the heights given by the first and last values and uses these for the result if the argument is respectively less or greater than the table range. This allows one to deal with curves with horizontal tails without having too punch a long T card (see Chapter 5).

OUTPUT

DYSMAP outputs tabulated and graphed variables. Ratnatunga gives full details and the best method of learning is by practice. It is a good plan to tabulate a large number of variables as a check on the internal working of the program, and to plot those which are most important, given the purpose of the model.

Printing and plotting are done at intervals defined by the user inserting, for example:

C PRTPER = 1

C PLTPER = 1

which would print and plot the chosen variables every time unit in the model, *not* every DT.

In practice it reduces the bulk of the output to define PRTPER as a variable e.g.

A PRTPER.K = 1 + STEP(9,11)

This prints the values at TIME = 1, 2, 3 . . . 10, 20, 30, 40 etc.

If the program gives trouble, PRTPER and/or PLTPER could be reduced to DT, with a shorter LENGTH, to show up the fine detail of what is happening. The levels, auxiliaries and supplementaries are output with their values at the stipulated time, but the values of rates are for the DT *commencing* at that time.

RERUNS

One of the most powerful features of DYSMAP is the ease with which a rerun can be carried out.

At the end of the model a card with RUN and a name or number is placed. This can be followed by any number of cards giving new values of constants or for tables, together with another RUN card, and a suitable title for the rerun showing which parameters have been altered. This automatically runs the model again with those new values in place of the old ones, thus enabling the analyst to experiment with different parameter values and policies at one run.

After a rerun a constant keeps its new value and does not revert to that given in the original model. This can be overridden by defining the constant, in the main program, on a card with CP (or TP for a table) in columns 1 and 2. The constant then reverts to its original value after any subsequent reruns.

4.5 Initial Conditions

All levels must have an initial condition, and so must some rates as described in Section 5.6. These values are needed to get the simulation started as shown in Fig. 4.2 and discussed in Section 4.1. They are put on an N card which usually follows the L card itself for ease of reference.

At least one initial value in the model must be a pure number e.g.

```
N  LEV=150
```

but it is often convenient to use a parameter in the N equation, defining the parameter on a separate C card so that it can be changed in a rerun e.g.

```
N  LEV=ILEV

C  ILEV=150
```

Although a number must be provided somewhere the other N conditions can usefully be specified in terms of that number, or each other, e.g.

```
N  LEV2 = LEV
```

or

```
N  LEV3 =AUX1*AUX2+LEV4
```

where AUX1 and AUX2 denote auxiliary equations which exist in the model. This procedure saves an immense amount of work and, without it, the rerun facility would be nearly useless. Expressions can be as complicated as required, but must not contain postscripts.

DYSMAP automatically sorts these N equations into a computable sequence and then uses the specified number to calculate the numerical value for the other levels.

If a feasible computable sequence cannot be found, DYSMAP outputs the information that simultaneous equations exist among the N equations. For example, if the user has put

N LEV1 = LEV2

N LEV2 = LEV1

N LEV3 = 100

the calculation cannot be started for LEV2 and LEV1. The reason may be a punching error e.g. that the first of these conditions should read

N LEV1 = LEV3

Another possibility is that the user may have tried to be too clever at defining initial conditions in terms of each other. In that case another number has to be inserted, e.g.

N LEV1 = 200

Compare this with simultaneous equations among *auxiliaries*, as discussed in Section 5.7.

Where an auxiliary appears in an N equation it does so without a postscript. The auxiliary itself is defined on a perfectly ordinary A card, following all the normal conventions for equation syntax and obeying the rules for equation structure which are discussed in Chapter 5.

Although an auxiliary or rate could be initial conditioned itself on an N card there is no need to do so. DYSMAP, in computing the initial conditions, will automatically pick out the appropriate auxiliary or rate equation and use it as though it were an N equation. Since auxiliaries and rates are ultimately dependent on levels, which will have N cards of their own, this simply amounts to a more sophisticated procedure for sorting the N conditions to get a computable sequence. When this is done, DYSMAP prints out the message INITIAL CONDITION UNDEFINED. DYNAMIC EQUATION BY DEFAULT followed by the name of the auxiliary or rate in question.

If an initial condition is not provided at all for a variable which needs one, DYSMAP assumes it is zero for the first run and any reruns are aborted.

It is essential to check that the values of variables at TIME=0 are consistent with each other and with what was intended by the initial values which were defined. In a complicated model it is very easy for inconsistencies to creep in and for false dynamics to be generated, as discussed in Section 4.12, *Incorrect Initial Conditions.*

4.6 Layout of a DYSMAP Program

A dynamic model is usually simple to start with, but rapidly becomes more complex as more and more detail is incorporated. In fact, the list extension methods described in Chapter 3 is intended to make this happen in a controlled fashion.

In writing and developing a program it is essential to make provision for complexity to increase in order not to lose track of what is going on.

The best way of doing this is to group the model equations into sectors, according to the part of the system which they describe. The name and purpose of the sector is punched on NOTE cards which precede it.

It is very important to know what variables have been used in a program and this is done by including in it a documentation sector. This has cards such as

D INV=(U) SALES INVENTORY

The units of measure, or dimensions, being stated in brackets. DYSMAP will automatically check these for consistency as discussed in Chapter 5. From this information DYSMAP automatically produces a documented breakdown of each equation, its constants and the definition of its variables, many examples of which appear in the book. It also produces a MAP showing where the variable is defined in the program and naming the variables in whose equations it is used. It also identifies variables which have been used but not documented or documented but not used, and this is very useful in keeping abreast with amendments to a model.

4.7 The Choice of DT

The proper value must be chosen for the important parameter DT. Basically, there are three factors to consider, delays, numerical accuracy, and rounding errors.

If DT is too large the delays become unstable because the internal levels acquire or lose too much of the quantity being delayed and this leads to large fluctuations in the output rate. Forrester has shown that, to prevent instability from delays

$$\mathrm{DT} < \underset{i}{\mathrm{MIN}} \left(\frac{\mathrm{DEL}_i}{2n_i} \right)$$

where the ith delay is of magnitude DEL and order n.

However, it is also necessary to ensure that the output from the model delays approximates reasonably closely to that of the system delays. It is shown in the Annexe to Chapter 2, that this gives the condition

$$\mathrm{DT} < \underset{i}{\mathrm{MIN}} \left(\frac{\mathrm{DEL}_i}{10n_i} \right)$$

Numerical accuracy arises because we are approximating differential equations by difference equations. For example, if DT is set at 0.5, for the system in Section 2.12, by applying the first of the above rules, its transition from the explosive to the sustained oscillatory mode takes place at, approximately, TAI = 2.3. The value found by the differential equations of Appendix A, Section 8, is TAI = 1.5, and if DT is set to .03 the simulation yields very nearly the correct critical value of TAI. This suggests that DT has to be very small for great accuracy.

Rounding error can occur when DT is successively incremented to produce the passage of TIME. This may lead to uncertainty about the exact times of occurrence of time-related events such as SAMPLE, PULSE etc. and to some very irritating printing and plotting at TIMEs which are not exactly those specified. It can be avoided by making DT the decimal equivalent of an exact binary fraction such as 1/4, 1/8, 1/16, etc.

Since computer cost is not usually a significant factor in the overall cost of a study it is as well to make DT very small, e.g., .125 or .0625, providing it is still an integer divisor of sampling, pulsing, printing and plotting periods. When the model is fully developed DT can be increased to reduce computer cost and time, providing an occasional check is made with a small DT.

4.8 Special Techniques in DYSMAP

The standard functions can be used in many advanced ways. We cannot cover every possibility, but the examples should promote the user's own ingenuity. Small programs should be written to demonstrate the methods described. Chapter 5 draws on this material in developing typical equation forms for systems.

A Calculation of Most Recent Values

In real systems, Cumulative Sales, for example, increase continuously, but the only information available to the decision-maker is the value at the end of last month, and, during the coming month, decisions will be made on the basis of progressively older information. Even though such decisions may lead to actions which will affect future sales there is nothing for it but to use the old data. If this is what happens in the system then it must be modelled.

The discontinuity is achieved by using the SAMPLE function which converts a continuous variable into a series of discontinuous steps. For the sales data example, we may put:

OTOT.K=SAMPLE(CSALES.K,PERD,0)

OTOT	Cumulative Sales total at last sample point. Units.
CSALES	Actual Current Cumulative Sales. Units.
PERD	The interval between sampling instants. Time units.
0	Value of OTOT during the first PERD of simulated time.

This will make OTOT follow a stair-step pattern, touching CSALES at PERD, 2 x PERD etc.

If PERD is not an integer multiple of DT some rather curious results are produced, so that it is advisable to ensure that it is.

B Calculation of Two or More Old Values

(This procedure is available in some versions of DYNAMO as the BOXCAR function).

The method applicable to the calculation of the most recent old value will not work when two or more are required. If one wrote

OVAL1.K=SAMPLE(CVAL.K,PERD,0)

and

OVAL2.K=SAMPLE(OVAL1.K,PERD,0)

to give the old values of some variable, the current value of which is CVAL, at the ends of the previous two periods, the auxiliary sorting routine in DYSMAP would carry out the sampling in the order given here, so that OVAL1 and OVAL2 would be identically equal.

If OVAL1 and OVAL2 are levels, the PULSE function can be used to make rapid changes to them during the DT immediately following the end of a period.

Before giving the equations it is necessary to point out that although OVAL1 and OVAL2 are values of CVAL at the past time points t_1 and t_2, they have been carried forward in the system and are current values at the time point labelled NOW. This is shown, in Fig. 4.3, by the horizontal line prolonging the level of OVAL2 from t_1 to NOW.

The equations are based on using VDIFF, the difference between CVAL and OVAL1, to generate a rapid change in OVAL1 during the DT following the end of each PERD. Since the PULSE only operates once in each PERD there is no need to sample VDIFF in calculating the change to be made.

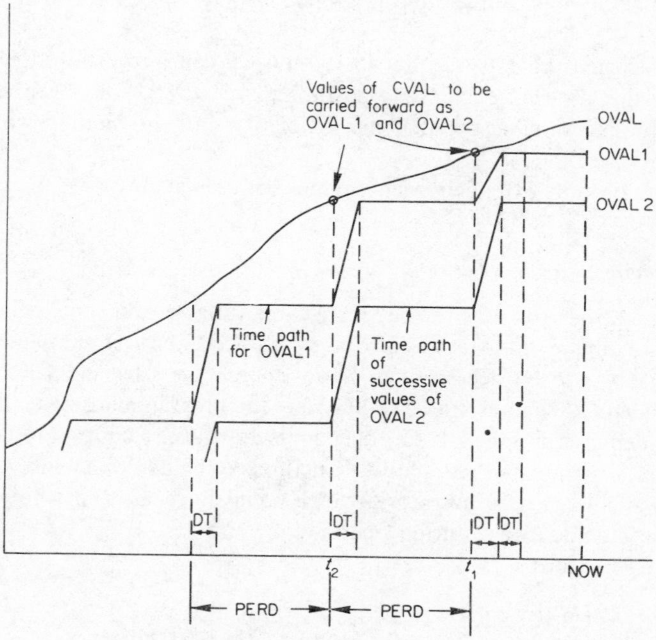

Fig. 4.3 Calculation of old values of a variable

This leads to:

VDIFF.K=CVAL.K–OVAL1.K 4.1,A

and then to:

OVAL1.K=OVAL1.J+DT*PULSE(VDIFF.K/DT,PERD,PERD) 4.2,L

(The convention for equation reference is to number the significant equations according to their position in the chapter and to add a letter denoting the type of the variable on the left hand side of the equation where it is necessary to make the type explicitly clear.)

To get OVAL2 we require:

OVAL2.K=OVAL2.J+DT*PULSE((OVAL1.J–OVAL2.J)/DT,PERD,PERD) 4.3,L

Comparison of equation 4.3 with 4.1 and 4.2 shows that VDIFF is computationally completely superfluous and is only introduced here to show the method of approach.

Both OVAL1 and OVAL2 require initial values which will depend on the nature of CVAL. If CVAL is a cumulative variable, as in example A, the old values are initialized to zero.

Fig. 4.3 shows that OVAL1 and OVAL2 have reached their new values by the end of the first DT after t_1 and t_2 and are available for use from that time throughout the rest of PERD. Since OVAL1 and OVAL2 are levels, this is correct and the change in their values during the DT is perfectly normal behaviour for a level.

It will be clear from Fig. 4.3 that PERD must be an integer multiple of DT. If PERD = DT the procedure still works. In Fig. 4.3, CVAL has been shown as a steadily increasing variable for the sake of clarity, but the method works even if CVAL fluctuates.

The method can be extended to any number of old values.

C Total Change during a Period

One often wishes to know the net change in a variable over some period of time. In the sales example this would be total sales during the last complete month. It would probably also be necessary to know cumulative sales and sales during this month so far, so we shall use these to illustrate the programming steps, the problem itself being depicted in Fig. 4.4. The diagram is drawn for a continuously increasing variable such as cumulative sales but the methods to be described will work equally well for a variable such as Inventory, which could fluctuate and where one might wish to compute net change during a period.

The basic equations are:

SOFAR.K=CSALES.K–OTOT1.K 4.4 A

SPERD.K=OTOT1.K–OTOT2.K 4.5 A

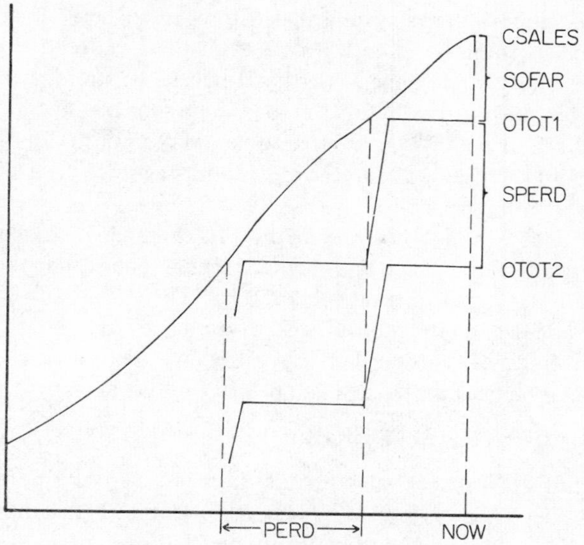

Fig. 4.4 Total change during a period

The method of example B being used to compute OTOT1 and OTOT2. Extension to previous old totals, OTOT3 etc. would provide the net changes during earlier periods, which could serve as a trend-indicating or forecasting device in a system model.

4.9 User-defined Macros

A macro is a piece of program which can be used again and again within another program to do the same type of calculation for different cases. For example, a corporate model might include production flows at each of several factories – all basically the same but having different numerical values for the flows themselves and for the factory parameters. To save multiple rewriting of the same piece of program, we write it once, using dummy variables, give it a name, and call it each time it is needed using the actual variables in each case as socalled ARGUMENTS in the macro.

Note: Both DYNAMO and DYSMAP are undergoing continual evolution so the following information may change. See Users Manuals.

In DYNAMO and in DYSMAP, macros can be written in simulation language syntax, and in DYSMAP they can also be written in FORTRAN and inserted within the DYSMAP program. Details of this latter procedure are given in Ratnatunga. For our purposes we concentrate on DYSMAP-Syntax macros, which are most likely to be of use in the average modelling situation.

Each macro is given a name, preferably indicative of its purpose, and it usually has one or more arguments. These are the quantities which are to be handled by the macro, or the constants in it.

For example, suppose that a system makes forecasts of several variables, such as sales, for a point 12 months in the future. After the 12 months have passed the system compares the actual value with that forecast 12 months previously and, perhaps, modifies the forecast for the next 12 months in the light of the comparison. This is a rolling procedure, done every month for the coming 12 months in the light of what was forecast 12 months ago; it is not something done only every 12th month.

If only one variable is involved, the method of example B can be used to carry the forecast forward through 12 intermediate stages. The reader should write the equations for PERD = 1 month as an exercise.

For several variables, or for multiple use of the same old forecast, this would be very tedious, so that a MACRO called, say, CURVAL (for current value) could be written. This could be used in the program proper by putting, say,

CVSFC.K = CURVAL(SFCST.K,SFIV) 4.6,A

where CVSFC Current value of Sales Forecast made 12 months ago

SFCST Sales forecast. (The value from 12 months ago is carried forward automatically by CURVAL.)

The sales forecast made now, but relating to 12 months hence, is injected into CURVAL and will come out in 12 months time. CURVAL contains the last 12 forecasts made 12, 11, 10 . . . months ago and these will come out in each succeeding month, being replaced, in each case, by the forecast made then and relating to 12 months from that date.

SFIV Initial Value for Sales Forecast, i.e. the forecast made 12 months before the start of the simulated period.

In this case SFCST and SFIV are the arguments, and mean that CURVAL is, in this case, being used to carry forward old sales forecasts. The whole point of the MACRO is that it could be used, in the same model, to carry forward old forecasts of anything else, e.g. order backlog, available labour or whatever, simply by using the appropriate arguments.

The macro itself appears at the very beginning of the program and, for this example, we give a few lines of the example the reader was invited to try three paragraphs earlier.

```
MACRO    CURVAL(OFCST, IVAL)

L £OF11.K=£OF11.J+(DT)(PULSE((OFCST.J–£OF11.J)/DT,£D,£D))
N £OF11=IVAL
L £OF10.K=£OF10.J+(DT)(PULSE((£OF11.J–£OF10.J)/DT,£D,£D))
N £OF10=IVAL etc.
L CURVAL.K=CURVAL.J+(DT)(PULSE((£OF1.J–CURVAL.J)/DT,£D,£D))
N CURVAL=IVAL
N £D=1
```

MEND

The following points should be noted.

1. The MACRO appears at the start of the program, immediately after the program title.
2. It starts with a card bearing the word MACRO, the name of this particular one, and typical cases of its arguments using dummy names *without* time postscripts.
3. The dummy arguments have to appear at least once on the right-hand side of an equation within the macro in order to specify what it is supposed to do with them.
4. Since this macro is supposed to carry an old forecast forward for 12 months there are 11 intervening variables OF11, OF10, etc., which represent intermediate stages in the process, but which are simply part of the macro's internal workings and not of its output. The names of these internal variables are always preceded by a £ or $ sign. DYNAMO actually recognizes $ but American computers used in Britain usually treat £ as equivalent to $, at least as far as card punching conventions are concerned.
5. The name of the macro has to appear on the left-hand side of one equation, as though it was a variable name, in order to create the output the macro is intended to provide. The macro name may also appear, as a variable, on the right-hand side of equations in the macro as often as required.
6. The usual identifiers of equation type appear in column 1 and the standard rules of equation format apply to macro equations, with the exception that constants have to be specified as though they were initial conditions, as for £D, the pulse parameter.

   Notice that in this case we have, for simplicity, assumed that the forecasts have been steady for the previous 12 months so that all the levels for the dummy forecasts OF1–OF11 have been initialized to IVAL. This is optional and whatever initial conditions are desired can be used, provided the macro has the proper number of arguments.
7. The macro has to end with a MEND card and can then be followed by as many other macros as are needed.
8. The initial value is, in this example, specified by the name of a constant, such as SFIV in the specific usage of the macro in equation 4.6 and, generically, by IVAL in the macro definition.
9. For each specific usage of the macro in the model an actual numerical value must be provided for the initializing constant.

In general, macros provide great flexibility in programming, especially where there are complicated decision rules. The novice should, however, avoid them as an apparent need for them may indicate that he has overcomplicated his model.

4.10 Flow-charting Conventions

Although the influence diagram is the more useful for model building, at the programming stage it can be helpful to draw a program flowchart. The drawback is that they are rather tedious to produce and are difficult to alter and update. Details of the conventions and examples of use are given in Pugh and in the books by Forrester and Jarmain listed in the Bibliography. For the purposes of this book we shall rely exclusively on influence diagrams, but the serious student should familiarize himself with flow diagramming by studying Pugh's Appendix G and the examples in Jarmain.

4.11 Model Testing

Once the model has been written and programmed it has to be tested. The testing is intended to fulfil three main needs:

1. Detection of errors in the model apart from syntax errors which are automatically located by DYSMAP.
2. Determination of the basic dynamic properties of the system.
3. Assessment of the adequacy of the model as a representation of the system.

Sections 4.11 and 4.12 deal with the first step in this process, the others being treated separately in Chapter 7.

Errors in the model are generally of two types: mistakes in the use of the programming language, and faulty equation writing. Even when both of these have been eliminated, the model will not necessarily be an acceptable description of the system, since some part of the system may not have been included, or the equations used, though not containing any actual errors, may not represent the way the system actually works, or the parameter vaues may be wrong. All these matters are dealt with in Chapter 7 under the umbrella of assessing the adequacy of the model. For the present we confine ourselves to tests which are intended to ensure that the influence diagram has been properly translated into equation form. Dimensional analysis is treated separately in Chapter 5, and automatically by the DYSMAP dimensional analyzer.

Many non-syntactical errors can be found from the output of the program when the system is driven by a test input. Test inputs are of two types: standard inputs and time series. A standard input is one which takes a very simple form and is designed to elicit information about the system which is not cluttered up by too much detail. This is particularly useful in checking for errors. A time series input is taken from historical data about the system and is mainly useful for assessing model adequacy. For the present we consider standard inputs only.

The simplest standard input is the step, which entails changing the driving input from one steady value to another, about 10 or 20% greater, after a predetermined time and then allowing it to remain at the new level. Any dynamics observed before the step will be false and arise from errors in the model. Dynamics after the step will be real, and indicative of the system's inherent characteristics, providing there are no false dynamics.

It is sometimes appropriate to produce a model in which there are no external inputs and the system's behaviour arises purely from its initial conditions and parameter values. This models a system which generates its own growth – for example in a corporate planning model. This is inherently more difficult to test for non-syntactical errors as there are no reference values against which to compare the output. For such a case the testing process would require a very careful examination of the model's outputs with a close watch for any 'odd' behaviour. It is practically impossible to give a general definition of 'odd' so it is hard to give any general guidance and each case has to be treated on its merits.

4.12 Examples of Detection and Correction of False Dynamics

Since there are almost endless opportunities for making mistakes in any kind of model building (something which is not usually mentioned), it is better to consider some specific examples of the detection and correction of errors, rather than give vague guidelines.

Incorrect Initial Conditions

The idea of the step input is to start the system in a stable state and then shock it to see what happens. For this tactic to work the initial conditions have to reflect a truly stable state and if they do not, false dynamics will be generated. The first example is of a system in which the initial conditions are incorrect and shows the false dynamics thus generated. The influence diagram for one of the faulty parts of the model is shown in Fig. 4.5. (This example is very similar to the IB rules for the

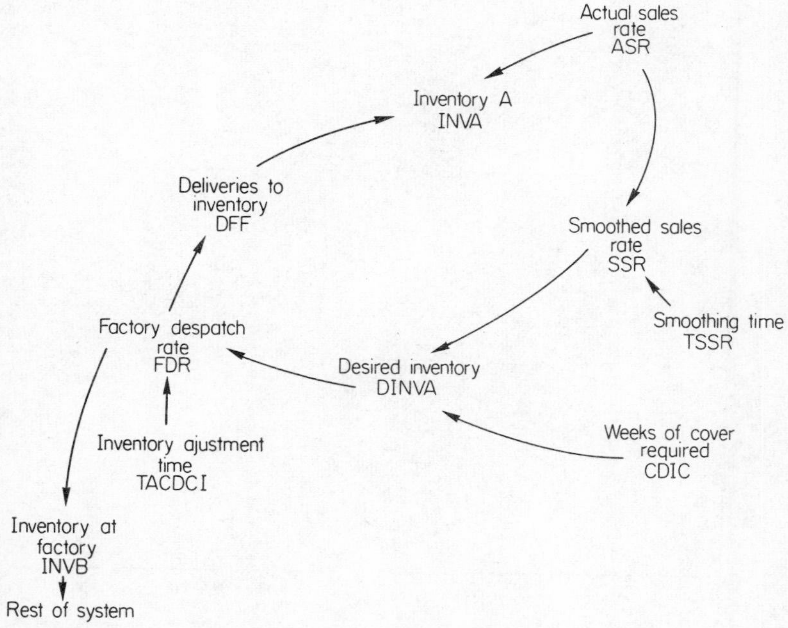

Fig. 4.5 Influence diagram for system showing false dynamics

sample system in Section 2.12.) The variable names are given under the description in the diagram.

The Actual Sales Rate ASR is the input variable and starts at a value denoted by ISSR (75 units/week) and is changed at the tenth week to 85 units/week. The Actual Sales Rate is averaged to give SSR, the Smoothed Sales Rate, which determines the Desired Inventory, DINVA, as a number of weeks of sales.

A DINVA.K=CDIC*SSR.K

Goods to replenish Inventory A come from the factory at a rate called the Factory Despatch Rate and determined by

R FDR.KL=(DINVA.K–INVA.K)/TACDCI

The goods from the factory arrive at Inventory A at a rate DFF which is simply a delayed version of FDR. The equations for Inventory A can thus be written:

L INVA.K=INVA.J+(DT)(DFF.JK–ASR.JK)

N INVA=ISSR*CDIC

In the second of these equations, INVA has been initialized to the Desired Inventory, so that the system appears to be correctly set up.

The step response is shown in Fig. 4.6. This shows the step in ASR and, before the step takes place, fluctuations in Inventory A.

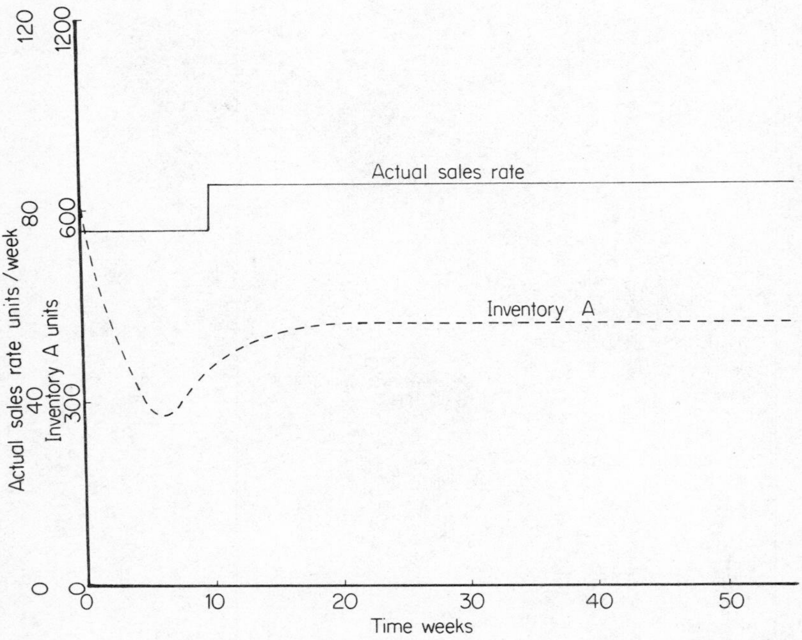

Fig. 4.6 False dynamics generated by incorrect initial conditions

Since the system shows dynamic behaviour when it ought to be stable there is an error. If Inventory A varies when ASR is constant, then DFF must be varying and, therefore, so must FDR. However, FDR should be constant and equal to ASR for the system to be stable, and applying this to the equation for FDR we find that

ASR=(DINVA–INVA)/TACDCI

or

INVA=DINVA–TACDCI*ASR

is a condition for stability of INVA.

When the system is steady, SSR = ASR and, using the equation for DINVA, Inventory A would have to obey the equivalent condition,

INVA=(CDIC–TACDCI)*ASR

for there to be no fluctuations in INVA.

When the system starts, ASR = ISSR, but

INVA=CDIC*ISSR

instead of

INVA=(CDIC–TACDCI) * ISSR

Clearly Inventory A is too large and it is easily seen that it is so large that initially FDR is zero so that INVA is being depleted but not replenished and Fig. 4.6 shows this quite plainly. The correction is to put

N INVA=(CDIC–TACDCI)*ISSR

as the initial condition for INVA.

False Dynamics due to Mistakes in Timing

The PULSE and SAMPLE functions are fruitful sources of programming errors and our second example of false dynamics concerns such a case.

The purpose of the model was to study the detailed dynamics of a production system in coping with variations in demand. The company had a policy of deciding at the start of each month what the production level was to be for the month, and relied on the stocks held by its dealers and in its own warehouses to absorb the fluctuations in public demand. The point in the model at which the error occurred was the dealer sector.

The key variable in the dealer sector is the Dealer Order Rate, DOR. Dealers tend to bunch their orders in the last week of the month and, typically, DOR might be 50 units per week in the early weeks of the month but rise to 100 units per week in the last week of the month. Since a month was taken to be 4.3 weeks this would lead to a total quantity ordered of 3.3 x 50 + 100 = 265 units. In any given month, dealers might order more or less than this, depending on their forecast of demand. The Dealer Forecast depends on the Dealer's assessment of demand in units/week and his observations of his own sales rate. For simplicity (since we are making a point about programming errors and not about system behaviour) we suppose that the dealers forecast demand perfectly, and, in that case the dealer forecast would be steady when demand and sales are uniformly level. The dealer forecast, DFCAST.K, multiplied by the number of weeks in the month, gives the dealers' estimate of monthly demand at time K. This quantity, divided by 265, is the ratio by which the basic dealer order rate of 50 or 100 has to be multiplied in order to get the actual dealer order rate in the early part of the month or during the last week respectively. Thus, if time K is in the first 3.3 weeks of the month, the Dealer Order Rate will be given by

```
R DOR.KL=((4.3*DFCAST.K)/265)*50
```

and in the last week it will be

```
R DOR.KL=((4.3*DFCAST.K)/265)*100
```

Since we cannot have two equations for the same variable, the terms 50 and 100 were replaced by FACTOR.K, which has the value 50 in the early part of the month and 100 in the last week. A dummy time, DTIME.K, represents the point in the month. Using a value of DT = 0.1 (not an exact binary fraction but chosen to simplify the arithmetic), the equations needed are:

```
A OTIME.K=SAMPLE(TIME.K,4.3,0)
```

so that OTIME.K is the chronological time at which the current month started, together with

```
A DTIME.K=TIME.K–OTIME.K
A FACTOR.K=TABHL(TDOP.DTIME.K,3.3,3.4,0.1)
T TDOP=50/100
A DOR.KL=((4.3*DFCAST.K)/265)*FACTOR.K
```

The effect of this should be that as long as demand is constant, the variations in Dealers Stock, DSTOCK, should be about some constant level.

In practice, when the model was run, the variations shown in Fig. 4.7 were observed. The Dealer Order Rate shows the right kind of spiky behaviour, and dealer stock fluctuations, but has a downward trend even when DEMAND is steady.

Clearly, the model is wrong, the error being that Dealer Order Rate is always slightly too low, otherwise the stocks could not decline when everything else is steady. The error is in the equation for FACTOR.K, which makes FACTOR.K = 50 when DTIME.K is 3.3 or less and FACTOR.K = 100 when it is 3.4 or more. From 3.4 to the end of the month at 4.3 is only 0.9 weeks and not 1.0 weeks as it should be, so that FACTOR is 50 at 3.3 when it should be 100. The effect is to make DOR.KL slightly less than it should be for stability, so the Dealer Stock slowly declines. The correct equation is

```
A FACTOR.K=TABHL(TDOP,DTIME.K,3.2,3.3,0.1)
```

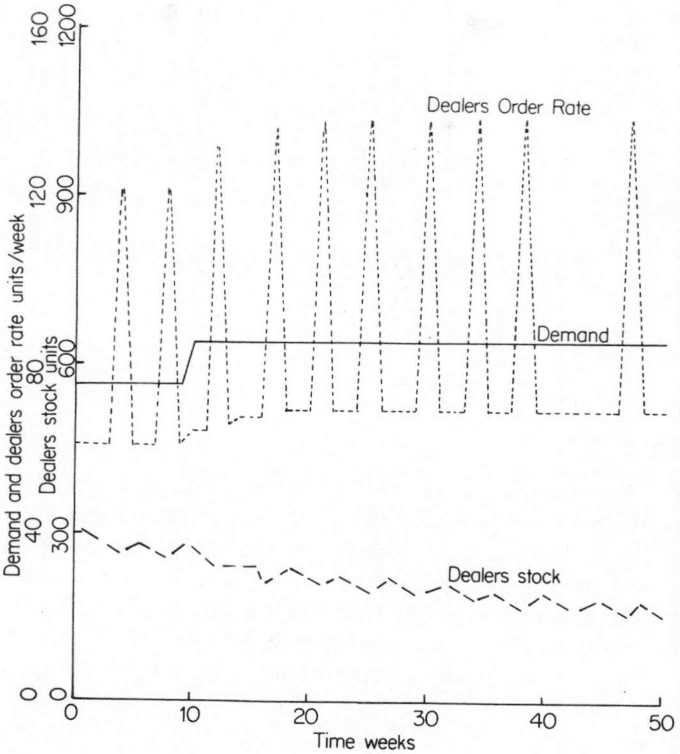

Fig. 4.7 False dynamics generated by errors in timing

The reader should run this little bit of program to see that it does give the required correction.

In general, the times at which events based on changes in values actually takes place requires careful attention during programming, but a step input is a very useful error detector.

Chapter 5

Equations for Systems

5.1 Introduction

The earlier chapters dealt with the basic ideas of modelling and simulation programming. We now turn to the difficult probem of formulating equations for systems. We cannot give suitable equations for every possibility in such a wide-ranging field as System Dynamics, but this chapter discusses some of the major aspects. The reader should carefully study the programs in Appendix B which contain many more examples of equation types for different situations.

Before we can write equations we must, however, be able to deal with units of measurement.

5.2 Dimensional Analysis

The units in which a variable is measured are called its DIMENSIONS and complications arise from the different variables used in the same model. For example, the size of oil tankers is expressed in Deadweight Tons, DWT, which is the amount of crude oil which can be carried in the ship at any one time. The price paid for chartering oil tankers is usually expressed in £/DWT/Month. However stocks of crude oil would be measured in Crude Tons, CT, and the rate at which the ships can deliver oil would be in CT/Month. Eventual sales would, however, be measured in Product Tons, PT. A model of an oil company's tanker-chartering, production, and distribution policies would involve all these variables.

As we remarked before, in management systems we have very great freedom to choose how the control actions are to be arrived at. For example, a production rate, in units/week, can be based on whatever other variables and parameters we choose, assembled into an equation of whatever form we wish. The factors in the equation might be measured in a mixture of units, £, weeks sq. metres, °F, £/Unit/Week or anything. We do not even insist that the factors and their equation should give a sensible outcome for the system performance, but we have to insist that, when the factors have all been assembled, the final result is measured in the same units as the variable on the left-hand side is intended to have.

Each equation has to do two things; to transform the numerical values for the quantities on the RHS into a numerical outcome for the LHS; to transform the individual dimensions on the RHS into a resultant dimension for the LHS quantity. We require that the dimensional transformation be consistent, in that its result should be the dimension that the LHS is supposed to have. We must avoid situations in which this does not happen, i.e. we must avoid not comparing like with like.

DIMENSIONAL ANALYSIS helps us to do this. Although DYSMAP can check that the equations in a model are DIMENSIONALLY CONSISTENT it cannot correct an inconsistency and it is, therefore necessary to understand dimensional analysis. There are no standard rules for correcting dimensional inconsistencies, and each case is dealt with on its merits. However, the techniques of dimensional analysis can be used to create the *dimensionally* correct equation directly (Section 5.3).

Dimensions are written in square brackets, using abbreviations. Negative exponents represent division by another dimension. Thus, the dimensions of an inventory, which is measured in Units (i.e. number of articles or packages) would be written as [U], and those of the production rate which replenishes it, and which are units per period of time, would be written $[U \times T^{-1}]$. This is called a COMPOUND DIMENSION. Note the use of the multiplication sign to separate the components. This avoids confusion with dimensions which have a multiple-character abbreviation, such as DWT. In this context 'units' does not mean simply 'number', as in arithmetic, but 'number of *something in particular*', such as diesel pumps, or washing machines.

Care is needed in choosing abbreviations, for example the production capacity of a coke-processing plant would, perhaps, be measured in tons/year. This could not be written as $[T \times T^{-1}]$ as confusion would arise between T as an abbreviation for tons and for time. One might write $[T \times Y^{-1}]$ and, in system dynamics, T does not necessarily mean the time unit, as it usually does when dimensional analysis is used in engineering and physics.

More complex compounds can be expressed, using the ordinary rules of algebraic manipulation of the exponents. (See, for example, Chapter 1 of my *Mathematics for Business Decisions*, Thos. Nelson and Sons and Barnes and Noble Inc.) For example the rate of increase in coke-oven capacity has to be measured in Tons per year per year, i.e. $[T \times Y^{-1}] \times [Y^{-1}]$ or $[T \times Y^{-2}]$.

Dimensionless quantities such as ratios and multipliers, often arise. Their dimensions, or lack of them, are expressed as [1]. If a dimensionless quantity appears on the RHS of an equation, it may be ignored in any dimensional checking. If it appears on the LHS, the dimensions of the variables on the RHS must cancel out to [1] for the equation to be dimensionally valid.

Different time units, such as years and months, often arise in the same model, simply because they are relevant to the system. Their occurrence in an equation, which thereby apparently becomes dimensionally inconsistent, helps in the proper use of the appropriate scaling factors for converting one time unit into another, such as

$$\text{Years} = \text{Months}/12$$

or

$$[Y] \equiv [M \times 12^{-1}]$$

We now illustrate with more examples drawn from the tanker-chartering model, of Chapter 11 and Appendix B.11.

A Compound Dimension

Because the sizes and speeds of individual oil tankers vary quite widely one from another, it is necessary to define a 'notional tanker'. This is a ship of 19450 DWT which can carry 99580 tons of crude oil per year between Bahrein and Rotterdam, via the Cape. For an actual ship on any route, the carrying capability can readily be found and this is very useful in settling prices in the ship chartering markets. There is, therefore, a shipping constant, CONS, which converts DWT into CT/month or gives the shipping tonnage needed to move crude oil at a given rate. Since the basic unit of time is the month, we define CONS by

$$\text{CONS} = \frac{19450}{(99580/12)}$$

but what are its dimensions?

We write [CONS] to denote the unknown dimensions of CONS and then for consistency we have

$$[\text{CONS}] \equiv \frac{[\text{DWT}]}{[\text{CT} \times Y^{-1}] \times [M \times Y^{-1}]^{-1}}$$

simply replacing each of the numbers in the defining equation, by its dimensions. Hence,

$$[\text{CONS}] \equiv [\text{DWT}] \times [\text{CT} \times Y^{-1}]^{-1} \times [M \times Y^{-1}]$$

We now expand the RHS to get

$$[\text{CONS}] \equiv [\text{DWT} \times \text{CT}^{-1} \times Y \times M \times Y^{-1}]$$

$$\equiv [\text{DWT} \times \text{CT}^{-1} \times M]$$

and these are the dimensions of CONS. Now CONS is the DWT needed to carry oil at so many crude tons per month. It should, therefore, have dimensions which, verbally, would be 'DWT per (crude ton per month)'. The brackets express what DWT relates to but they cannot, of course, be articulated. The 'per' before the brackets means that whatever is within them is to $^{-1}$ and the ordinary steps of exponentiation lead to the given dimensions for CONS.

In formulating dimensions for models it is necessary to be on the look out for variables which appear to be dimensionally similar to quantities like CONS but which are, in fact, very different. Consider for example, the price, P, paid to charter oil tankers. This is expressed in '£ per month per DWT'. The best way of sorting out these similar but different situations is to imagine brackets round the part of the dimensions which represent the actual flow of, in these examples, oil and money, and then to interpret 'per' as meaning $^{-1}$. In these two cases we get:

'DWT per (crude ton per month)'

or

$$[DWT \times (CT \times M^{-1})^{-1}] = [DWT \times M \times CT^{-1}]$$

and

'(£ per month) per DWT'

or

$$[(£ \times M^{-1}) \times DWT^{-1}] = [£ \times M^{-1} \times DWT^{-1}]$$

A little practice and care are called for.

Error Detection by Dimensional Analysis

In building complicated models errors of equation formulation occur and, unless they generate detectable false dynamics, they may be very hard to spot. Dimensional analysis provides a way of doing this; automatically when using DYSMAP.

The oil company model contained an equation for the rate of inflow of crude oil to the company's oil stocks in Europe. This rate, COAR, should have dimensions of $[CT \times M^{-1}]$, and the equation was, essentially

$$COAR = SCR/CONS$$

where SCR was the rate of chartering a particular category of ships, and had dimensions of $[DWT \times M^{-1}]$. The dimensions of COAR would thus be

$$[DWT \times M^{-1}] \times [DWT \times CT^{-1} \times M]^{-1} = [CT \times M^{-2}]$$

which is wrong.

With hindsight, it was obvious that the variable SCR, the rate of adding to chartered tonnage, should have been SCT, the total tonnage on charter, which has dimensions [DWT].

In summary, the basic technique of dimensional analysis is to copy the form, with its multiplications and divisions, of the equation being analysed, into a DIMENSIONAL IDENTITY and to obtain the RESULTANT DIMENSIONS. The principle of dimensional consistency requires that these be identical to those of the variable on the LHS.

Special Points in DYSMAP

The TABLE function is a dimensional transformation. For example (ignoring formalities),

DEMAND = TABLE (PRICE)

appears to be dimensionally inconsistent because it equates $[U \times M^{-1}]$ with $[£ \times U^{-1}]$. This is, in fact, acceptable because the graph from which the table has been derived makes the dimensional transformation by its own definition.

The other standard DYSMAP functions do *not* act as dimensional transformations. The CLIP function needs especially close checking. In it usual form of

V = CLIP (A, B, P, Q)

it provides that $V = A$ if $P \geqslant Q$ and $V = B$ if $P < Q$. It is, therefore, crucial to ensure that V, A and B are dimensionally mutually consistent, *and* that P and Q are. The dimensions of P and Q need not be consistent with those of V, A and B. The arguments of trigonometric functions have to be [1] so that, to write

V = A*SIN((6.283 * TIME)/PERD)

it is necessary that

$$[V] \equiv [A]$$

and

$$[TIME] \equiv [PERD]$$

5.3 Advanced Uses of Dimensional Analysis

Dimensional analysis can sometimes be used to infer an equation structure from the dimensions of its variables. This rather useful approach will be explained by means of examples, starting with one in which the answer is obvious, in order to show that the dimensional analysis approach agrees with common sense and first principles.

In modelling biological systems the birth rate of a species can be written as the product of the size of the breeding population, a normal fertility which is the average number of offspring per year per adult female, and a multiplier to reflect the effects on fertility of changes in the food supply. It is fairly obvious that

$$BR = POP \times FERT \times MULT$$

where the symbols indicate Birth Rate, Population, Fertility, and the Multiplier respectively and the dimensions of the variables are, from left to right, and using A to denote the number of animals, $[A \times Y^{-1}]$, $[A]$, $[A \times A^{-1} \times Y^{-1}]$ and $[1]$. The dimensions of FERT could be simplified, but we shall leave it in the long form. These dimensions are correct and we now show that formal application of DIMENSIONAL INFERENCE gives the same equation.

We start by assuming that BR is an unknown function of the other three variables. We express this as

$$BR = POP^a \times FERT^b \times MULT$$

where a and b are exponents to be determined. (The reader should think out the reason for MULT lacking an exponent.) Inserting the dimensions in the usual way

$$[A \times Y^{-1}] \equiv [A]^a \times [A \times A^{-1} \times Y^{-1}]^b$$

or

$$[A \times Y^{-1}] \equiv [A^a \times Y^{-b}]$$

The principle of dimensional consistency requires that the exponents of each dimension must be equivalent, and this enables us to write two equations relating the exponents of A and Y on the RHS to their values on the LHS. Thus, for A, $1 = a$ and, for Y, $-1 = -b$ or $a = 1$ and $b = 1$ are the correct exponents and the equation form is as derived from first principles.

We examine a more complicated example, again drawn from the oil-company model.

The rate at which ships are chartered, SCR, is $[DWT \times M^{-1}]$ and depends on the Expected Shipping Capacity Deficit, ESCD, which is $[CT \times M^{-1}]$, CONS, and the Deficit Correction Time, DCT, which is [M].

We write

$$SCR = ESCD^a \times CONS^b \times DCT^c$$

where a, b and c are to be found. The problem can be solved by reasoning from first principles, but that is a fatiguing and error-prone business. We therefore put

$$[DWT \times M^{-1}] \equiv [CT \times M^{-1}]^a \times [DWT \times CT^{-1} \times M]^b \times [M]^c$$

whence

$$[DWT \times M^{-1}] \equiv [DWT^b \times M^{-a+b+c} \times CT^{a-b}]$$

Equating exponents leads to:

for DWT $\quad 1 = b$

for M $\quad -1 = -a + b + c$

for CT $\quad 0 = a - b$

Hence:

$b = 1$

$a = 1$

$c = -1$

and the correct equation is

$$SCR = \frac{ESCD \times CONS}{DCT}$$

Being sure when to multiply and when to divide by a conversion factor such as CONS, is a great help in building complex models.

The modeller should be very careful indeed when using dimensionless quantities. They are tempting because they evade the labour of dimensional analysis, but this may lead to error because of the numerical effect they have and also because in a complicated model it is by no means easy to see whether one should multiply or divide by the dimensionless factor, or even whether one should use it at all. In all cases it is essential to check that the factor really is dimensionless, by careful reasoning from first principles.

In the earlier example on animal population it was evident that MULT had to *be* a multiplier, but in other cases it is by no means so obvious whether one should multiply or divide by an apparently dimensionless factor or even whether it should be in the equation at all. A more rigorous application of dimensional analysis is needed and the tanker model provides an example.

Oil can come from two sources *long-haul* and *short-haul*, in amounts C1 and C2 respectively in some period of time. The short-haul ratio, SHR is defined to be $C_2/(C_1 + C_2)$. If C_1 and C_2 are simply measured in crude tons, $[\mathrm{SHR}] \equiv [1]$. The size of a ship is in [DWT] and a standard ship can carry oil at a rate of DWT/CONS from a long-haul source. If the ships are used on a mixture of long and short-haul work, it is necessary to introduce a short-haul weighing factor, SHWF, which reflects the increased carrying capacity on the short-haul routes. It can be shown that:

$$\mathrm{SHWF}=(1-\mathrm{SHR})+\mathrm{SHR}\times\frac{\mathrm{SHVT}}{\mathrm{LHVT}}$$

where SHVT and LHVT are the respective voyage times in [M]. If $[\mathrm{SHR}] \equiv [1]$, then apparently $[\mathrm{SHWF}] \equiv [1]$ and one has to reason through whether to multiply or divide by SHWF, or whether to use it at all.

For example, it is not easy to work out from first principles whether we should put carrying capacity, CC, which has to be $[\mathrm{CT} \times \mathrm{M}^{-1}]$ as

$$\mathrm{CC} = (\mathrm{DWT/CONS}) \times \mathrm{SHWF}$$

or

$$\mathrm{CC} = (\mathrm{DWT/CONS}) \times (1/\mathrm{SHWF})$$

With $[\mathrm{SHWF}] \equiv [1]$ both forms are dimensionally correct, but so is CC=(DWT/CONS), and we do not know whether to drop SHWF from the equation, or to multiply or divide by it. As SHWF is approximately 0.7 this is a serious problem which can be solved by more careful definition of dimensions. If $C_1 \equiv [\mathrm{LHT}]$ and $[C_2] \equiv [\mathrm{SHT}]$ (for long-haul tons and short-haul tons) and $[C_1 + C_2] = [\mathrm{CT}]$, for ordinary crude tons, we now have

$$[\mathrm{SHR}] = \left[\frac{C_2}{C_1 + C_2}\right] = [\mathrm{SHT} \times \mathrm{CT}^{-1}]$$

and

$$[(1 - \text{SHR})] = \left[\frac{C_1}{C_1 + C_2}\right] = [\text{LHT}\times\text{CT}^{-1}]$$

so

$$[\text{SHWF}] = \left[1 - \text{SHR}\left(1 - \frac{\text{SHVT}}{\text{LHVT}}\right)\right] = [1] - [\text{SHR}]([1] - [1])$$

$$= [1] - [\text{SHR}]$$

$$= [1] - [\text{SHT}] \times [\text{CT}^{-1}]$$

$$= \frac{[\text{CT}] - [\text{SHT}]}{[\text{CT}]}$$

$$= \frac{[\text{LHT}]}{[\text{CT}]}$$

If we now put

$$[\text{CT} \times \text{M}^{-1}] = [\text{DWT}] \times [\text{DWT} \times \text{LHT}^{-1} \times \text{M}]^{-1} \times [\text{SHWF}]^{a}$$

using LHT in [CONS] instead of CT because that is implicit in the definition of the standard ship, we now see that $a = -1$ is the correct exponent, and our second equation for CC is the correct one.

In studying this example, the reader should distinguish between a variable and its dimensions. Thus in the statement that [SHWF] = [LHT]/[CT] we are not saying that the numerator *is* the long-haul tonnage, merely that it *has the dimensions of* long-haul tons.

The key to this example was the statement that 'it can be shown that SHWF = etc. This was an algebraic derivation for which the dimensional analysis provides a useful check that the numerical result is also dimensionally appropriate.

Having examined dimensional analysis, we now return to equation forms.

5.4 Level Equations

There are two kinds of level equations: the conserved level in which the flows into the level are, by their nature, indestructible, e.g. product, cash, orders etc; and the smoothed level, which operates only on information and where the requirements of conservation do not apply.

The Conserved Level

The basic form of the conserved level is:

LEV.K=LEV.J+DT*(±RATE1.JK±RATE2.JK+ . . . .) 5.1,L

The equation may contain auxiliaries:

LEV.K=LEV.J+DT*(RATE.JK±AUX.J) 5.2,L

(Note the use of the J postscript on the auxiliary.)

This use of auxiliaries contravenes the strict rules on coherence put forward in Chapter 3. DYSMAP will not, however, reject such an equation as invalid, and the usage is fairly widespread in system dynamics models. In general, however, the practice should be avoided, especially by the novice. An auxiliary is a step in developing a rate, and not a rate itself. If the auxiliary is used in a level equation a rate could, and should, have been used. Thus, although using an auxiliary in a level equation may appear to be a convenient shortcut, it may also conceal sloppy thinking about the system and, in a complicated model, this can lead to a very great deal of difficulty.

One situation in which auxiliaries are used in levels, is in the modelling of two or more streams arising from one source.

For example, the firm sets a price and, in response, the market orders at a certain rate. This sets in train a process of inventory depletion, credit control and cash flow, shown in Fig. 5.1. The links into and out of Price, Inventory, Cash Balance, and Accounts Receivable, show that these variables interact with other parts of the system in ways which need not concern us for the purposes of this example. 'Receivables' are the amounts of money owed to the company for goods supplied and charged to the customer, but not yet paid for.

We now write the equations for this system, leaving our superfluous arguments in functions, and some of the irrelevant flows from the levels.

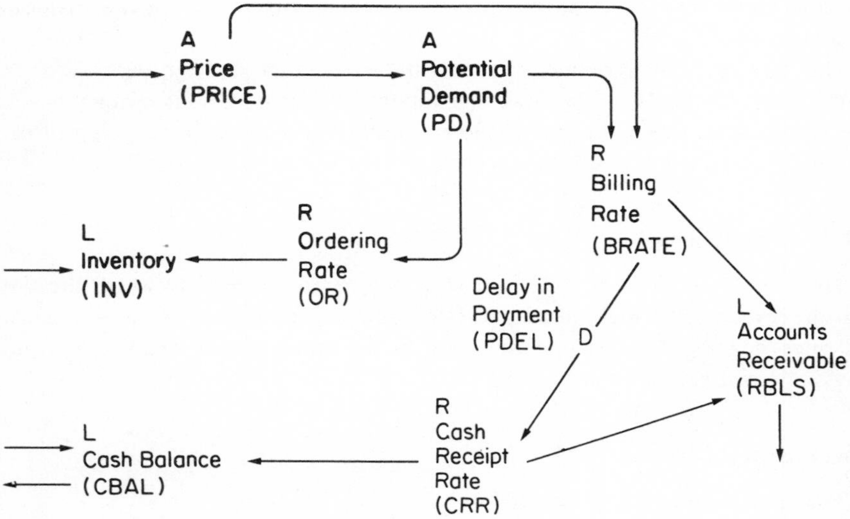

Fig. 5.1

PD.K=TABLE(PRICE.K)

PD Units/week

PRICE £/unit

OR.KL = PD.K 5.3,R

OR Units/week

PD Units/week

BRATE.KL=PD.K*PRICE.K

CRR.KL=DELAY3(BRATE.JK,PDEL)

RBLS.K=RBLS.J+DT*(BRATE.JK–CRR.JK) 5.4,L

INV.K=INV.J+DT*(Other Rates–OR.JK) 5.5,L

It is fairly clear that, as long as DYSMAP will accept PD.J instead of OR.JK in equation 5.5, no useful purpose is served by equation 5.3. The link from Potential Demand to Inventory would, therefore, be continuous, with OR being treated as an IMPLIED RATE, which is not explicitly stated. However, this uses a programming trick to get out of thinking about the forces governing the system.

The correct treatment would be

BRATE.KL=TABLE(PRICE.K)*PRICE.K

OR.KL=TABLE(PRICE.K)

The advantage of using PD is that constraints can be modelled, for example

AD.K = MIN(PD.K,CAPAC.K)

AD Actual Demand

CAPAC Production Capacity

which can be written as

BRATE.KL=MIN(TABLE(PRICE.K),CAPAC.K)*PRICE.K 5.6,R

OR.KL=MIN(TABLE(PRICE.K),CAPAC.K)

The Smoothed Level

The other permissible form of a level equation is

LEV.K=LEV.J+(DT/ATC)*(RATE.JK–LEV.J) 5.7,L

where RATE An input rate
LEV A Level which is the smoothed, or averaged value of RATE
ATC An adjustment time constant, called the SMOOTHING TIME

The second bracketed term *must* contain a *rate and the level which appears on the LHS of the equality*.

Now, if [RATE] ≡ [X], where X can be any dimension or compound, no matter how complex, and if [LEV] ≡ [Z], then what is [Z]?, given that [DT] and [ATC] are both, say [Months]. Dimensional analysis gives

$$[Z] \equiv [X]$$

Thus, *a smoothed level always has the same dimensions as its input rate.*

The basic function of equation 5.7 is to smooth out fluctuations in the inflow, RATE. Thus, when RATE rises and falls, so does LEV, but more slowly, and the shorter-term fluctuations in RATE are filtered out.

It is also the case that, if RATE rises from one steady value to another, LEV will eventually rise asymptotically to that new value. This is the first-order exponential delay referred to in Chapter 2, and used in the example in Section 2.13.

The first-order smoothing equation, 5.7, can be used to model what Forrester has called *psychological smoothing*. This involves saying that managers do not always respond immediately to new forecasts, but weight them according by the old forecasts and a factor representing the rapidity with which they are prepared to abandon their old beliefs about the future.

For example, let AFCST.K be an actual forecast made at K and relating to some future point in time, and let SFCST be a smoothed forecast, relating to the same point in the future. We write:

SFCST.K=SFCST.J+(DT/FAT)(AFCST.J–SFCST.J)

Where FAT is the Forecast Adjustment Time. (Note the auxiliary in a level again.)

This equation models the process by which, in effect, the firm does not immediately believe the new forecast AFCST but weights it by the previous forecasts which have been produced. In this case SFCST would not vary as rapidly as AFCST and, furthermore, would lag behind it.

The smoothing level delays as well as smooths. For a step increase in the input the level will reach 95% of its new steady state after 3 times the time constant; for a ramp input the level will lag the ramp by the time constant; and for a sinusoidal input the level will lag by approximately one quarter of the period of the input providing the smoothing time is at least as large as the period. If the smoothing time is appreciably less than the input period, for example in quarterly smoothing of annual fluctuations, the lag will be very slight. The reader should conduct simple simulation experiment to demonstrate these statements.

The smoothing level can be used on any information stream. Common examples are average sales, average production, average cash flow etc. This use of the word ‘average’ is intended to stress that an average value *of a rate variable* is a *level*. This is an example of the distinction between a rate as an instantaneous flow and a level as an item of conserved information which persists even if the flow ceases.

The smoothed level also has the useful property that ATC can vary without changing the smoothed output. Now, in the normal first-order delay of physical items

LEV.K=LEV.J+DT*(IN.JK–OUT.JK) 5.8,L

and

OUT.KLLEV.K/DEL.K 5.9,R

If IN is constant, LEV = DEL * OUT. Now, if DEL suddenly increases, the quantity in the level must increase even if IN is still constant. This can only happen if OUT falls for a time and equation 5.9 will ensure that this will happen.

In the smoothing of information, we do not wish the value of, say, Average Sales to change simply because we have started to use a larger value of ATC. Equation 5.7 shows that, in the steady state, RATE = LEV and the bracketed term is zero. If ATC is changed, LEV will be unaltered, which is the behaviour we require for smoothed information.

The reader should note carefully the distinction between delays in, say, the flow of orders *themselves* which are properly modelled by equations 5.8 and 5.9 and to which the arguments about the effects on the quantity delayed if changes in DEL *do* apply, and the smoothing of information *about* orders for which the smoothing equation, and its consequences, are appropriate model forms.

The output from a smoothing equation can be smoothed in its turn and there are, as we shall see later, uses for such a formulation. Thus one could have

SSALES.K=SSALES.J+(DT/AT1)(SALES.JK–SSALES.J)

and

DSALES.K=DSALES.J+(DT/AT2)(SSALES.J–DSALES.J) 5.10,L

in which SALES represents a sales rate, and SSALES and DSALES are its first and second smoothing. There is no requirement for the two smoothing constants to be identical.

Notice that in equation 5.10 we have a level feeding directly into another level which is a fine breach of the rules of coherence. In fact it is easier to regard SALES as an input to DSALES, with SSALES as an additional piece of information produced en route.

5.5 Rules for Rate Equations

The rate equations are the policy statements in the system. They reflect the rate at which the system will change during the forthcoming DT. This includes changes brought about by managerial choice (e.g. a production rate or an R and D expenditure rate) and those arising from the behavioural responses or fixed processes. Consider, for example a rate of inflow of orders in response to a choice of price, or the rate of development of new products in response to an earlier R and D expenditure rate. The reader should think out what is meant by 'the system's behavioural responses or fixed processes' in each of these cases, in the light of the discussion of controller, environment and complement in Chapter 2.

Because of these considerations, rate equations *must not contain DT*. The exceptions to this are treated in Section 5.9.

We have stressed that the rates are the flows in the system and, as such, are not directly observable. The only way that a rate can be measured is by inference from the effect it has had on a level, or levels. This is a general statement about the nature of all systems and it is not something forced on us by DYSMAP, but rather it is a proposition to which that language conforms. It is true for physical systems as well as for managed systems. We cannot observe the actual rate of inflow of orders but only the orders which have accumulated during the day, which enables us to infer the average rate at which orders arrived during that day. Similarly, those instruments which purport to measure physical rates do so by averaging them and translating the result into the position of a needle on a scale, i.e. by creating a level from the rate.

Now, if the equation for a rate contains other rates on its RHS, the foregoing argument implies that we are claiming to use a quantity which is *inherently* unknown in our policy choices. Although DYSMAP may allow the programmer to get away with it, *it is a mistake to use other rates in the equation for a rate*. Although the mistake appears not to be serious, because the program still gets executed, it is, in fact, very serious because it indicates that the system has probably not been properly understood. In short, the only case in which a rate may validly appear in a rate equation is when the two are connected by a delay. In this case, of course, the rate on the RHS is simply convention to remind us of what has been delayed and the delay is a model of the workings of the physical or behavioural processes of the system, not a statement of its policy choices or behavioural responses.

It is, however, very tempting to use rate-dependent rates in a model. For instance one may write

CFLOW.KL=PRICE.K*ORATE.KL

to model the flow of cash in response to a flow of orders.

In one sense there is nothing wrong with this because, as we remarked in Chapter 2, there is no mathematical difference between rates and auxiliaries and, since PRICE and ORATE both relate to the the same time there are no arithmetical errors as there would be if we put PRICE.K*ORATE.JK (as one would in DYNAMO) in which case the postscripts relate to different times. The real error, however, *is in the way the system has been interpreted* and the mixing up of instantaneous and unobservable flows with descriptions of the system's processes. Rate-dependent rates must be avoided, not because they are mathematically wrong but because they indicate sloppy interpretation of system structures.

Rate-dependent rates are readily avoided by giving a little thought to what actually happens in the system.

Consider, for example, orders being received and being despatched after a delay. This could be written

PD.K=*f*(PRICE.K, other factors e.g. capacity)

PD = (U/WK)   Potential Demand

with

ORATE.KL=PD.K

ORATE = (U/WK)	Rate of inflow of orders if all the demand leads to orders being placed.

DRATE.KL=DELAY3(ORATE.JK,PDEL)

with DRATE being the rate at which the goods are despatched.
If BRATE is the billing rate in £/week, the correct formulation is to use

CRATE.KL = PRICE.K*f(PD.K, etc)

to get the rate at which future bills are being accumulated in the Accounts Department for later release to the customers. We then write

BRATE.KL=DELAY3(CRATE.JK,BDEL)

with BDEL as the delay in the accounting system, assuming for simplicity that 3 is an appropriate delay order.

This formulation avoids rate-dependent rates, and it also allows the creation of an additional level, BAL, the amount of cash which will be received when goods can be despatched, i.e. the cash equivalent of the order backlog.

BAL.K=BAL.J+DT*(CRATE.JK–BRATE.JK)

Obviously BAL is connected with the system's financial liquidity. This illustrates how avoiding improper formulations leads to improved modelling of other aspects of the system.

The strict rules for rate equations are sometimes bent in actual programming. For example, raw materials are used to manufacture product and have to be ordered to replenish raw material stocks. It is very tempting to write:

RMOR.KL=RMUR.KL+(DRMS.K–ARMS.K)/TAMS   5.11,R

where

RMOR	Raw Material Order Rate
RMUR	Raw Material Use Rate
DRMS	Desired Raw Material Stocks
ARMS	Actual Raw Material Stocks
TAMS	Time to Adjust Material Stocks

and where RMUR could be found from the equation for the complete item production rate.

This formulation is incorrect and should not be used. It implies that a rate can be observed, which it cannot. A better approach is

ARMUR.K=ARMUR.J+(DT/TARMU)(RMUR.JK–ARMUR.J)

where	ARMUR	Average Raw Material Use Rate
	TARMU	Averaging time

and use ARMUR instead of RMUR in equation 5.11.

Although this is a little more laborious, it allows TARMU to be used as a system parameter, and, therefore, gives greater opportunities for system control than the incorrect approach.

Rate-dependent rates are occasionally used in DYNAMO to model very short delays by putting

OUT.KL = IN.JK 5.12,R

Again, this is not correct. If the delay is so short it is hardly significant to the dynamics and should not be modelled. DYSMAP would not accept this and would replace it, with a warning, by

OUT.KL = IN.KL

but it is still incorrect, for the reasons given earlier.

5.6 Initial Values for Rates

All rates which occur on the right-hand side of a rate delayed equation require an initial value, the most typical case being

OUT.KL = DELAY3(IN.JK,DEL)

A third-order delay has three internal levels and these have to have values in order to get the value of OUT in the first DT after TIME = 0, and also to keep track of future changes arising from the dynamics of IN. In DYNAMO and DYSMAP the provision for this is made automatically in that, if the programmer adds an equation,

IN = X 5.13,N

where X is a numerical value or an expression, they automatically set up three dummy levels each of which is set to a value of X x (DEL/3). The last of the dummy levels then controls OUT to be (X x (DEL/3))/(DEL/3), i.e. so that OUT and IN are initially equalized.

If the value of IN is known, as a *rate* (though possibly averaged over the DEL prior to 0) then no problem arises. Usually, however, the data is in such a form that we know the total quantity put into the delay during DEL and it is necessary to work out the initial value of IN which would give the same total quantity in the three dummy levels.

If L1, L2, L3 are the three internal levels and D1 and D2 are the two intermediate rates in the DELAY3 and we know that L1+L2+L3=K then

OUT = L3/(DEL/3)

DR = L2/(DEL/3)

D1 = L1/(DEL/3)

and we need to have

OUT = D2 = D1 = IN

Clearly

IN = L1/(DEL/3)

= (K/3)/(DEL/3)

or

IN = (K/DEL) 5.14

For an example of this for a complicated high-order rate see Appendix B11 lines 565–594. See the 'DYSMAP User's Manual' for more detail.

5.7 Rules for Auxiliary Equations

Auxiliary equations form the fine structure of a rate. There are no restrictions on the form of an auxiliary equation, except that the RHS should not contain DT. Auxiliaries can always be successively substituted into the rates of which they form part so that, collectively, they should conform to the rules for rates. In general, an auxiliary should depend only on levels and other auxiliaries. At this point, the reader should review the treatment of coherence presented in Chapter 3.

A common mistake in modelling is the generation of a set of simultaneous auxiliary equations which occur if a loop without a level has been created e.g.

AUX1.K = AUX2.K

AUX2.K = AUX1.K

This example is obvious and usually loops are much more involved, but DYSMAP and DYNAMO both identify and output such loops. Since DYSMAP sorts rates and auxiliaries simultaneously, a DYSMAP loop may include both variable types if the modeller has used rate-dependent rates. DYNAMO sorts rates and auxiliaries separately and could not detect such an eror but would permit a level-less loop involving a group of auxiliaries and an equation such as RATE1.KL = RATE2.JK. As we have seen in Section 4.1, this would produce completely spurious dynamics.

If a loop occurs, the only solution is to write out the equations and, by applying the type assignment method, find a level or levels and any corresponding rates. There are three options:

1. An ordinary conserved level
2. An averaging process to model a delayed information flow
3. A time delay e.g. DELAY3, whose internal levels will provide those required for the loop.

The first and last should have been found in the original type assignment, and option 3 is the most common. DO NOT put in an average simply to get out of a loop i.e. don't use a modelling dodge to get out of a situation which calls for careful thought. This will often simply confuse the issue by creating fake dynamics or even shifting the problem somewhere else in the model. Always, therefore, rely on type assignment, not on programming tricks.

5.8 Level–Rate Diagrams and Equation Formulation

In arriving at the equations in a model it is not sufficient to regard the levels, rates and auxiliaries as separate components of the system. It is necessary, in short, to study how the variables will combine to give a dynamic effect in order to be sure that the equation forms to be used are both formally valid and dynamically correct.

This desirable result is achieved by the use of LEVEL–RATE DIAGRAMS. These are simply graphs, with time as the horizontal axis, and two vertical axes – one for the level and the other for a rate, or the net value of a sum of rates.

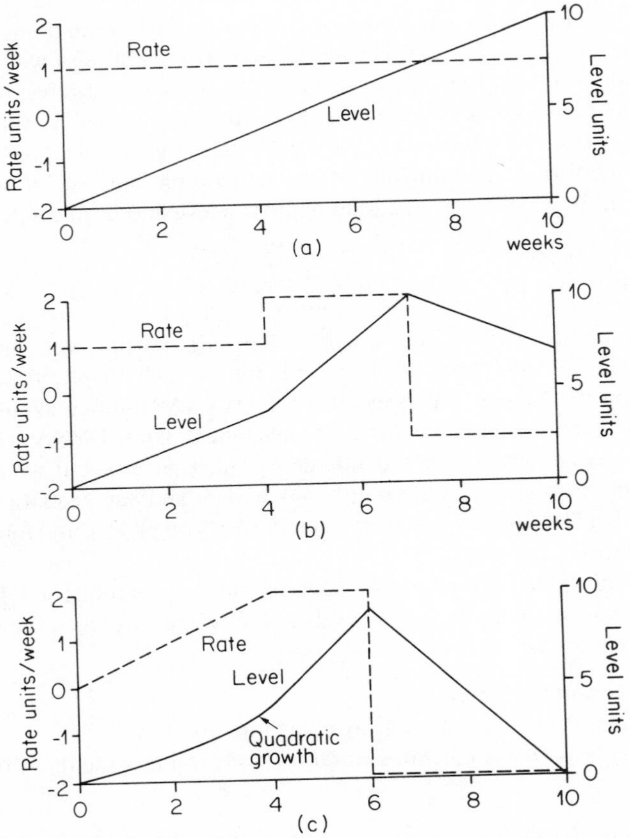

Fig. 5.2 Level-rate diagrams

The basic idea is to use the rate equations, and any auxiliaries which feed them, to deduce the time-form of the input to the level and thus to infer the dynamic behaviour of the level. Conversely, one can start from the level's dynamics and work backwards to see what kind of behaviour the rate would have to have had in order to generate these dynamics. This can then be checked against the rate and auxiliary equations, either mathematically or by a small simulation, to see if those equations produce the right dynamics.

For example, if a rate has a constant value of 1 unit/week how would the level it feeds behave? Fig. 5.2a shows the result. The best way of grasping level rate interactions is by some examples, the solutions being shown in the corresponding parts of Fig. 5.2.

Referring to Fig. 5.2b, if a rate is 1 unit/week for 4 weeks, 2 units/week for 3 weeks and −1 unit/week for 3 weeks what will the level's behaviour be?

Referring to Fig. 5.2c, a rate increases uniformly from zero to 2 units/week in 4 weeks. It remains steady for 2 weeks and then suddenly switches to −2 units per week for 4 weeks. What does the level do?

The reader should construct similar examples for himself, and, depending on his level of mathematical expertise, should experiment with equation forms to produce the required behaviour in the rate. Forrester's *Principles of Systems* gives more detail on level-rate diagrams.

5.9 Valid Uses of DT in Equations

As a general rule DT should not appear in equations, except for levels. The reason is that DT is part of the technique of simulation. Its numerical value is chosen, as shown in Section 4.7, to ensure that the simulated difference equations approximate sufficiently well to the differential equations of the system; to ensure stability in delays; and to avoid rounding error in TIME incrementation. The occurrence of DT in an equation other than a level implies that it has somehow become part of the decision process. In general, therefore, *a DT on the RHS of an equation is a mistake*.

There are some exceptions to this which we shall now discuss. The reader must be very careful, in using these exceptions, to avoid the trap of using 'legitimate' programming tricks as devices for getting himself out of trouble caused by a poor understanding of the system, or by mistakes elsewhere in the program. As usual, we give only illustrations, rather than definitive rules for permissible uses of DT.

'Year-end' Modelling

In Chapter 4 we used the PULSE function to generate rates which were capable of emptying a level very rapidly. These involved equations such as:

```
RATE2.KL=PULSE(LEVEL.K/DT,PERD,PERD)
```

and then

```
LEVEL.K=LEVEL.J+DT*(RATE1.JK–RATE2.JK)
```

This is an example of the injection of discontinuity into an otherwise continuous system. Such discontinuities often cause appreciable dynamics in a system, for example, where a firm, or an economy, uses financial-year budgets, rather than the rolling 12 months variety. The use of DT enables one to model the transition from one period to the next.

Time-series Modelling

It is sometimes useful to be able to record the time-series generated within a system. For example, successive monthly profits, or other performance figures, for use in modelling some forecasting situations, as discussed in Section 5.12.

In Chapter 4 we showed how such old values could be recorded, using the PULSE function and DT. This is essentially the same as the year-end case, and for the same reasons.

Constraints

Certain types of constraint require DT to be used on the RHS, in order to model them effectively. We shall show that this is a legitimate use of DT, but it should be assessed critically in each case to ensure that it is being used appropriately, and not as a fudge-factor to cover up inadequate modelling.

Consider a DEMAND, units per week, to be met from STOCK, units, with SALES being the actual rate of sale. Intuitively,

SALES.KL = MIN(DEMAND.K,MAX(0,STOCK.K))

The MAX avoids negative sales if stock has gone negative but this is incorrect, both dimensionally and operationally. Obviously, STOCK will vary from one DT to the next and, since DEMAND is in units per week we have to consider the amount which could be despatched from stock during a DT. Thus we put

SALES.KL = MIN(DEMAND.K,MAX(0,STOCK.K/DT))

Chapter 10 has this formulation for the case where unmet DEMAND is lost. The reader should work out the case where it accumulates into an order backlog, considering whether new demand, or the backlog, is supplied first. The reader should also think of several more applications of this constraint form.

Having dealt with the principal requirements for equation formulation, the rest of the chapter covers some of the more advanced aspects of equation structure.

5.10 The Three-term Controller

The general form of the rate equation is

RATE = *f*(DESIRED–ACTUAL states)

or

RATE = *f*(&)

where $\mathcal{E}$ is the error or discrepancy in the system. Classical control theory recognizes three forms for the function *f*, together with their combinations. These are:

1. RATE = $K_1\ \mathcal{E}$ PROPORTIONAL CONTROL
2. RATE = $K_2\ \dot{\mathcal{E}}$ DERIVATIVE CONTROL

where the derivative is usually taken with respect to time

3. RATE = $K_3 \int \mathcal{E} dt$ INTEGRAL CONTROL

The combinations are:

4. RATE = $K_1\ \mathcal{E} + K_2\ \dot{\mathcal{E}}$ PROPORTIONAL PLUS DERIVATIVE (PPD)
5. RATE = $K_1\ \mathcal{E} + K_3 \int \mathcal{E}\ dt$ PROPORTIONAL PLUS INTEGRAL (PPI)

and the Ks are chosen to meet control objectives and might even be functions rather than constants.

We now give examples of equation forms for these.

Proportional Control

This is by far the most common case in *stated* managerial practice, because it is easy to formulate practical decision rules based on it. In fact, simple proportional control is rarely used in its pure form, and most managers will, if pressed, admit to using something equivalent to cases 4 or 5 if case 1 does not give results soon enough.

Proportional controllers are however often used in System Dynamics models, mainly because they are often reasonably close to actual industrial practice. The basic form, for example, in the inventory case is

ORATE.KL=CONS*(DINV.K–INV.K)

where the proportionality constant, CONS, is always expressed as the reciprocal of an adjustment time, e.g. TAI, the time to Adjust Inventory. This gives:

ORATE.KL=(DINV.K–INV.K)/TAI

This gives the controller an intuitively acceptable form, as TAI has an operational meaning, and it also behaves as a first-order delay in its adjustment of ORATE to a step in DINV.

Naturally, TAI can be a variable so that the response can be varied according to the magnitude of $\mathcal{E}$. In that case, the controller is no longer strictly proportional.

As we have seen, this controller can only operate if, say, DINV > INV, and it will always fail to achieve its purpose of regulating, in this case, inventory *and* production. An alternative is the 'proportional plus inertia' form in which we put, say,

ORATE.KL=(DINV.K–INV.K)/TAI+AVCON.K

where AVCON is average consumption. We could equally well have used the average of ORATE itself, or even something else. The effect is to introduce a damping or inertia effect which says that ORATE should be based on the error *and* the previous level of activity. This damps down error response, but may make the system as a whole more sluggish.

Derivative Control

The first essential is a means of approximating the derivative. The crudest way of doing this is to put (using some superfluous variables to simplify the presentation)

ERR.K=DINV.K–INV.K

and

OERR.K=OERR.J+DT*PULSE((ERR.K–OERR.J)/DT,DT,DT)

This gives two values of the error, DT apart, and then

DERIV.K=(ERR.K–OERR.K)/DT

The value of DERIV.K can be used directly in the control equation as it is dimensionally correct.

In practice, this whole approach is generally unsatisfactory as DT will generally be very small and, as a result, DERIV will then be very sensitive to small changes in ERR. This could lead to instability in ORATE, which can be reduced by calculating OERR for PERD, using the methods of Section 4.8, where PERD = nDT and n has to be found by experiment.

A much more satisfactory approach is to estimate the derivative of ERR by a quadratic approximation. Let ERR be the error at the current moment i.e. at TIME, and let OERR1 and OERR2 be two old values for TIME–P and TIME–2P respectively. Let a, b and c be the unknown coefficients of a quadratic such that

$$aT^2+bT+c=E$$

$$a(T-P)^2+b(T-P)+c=O1 \qquad 5.15$$

$$a(T-2P)^2+b(T-2P)+c=O2$$

where T,E, O1 and O2 are abbreviations for TIME, ERR, OERR1 and OERR2 respectively and we drop the .K postscript.

It is easy to show that

$$a=\frac{E-2O1+O2}{2P^2}$$

and

$$b=\frac{E(3-+2T)+4O1(T-P)+O2(2T+P)}{2\ P^2}$$

and the estimate of the derivative of ERR will

DERIV.K = 2a*TIME.K+b

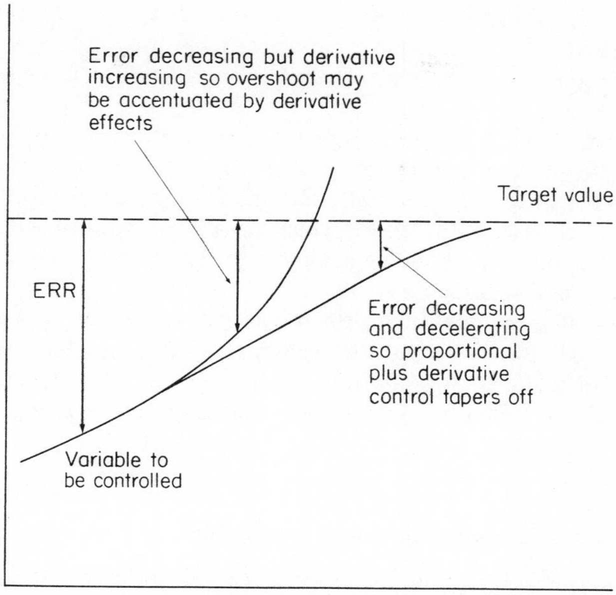
Error decreasing but derivative increasing so overshoot may be accentuated by derivative effects
Target value
ERR
Error decreasing and decelerating so proportional plus derivative control tapers off
Variable to be controlled

Fig. 5.3 To illustrate derivative effects

Obviously a, b and c are themselves variables as they satisfy equations 5.15 only at some particular value of TIME.

The idea of derivative control is that as ERR $\rightarrow 0$ it should slow down, so that both ERR and its derivative ought to be decreasing. Ideally then control action should be tapering off, and there should be less chance of overshoot. In practice, because most practical business control systems are rather poor in the first place, although ERR may be decreasing, its derivative may increase and the derivative component in the controller then applies more control action, not less, and overshoot occurs. These two cases are shown in Fig. 5.3.

In general, therefore, derivative regulation is hard to apply effectively, and a better solution is usually to attempt a straight proportional, or proportional plus inertia, control elsewhere in the system to try to inhibit the underlying instability.

Integral Control

This accumulates the error so that, as time passes, an even stronger control is applied if the situation is not improving. Any delays in the system may create the danger of overshooting and even of explosive oscillation being set up. The form is

INTERR.K=INTERR.J+DT*ERR.J

and the integrated error, INTERR, has dimensions of 'unit-days'. This make the control equation into:

ORATE.KL=INTERR.K/INTM

where INTM, the Integral Correction Time, has dimensions of $TIME^2$, and is analogous to TAI. Finding a suitable numerical value for INTM may not be easy. See below Section 5.11, *Ratios as Dynamic Controllers*, for a use of an integral controller.

The advantage of integral control is that it tends to eliminate any long term final error in the system, but the drawback is that it is sometimes not easy to translate into intelligible managerial terms. In addition, the long delay in many managed systems may disguise control actions which have been initiated but have not yet had a chance to take effect. Adding more control action to that already taken may well lead to very serious instability.

In general, although the three-term controller exists and is widely used in engineering control, the best results for *managed* systems can normally be obtained by relying on proportional or proportional plus inertia control, allied to a very close examination of the structure of the system, with appropriate changes, as we shall exemplify in Chapter 8.

5.11 Miscellaneous Points on Equation Formulation

Having discussed the formal rules of equation structure, the influence of dynamic relationships via level-rate diagrams, and the possible types of control law, we now cover some detailed points on equation formulation. The reader who wishes to do some modelling should, however, carefully study the large number of examples of equations contained in the computer programs of Appendix as he works through the case studies in Chapters 10 to 12.

Modelling of Ratios

Ratios are common in System Dynamics models, particularly as indicators of the need for control action. For example, in the corporate growth case study in Chapter 12, the firm's financial policy is regulated by the Financing Ratio, FRAT, defined in terms of Achievable Rate of Spending, ARS and the Required Rate of Spending, RRS, by

FRAT.K=ARS.K/RRS.K

Now, if the firms sees no need for expansion, RRS=0, FRAT$\rightarrow\infty$ and the simulation program fails because of an attempt to divide by zero. One solution is to put

FRAT.K= ARS.K/(RRS.K+CLIP(0,1,RRS.K,0.1))

This replaces the denominator by a positive value when RRS=0 and leads, of course, to a large value of FRAT, which would more usually be in the region of 1.0 when the firm is just about financing its own growth. This shows the need to understand the significance of a modelling situation when using a programming trick to avoid trouble with a model.

Ratios as Dynamic Controllers

In the last subsection, FRAT was instantaneous in that changes in RRS and ARS would operate immediately. A more complex case arises when we have to allow for a ratio to progress from its current value to a new one. For example, a firm makes two categories of product, A and B and the production of A as a fraction of total output is measured by the A-B ratio, ABRAT. Over the long term, for plant utilization reasons, the firm wishes ABRAT to keep to a target value, LTTAB. Its present value is ABRAT and we shall suppose for illustration, that ABRAT > LTTAB. In a short-term, the firm moves ABRAT up or down as needed in response to market indications and the present requirement is for a value denoted by STABR and this is, we imagine, even larger than ABRAT. As shown in Fig. 5.4 the firm wishes to move ABRAT up towards STABR, but at the same time would like to see ABRAT eventually trend downwards towards LTTAB.

The extent to which the firm is *willing* or able (because the ratio may take time to change) to react to these forces may be important to system behaviour, and we can model this by using the rate of change of ABRAT, RCABR and writing

RCABR.KL=(STABR.K–ABRAT.K)/STAT–(ABRAT.K–LTTAB.K)/LTAT

5.16,R

where STAT and LTAT are short and long-term adjustment times, not necessarily unequal, and ABRAT is a level of course.

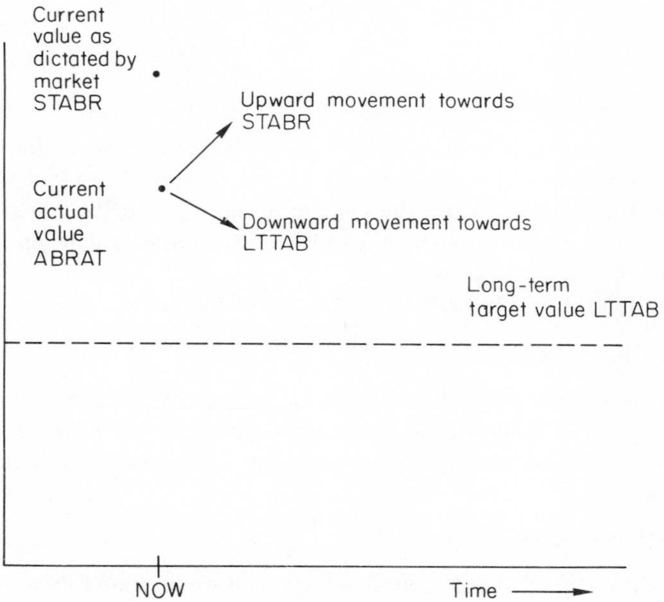

Fig. 5.4 Dynamic evolution of a ratio

If STABR AND LTTAB remain constant, ABRAT will stabilize at a point between them and the reader should work out where this will be. If, therefore, it is an absolute requirement that ABRAT→LTTAB, eventually, one could take the error (ABRAT–LTTAB) integrate it, and add it to equation 5.16 paying attention to dimensions. Thus we have

INT.K=INT.J+DT*(ABRAT.J–LTTAB.J) 5.17,L

and

[INT] = [WK]

and if we add a term, –INT.K/INTECT, where INTECT in the Integral Error Correction Time, RCABR will not be dimensionally valid. We therefore average the integration by

AVINT.K=INT.K/TAVINT 5.18,A

with TAVINT as a suitably chosen period to reflect the importance management attach to the need to have the ratio at LTTAB. AVINT will be [1] and we cannot smooth INT in the usual way (e.g. equation 5.7) because the result would have the same dimensions as INT. We now add a term –AVINT.K/INTECT to RCABR to obtain a dimensionally valid, and operationally correct, result for ABRAT, as the reader should verify by dimensional analysis and simulation. This is different from, but equivalent to, the method of Section 5.10. We could complicate matters still more by only allowing the term with INT in the equation for RCABR to come into operation if the error (ABRAT–LTTAB) has persisted for some given length of time.

Multiplicative Relationships

Multipliers are commonly used in SD models to reflect interactions between variables which are being controlled separately. For example, Sales Potential, SP, in a chain store firm may depend on the number of shops, NSHOPS, and a multiplier EPSM to reflect the employees per shop, EPS, on the grounds that, up to a point, increased staffing will increase sales.

If EPSM is defined by a suitable table function of EPS we could have

SP = NSHOPS*EPSM

If we now want to consider the effect of the stock per shop, SPS, and argue that increasing stock availability in the shops should increase the sales potential, we might introduce SPSM, a stock per shop multiplier, defined as a table function of SPS, and have

SP=NSHOPS*EPSM*SPSM

This raises the risk of double counting because if we increase EPS so that EPSM becomes 1.2, *and* increase SPS so that SPSM also becomes 1.2, SP would increase to 1.44 of its earlier value. However, the stock will only be effectively deployed if the employees are there to sell it and they will only be effective if they have something to sell. In short EPSM and NPSM are not independent, as we have written them as being, and we have double-counted.

Great care and thought is needed for this situation. Either a function must be worked out which correctly relates EPSM and SPSM to the *joint* effects of SPS and EPS or recourse must be had to the two-dimensional table functions which are available on later versions of DYNAMO and DYSMAP. We stress, however, the need for care and thought rather than the glib use of programming devices.

Setting-up Table Functions

Table functions usually present little difficulty. One simply draws the curve to be input, chooses a series of equally spaced points on the horizontal axis and reads off the values from the curve, entering them, in order, in the table. There are many examples of this in Appendix B.

The equally spaced points have to be chosen so that straight lines drawn through the points they define on the curve would be a reasonable appropriation to it. (Some version of DYNAMO and DYSMAP use a polynomial approximation to the curve, but earlier ones use linear interpolation.)

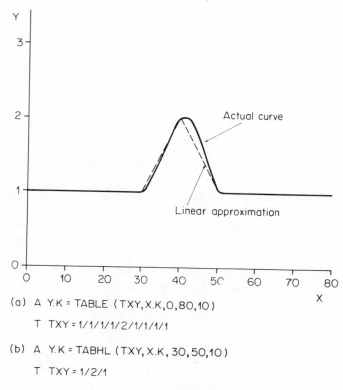

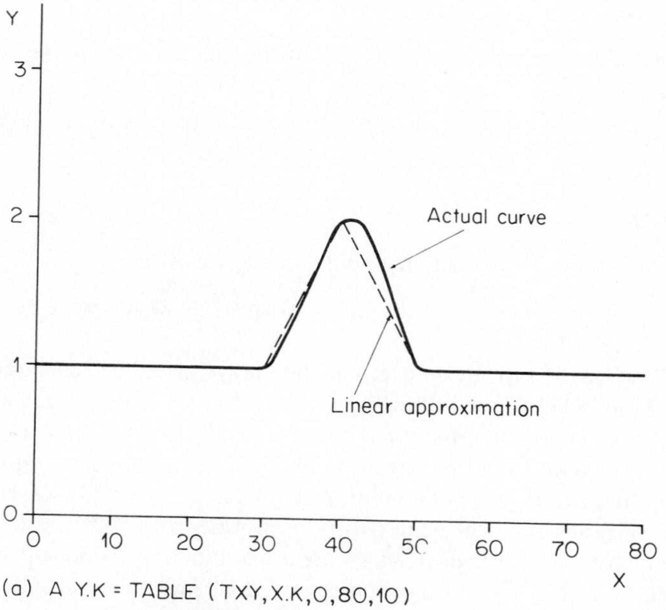

(a) A Y.K = TABLE (TXY,X.K,0,80,10)

T TXY = 1/1/1/1/2/1/1/1/1

(b) A Y.K = TABHL (TXY,X.K,30,50,10)

T TXY = 1/2/1

Fig. 5.5 To show TABLE and TABHL

If the curve contains sharp discontinuities the points have to be close together at the discontinuity and this leads to a long and troublesome table. This can be avoided by using a TABHL function which takes the end values of the table if the variable used as argument exceeds the stated range. If this happens with a TABLE function, the program would fail. For example, the curve in Fig. 5.5 can be defined using three values in a TABHL but would require nine with TABLE.

Really sharp discontinuities are better injected using STEP or PULSE functions, perhaps adding or subtracting them to get a step-wise pattern. In this way the heights of the steps are easily changed in a rerun simply by changing one or two parameters, rather than having to repunch a table.

If a TABHL function is used to inject a time series pattern of demand, say, the flat tails it produces may be unrealistic at the start and finish of the run, as they imply that the variable demand experienced throughout the run was suddenly switched on and, as suddenly, off, at the start and finish respectively. This is easily solved by simply defining LENGTH, and the table, to be, say, 120 when one only really needs it to be 100 to show the dynamics, and then simply ignoring the output before 10 and after 110.

Repeating Cycles

Repetitive cyclical behaviour is easily input by the standard trigonometric function SIN and COS, e.g.

```
A DEMAND.K=100+SEAMP*SIN(6.283*(TIME.K + PERD/4)/PERD)
C SEAMP=50
C PERD =52
```

This example shows several detailed points in one equation:

1. DEMAND will oscillate steadily between 50 and 150 with a period of 52 weeks i.e. annual seasonality.
2. If SEAMP was made variable an irregular 'sine' wave would be generated.
3. By using TIME.K+PERD/4 within the SIN instead of TIME, the sine wave starts at its peak, rather than its mid-point. This is usually a good thing as the model starts in a loaded-up condition which should help to disguise any minor errors of initial conditions thus getting more information from a given LENGTH.

   The same result could be achieved by using COS with TIME as its argument but, as shown later, we might have to use a sine function to model forcasting in the system and it is less error-prone to use the same function with different arguments than to mix SIN and COS in the same model.

A non-uniform shape for a seasonal pattern can be put into a table with appropriate time parameters, e.g. TABHL(TSP,TIME.K,0,52,4) could be used to define a Table of Sales Pattern over a 52 week period. If LENGTH=156 we need TSP to be three times as long, or we have to make it repeat itself three times. This can be done by writing

A DEMAND.K=TABHL(TSP,TDIFF.K,0,52,4)

T TSP=suitable values

A TDIFF.K=TIME.K–YEND.K

L YEND.K=YEND.J+DT*PULSE(52/DT,52,52)

N YEND = 0

The reader should work out why the year-end variable, YEND, and TDIFF, produce the required behaviour.

5.12 Forecasting Equations

Forecasting is very common in managed systems and therefore must be modelled. There are two situations:

1. Exogeneous. The controller is affected by an input, which it cannot influence, from the complement, e.g. a demand pattern. The controller uses forecasts of this input and these must be modelled, without necessarily specifying how the forecasts are made, or how any errors involved actually arise.
2. Endogeneous. The controller records past states of itself, the complement, or the environment, and projects them into the future for use in decision-making.

We now discuss DYSMAP equations for these, examining some of the implications in Chapter 9.

Exogeneous Forecasting

The system is driven by, say, Demand, but one of the factors affecting its performance is the controller's demand forecasts (or the environment's) (see Fig. 5.6).

Normally Demand might be represented by a table function, TDMD, which gives the values which Demand is going to have at each point in time, e.g.

DEMAND.K=TABLE(TDMD,TIME.K,n_1,n_2,n_3)

generally n_1=0, n_2=LENGTH. In the simulation, TIME.K represents the current time and if FHOR is the forecasting horizon, i.e. the distance ahead that one is trying to forecast, then a perfect forecast made at TIME.K and representing what

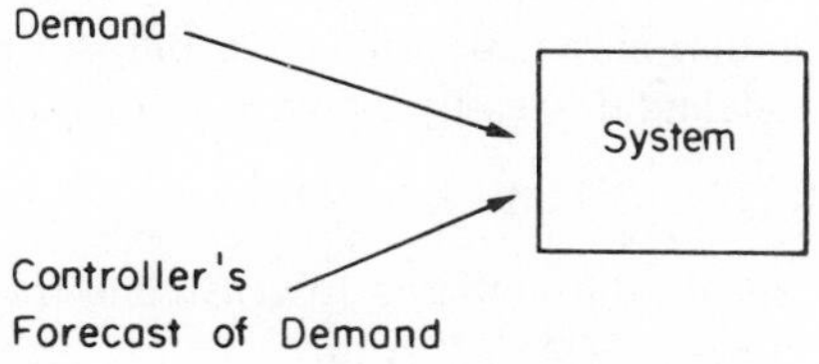

Fig. 5.6 Exogeneous forecasting

is currently the forecast for the state of demand at TIME.K+ FHOR is going to be:

$$FCAST.K=TABHL(TDMD,TIME.K+FHOR,n_1,n_2,n_3) \quad 5.19,A$$

This is the same table name, with TABHL to prevent FCAST being undefined towards the end of the run when TIME.K+FHOR > LENGTH. This leads to horizontal forecasts in the last FHOR time units of LENGTH. This is rather illogical if TDMD has been designed to produce a fair amount of variation during LENGTH. This can be avoided by making n_2=LENGTH+FHOR and putting suitable extra values towards the end of TDMD.

We could represent the forecast errors by ignoring equation 5.19 and using something entirely different. This would not make a great deal of sense and it is far more useful to relate the forecast to future Demand and specifically to model forecasting errors. This allows one to test the controller's sensitivity to forecasting errors, whether some errors are more serious than others or, indeed, whether some errors could be beneficial to system performance, e.g. does it pay to be optimistic about the future?

This can be achieved by putting:

$$FCAST.K=TABHL(TDMD,TIME.K+FHOR,n_1,n_2,n_3)*EBF.K$$

where EBF is called the ERROR AND BIAS FACTOR.

The EBF can be used in a variety of ways:

1. When EBF=1, the forecasts are perfect, and it is often useful to see just how much difference this unattainable, but much-sought, end would make to system performance. Often, the answer is 'very little', as in Chapter 10, which has some striking implications.
2. If EBF>1 the forecasts are optimistic, and conversely if EBF<1. EBF can also vary about 1.0 to reflect alternate optimism and pessimism, by relating it to TIME in a table function.
3. More sophisticated possibilities are to relate EBF to a comparison between Forecast and Outcome using an old-value MACRO such as that described in Section 4.9. There are objections to this procedure which are discussed in Chapter 9.
4. If forecasts are high, when compared to previous values or to productive capacity, many firms bias the forecast downwards, and conversely when the forecast is low. This may be caused by an unwillingness to believe evidence which is too different from recent experience, or it may be in anticipation of inability to sell more than can be made or pressure to sell what is going to be made. The EBF formulation enables us to avoid stating the cause of the phenomenon, but to model it fairly easily by writing, for example

   FCAST1.K=TABHL(TDMD,TIME.K+FHOR)

   FCAST1 being, so to speak, the unbiased forecast, and then getting the forecast which is going to be believed, FCAST2, from

   FCAST2.K=FCAST1.K*EBF.K

EBF is modelled by observing the difference between FCAST1 and, say, productive capacity, PCAP, as a fraction of PCAP (to scale it)

DIFF.K=(FCAST1.K–PCAP.K)/PCAP.K

and, perhaps

EBF.K=TABHL(TEBF,DIFF.K,–.4,.4,.2)

TEBF=1.2/1.05/1.00/.95/.80

the values in TEBF model a progressively heavier biasing as FCAST1 departs more severely from PCAP. In practice one could probably get a fair idea of the values in TEBF from past data and, of course, use sensitivity analysis. Recall that these equations are not intended to *describe how* forecasts are made but to *model the results* of informal behavioural processes which have not been made explicit.

Exogeneous forecasts may also be made for seasonal-variable demands. The seasonal demand can be modelled by a sine wave:

DEMAND.K=BLEV.K*(1+SEAMP*SIN((6.283*TIME.K)/PERD))

where	BLEV	Base Level of Demand
	SEAMP	Seasonal Amplitude
	PERD	Periodicity of Demand Cycles (usually 12 months)

FCAST can be seasonally corrected by replacing TIME.K by (TIME.K+FHOR) in the argument of the sine function, with suitable changes to BLEV if it represents a growth process.

Repetition of table functions, as discussed in Section 5.11, *Repeating Cycles*, leads to problems in that TDIFF and FHOR will be off the range of the table and into the early part of next cycle as soon as TDIFF+FHOR=52. This can be dealt with by putting

FCAST.K=TABHL(TSP,TVAR.K,0,52,4)

TVAR.K=CLIP(TDIFF.K+FHOR–52,TDIFF.K+FHOR,TDIFF.K+FHOR,52)

with the other equations as before.

The method of producing a forecast by using TIME.K+FHOR as the argument in a time-dependent TABLE or periodic function gives incorrect answers when used in a STEP or other function which only acts at fixed points in time. Thus, if

DEMAND.K=100+STEP(MAG,STM)

we would want a perfect forecast to go from 100 to 100+MAG as soon as TIME.K+FHOR<STM. This can be achieved by putting, say,

FCAST.K=100+CLIP(MAG,0,TIME.K+FHOR,STM)

or more simply

FCAST.K=100+STEP(MAG,STM–FHOR)

Similar considerations apply to the time of the first of a series of PULSE functions. Thus, for continuously-dependent functions such as TABLE and SIN use TIME.K+FHOR, and for fixed time processes such as STEP use TIME.K–FHOR.

The methods discussed so far apply to the forecasting of the height of the demand curve at some future point, i.e. to forecasts of the potential *rate* of sales. However there are many cases where the firm is more interested in the potential *amount* of sales during the next FHOR months. This forecast would then be compared to available stock and scheduled production to see whether or not there was a rough match or whether extra or less production was appropriate. In such a case demand at each moment in time will continue to be governed by the height of the TABLE but the forecast has to be based on the area under the curve between TIME.K and TIME.K+FHOR. In order to make the most of the DYSMAP rerun facility we have to provide for changes in FHOR. In the following equations FHOR=2 but provision is made for FHOR to be as high as 3.

The forecast area is taken by considering successive suitable slices of area and adding up those which are to be considered. If TABHL does not vary too sharply we can put

FDUM1.K=TABHL(TDMD,TIME.K+PERD/2,n_1,n_2,n_3)*PERD

FDUM2.K=TABHL(TDMD,TIME.K+1.5*PERD,n_1,n_2,n_3)*PERD etc.

i.e. we take the mid-point values of a series of slices each PERD units wide and multiply by PERD to get the area of that slice. Then

FCAST.K = FDUM1.K*CLIP(1,0,FHOR,PERD)

+FDUM2.K*CLIP(1,0,FHOR,2*PERD) etc.

The reader should work out an equation for forecasting the area under a STEP demand pattern.

Endogeneous Forecasting

In endogeneous forecasting we model explicitly the process by which the controller attempts to forecast the future state of some variable which it has recorded in the past. The procedure allows us to study the dynamic *effects* of different forecasting techniques, but see Chapter 9 for a discussion of forecasting *accuracies*. The variable may be an externally imposed driving force, such as consumption, or it may be an internally generated system component, for instance the order backlog.

The first of these cases appears to be the same as exogeneous forecasting. The difference is that then we were modelling the system predicting the future by looking *at* the future, we are now treating the case in which records of the *past* are used to predict the future.

The situation may be shown most simply in Fig. 5.7. This shows a variable V, of which we wish to predict the value at NOW+FHOR. In the diagram, V has been shown with a smooth trend, merely as a diagrammatic simplification.

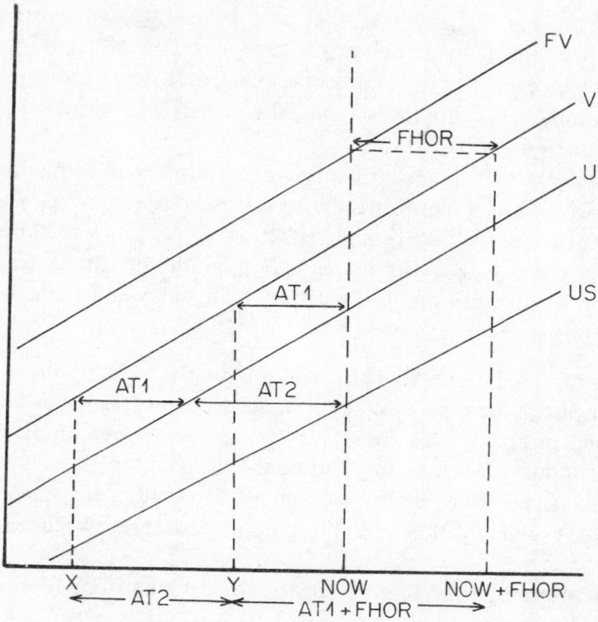

Fig. 5.7 To show simple endogeneous forecasting

The graph of V has to be paralleled and preceded by a graph of its forecast value, FV, displaced to the left by FHOR. The problem is to find some way of moving V to the left without knowing V in advance. There are any number of progressively more sophisticated methods of doing this which involve recording one or more old values of V, fitting curves to them, and then extrapolating the curve FHOR into the future. Since the extrapolation is done at NOW this is tantamount to shifting the V-curve to the left. For ease of explanation we shall illustrate the simplest way of doing the extrapolation. More complex methods can be programmed in much the same way but at greater length.

The simplest procedure is to smooth V twice, using different time constants if required, to get VS1 and VS2. These will lag V of course. The equations are

VS1.K=VS1.J+(DT/AT1)(V.JK−VS1.J)

VS2.J=VS2.J+(DT/AT2)(VS1.J−VS2.J)

The slope of V between time instants X and Y, as estimated at NOW, would thus be

$$\frac{\text{VS1.K} - \text{VS2.K}}{\text{AT2}}$$

and the forecast would be

FV.K=(VS1.K−VS2.K)*((AT1+FHOR)/AT2)+VS1.K

The geometry of Fig. 5.7 suggests that this procedure will work fairly well providing V is fairly regular in its behaviour and providing AT1 is fairly small and AT2 fairly large. However, the dynamics of a first-order delay indicate that if V is liable to fluctuation, AT2 should not be larger than one third of the interval between significant changes.

Obviously, there are any number of refinements which can be made to this very simple scheme. One improvement would be to use old values of V rather than smoothed data, but this is not very satisfactory if V is noisy. More complex formulae can be used, e.g. a linear regression could be fitted to 12 old (or smoothed) values, a quadratic could be fitted to 3 old values, and so on. Three points should be borne in mind:

1. The dynamics of the system will be heavily influenced by the policies used to act upon the forecast values, so that the system decision structures are at least as important, and probably far more so, as an avenue to improving system behaviour, as 'improved' forecasting formulae (see Chapter 9);
2. The system may have loops which correct for forecasting errors (or such loops may be created) which will render the pursuit of improved forecasting largely illusory;
3. It may be far better for system performance to make poor forecasts of some other variable than to make 'better' ones of V.

We shall, however, have more to say about forecasting in Chapter 9.

5.13 Time Errors in Forecasting

So far we have considered cases where the error in the forecast is one of magnitude. For example, if the exogeneous demand is going to be 100,FHOR from NOW, and the forecast is 120 the system is being optimistic about the magnitude.

Now, consider a step input

DEMAND.K=100+STEP(SMAG,STM)

then the appropriate forecasting equation would be

FCAST.K =(100+STEP(SMAG,STM–FHOR))*EBF

and FCAST may get the size of the step wrong but it will correctly predict its timing, FHOR beforehand.

This might easily be unrealistic and the way out would be to put

FCAST.K=(100+STEP(SMAG,STM–FHOR–TERR))*EBF

where TERR is the timing error, which could be as complicated as one wished. If TERR is positive, the – sign in front of it will mean that the STEP is forecast too soon, and vice-versa.

Alternatively we could have:

FCAST.K=(100+STEP(SMAG,STM–FHOR*TERFH))*EBF

where TERFH is a multiplier for the timing error as a proportion of the supposed forecasting horizon. If TERFH>1, events are forecast too soon.

Similar methods can be used to get timing errors in repeated demand cycles.

5.14 Shape Errors in Seasonal Forecasts

In seasonal situations, particularly agricultural ones, the shape of the demand pattern for, say, fertilizer may be quite different from the expected shape because of the weather. This is more than a matter of timing errors being made; the whole pattern of events is different.

This can be dealt with by having two patterns, a forecast pattern, TFP, and an actual pattern, TAP. TFP would be based on the normal course of events and TAP would be a different shape. Actual demand could then be, say,

DEMAND.K=TABHL(*TAP*,TDIFF.K,0,52,4)

in the usual way. Forecasts would then be generated from

FCAST.K.=TABHL(*TFP*, TVAR.K,0,52,4)*EBF

with TVAR defined as in Section 5.12, *Exogeneous Forecasting*, except that TDIFF.K+FHOR could be replaced by TDIFF.K+FHOR–TERR to bring in timing errors. Note that the values in TFP and TAP refer to the same points in time and that the time-shift due to forecasting is brought in by FHOR–TERR, not by altering the time location of the entries in TFP. In general, the average values of TFP and TAP (or the area under them) should be equal, as magnitude errors are brought in by EBF.

The reader should think out, with the aid of graphs of TFP and TAP, precisely what he is doing before using this approach.

Chapter 6

Properties of Feedback Loops

6.1 Introduction

In Chapter 2 we studied the behaviour of a simple system as affected by its structure and control policies. We now turn to the central question in system dynamics – what is the connection between structure and performance or, conversely, between observed dynamics and underlying structure?

The behaviour of the simple system in Chapter 2 reveals some interesting aspects, as shown in Fig. 2.19:

1. ORATE is the same shape as CONS, but is much larger in amplitude and lags slightly;
2. Comparing COMRATE with ORATE, we see that the lag has become much greater and that COMRATE is smaller in amplitude than its predecessor, ORATE;
3. Comparing CONS with INV we again see changes in amplitude, but a complete reversal of the peak and trough pattern.

These comparisons show that the system is doing three things in response to the input:

1. Transforming a step input into outputs of a completely different shape;
2. Changing inputs of one magnitude into outputs of different magnitudes;
3. Creating lags between outputs and inputs.

We, therefore, need a theory of how systems do this kind of thing in order to tackle the problem of improving behaviour. We start from the axiom that a system's feedback loops (FBLs) and their components determine its dynamics and, in improving its behaviour, the feedback loop is the basic unit of analysis. This is the salient feature of System Dynamics. *Loops govern dynamics*, and understanding the properties of the loops enables dynamics to be explained and changing the loops permits us to suppress undesirable dynamics or reinforce desirable behaviour. The basic method of dynamic analysis is, therefore, to analyse the connection between a system's loops and its dynamics, *checking* the results by simulation. It is NOT a matter of simulating every possible change to the system in the hope that something will be found.

This chapter introduces a trial-and-error approach to system analysis which calls for a very challenging blend of skill and and experience. The methods are less rigorous than mathematical analysis, but they can be more speedily learned and are easier to apply to the fairly complicated models which often emerge from practical studies. Experience has shown that these methods work very well for most problems, particularly where there are severe non-linearities, perhaps because, for some psychological reason, most people find that they stimulate a more creative approach to redesign of the system than would be the case with more formalized, mathematical approaches.

In studying FBL properties we first discuss stability, and then we study dynamic behaviour and loop structures.

6.2 Stability

The idea of stability is important in system dynamics, but it is difficult to give an easy definition of it, because stability is relative and is not necessarily a good thing. We therefore need a concept of stability which will reflect these factors.

Consider the way in which a system responds to shocks from its inputs. For a step input the system is said to be STABLE if it eventually settles to a steady state. The time taken to settle, and the degree of oscillation undergone are measures of relative stability, and if both can be reduced the system has been made more stable, or more heavily DAMPED. Section 2.12 shows that it is usually a matter of trading-off settling time against oscillation.

If the system settles back to its *original* operating condition after a step change in the input, it is ULTRASTABLE.

It is easy to see that ultrastability is not necessarily desirable in a business system. Consider two variables, total Market Size MS, and the company's Sales Volume SV, and suppose that MS rises. The ultrastable, and two versions of the stable, cases are shown in Fig. 6.1. The ultrastable transients are almost irrelevant to system performance as, in this case, it is a notably bad response to the market. Ultrastability is, however, easily generated within a system. (A system OUGHT, usually, to be ultrastable to an impulse shock.)

Internal variables, such as a discrepancy between a target state and an actual state should, at first sight, be ultrastable if the control loops are doing their job. However, recall Section 2.12 with the IB rule where the system could not create an ultrastable error between desired inventory and inventory and, when it did (with the IOB rule) performance could be said to be worse because stock levels were much higher. Compromises like this do happen, but there are plenty of cases where improving the control improves all the other variables without trading improvement in one against deterioration in another.

A system's reaction to periodic inputs is important, and it is CYCLICALLY STABLE if it eventually exhibits sustained periodic behaviour, not necessarily

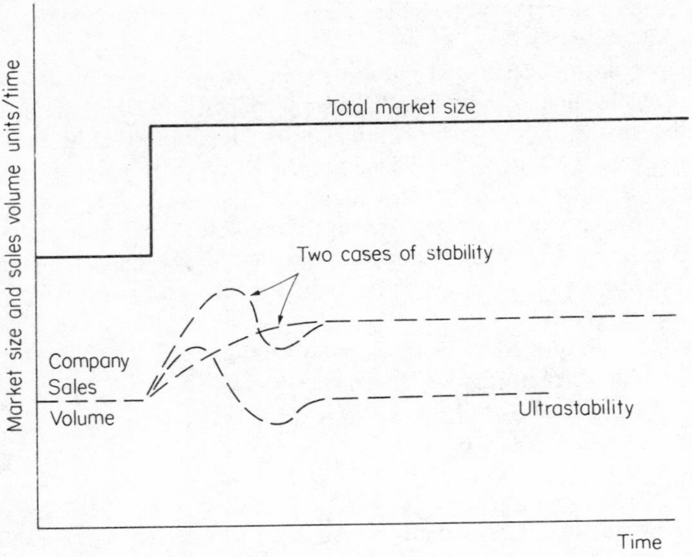

Fig. 6.1 To illustrate ultrastability

sinusoidal. In this case, stability is improved if the amplitudes of the outputs can be reduced (recall Figs 2.20 and 2.21).

Finally a system is EXPLOSIVELY UNSTABLE if it generates explosive oscillations in response to an input, or as a result of its own initial conditions if there is no exogeneous input. If a system explodes for *all* reasonable values of its parameters it is INHERENTLY UNSTABLE. Usually explosion, in a given structure, can be corrected by proper choice of parameters (see Section 2.12) and inherent instability is rather rare. In any case of explosion the first thing to check is the value of DT (Section 4.7).

These definitions are conceptual and relative, because lack of damping in a managed system is not necessarily bad. For instance, in a production system we should probably prefer to see inventory become less damped if there was a compensating improvement in production damping. Similarly, we might be prepared to tolerate a poorly-damped cash balance in return for a greater degree of damping in, say, long-term debt, or earnings. Usually, one can only weight these trade-offs fairly arbitrarily so it is not necessarily bad to have qualitative measures of stability such as are furnished by the system graphs.

If the dynamics are obviously terrible, the qualitative concept of stability is the proper point of view since the priority is to achieve basic control of the system. Chapter 10 illustrates such a problem. If, however, the system is already fairly well controlled the problem is more one of fine tuning and the equations for system performance indicators in Chapter 7 may be appropriate. However, the trade-off between the relative stability of different variables will nearly always be fairly arbitrary, as it is very hard to arrive at the real economic costs of stock-holding and adjustments to production rates etc. See the case study in Chapter 11 and the discussion in Rivett's *Principles of Model Building*, Chapter 5.

Thus far we have only looked at stability in cases of negative feedback. Positive feedback is supposed to produce growth so that it would be misleading to refer to positive stability in the terms which we have so far used. We shall, therefore, say that a positive loop is GROWTH STABLE if it produces steady growth and 'damping' will imply the extent to which the system oscillates about the exponential growth pattern.

For a positive process there are two special cases. A system is MORBIDLY STABLE if its positive growth process has failed and is producing decline instead of growth. A stable positive loop which produces enough growth to force the system into the overshoot and collapse mode, is OVERSHOOT UNSTABLE.

Clearly, where a system contains positive feedback, we want to find and prevent the conditions for morbid stability and overshoot instability.

6.3 Describing Dynamic Behaviour

Dynamic behaviour has to be described in two ways in order to get a full picture of the system's response to an input. The TIME DOMAIN is simply what happens as time passes, i.e. it is true dynamic behaviour. In the FREQUENCY DOMAIN the system is driven by a sinusoidal input and we observe the way in which its eventual steady state response changes as the frequency of the driving sine is altered. The frequency domain is very important because it shows how sensitive the system will be to the noise which is always present in real world inputs.

Time Domain

In the time domain the behaviour of any single variable is described by its:

1. Speed of Response for positive loops and the initial step response of a negative loop.
2. Oscillations by their period, amplitude, and damping, or degree to which each oscillation is lower than its predecessor (these are the detailed components of instability).

This is treated more precisely in Chapter 7. For two or more variables, their relative dynamic behaviour is described by their lead/lag relationship as in Chapter 2, while their individual dynamics are characterized as above.

In the time domain the most important aspect is the occurrence of DOMINANCE. This means that, at any one time, the system is under the sway of one, or perhaps a few, loops, and they are overriding the behaviour of the other loops in the system. As time passes, the dominant loops may cease to be so and control passes to other parts of the system which then impress their characteristics on it. Changes in dominance are generally undesirable, and we usually seek to ensure the continued dominance of loops which can produce desirable behaviour. This implies that we must seek to prevent loops which produce bad behaviour from becoming dominant or at least to ensure that their dominance is as short-lived as possible.

We shall have more to do with dominance later. For the moment we simply introduce the concept and remark that it arises out of the presence of non-linearities in the system.

Frequency Domain

The time domain is simply the behaviour pattern of the system as time goes by, for a given input. However, some inputs, especially noise, can be broken up into combinations of sinewave inputs of different frequencies. We therefore need to know how the behaviour pattern, when all transients have died away, depends on the input frequency.

In the frequency domain, behaviour is characterized by the POWER SPECTRUM. This is found by taking two variables in the system, injecting into one a sine-wave and measuring the amplitude of the resultant oscillations in the other variable.

The ratio of the amplitudes is called the SYSTEM GAIN. When this is plotted against the input frequency a curve such as Fig. 2.20 is obtained, and this is the power spectrum.

Another important frequency domain characteristic is the PHASE PLOT. The injected sinusoid will, in a *linear* system give a sinusoidal output which lags behind the input by an amount called the PHASE SHIFT. This is shown in Fig. 2.19 in which the phase shift between ORATE and COMRATE is approximately 0.15 of the period. The phase plot shows how the phase shift depends on the input frequency.

6.4 Feedback Loop Structure

Our approach to the relationship between loops and behaviour is to examine the components of the loops, and to consider how they affect behaviour.

A loop has five characteristics:

1. Polarity; whether it is positive or negative i.e. what it is supposed to do.
2. Gain; the amount by which it amplifies or attenuates an input in transforming it into an output.
3. Delay; the extent to which it shifts an output so that its peaks and troughs do not coincide with those of the input.
4. Pure integration; the number of ordinary levels which it possesses, because these have important effects on gain and delay.
5. Linearity; if the loop is non-linear its behaviour may be markedly affected.

6.5 Polarity or Sign

Loops are either positive or negative and this refers to what the loop is supposed to be capable of doing.

NEGATIVE loops are the goal-seekers or error-correctors in the system. Their function is to adjust some part of the system to a goal, which may be fixed or moving.

POSITIVE loops are the growth-generators in a system. In practical systems they are always found in conjunction with negative loops.

A loop's polarity can be found by following it round and working out whether an increase in the value of an arbitrarily chosen starting variable leads to a further increase or to a decrease in the same variable. This gives the polarity as positive or negative respectively.

Alternatively, count the number of – signs on the influence links and *if there is an EVEN number of – signs the loop is POSITIVE.* (Another method which has been proposed is to count the number of *changes* of sign going round the loop. This does NOT always give the correct result and should not be used.)

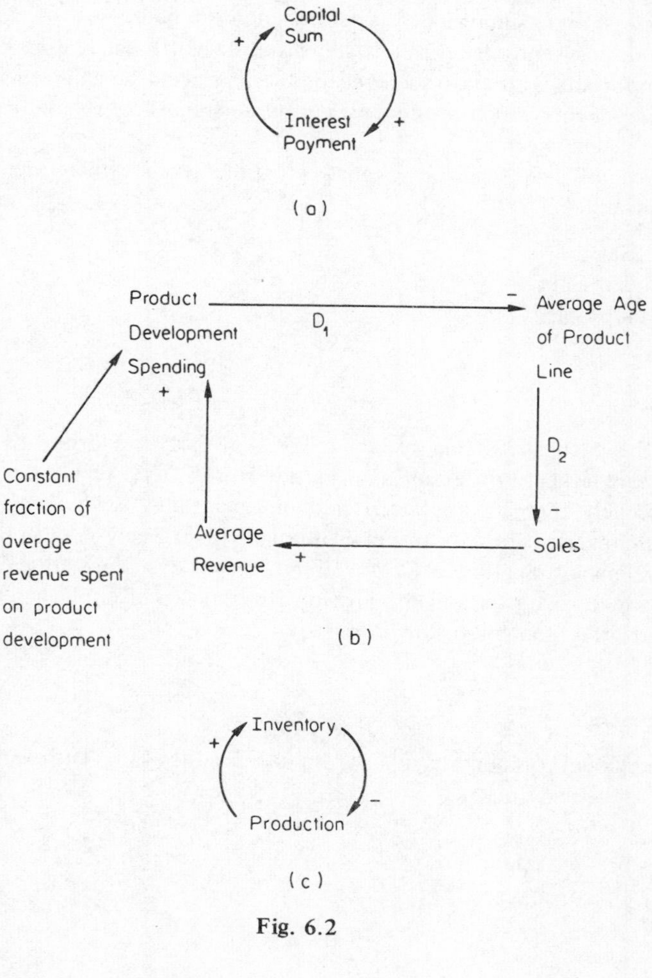

Fig. 6.2

For example:

1. Compound Interest (see Fig. 6.2a). In this diagram, there are no negative signs and loop is positive.
2. Product Development (see Fig. 6.2b). In this diagram, there are two negative signs so the loop is positive.
3. Production–inventory (see Fig. 6.2c). This has one minus sign, and is negative.

6.6 Gain

This is the ability of a loop to alter the magnitude of the input in creating an output.

Gain may be greater than unity, in which case the input is magnified, or conversely, when the input is attenuated. The simple system in Section 2.13 was of the former type and we implied that that was undesirable. This is by no means always the case and gain, of itself, is neither good nor bad.

In practical system analysis it is almost never worth calculating gain precisely. What is important is to grasp the concept and its effects, be able to recognize those system components which produce it, and understand how to alter components to increase or decrease gain.

The idea of gain can be applied at three levels which are, in order of comprehensibility:

1. System Gain
2. Component Gain
3. Open Loop Gain

System Gain

When the cyclically stable system of Section 2.12 was driven by a sine wave, ORATE was markedly different in magnitude from CONS. This is the essential idea of gain; namely that an input is altered in magnitude, *and possibly also in sign*, when it is converted into an output. Although this is easily seen in the case of the overall system of Section 2.12 *any* input/output transformation is potentially capable of involving gain, and the idea of gain can also be applied to the individual components of a loop and to the loop itself.

Component Gain

The individual components in a system also produce gain. The example shown in Fig. 6.3 has the decision rule

$$\mathrm{OR} = \frac{\mathrm{D-I}}{\mathrm{TAI}} = \frac{\mathrm{D}}{\mathrm{TAI}} - \frac{\mathrm{I}}{\mathrm{TAI}} \qquad 6.1$$

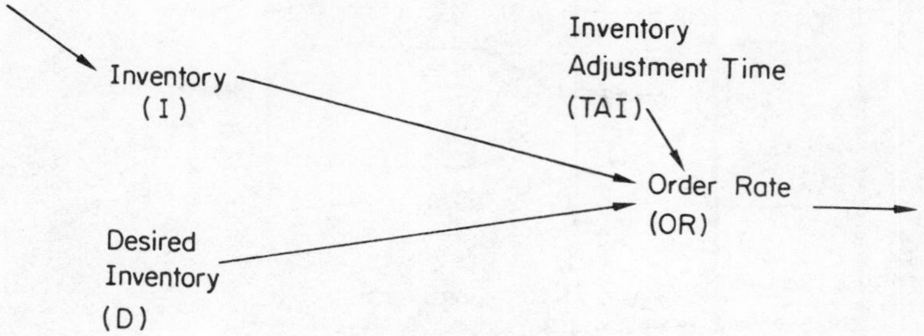

Fig. 6.3

For the *link* between inventory and order rate, a unit increase in inventory leads immediately to 1/TAI units *decrease* in order rate. Thus the gain *between inventory and order rate* is –1/TAI or as shown in Fig. 6.4.

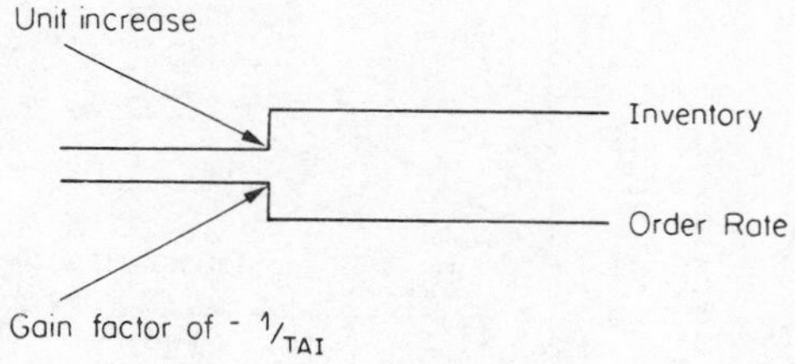

Fig. 6.4

In general, *the control of a rate by a level is a component gain*, and, in this case, increasing TAI decreases the component gain.

Table functions for non-linear policies are an important source of gain, because they help to produce the phenomenon of dominance. For example, we wish to model the modification of Desired Product Development Spending by financial constraints:

APDSR=FCM × DPDSR

where

DPDSR Desired Product Development Spending Rate

FCM Financial Constraint Multiplier

APDSR Actual Product Development Spending Rate

and FCM is a gain operator. This might appear as in the influence diagram Fig. 6.5, where the dotted lines show other parts of the system. Since the graph relating FCM to the Financial Reserves might be as Fig. 6.6, the component gain varies as Financial Reserves rise or fall, relative to Required Spending.

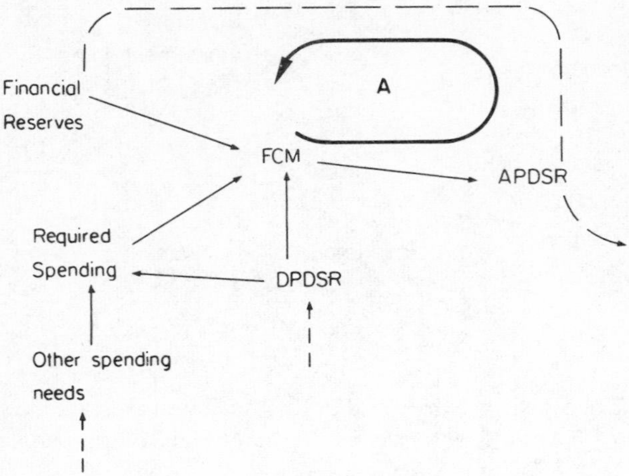

Fig. 6.5

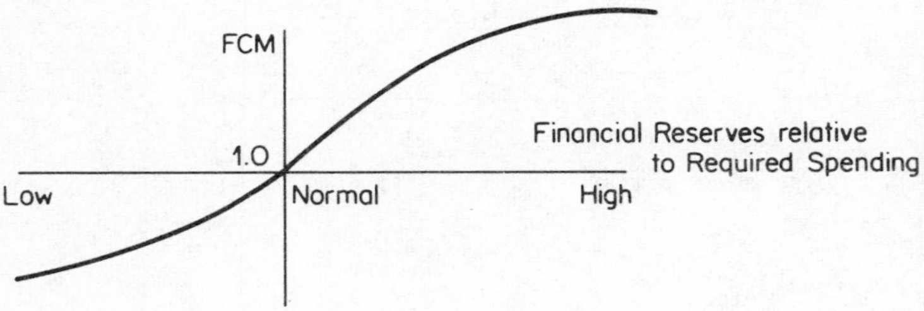

Fig. 6.6 Relationship between FCM and financial reserves

The varying gain affects loop A and could drive it into oscillation, or at least reduce its damping and, therefore change its effect on the system. One solution might be to flatten the curve to make the gain less sharply variable.

Open Loop Steady-state Gain

The loop is the basic unit of analysis in a system and it is, therefore, useful to be able to identify loop gain. This can be done by imagining the loop to be pulled out of the system, cut and stretched out, so that it starts at an arbitrarily-chosen component and comes back to it (see Fig. 6.7). The open-loop steady-state gain is then defined to be the product of the component gains.

6.7 Recognizing Gain-producers

Pure gain can occur in four main ways:

1. Multipliers

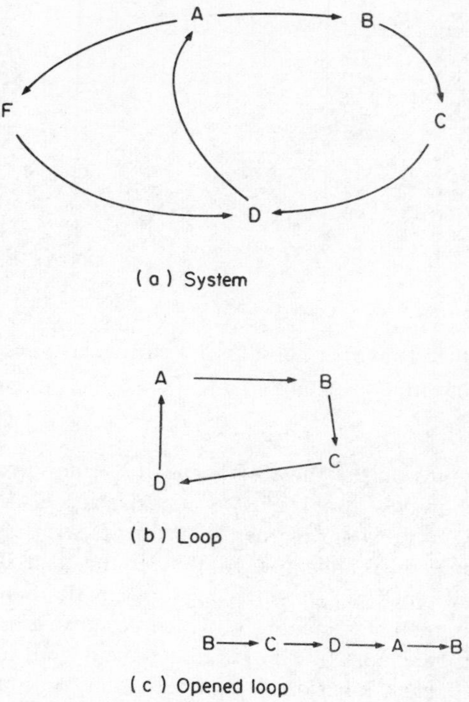

Fig. 6.7 Open-loop steady-state gain

2. Difference Relationships (Proportional Controllers)
3. Functional Relationships
4. Table Functions

Multipliers

For an equation form such as

Y.KL=A.K*B.K*X.K

the gain between X and Y is A.K*B.K.

This example immediately implies the possibility of trouble in the loop. As A and B vary so will the gain. The 'vigour' of this loop (which is a good way of thinking about gain) will therefore depend on the state of the rest of the system, which it will in turn, affect.

We cannot say whether this would be a good thing or not. If A and/or B were functions of some error condition and the loop was intended to correct this error, then it might well be useful to increase the loop gain as A or B got too large. The drawback would be that damping might be decreased and there would be a risk of setting undesirable oscillation in the system.

Difference Relationships (Proportional Controllers)

If

$$Y.KL = \frac{D.K - X.K}{T}$$

the gain from X to Y is $-\frac{1}{T}$.

Functional Relationships

Multipliers and Difference Relationships are particular types of functions. Let us now consider functions more generally, e.g.

$$Y.K = f(X.K)$$

The gain, G, depends on the value of X and for some value, X_0 say, would be $G = f(X_0)/X_0$. It is however also useful to consider the GAIN DERIVATIVE or $dG/dX_0 = (f'(X_0) - G)/X_0$, where the derivative of f is with respect to X. Its main use is as a dominance-shift indicator, so that if the gain derivative is large in absolute magnitude when $X = X_1$, large changes in gain are to be expected whenever X is near X_1. This is the phenomenon by which dominance in the system is likely to be shifted from one loop to another. Large gain derivatives therefore often indicate a need to change a function so as to reduce any abrupt changes in Y which it may create, i.e. to make the curves of functions as smooth or flat as possible.

Table Functions

Since these express functional relationships which cannot conveniently be reduced to analytic form the foregoing ideas apply. The table function is, however, easily visualized and, therefore, more readily changed. For example, consider a graph such as that in Fig. 6.8. To fix ideas, suppose that the curve represents the policy by which overtime worked, Y, is related to order backlog, X. Obviously, the region from X_2 to X_3 does not have constant gain because Y_1/X_3 is less than Y_1/X_2, and similarly for the region X_0 to X_1.

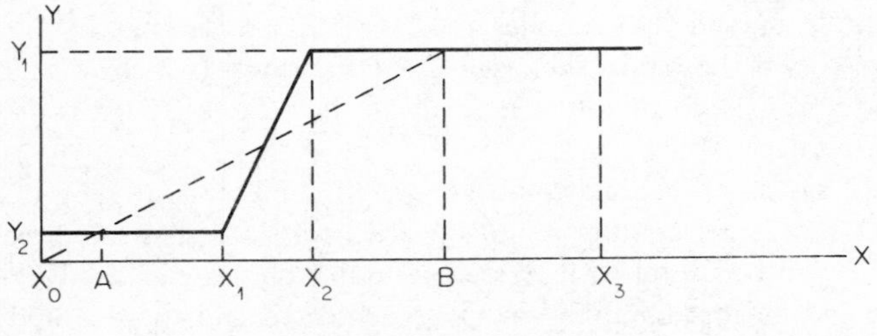

Fig. 6.8

There are, however, pairs of points, such as A and B, at which gain is equal.

Obviously gain varies along the curve and, because the function is not analytical it would be tedious to calculate a curve relating gain and X. Elementary trigonometry shows, however, that the region from X_1 to X_2 produces a relatively major discontinuity in gain in which a small change in X leads to relatively large changes in Y. If X usually moves in the range X_1 to X_2 there will be fairly rapid movements in Y and, hence, instability in Y, and one cure for this would be to change the overtime policy by pushing the curve to the right or left to take the region of discontinuous gain out of the range X_1 to X_2.

This illustrates the difficulty of this kind of analysis. In this example, there is a fairly obvious feedback between X and Y and this may simply push the system into the new range for the discontinuity in X. One solution might be to use a linear curve with constant gain but the presence of constraints on, say, maximum overtime still leaves the possibility of a gain discontinuity.

6.8 Delays

Delays are important because of the effect they have on a total stream of information or entities, and not because they affect individual members of that stream.

If the input stream is steady, the output has the same magnitude as the input and is also steady. Individual entities in the input are of course delayed by the delay magnitude, DEL, but the total output stream is exactly the same shape as the input.

Dynamics occur when the input rate varies. For a sinusoidal input the output differs in magnitude from the input and lags behind it. The gain and PHASE SHIFT depend on the order of the delay and the period of the input as in Fig. 6.9.

This shows that if the period is very long compared to DBL, the output and the input are very similar – long period inputs being fairly close to the steady-state. As the period gets closer to DEL, the output is further attenuated and the lag increases.

When the period is the same as DEL the phase shift is approximately DEL/2, for the third-order delay commonly used in modelling. Thus the output is just reaching its maximum when the input is at its minimum, i.e. the output is doing the opposite from the input. This is a potential source of poor system behaviour if the input contains noise of period close to DEL.

While delays are important components of systems they are in a sense, artificial, as they arise from the interaction of gain and integration through a feedback loop. This is easily seen for a first-order delay which (Section 2.8) has the equations:

IN.KL-*derived from rest of system* 6.2,R

DLEV.K=DLEV.J+DT*(IN.JK–OUT.JK) 6.3,L

OUT.KL=DLEV.K/DEL 6.4,R

with DLEV being an internal or dummy level which acts as a reservoir for the delayed entities. This is shown in influence diagram form in Fig. 6.10. 1/DEL is a

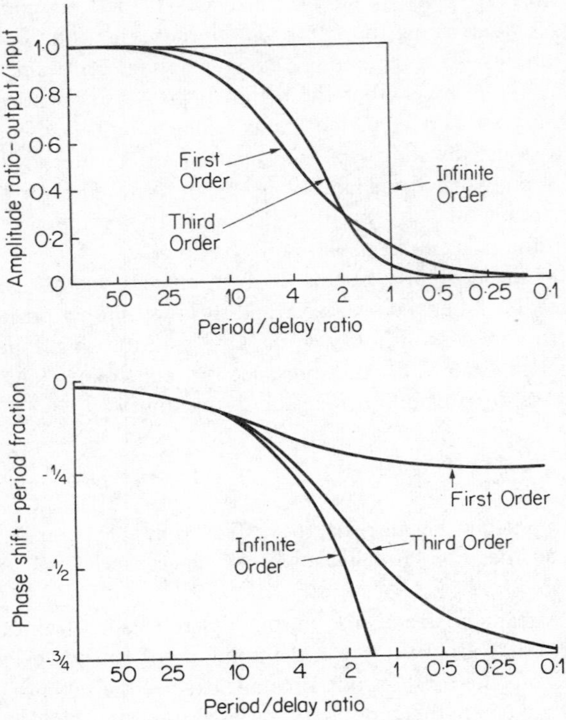

Fig. 6.9 Gain and phase shift in delay F (adapted from Forrester)

Fig. 6.10

gain component between DLEV and OUT, as is seen from equation 6.4, and it is the *closing of the feedback loop* from OUT to DLEV which transforms the gain effect into a delay.

The important and difficult idea is that *the point of view determines whether a component acts as a gain or delay*. For the link from DLEV to OUT there is a gain of 1/DEL. When we broaden the view from that link to the subsystem between IN and OUT the presence of the closed loop and its constituent integration, transforms the 1/DEL gain *between DLEV and OUT* into a delay of DEL *between IN and OUT*.

This is why the constant in a proportional control rule is expressed as the reciprocal of a smoothing or adjustment time, apart from the operational meaning

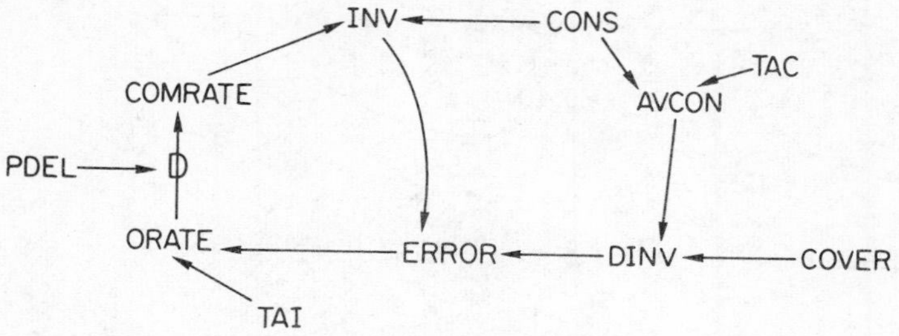

Fig. 6.11 Production–inventory example

of the adjustment time as the time within which the discrepancy would be eliminated if no further control actions were initiated.

Consider a common proportional control situation, as in the production–inventory example in Fig. 6.11, where the terms were defined in Section 2.13. We list the salient points:

1. PDEL is a perfectly ordinary physical delay, for example of third order.
2. From ERROR to ORATE, 1/TAI is a gain component because

$$\text{ORATE}=\frac{\text{ERROR}}{\text{TAI}} \qquad 6.5$$

   Between INV and ORATE the gain is −1/TA1 as in equation 6.1, *for this proportional control rule*.
3. The closed loop converts TAI into a first-order delay, exactly as in equations 6.2–6.4. The delay magnitude will be TAI.
4. The total delay in the loop is PDEL+TAI, and *loop delays are always additive*, as opposed to gains which are multiplicative.
5. The total delay characteristics for the loop are not usually easy to determine. If it happens that PDEL is 3 x TAI then the cascading properties of delays would make the total delay exactly the same as a single fourth-order delay of magnitude PDEL+TAI. This will be unusual, and there is no virtue in making TAI conform to PDEL to ensure a neat fourth-order representation.
6. PDEL is fixed, short of altering the production technology, so controllability in the loop can only be improved by selecting TAI to give a small phase shift and near-unity gain for the dominant frequencies in CONS transmitted via DINV, using Fig. 6.9 as a rough guide, but this may affect the damping of the step response, as in Section 2.13.
7. TAC also acts as a delay, because there is a loop created by the smoothing equation. This may be shown by detailing that part of the influence diagram (see Fig. 6.12) and writing

$$\text{AVCON.K=AVCON.J+(DT/TAC)(CONS.JK}-\text{AVCON.J)} \qquad 6.6,\text{L}$$

8. COVER is a pure gain operator which may be large enough to overcome the gains of less than unity which are produced by the delays. For example, if CONS

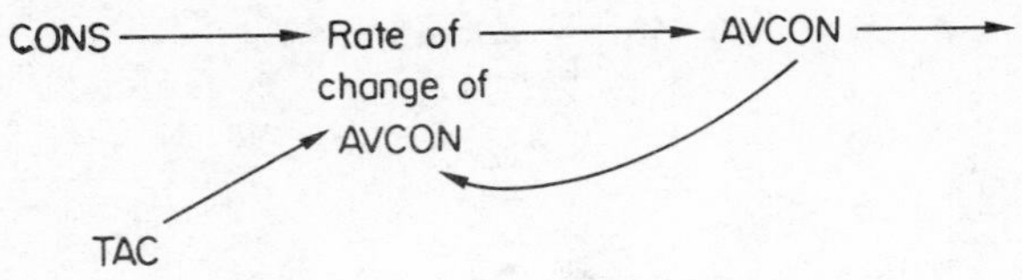

Fig. 6.12 Influence diagram

has a period of 26 weeks and if TAC=4, TAI=6, PDEL=4 and COVER=10, as in Section 2.13, and using a more accurate graph than Fig. 6.9 we can try to analyse the system as follows.

For the TAC delay, the PERIOD/DELAY ratio is 23/4 or 5.75, so the gain between CONS and AVCON will be 0.7. Since COVER = 10, the amplitude of DINV will be 10. x 0.7=7 times that of CONS.

We can use Fig. 6.9 in this way for components, but not for a system as a whole, even for such a simple case as this. For example, if we assume that cover is a gain on (AVCON–INV) we can find that the system gain is 3.64 between CONS and ORATE at his periodicity, but Fig. 2.20 shows that it is, in fact 4.95 so we are way out.

It is, however, useful to use the simulator to draw graphs like Fig. 6.9 for the system as a whole using say, CONS, as an input variable driven by a sine wave and measuring the resultant gain and phase shift at, say, COMRATE for different values of the input period. It is also useful to draw a similar graph showing settling time against the period as in the lower diagram in Fig. 6.9.

Table 6.1. *Loop Characteristics and Dynamics*

	Loop Polarity (Type)	
	Positive	Negative
Gain	*Increase* leads to more rapid growth	*Increase* leads to decreased damping
Delay	*Decrease* leads to more rapid growth	*Decrease* leads to decreased damping

These rules are relatively simple statements of some rather complex phenomena and they have to be applied with care in each particular case, particularly where the control rules are not simple proportional controllers.

The best approach is, therefore, as we remarked, to grasp the nature of gain and delay and to use the summary presented in Table 6.1, coupled with a careful study of the dynamics of the model being analysed.

6.9 Pure Integration

The ordinary level equation

L.K=L.J+DT*(IN.JK–OUT.JK) 6.7,L

is a pure integration of, in this case, two rates. A smoothed level is *not* pure integration, but a delay, in which the level controls the output. In a pure integration the output is not controlled by the level but by some other part of the system.

The difference between pure integration and first-order delay can be seen from influence diagrams and their associated level-rate diagrams.

Pure Integration

For pure integration, if the Depletion Rate is zero, for simplicity, and the Addition Rate changes from zero to a positive value, the level-rate diagram will be as shown in Fig. 6.13.

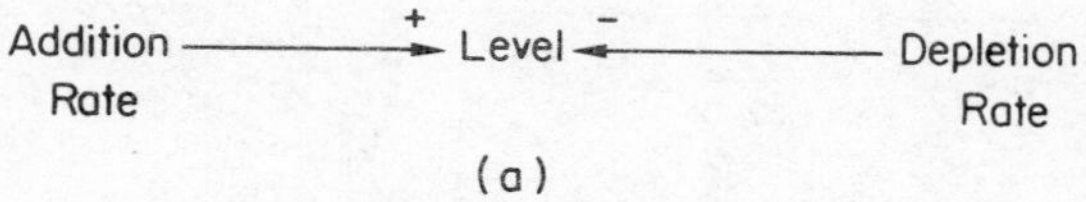

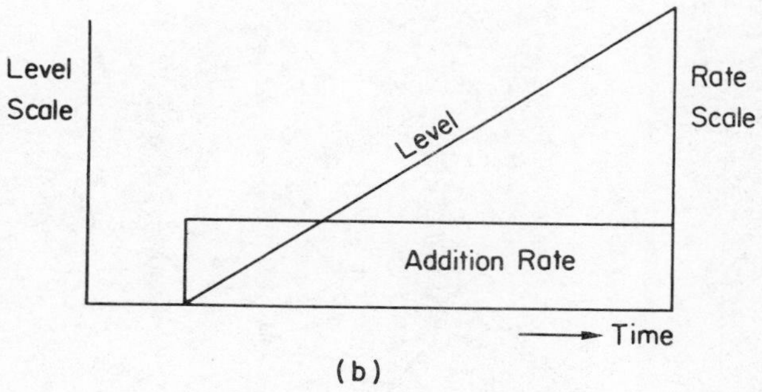

Fig. 6.13 Pure integration

First-order Delay

The first-order delay influence diagram is shown in Fig. 6.14a. This shows the structural difference, the level-rate behaviour for the same change in Addition Rate is shown in Fig. 6.14b. The Depletion Rate cannot remain at zero, because the Level rises to a value T times as large as the step in the Addition Rate.

Pure integrations act as phase shifters with respect to the net input rate (IN–OUT), the shift being 1/4 period. This can easily be shown analytically, but it

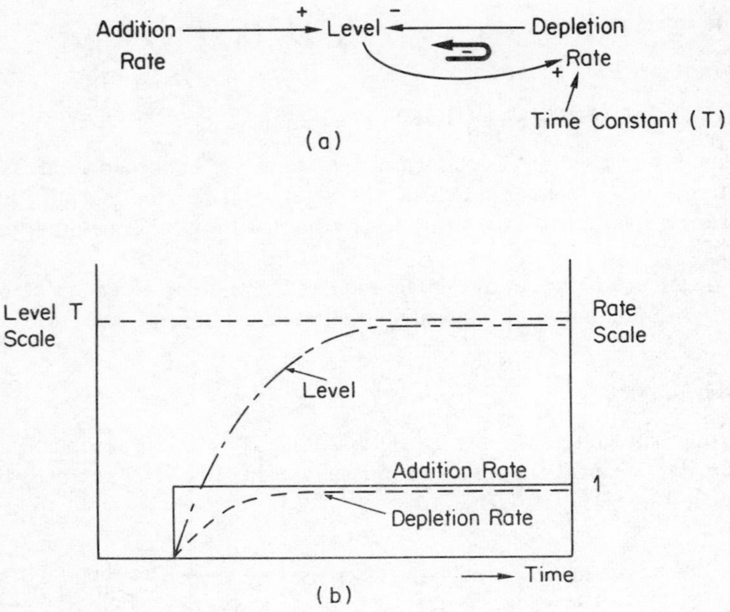

Fig. 6.14 First-order delay

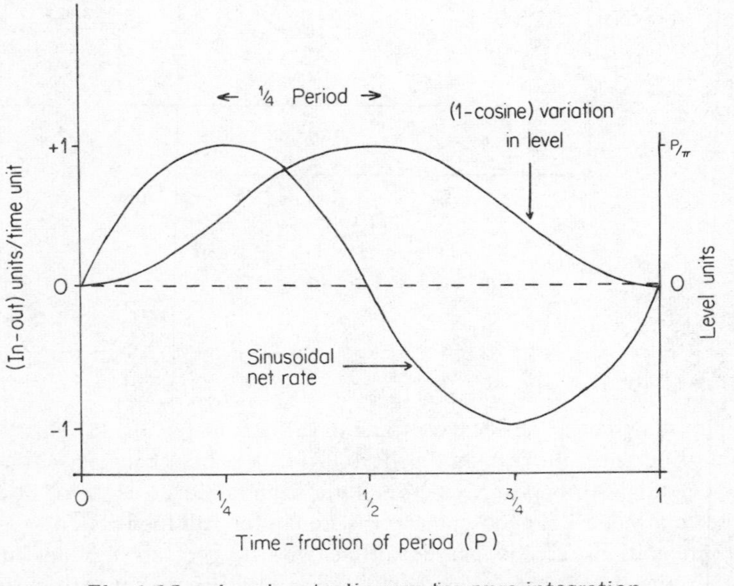

Fig. 6.15 Level–rate diagram for pure integration

is also shown in the level-rate diagram in Fig. 6.15. Note the phase shift, which is always 1/4 of the period for any regular net input rate, also the scales for the net rate and the level. For pure integration, gain is proportional to the period of any regular input. For the sinusoidal input the gain is $P/2\pi$, because the level goes from P/π to 0 while the rate goes from +1 to −1.

In general, the phase shift effects of delays or of pure integrations and the consequent reduced damping, can be most easily attacked by the use of derivative control. Unfortunately derivative control tends to reduce critical periodicities and can amplify noise responses which we may have been at pains to reduce.

This drives home the point that system design is a compromise and perfect control is not possible. The various performance criteria discussed in Chapter 7 may help in discriminating between one control action and another by numerical criteria, but even these involve arbitrary, judgement, weightings of performance values.

6.10 Gain, Delay and Integration in Negative Feedback

Armed with these concepts of gain and delay, it is now useful to study feedback round a delay in a single loop, when it is subjected to an exogeneous input. The whole point of system dynamics is that it deals with systems of many loops which interact with each other, so that a variable in loop A is driven by being a member of loop B and, in turn, drives loop B by the way in which loop A reacts to an original change in the variable. It is therefore useful to understand the way in which loops respond to impulse inputs. The reader should use the program in Appendix B.6 to confirm what follows.

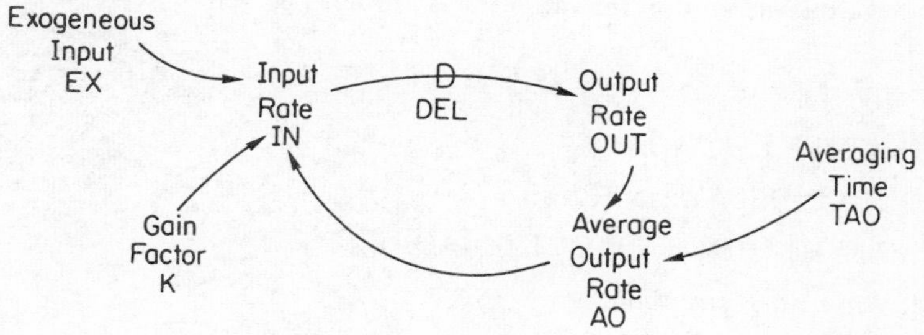

Fig. 6.16 Simple loop with delay but no pure integration

Negative Loop without Pure Integration

Consider a simple loop with a delay but no pure integration, as in Fig. 6.16.

The loop will be positive or negative, depending on the sign of the gain factor K but positive feedback is dealt with separately in the next section.

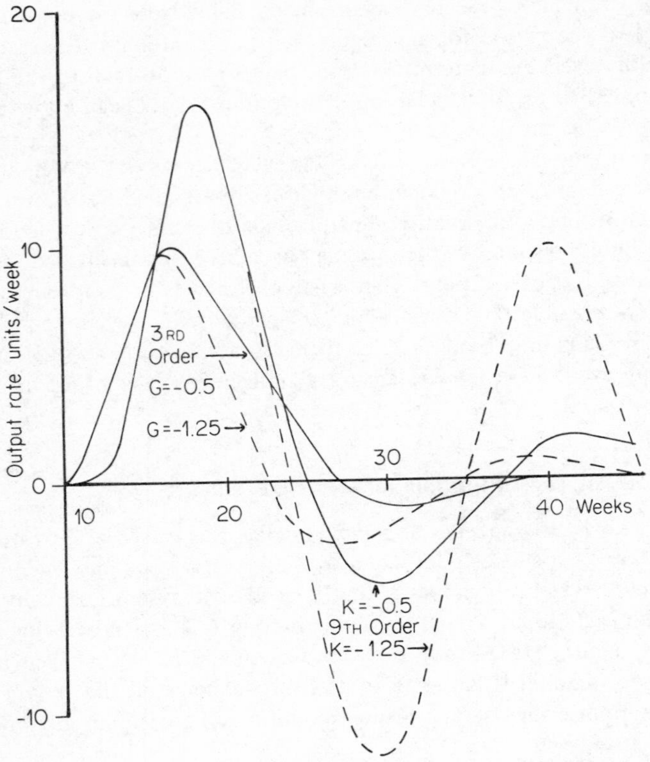

Fig. 6.17 The impulse response of a negative loop without integration for various delay orders and gains

The equations for the system are:

IN.KL=EX.K+K*AO.K 6.8,R

OUT.KL=DELAY*n*(IN.JK,DEL) 6.9,R

AO.K=AO.J+(DT/TAO)(OUT.JK–AO.J) 6.10,L

K = *some negative value*

The impulse response of this loop is shown for various values of the gain and the order of the delay in Figs. 6.17 and 6.18. They should be examined carefully as they show that:

1. An increase in the gain (i.e. making K more negative) decreases the damping and decreases the period of oscillation, though the second effect becomes less marked the higher the order of the delay.
2. Increasing the order of the delay for a given gain decreases the damping but does not materially affect the natural period of the loop.
3. Reducing the magnitude of the delay reduces the damping.

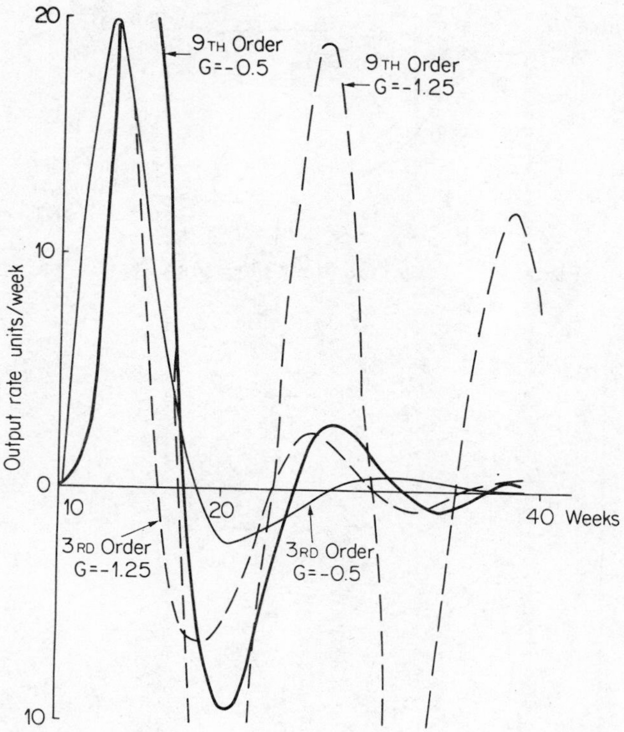

Fig. 6.18 The system of Fig. 6.17 with delay magnitude halved

Negative Loop with Pure Integration

Consider a loop which contains one pure integration, as opposed to the smoothing integration in the previous case; the system is shown in Fig. 6.19.

The equations for the system are:

IN.KL=EX.K+G*INT.K 6.11,R

OUT.KL=DELAY*n*(IN.JK,DEL) 6.12,R

INT.K=INT.J+DT*OUT.JK 6.13,L

G = *some negative value*

This is a very common type of loop. It is similar to the production–inventory system previously studied, if the exogeneous factor is regarded as the Desired Inventory.

Some examples of the response of the system to an exogeneous impulse are shown in Fig. 6.20.

The general conclusions for a loop with integration (the student should check

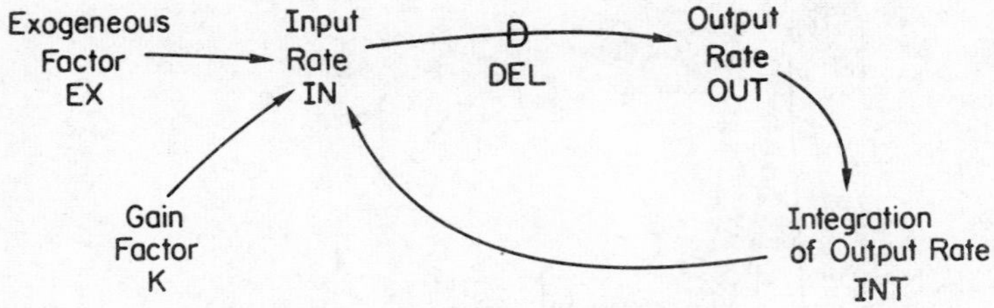

Fig. 6.19 Loop containing one pure integration

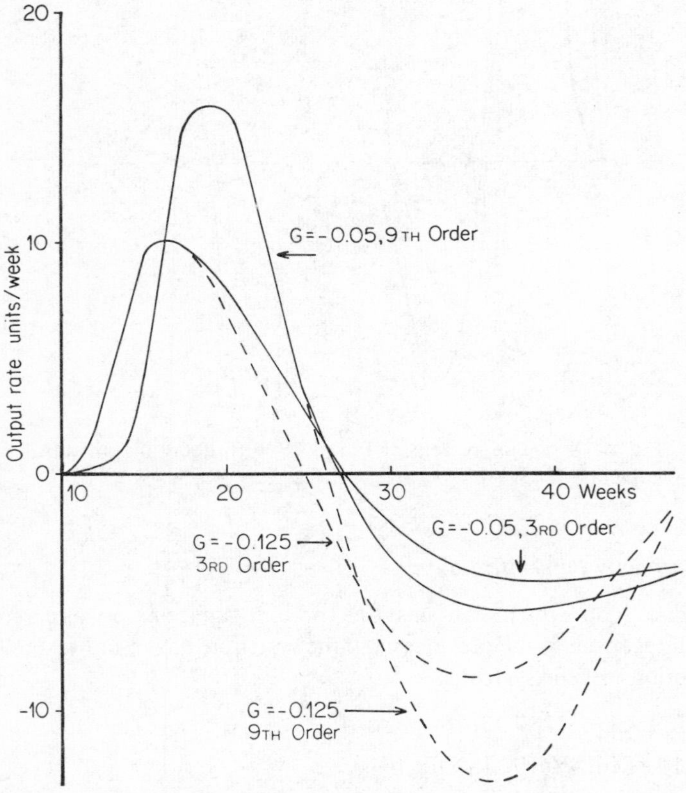

Fig. 6.20 Impulse response of a negative loop with pure integration

this by simulation) are:

1. In a system with pure integration, the natural period is larger than for a system without integration.
2. If a loop possesses integration the gain has to be much less than in an ordinary loop to preserve stability.

3. Increasing the gain or reducing the delay or increasing the order of the delay all tend to decrease the damping in the loop and the natural period of oscillation, though this last is not usually appreciable.

6.11 Gain, Delay and Integration in Positive Feedback

The parameters in a positive loop affect performance in a different way from those of a negative loop.

Positive Loop without Pure Integration

Consider first a positive loop without *pure* integration.

A company re-invests part of its average investment income in external funds which are repaid, with interest, after a period. For simplicity we assume that the interest is paid at the same time as the repayment of the principal. This is shown diagrammatically in Fig. 6.21, and the equations are

AVINC.K=AVINC.J+(DT/IAP)(NINV.JK–AVINC.J) 6.14,L

RINV.KL=PIE*AVINC.K 6.15,R

NINR.KL=NIF*DELAY*n*(RINV.JK,IPERD) 6.16,R

(Obviously NIF > 1 to model the repayment of the loan.)

There are two gain operators, PIE and NIF, and the open loop gain is

OLG=PIExNIF 6.17

The delay in the loop is IPERD+IAP and the delay order is (n+1), because AVINC is a first-order delay.

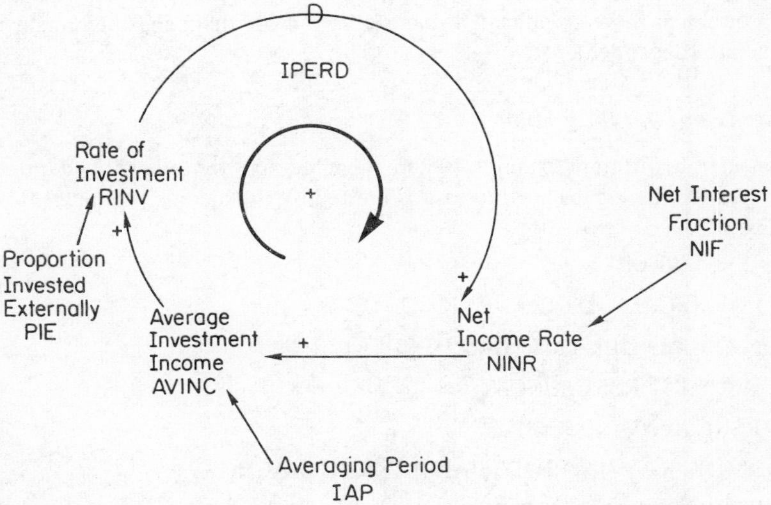

Fig. 6.21 Example of a positive loop without pure integration

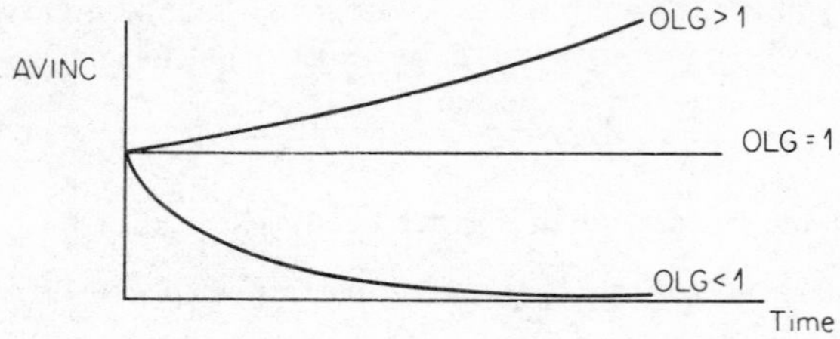

Fig. 6.22 Gain vs growth for positive loop without integration

When NINR > AVINC the system is growing and substituting equation 6.15 into 6.16 shows that this can only happen if OLG > 1.

If the OLG < 1 because the company is not reinvesting enough, i.e. PIE is too low, then the loop will decay asymptotically. (Morbid Stability, Section 6.2, can only arise if negative quantities are allowed, which would be nonsense in this example as it stands, but it could happen if losses were modelled.) This can be shown as in Fig. 6.22.

If RINV is subject to an impulse, as in Section 6.10 on negative loops, the system is ultrastable if OLG < 1 and growth stable if OLG > 1.

Reducing the delay in the loop would also increase the growth rate. Thus, if IAP was reduced then, providing growth was taking place, it would be recognized more quickly and more rapid increases of RINV would lead to still further growth. If OLG < 1, reducing the delay would produce more rapid decay.

The higher the order of the delay, the higher the growth rate because the reduced dispersion is equivalent to a more rapid response. (Note that in a practical modelling case, as opposed to an example on system properties, IPERD and NIF would be much more complicated than we have made them and they might well, in fact, be interdependent.)

Positive Loop with Integration

Consider a different example where demand has far outstripped production capacity and where capacity increase is limited by the income generated from sales (see Fig. 6.23).

The equations are

CAP.K=CAP.J+DT*RANC.JK	6.18,L
RANC.KL=DELAY*n*(RNCS.JK,CDEL)	6.19,R
RNCS.KL=RSNC.K/COST	6.20,R
RSNC.K=AVINC.K*FISC	6.21,A
AVINC.K=AVINC.J+(DT/IAP)(NINR.JK–AVINC.J)	6.22,L
NINR.KL=SALES.K*NPF	6.23,R
SALES.K=CAP.K	6.24,A

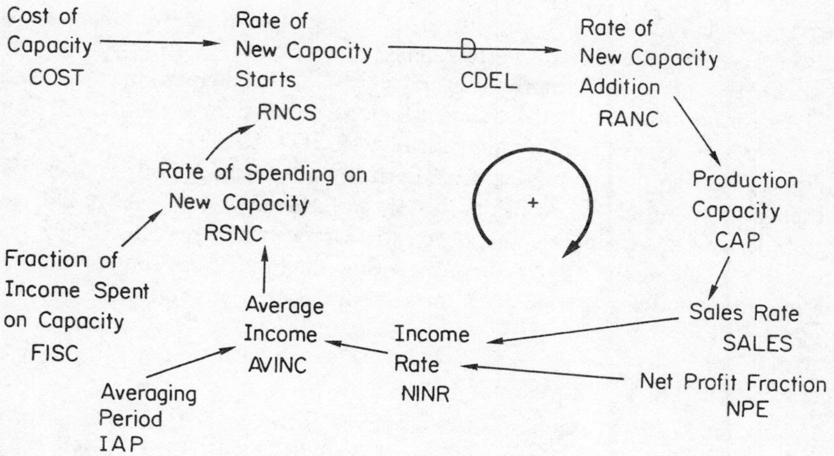

Fig. 6.23

Note the modelling of SALES and RSNC as auxiliaries, and the *essential* difference between this loop and the previous one is the pure integration at CAP, equation 6.18

There are three gain operators, NPF, FISC and COST and

$$\mathrm{OLG} = \frac{\mathrm{NPF} \times \mathrm{FISC}}{\mathrm{COST}} \qquad 6.25$$

Clearly, OLG > 0 but OLG may very well be less than 1.0.

The delay in the loop is the *n*th-order construction delay, CDEL, plus the first-order smoothing at IAP and the order of the delay is (n+1). In practical modelling, n might be no higher than 6 or 9 to reflect the dispersion caused by the running-in of new plant and the likely variability between completion of the various

Table 6.2. *Summary on Positive Feedback*

	No pure Integrations	Pure Integrations
Gain	G > 1 Growth G = 1 Static G < 1 Decay	G > 0 Growth G = 0 Static G < 0 Decay
Delay	*Magnitude* Mag. Decreases, Growth Rate Increases *Order* Order Increases, Growth Rate Increases	

stages of construction. It is nearly always very unrealistic to model capacity addition by an infinite order or perfect delay.

Since OLG > 0 it will be fairly clear that RANC > 0 and equation 6.18 therefore provides for continued growth in the system, as opposed to the situation without pure integration where the growth condition was OLG > 1.

Fig. 6.15 shows that pure integration possesses attributes of both gain and delay, so there are no simple relationships for the effects on growth rate of changes in magnitude and dispersion of the delays. However, the same qualitative results apply, that reducing the magnitude or increasing the order increase the growth rate.

These results on positive feedback may be summarized in Table 6.2.

6.12 The Limits to Growth

The characteristic of stable positive feedback is that all the variables in the loop double their value at a constant interval called the DOUBLING TIME (Section 2.19). The doubling time is easily determined for any positive process. For example, the macro-dynamics of RTZ in Fig. 1.5 are not precisely positive feedback from 1962 to 1970 but, for simplicity, assume that they are.

The turnover goes from £50 million to £450 million in 9 years (in round figures). Exponential growth is described by the equation

$$y(t) = ae^{kt}$$

At t=0, $y(t)$=50, so a=50. At t=9, y=450, so

$$450=50\ \mathrm{e}^{9K}$$

and

$$k=\frac{1n9}{9}=0.244$$

Let T be the time at which y=100, i.e. at which it doubles its initial value. Then:

$$100 = 50\ \mathrm{e}^{.244}$$

or

$$\mathrm{T} = \frac{\ln 2}{.244} = 3.2$$

In general, the doubling time, T, is given by

$$\mathrm{T} = \frac{\ln 2}{k} = \frac{.693}{k}$$

Now consider a general positive process in which T=1 time period and a=1. Then the value of the variable at the first few time points will be:

Time	0	1	2	3	4	5	etc
Value	1	2	4	8	16	32	

In any real situation, there must be some limit to the growth which can take place. Very often, in socio-economic systems, the limit is not a physical one as it might be in the case of a falling object dashing itself to pieces when it hits the ground. Socio-economic limits are usually of the type which, if significantly exceeded, will cause the system to collapse from 'starvation.' For example, a firm grows beyond the market's real limits and satisfies demand now which should have been reserved for the future. As the future unfolds there is little demand to be satisfied and the firm has, so to speak, eaten its seed corn. Forrester and Meadows have discussed similar processes in their works on the limits to world economic growth (Appendix C).

Let us imagine that, in the above case, the limit happens to be 32. At the start of the positive growth, the system is at 1/32 of its limit and at 1 time period, the doubling, which would undoubtedly be the cause of much gratification, has taken the system to 1/16 of its limit.

The comparatively large increase of four units between time 2 and 3 still leaves the system at only 1/4 of its limit, but within two more doubling times it reaches the limit, whereas it only went from 1/16 to 1/4 in the previous two periods.

Even more ominously, the system goes from 1/2 its limit to the limit itself in 1 doubling time from 4 to 5.

Because of the delays in positive FBLs and the fact that most socio-economic systems are very poorly controlled, it is clear that nearly all positive processes carry with them the seeds of very serious trouble which will come upon the system very rapidly. There is therefore a very strong chance that any positive process could easily become overshoot unstable. In general, the occurrence of a positive loop in a model should be a warning signal to the analyst, first to check that it really is a positive loop, secondly to make sure that it really exists in the system, and thirdly to design control actions which will prevent overshoot and collapse.

6.13 Pseudo-positive Loops

A fairly common phenomenon in modelling is a loop which is, to all appearances, positive, but which never acts as though it were. (Ted Sankey and Irwin Dawson drew my attention to this problem.) These PSEUDO-POSITIVE loops have to be identified as such and not confused with the genuinely positive loops to which the remarks of Section 6.12 apply.

In the chartering of oil tankers, a discrepancy between the Charter Tonnage available and that required leads to a decision about the rate at which ships are to be chartered, the Charter Rate. Those ships become available after a short delay and are then part of the Charter Tonnage.

The rate at which ships complete their charters depletes the Charter Tonnage and is most appropriately modelled as a high-order, long duration, delay of the Charter Rate. This is shown in Fig. 6.24.

By counting signs, loop A is negative and B is positive and it might therefore appear that an increase in Required Tonnage leads, via loop B, to an ever-increasing charter rate. This would be disastrous for the system were it not nonsense in the modelling.

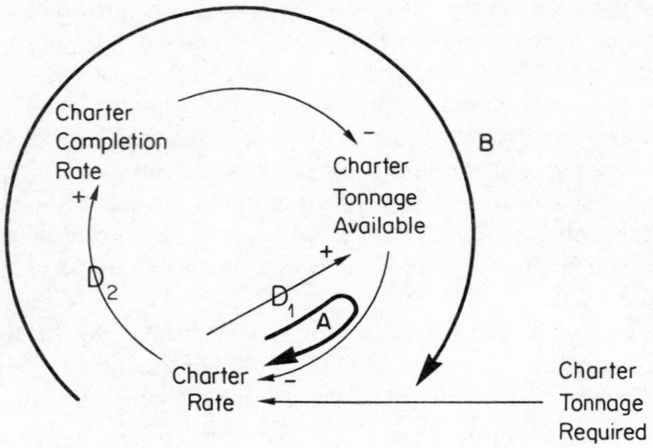

Fig. 6.24

The solution lies in the relative delays. Obviously D_1 has to be a good deal less than D_2, otherwise there would be no point in chartering at all. Because the delay in loop A is so much lower than that in B, the former effectively short-circuits the latter and loop B never operates as a positive loop in the usual sense.

Chapter 7

Validation, Performance and Analysis

7.1 Introduction

Having examined how a model is formulated and programmed, and having looked briefly at the theory of single feedback loops, we now attempt the rather daunting task of applying these ideas to models of several loops. In this chapter we shall attempt to offer some general guidance, but there is no question that practice is the preferred instruction. In Chapter 8 we shall therefore study an example problem and, in later chapters, examine real industrial situations.

We first discuss model validation and system design objectives so as to create the background against which system design is to be studied.

7.2 Model Validation

Unquestionably one of the most difficult areas in management science is that of trying to establish whether the model is 'valid'. A great deal has been written about the problem, but no really satisfactory solutions have been proposed, and many writers take refuge in philosophical abstraction or statistical mathematics. We shall attempt a middle course, appealing more to common sense than to anything else. In general we follow the reasoning of Forrester's excellent Chapter 13, to which the reader should refer for a fuller discussion.

It is often the case that validity is confused with truth and attempts are made to 'prove' that a model is 'true'. Validation means something quite different. The Concise Oxford Dictionary defines valid to mean 'sound, defensible, well grounded', and we shall refer to model validation as meaning the *process by which we establish sufficient confidence in a model to be prepared to use it for some particular purpose.*

To fix our ideas, recall that system dynamics models are policy-design tools. We build them in order to know what changes to make in a system so as to improve its behaviour. We only bother about validation to improve our confidence in what we have done to the point where we will take the risk of making a change to the system. Thus the only true test of a model's validity which is theoretically possible is to observe the actual system at a suitable time after the change to make sure that behaviour has been improved. Even this test is, in practice, very difficult because it will be very hard to disentangle the effects of policy changes from all the other things that have happened to the system, and in any case, an *ex post facto* check like this is irrelevant to the decision to implement a recommended design change.

Strictly speaking, therefore, true validation is impossible and we are left with a confidence-boosting exercise. Unfortunately, it has become accepted to use 'validation' to mean 'confidence-boosting', and we shall have to conform to the convention. This approach should, however, put the validation issue in the perspective in which it should be viewed; namely that models are built for a purpose, or to answer questions, and validation only has meaning in relation to whether the model meets its purposes.

This leads us to thinking along lines which can be phrased as a series of questions which can be asked about a model in order to guide ones judgment in assessing its validity and utility. Note that there are no such things as models which are absolutely valid or completely invalid, except in relation to a particular purpose, and a model which is good for one purpose may be so poor as to be misleading for another.

1. Is the system boundary right? If the model does not include those parts of the system, especially its controller, which can be changed to improve behaviour, then it is useless and therefore invalid. Thus, a model might be an excellent treatment of production distribution decisions, but be largely irrelevant to corporate planning issues.
   A corporate planning model might (not 'must') have a production sector and it would probably not add to the model's validity as a corporate planning tool to improve that sector. Indeed, it might well detract from it by obscuring the real problems in a lot of detail.
2. Are there any gross errors? A model which produced, say, negative product prices, or a cash flow of £10^{26} per month would obviously be not particularly valid because it is producing behaviour which is conceptually impossible, or which is beyond all sense and reason for the system.
   Errors of this type may be due to simple mistakes, such as forgetting to divide by 100 when using percentage values. Alternatively they may arise from failure to model constraints properly or to represent decision functions realistically, or, particularly, from dimensional errors.
   Note that we are already using words such as 'properly' and 'sense and reason'. This should imply fairly clearly that validation is not just a statistical exercise in curve fitting, but also a matter of judgment, even when statistical procedures are employed.
3. Is there a correspondence between the model structure and the system? This is a matter of developing confidence in the model, on the part of the analysts building it and on that of the managers for whom it is being built, that the model represents the system as it is. One checks that the proper variables have been correctly interconnected, and that the decision functions in the model reasonably reflect those actually used.

This is very difficult to do. It is very rare indeed that data are available to verify that the modelled decision function reflects what was done in the past. Even when they are available, the best that can be done is to reject an obviously wrong formulation.

In practice, a good approach is to conduct a simulation session with the decision-maker. Ask him what he would do under various sets of circumstances and try to construct a function which does the same things, *for the same reasons*. For this purpose, the Working Memorandum technique described in Chapter 13 is very effective.

If the problem is hard enough for controller elements, it is doubly so for the environment. Review the discussion in Chapters 2 and 3.

4. Are the parameter values correct? This is, in many ways, not a very important question. The dynamics of a system are usually not affected very much by most of the parameters, providing they are within a fairly broad range. However, some of the parameters will be more critical, and changing their values may change the behaviour mode of the system from, say, sustained to damped oscillation. An example of this is the parameter AI in Section 2.12 which has three critical values which determine whether the system explodes, just oscillates, has damped oscillations, or has a smooth S-curve response.

   The critical values for a system can usually be found by inspecting the loop structure and testing the model by simulation runs. However, even this may not be worth the effort because what matters is not so much what the value used to be, but what it is going to be in the new system. This is the first inkling of the concept of ROBUSTNESS which will be examined later and in Chapter 9.

5. Does the model reproduce the system behaviour? This implies that we have time series from the system which are compared to series for the same variable from the model, and the model 'fails' if its values do not agree 'sufficiently well' with the actual history. This is a classical approach which is often allied to very sophisticated statistical procedures but there are some very serious difficulties in actually applying it.

   a. In practice it is rarely possible, as the data are usually not available. In any case, even when data can be found, they relate only to the system states, and rarely are the policies by which those states were controlled known. Comparing model series to actuals is meaningless unless one also knows that the policies were identical and were consistently applied.

   b. The available data will never cover all the model outputs, so many of the variables must remain untested. It seems hard to reject a model because one or two of its outputs do not match an uncertain past history, and another 10 or 20 or 200 are completely untested.

   c. The purpose of the model is to design a control system which will function effectively in the face of a variety of shocks in the future. It seems unreasonable also to expect it to summarize past history, and it is not easy to see any good reason why it should.

      The only reason which holds water is that, if the model purports to represent the system's structure *and* policies, and if neither of these have changed during some past period, then the model obviously should reproduce the same data, to some degree of exactness. Unfortunately, most of the statistical tests for the agreement between two time series (the model's and the data) require about 30 data points. Even with monthly data it is unlikely that one could find a 'representative' period 2½ years long during which there were no system changes and for which the data can be found. For quarterly data there is practically no prospect of doing so.

It sometimes seems that an analyst has sought some kind of statistical verification of his model for no discoverable reason other than that it is somehow 'scientific' to do so. This temptation should be resisted. To 'fit' a model to data comes very close to asserting that only parameters matter, and that policy produces only noise in the system. This is a very poor basis for building a model for the purpose of policy design.

d. It is not immediately clear that there is any justification for choosing any particular statistical confidence interval, other than by judgment. Much statistical testing therefore reduces to somebody's opinion covered in a layer of advanced mathematics.

In general, therefore, we feel that the best test of confidence is the knowledge that the model has been carefully built up in conjunction with management. This is perhaps best exemplified by the model in Chapter 10, which was considered to be 'validated' when the Marketing Director jumped up (literally) and exclaimed, 'but that is exactly what happens'.

In this case the Marketing Director could produce no statistics to support his feelings, but it was also known that the model had been carefully put together by an analyst and a member of management, with a clear purpose in mind.

Undoubtedly the most convincing argument of all, however, is the sheer unavailability of data on industrial systems. In only one case has this writer been able to test a model against historical data. Even that only applied to one model variable out of several, and involved some very difficult digging to show that the historical policies did accord with those in the model throughout the data period, which was only 18 months.

7.3 Assessing System Performance

The first stage in model analysis is assessing the performance of the system. There are three aspects to this: choosing the basis for judging performance; assessing how well the system performs; deciding on performance targets and criteria and what, if any, deterioration could be tolerated in one performance indicator for the sake of improvement in another. We shall study the first two points in the next two sections, leaving the third to emerge by example as we proceed through the book.

7.4 Qualitative Considerations in Performance Evaluation

Before calculating numerical values for model performance, we have to decide what is to constitute 'good' performance. This is what is meant by 'qualitative' considerations. The variety of situations is so large that the best we can do is offer some general guidelines, first for the time domain and, subsequently, for the frequency domain. The guidelines are in the form of ways of assessing system performance, and how they should be viewed depends on what the system is supposed to do. For instance it would not do to apply criteria of damping in such a way as always to produce very damped systems. If the controller was supposed to respond rapidly to signals from the environment or the complement we might want to reduce the damping. At any rate we might wish to reduce the damping of the response to, say, annual seasonalities and increase it for any other variations. In short it is the essential first step in model analysis to decide what the system should achieve before one considers how effectively it does it. We shall repeat this point at intervals.

Time Domain

1. AMPLIFICATION The purpose of many industrial systems is to insulate the controller's internal variables from shocks provided by the environment or the complement. Thus, if we see that production rate has an amplitude which is comparable to, or even greater than, the amplitude of sales rate, we should be entitled to say that the system was not very successful in achieving its fundamental purpose.
2. TARGET ACHIEVEMENT Any system which contains negative feedback has error signals which it is supposed to be eliminating. One way of assessing the effectiveness of the system is to study the dynamics of the error signals. For example, in Section 2.12 the system is supposed to keep inventory up to the

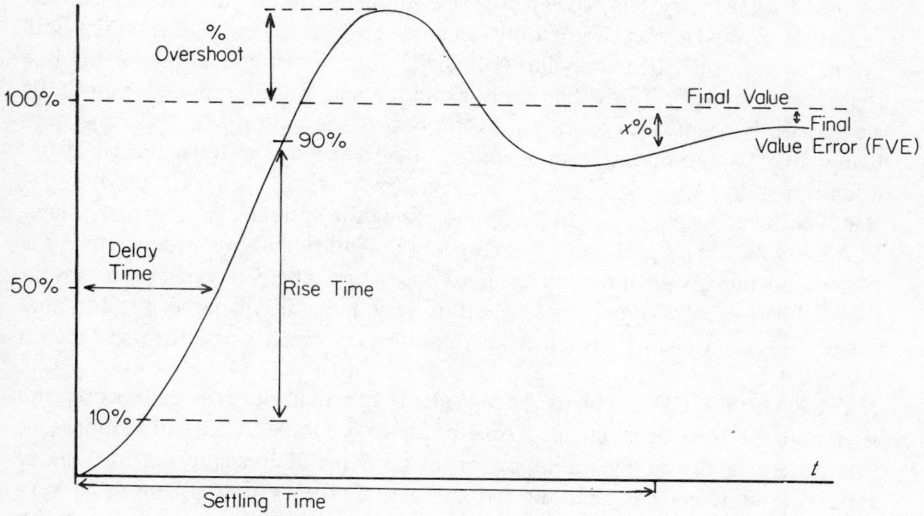

Fig. 7.1 Performance criteria in the time domain (after Elgard)

desired level, and one of the criticisms we made was that basing production simply on inventory discrepancy was basically unable to achieve that.

3. FLEXIBILITY For most business systems it is desirable to respond quickly and effectively to genuine and permanent (or, at any rate, semipermanent) changes in the environment or complement. We call this flexibility and it is measured by the RISE TIME i.e. the time taken by the system to make the first major movement to its new operating level – usually defined as the time needed to go from 10% to 90% of the new level.
4 DAMPING (% OVERSHOOT) Having reached its new level it is important that any overshoot should die away rapidly and this is measured by the extent to which successive oscillatory peaks are lower than their predecessors.
5. SETTLING TIME This is the time for the variable to settle to within $x\%$ of its final value.

Most of these ideas can be illustrated as in Fig. 7.1.

Frequency Domain

In the frequency domain there are several criteria which can be applied. The approach is to inject into the sytem, the loop, or the component, a series of sine waves of increasing periodicity and to plot the gain ratio for some output variable, exactly as we did in Section 2.12 (suitable DYSMAP equations can be found in Appendix B2.12). The most important criteria are:

6. BANDWIDTH Bandwidth is a measure of the ability of the system to follow an injected signal. Usually, in a business system, we want the bandwidth to be narrow i.e. we want the system to be insensitive to noise components. However, we might want the system to be responsive to long-period inputs such as seasonal variations or business cycles, if only because few micro-economic systems could afford to ride out such long-duration difficulties. We might also take the line that, as seasonality (providing it is genuine) and business cycles cannot be abolished, the controller ought to do the best it can with them. This may mean being able to respond to opportunities and make an 'orderly retreat' from downturns.

   Bandwidth can usually be reduced by increasing the system's information delays.
7. PEAK GAIN The peak gain is an aspect of stability and the more damped the system is, the lower the peak gain, to the point where it vanishes in a very overdamped system. Peak gain is often very large in managed systems and usually implies that the system is making life very much worse for itself than it needs to.
8. NATURAL PERIOD This is the periodicity at which the peak gain occurs and it is also the time between successive peaks of a damped transient response. If the system is driven by an input which contains appreciable proportions of variations with periods near the natural period, the system will respond very dramatically and will usually amplify those input components, thereby producing very unsatisfactory dynamics (recall Fig. 2.20).

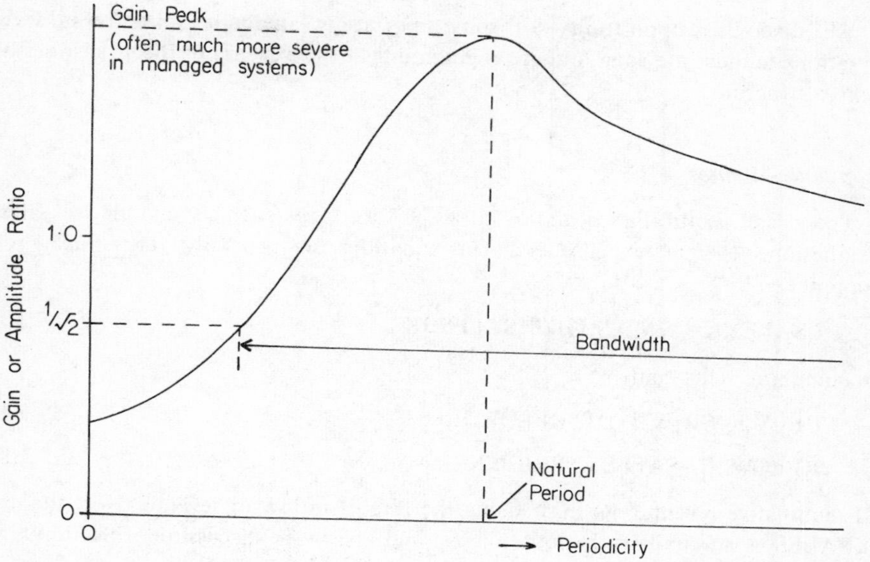

Fig. 7.2 Performance criteria in the frequency domain (after Elgard)

The frequency domain concepts are shown in Fig. 7.2.

Most of these concepts were derived by control engineers for linear systems. They do not apply exactly to non-linear systems, but close analogies can be found by simulation though the response curves may be much more complicated.

7.5 Simple Numerical Performance Measures

It is often useful to compute numerical values which provide an index of how well the system has performed during a simulation run. There are many possible measures and we illustrate some which have been found to be useful in practical modelling, together with equations for their calculation. The experienced analyst will, of course, be on his guard against using a performance measure (PM) just because it is easy to calculate. Defining a good PM is far harder than calculating it.

In practice a single PM is unlikely to be a sufficient index of performance to distinguish between one control strategy and another. A multiple performance measure, MPM, will be needed, which will be a function of several PMs. It is rare for this function to be any other than purely subjective, though there is nothing wrong with that if the opinions are given after serious thought by the people who will have to live with the results of the redesign.

All PMs discussed here are final values, i.e. they exist at the end of a simulation run, though they may be carried forward to the end from some intermediate point in the simulation. We now discuss some PMs, roughly in order of the complexity of the calculations needed.

We take this opportunity of sometimes using auxiliaries directly in level equations to illustrate economy of programming at the expense of the strict rules of system structure.

Cumulative Values

These are useful indicators of total performance, such as cumulative sales, production, revenue, expense etc. The equations are generally very simple, for example:

CSALES.K=CSALES.J+DT*SALES.JK 7.1,L

for cumulative sales, and

CREV.K=CREV.J+DT*RFLOW.JK 7.2,L

RFLOW.KL=SALES.K*PRICE.K 7.3,R

for cumulative revenue, RFLOW being the rate of inflow of revenue. Note the use of SALES as an auxiliary in equation 7.3 and review the discussion of Section 5.5, for which equation 7.3 is an abbreviation.

Point Values

The maximum values achieved by, say, inventory are indicators of performance, but they cannot be calculated by straightforward use of the MAX function, because of the time-shifts involved. However, if INV denotes inventory and OMAX is the previous highest value of INV, and if:

MAXINV.K=MAX(INV.K,OMAX.K)

we have the larger of the current inventory and the current previous maximum value. From this OMAX can be updated by

L OMAX.K=OMAX.J+DT*((MAXINV.J–OMAX.J)/DT)

and

N OMAX=0

This is another example of legitimate use of DT in an equation. Similar considerations provide minimal values, but care may be needed in some cases. Thus, if we wish to find the minimal value of some variable X we should write:

A MINX.K=MIN(X.K,OMINX.K)

L OMINX.K=OMINX.J+DT*((MINX.J–OMINX.J)/DT)

N OMINX = 10E6

say.

If, however, X is zero at TIME=0 then that will be the value carried forward and reported as OMINX at the end of the run. This will often be unrealistic, as, for example, when a step response is being studied, and it may be better to allow transients to die away before applying the minimal value calculation. This can be achieved by putting:

```
A MINX.K=MIN(X.K,OMINX.K)
R RCOMIN.KL=((MINX.K–OMINX.K)/DT)*D.K
A D.K=CLIP(1,0,TIME.K,CTIM)
L OMINX.K=OMINX.J+DT*RCOMIN.JK
N OMINX=10E6
```

This introduces a dummy rate, RCOMIN, which is held to zero until CTIM, at which time the initial transients are deemed to have ceased to matter. The value of CTIM would be found by a visual examination of the simulation graphs for an initial run. This technique is used in Appendix B, Section 2.12. The time, MTIME, at which, say, INV reached a maximum, or the time of the latest of a series of equal maxima, can be found by roughly analogous methods:

```
DI.K=CLIP(1,0,MAXINV.K–OMAX.K,0)
```

DI being a dummy variable which is 1 whenever the previous maximum is exceeded.

If OTIME is the time at which the previous maximum occurred then

```
A MTIME.K=DI.K*TIME.K+(1–DI.K)*OTIME.K
```

and

```
L OTIME.K=OTIME.J+DT*((MTIME.J–OTIME.J)/DT)
N OTIME=0
```

The amplitude of a variable's dynamics can be calculated from the maximum and minimum values. The true amplitude is the difference between a peak value and the following trough whereas the maximum and minimum calculated in the preceding equations are for the whole of the run and are not necessarily consecutive. If, however, we put:

```
AMP.K=MAXINV.K–MININV.K
```

and take the value of AMP when TIME=LENGTH we shall have the maximum spread of the variable. Using this method we can take a ratio of two amplitudes by:

```
RATIO.K=AMP1.K/AMP2.K
```

The last value of RATIO will be the maximum, but the dynamics of RATIO would tell us something about the way in which shocks are propagated through a system, especially if AMP2 relates to an input signal. This method allows us to plot frequency-responses such as Fig. 2.20.

Danger Zones

If Cash Balance, CBAL, falls below some minimum value MBAL the company is, in some sense, at risk and it would be useful to have a measure of the extent and duration of the risk. This can be achieved by:

DAREA.K=DAREA.J+DT*((MBAL.J–CBAL.J)*DI.J)

with

DI.K=CLIP(1,0,MBAL.K,CBAL.K)

The danger area, DAREA, has dimensions of £-Weeks. We could also write, say

RCDAREA.KL=((MBAL.J–CBAL.J)**2)*DI.K

and

DAREA.K=DAREA.J+DT*RCDAREA.JK

to penalize large departures from MBAL. This quadratic formulation is completely arbitrary and has only become standard practice in control engineering because it is *analytically* tractable, which is not a requirement in simulation.

Duration of Danger Periods

It is often useful to know how long a period of unsatisfactory performance has lasted during a total run. For example, Product Manufacturing Rate, PMR, is the greater of the Planned Production Level, PPL, and Production Sustainable with Available Component Stock, PSACS, i.e.

PSACS.K=CSTOCK.K/DT

PMR.KL=MIN(PPL.K,PSACS.K)

where CSTOCK is Component Stock. Switch for Production Cutback, SPC, would then be:

SPC.K=CLIP(0,1,PMR.KL,PPL.K)

(Note the use of PMR.KL and work out how this operates.)

Now SPC changes from 0 to 1 as soon as production cuts back so if TSC denotes the Time of Start of Cutback the duration of the cutback, DCB will be:

DCB.K=TIME.K–TSC.K

providing SPC only changes when the cutback starts and then holds that value. This can be achieved by asking, each DT, whether the production slowdown has just started or whether it was already in being. Clearly this involves the Old Value of SPC, OVSPC, which is

L OVSPC.K=OVSPC.J+DT*((SPC.J–OVSPC.J)/DT

and then

L TSC.K=TSC.J+DT*((TIME.K–TSC.J)/DT)*SPC.J*(1–OVSPC.J))

In this equation, TSC will be rapidly updated to current TIME from its old value (which would be the last time that a production cutback started) whenever SPC=1 and OVSPC=0, i.e. whenever another cutback starts. One can then define a plotted indicator of cutbacks, PIC as

PIC.K=CLIP(1,0 DCB.K*SPC.K,CRITV)

so that as soon as the cutback has lasted more than a certain length of time PIC changes to 1 and can be plotted.

The total duration of all slowdowns during LENGTH, TDAS is simply

TDAS.K=TDAS.J+DT*SPC.J

which is incremented by DT for every DT that SPC=1.

Many control systems actually have two mutually exclusive modes of operation e.g. normal and emergency. This technique can be used to record how much time is spent in, say, the emergency mode.

Control Errors

Many control mechanisms are supposed to keep a state variable at its desired value, e.g. DINV and INV in Section 2.12. One popular measure of performance in this respect is the Integral of Squared Error ISE, defined by, for example

RCISE.KL=(DINV.K–INV.K)**2

RCISE being a dummy rate, and

ISE.K=ISE.J+DT*RCISE.JK

The reader should review Section 2.12 to see whether this would be a good criterion in the absence of a danger zone PM.

A variation of this is the Integral of Time and Absolute Error ITAE. Where the Absolute Error, ABSERR, is written:

ABSERR.K=CLIP(DINV.K–INV.K,INV.K–DINV.K,DINV.K,INV.K)

DYSMAP has no absolute value function, but this use of CLIP provides absolute errors. Then

RCITAE.KL=ABSERR.K*TIME.K 7.2,R

and

ITAE.K=ITAE.J+DT*RCITAE.JK

It this is used to measure the response to a step input at TIME=STM then TIME.K in equation 7.2 should be replaced by (TIME.K–STM).

The ITAE criterion is fairly light on the transient response to a step, but bears down heavily on final value errors One particularly useful value is the Final Value

Error, FVE. If, say, INV is supposed to be controlled to DINV then a PM of the step response is simply to observe ERROR.K=DINV.K–INV.K when TIME= LENGTH. This indicates whether, as in Section 2.12, the system ever achieves its goal.

If the FVE is non-zero, both ITAE and ISE are non-convergent and attain such very large magnitudes, even for finite LENGTH, that they are poor discriminators between control rules.

7.6 Design Objectives and Robustness

When we apply dynamic modelling to corporate systems we are trying to achieve two things: improved performance and robustness. We have already mentioned several ways of assessing performance, so it is not hard to imagine that we can describe the ways in which the system is unsatisfactory and hence define objectives which the design exercise is to achieve. For example, we could say that response of the system to a step input was insufficiently damped, and that it amplified short-period noise far too strongly. This implies objectives, and a knowledge of the loops in the system will suggest changes to the system which might satisfy them. So far, so good, but what is ROBUSTNESS about?

A system is ROBUST is it can be made to work well regardless of what happens to it from the environment or the complement, by making suitable changes to its policies and/or its structure. It also means that the system can be made to perform effectively even though some of its parameters or relationships (especially those in the environment) are not known with certainty, or are even quite different from what has been assumed in formulating the model. We shall call these two aspects of robustness 'external' and 'internal' where it is necessary to distinguish between them.

This approach to the policy-design problem is, we believe, fairly novel, It seems to possess more ramifications the more one examines it, and we shall do so in a little more detail, because it seems to take us rather beyond its nearest analogy–sensitivity testing.

7.7 External Robustness

If a model is externally robust we mean that it produces good performance whatever the external shocks which fall upon it. For example, if we decide that the system ought to have a damped oscillatory response, then a policy will be robust if it produces such a behaviour *mode* whether there is an upward step of 10 or 100 in an exogeneous variable, and still produces that mode for downward steps.

On the other hand, suppose we were studying a corporate planning situation, where the aim was to respond flexibly to changes in total market size by generating good market share performance. In this case the policies would be robust if they increased market share when total demand rose, and defended it when it fell, regardless of the detailed time path during the rise and/or fall. This is very different from a classical approach, but we believe it to be more realistic in its relation to real managerial problems, and to be more useful for managers, and more stimulating for analysts.

The classical or conventional approach would be to attach probabilities to future patterns of demand or, in effect, to make a prediction that, say, a step increase in demand is the most likely thing to happen in the future. Given this prediction, one then makes some kind of plan for the future, i.e. an action is chosen, perhaps by methods of great sophistication.

Our approach is to say that a single plan based on a prediction, even if derived from probability distributions and optimal actions, concentrates attention within one or more arrows on an influence diagram. A useful, complementary, approach is to recognize that predictions always turn out wrong, and that corrective action will need to be applied. It is, therefore, valuable to examine the kinds of system policies which will always perform well and thus to design the policy framework within which conventional analysis can be applied. For example we should like to know how sensitive the system is to forecasting errors, what alternative controls are possible, and what performance criteria are most useful for the system. Is it, for example, better for the *system* for an optimizing procedure used in, say, production planning, to minimize costs, maximize throughput, or minimize average delivery delay?

These are all aspects of external robustness, but the concept will repay a good deal of thought and careful application to the case studies in Chapters 10–12.

It is, however, absolutely essential to realise that, in talking about robust performance, we are referring to modes of behaviour rather than to specific numerical values. Thus, when we mentioned the system still producing a damped response, whether the exogeneous step was 10 units or 100, we were not implying that the numerical values achieved by some internal system variable would not change whatever the step. Quite the contrary, the system's performance could be numerically quite different in the two cases, but because the response is qualitatively unchanged, robustness will have been achieved.

7.8 Internal Robustness

This is an even more profound concept than external robustness and it is very closely related to sensitivity testing. The latter, however, merely implies that we will see how much the system's quantitative, and perhaps qualitative, behaviour changes if parameter values are altered. This arises because there will often be many parameters, or functional relationships, in a model about whose value we are rather uncertain. For many of the parameters and functions the qualitative, and perhaps the quantitative, behaviour will not be much affected by quite large variations in the parameter value. It will, therefore, not matter very much if we make mistakes in assigning values to them. However, there will be at least some parameters to which the quantitative behaviour, and perhaps the qualitative will be much more sensitive. The aim of a sensitivity analysis is to conduct a programme of simulation experiments to identify the sensitive parameters. More effort can then be devoted

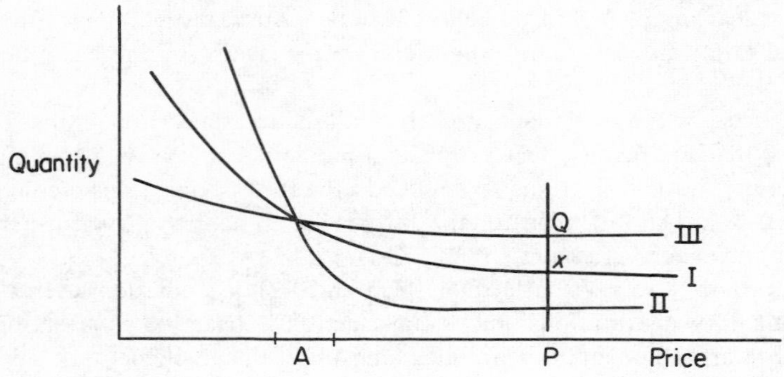

Fig. 7.3 Demand curves

to collecting data for accurate measurement of the parameter, and the model will, therefore, be numerically more 'accurate'.

Internal robustness goes beyond this. It recognizes that there will be parameters and functions which are imperfectly known. Indeed there may be functions which can never be known, no matter how much measurement is devoted to them. The obvious example is a demand curve, which might be like any of these shown in Fig. 7.3.

In Fig. 7.3, all three pass through the same point and data in region A would be hard to ascribe to curves 1 or 3, though 2 might be a little easier.

The idea of internal robustness is to say that, if a control policy depended on a parameter or relationship which was hard to estimate, then it would be essentially a poor policy. Far better to devise a means of by-passing the uncertainty and making the control depend on something which can be relied on more accurately.

The uncertain part of the system cannot be switched out or removed. If it is part of the environment, then that is that, and management cannot legislate it out of existence. However, it may be possible to circumvent it by some suitably-designed control system, so that the effects of the uncertainty on the numerical performance on the system are reduced, and on the qualitative performance are eliminated.

An example may clarify the concept. Management believe that the demand curve is 3 and they set the selling price at P, forecasting sales of Q. If the demand curve really is, or becomes, 1, then the actual demand will turn out to be X and difficulties may ensue. A robust policy would be one in which there were alternative control mechanisms which would adjust either the price set, and/or the quantity offered, drawing on other information such as orders received, inventory unsold, etc, so that the behaviour of the system, as measured by, say, stability of cash flows, was always good even if the actual numerical value was affected by the erroneous demand curve prediction.

Internal robustness is not an easy concept to grasp, and it is certainly not simple to achieve, especially where there is a large environment component in the system. This is to be expected, because systems with environments such as markets or competitors are hard to deal with in any case. It is therefore more fruitful to regard robustness as an insight and objective, rather than a technique.

Another example may illustrate the point about redesign. In a particular production system, the firm occasionally encountered large backlogs of orders but there was considerable disagreement within the firm about what the backlog meant. One group held that it was genuine business which might be lost if production was not increased but another argued that the backlog was a fake caused by customers placing orders which they had no intention of confirming, simply to get a place in the queue if delivery delay increased. Clearly one could not resolve this debate by any amount of testing of data but it had considerable implications for the way in which the system might be controlled.

Since both schools of thought agreed that backlog was undesirable, the answer was to redesign the way the production system reacted to 'hard' information on orders to make backlog occur much less often, thus rendering it irrelevant what one believed about the reality of the orders in it.

7.9 Methods of Model Analysis

The purpose of analysing the model is to find out what to alter to improve its performance. There are two main lines of approach; to take the model to pieces and study its component parts so as to get to know its internal workings intimately in the hope that this will trigger off ideas in the analyst's mind; or to apply to the model as a whole rigorous mathematical methods which 'guarantee' the finding of an effective solution. Naturally, these are *not* watertight categories, and there is rather a blurred distinction between them. However, we can divide the two approaches into four broad methodological categories:

Approach 1 a Loop analysis
b Sensitivity analysis
Approach 2 a Linearization and Eigenvalue analysis
b Optimal control

The ensuing chapters are rooted very firmly in Approach 1, but we first discuss all the methods before explaining Approach 1 in more detail.

1a *Loop Analysis*

This method works by looking at the overall structure of the loop pattern, to eliminate any obvious faults, and then analysing each loop in detail to see how it interacts with other loops. From this one hopes to suggest possible improvements to the model which have to be tested by simulation.

The first advantage of the method is that it is easy to learn the principles of it, which makes it accessible to analysts who have not had a substantial mathematical and control theory training, so it should be useable by analysts whose prior training has given them that knowledge of the actual system which is an essential for successful work.

The second advantage is that it puts a good deal of emphasis on the prospect of changing the whole structure of the system, and this is an avenue of approach which is very useful for the rather poorly controlled systems which are typical in business. In short, it permits the exercise of intuition and imagination.

The disadvantages are that, if badly done, it is time-consuming (though developments of DYSMAP should reduce this), that the use of the method, as opposed to the principles, is not easy to learn, except by practice, and that it is not mathematically rigorous, or 'optimal'.

Loop analysis has its origins in classical control theory and the advanced student would profit by a subsequent study of some of the standard control texts mentioned in Appendix C.

1b *Sensitivity Analysis*

On a simple view, sensitivity analysis involves making changes to the parameters and/or the structure of the model and seeing what effect it has on the performance. Structural changes are merely a special case of parameter changes if we imagine each link to have an extra parameter with a value of zero when it is out of the model, and 1 when it is in.

As a means of searching for a system redesign, sensitivity analysis on a trial-and-error basis is not practical. There are literally infinitely many possibilities to be simulated, so some sort of guidance has to be obtained, either by loop analysis or by mathematical treatment of a linearized model. The latter methods are advanced and are beyond the scope of this text, but are described in the thesis due to J. A. Sharp referenced in Appendix C.

An acceptable compromise is the use of sensitivity testing and loop analysis as complementary methods, and this is the approach we shall adopt.

2a *Linearization and Eigenvalue Analysis*

Standard methods of stability analysis have been evolved by control engineers which are based on the study of the eigenvalues of the state–space equation (see bibliography in Appendix C). The methods have all the advantages of mathematical rigour, but they have a number of disadvantages.

Firstly, they apply to linear systems and many, but by no means all, managed systems contain non-linearities. This may not be a serious problem, as the non-linearity may be ignored or it may be removed by linearizing the model about some chosen region of operation.

However, several drawbacks remain when this has been done.

1. The method requires considerable mathematical skill, which may not be possessed by the person who understands the system well enough to model it.
2. Psychologically, they focus attention on the system as it is and may obscure the possibility of structural changes (though that would be the fault of the analyst, not the method).

3. The system may move out of the range of validity of the linear approximation, though not, of course, if it was linear in the first place.
4. The method is based on the idea of ultrastability which, as we have seen, may not be appropriate for managed systems, especially if they contain positive feedback.

2b *Optimal Control*

The idea of optimal control is that if we wish to move a system from a state x_1 at time t, to a state x_2 at T then we can do so along an optimal path under control actions which can be calculated to give that optimality. Very many options are possible, for example we may not specify T, but simply move from x_1 to x_2 in a minimum time, or at minimum cost for a given T, or whatever. This method has obvious and powerful advantages, but it has disadvantages:

1. It requires advanced mathematical ability.
2. It is difficult to apply to the rather elaborate models which are often built of industrial situations.
3. The optimality criteria are usually rather simple, to prevent the mathematics becoming too difficult, though standard computer programs for optimization are a big help.

Summary on Methods of Analysis

It will now be clear to the reader that we shall use methods 1a and 1b, through we shall not often explicitly refer to them as separate methods, because they are hard to divide from one another. The main reason is that approaches 2a and 2b require such advanced mathematical knowledge, and ability, that to learn them will be beyond the time usually available to the management analyst, and certainly beyond the scope of this text. However, the reader who intends to do a lot of dynamic analysis would do well to obtain at least a basic familiarity with the more sophisticated methods. He should, however, always bear in mind that the advantages of mathematical rigour must not be allowed to override the real disadvantages of the assumptions which may have to be made in using the advanced methods.

7.10 Steps in Loop Analysis

In the next chapter we shall analyse a model in a fair amount of detail. In this section we shall try to summarize the essential steps in model analysis in a convenient form to which the reader can refer as he works through Chapters 8, 10, 11, and 12. It is important to understand that there is no set procedure to follow and no standard formulae to apply. At best there are hints and guidelines which leave the analyst free to use his own wit and insight, and which he should learn to ignore or break when it is convenient to do so.

There are three phases in analysis of a model of a practical system:

1. LOOP PRE-ANALYSIS. This is a search for gross faults in the model i.e. those parts of the model which are obviously out or place, or likely to cause trouble.
2. DETAILED ANALYSIS. Having eliminated, or at least noted for future attention, the gross faults in the model we analyse its loops in more detail.
3. REDESIGN. In this phase we try to bring about changes which will improve model performance.

In practice, of course, these phases, particularly the second and third, tend to run together. However, they are a useful explanatory device and the student will probably find it helpful to try to separate his work along these lines, at least until he has acquired some experience in system analysis.

7.11 Loop Pre-analysis

The first phase is to get to know the model as a structure. During the model-building work the analyst will have been trying to reproduce the real system as a simulation model. He now has to consider the model's attributes as a control mechanism, in the light of the objectives and nature of the system it reflects. First he must ensure that the work done so far justifies embarking on pre-analysis.

As an outline, we suggest the following steps:

1. Identify system purpose. The system is supposed to do something, or even several things. For example, the production system in Section 2.12 is supposed to ensure that production adjusts smoothly to changes in demand level and that inventory is kept within bounds. The system in Chapter 10 is supposed to do much the same thing. Chapter 11's system has to do about six things, which are described there, and what the system of Chapter 12 is supposed to do is the first thing the reader should work out, if he hasn't given up before then.
   Notice that we say 'supposed'. It may not do it very well, it may not be possible to do it, and there may be apparently irreconcilable conflicts in a multi-objective system. Be aware of these possibilities from the start and watch out for solutions to emerge from the model analysis.
2. Make sure the model is valid. This can (probably) be assumed for text-book problems but in practice, one has to check that the model is valid in the sense discussed earlier, and that it concludes the right parts of the system to throw light on whether the system does what it is supposed to.
3. Subdivide the model. Identify the areas of the model, using the influence diagram, which represent the controller, the environment, and the complement. The controller is the only part where changes can be made, so it is usually pointless to consider changes to be made elsewhere in the system (but see Chapter 10 for a partial counter-example to this).
   From the subdivision and a careful study of the dynamics, work out what the complement can do to the environment and how the latter transforms this in

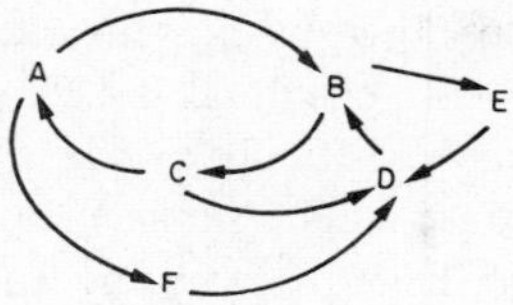

Fig. 7.4

passing on the shocks to the controller. Does the environment make matters worse? Can the environment be abolished e.g. by eliminating dealers from a distribution chain? If so, what has been identified as environment should strictly be controller but the original distinction between the two may be worth retaining as the abolition may be only partially successful e.g. dealers don't always do what they are told. Find out how the environment modifies signals from the controller in passing them back to it as signals from the environment. Try to assess, by sensitivity analysis and additional simulation runs, the internal and external robustness of the environment. Taking into account what the complement does directly to the controller, you should now have a very good picture of the kinds of shocks that the controller has to cope with.

4. Identify the Main Components of the Structure. Carefully examine the model, especially the controller, and look for the feedback loops, making a note of them for later reference. Try to see how the loops fall into groups, or subsystems. This may involve redrawing the influence diagram to make the loop patterns clearer. This can be made easier by using symbols such as I to denote a group of loops, which, in full, might look like Fig. 7.4. The direction of the arrow on the circle denoting subsystem I has no connection with the actual directions of the arrow within the subsystem. It merely serves to symbolize the presence of loops within I.
   There may not be any particularly noticable subsystems, but, if there are, the end result might look like Fig. 7.5. in which A and B are variables in the complement, I–IV are subsystems, and E and C indicate whether the subsystem is in the environment or the controller.
5. Common Sense. Before rushing ahead, switch off the computer (metaphorically!) and stop and think.
   This rather trite advice is worth following because most control systems encountered in practice are rather poor. They have evolved, rather than been

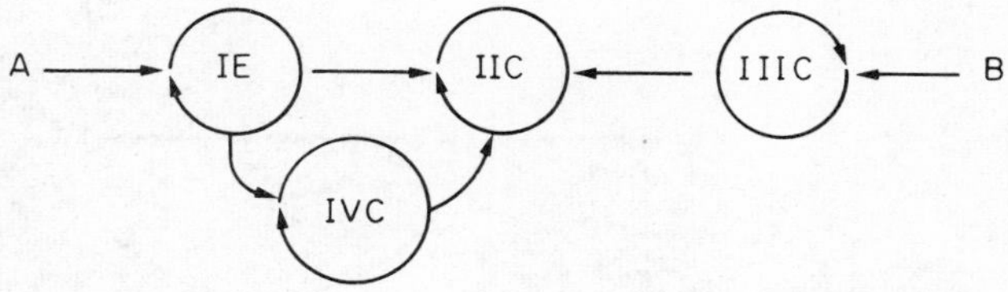

Fig. 7.5

designed. This is not a bad thing, by any means, because at least they work and their parameter values usually ensure at least a degree of stability. (The reader will find that it is remarkably difficult to select parameters for an hypothetical system for which the response will be stable.) However, practical systems tend to be very sluggish, to do very little to damp shocks and usually they have glaringly obvious deficiencies in the control structure. The quality of the analysis will be greatly improved if the serious faults can be picked out, and, generally, the system understood before a mass of computer printout obscures the issue. Look, therefore, for the following:

a. Unconformable loops. These are loops whose mode of operation is not consistent with the purpose of the system. For example, a positive loop in a system which is basically negative *in purpose*, is out of place and will almost certainly produce unsatisfactory dynamics. Notice that unconformability is related to the purpose of the system. A positive loop in a collection of negative ones is not unconformable simply because its sign does not match those of its neighbours, but because its presence is not compatible with the essential purpose of the system.

b. Poorly-connected subsystems. The whole point of the loop analysis method is that a system must possess feedback to achieve tight, effective, control. If it does not exist, then we have to introduce it. For example, a system which broke into subsystems as shown in Fig. 7.6 would have poor connection between the subsystems and would probably be hard to control effectively. In a case like this, there may be links from, say, F to C, which appear to connect subsystems I and II. These have to be checked to make sure that they really are control links and not simply constraints. See Chapter 11 for a case of this.

It is fairly evident that there is a good chance that subsystems I and III could amplify or, at any rate, fail to damp, shocks from variables A and L in

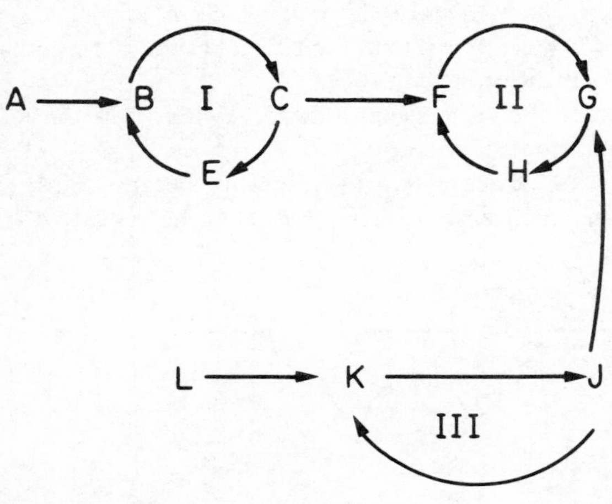

Fig. 7.6

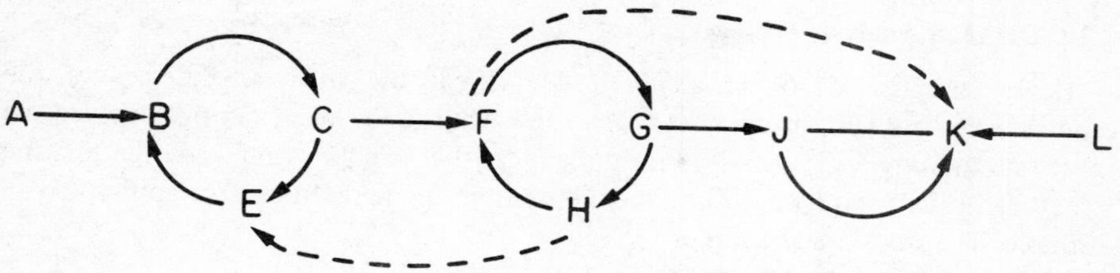

Fig. 7.7

transmitting them to subsystem II. This could be confirmed by careful study of the dynamics, using this simple analysis of the system's structure to guide one in what to look for in the dynamics.

There might be any number of ways of reconnecting the system to create additional loops, perhaps removing some existing links. The main thing is that looking at the loops suggests a train of thought which might, depending on what the variables are, lead to a structure such as that shown in Fig. 7.7, where new links are shown by dashed lines. Notice that the labels of the subsystems have vanished. The new loops connect them together, so that their separate identification would be rather artificial.

c. Amplifying input structures. In Section 2.12 we saw a case of an input, CONS, acting directly on a loop and also indirectly, with high gain and delay, through AVCON and DINV. This is a very common structure, which appears in Chapters 10 and 11. Usually it is a destabilizing influence.

d. Dominance shifting mechanisms. These are always non-linearities in the system which cause its behaviour mode to be dominated first by one loop, or group of loops, and then by another. The typical examples are stair-step growth and peaked oscillation. When the behaviour changes from one mode to the other there will be a variable somewhere in the system which has changed its position on a non-linear curve, often from one side to the other of a peak or elbow in the curve. This causes dominance to change, and keeping the preferred dynamic mode in the dominant role involves preventing these non-linear shifts, either by changing the non-linear curve, if it is a policy in the controller, or by not allowing the variable to move by introducing some other control action.

   This illustrates an aspect of external robustness. If the dominance-shifting variable is knocked, by an outside shock, out of the range in which it produces the preferred dominant mode, then the system will be made more robust if control policies can be devised which always restore the variable as rapidly as possible to its desired range.

6. Eliminate the Gross Faults. Having looked for any of the faults in the last section, eliminate them. For example change an unconformable positive loop to negative by suitable changes to the system.

7.12 Detailed Analysis

Having eliminated, or at any rate identified, the main weaknesses of the modelled control system we now embark on what is the most taxing, but also the most stimulating part of system dynamics–the analysis and redesign of the structure. At this stage we outline the method, in the next chapter we apply it to an example. The steps are as follows:

1. Carefully think over the purpose of the system and check that you fully understand what it is supposed to do. Review this with the managers involved in the real system. Arrive at a set of measures of system performance which translate the managerial realities of the system into purely technical aspects of dynamic behaviour. In particular, check that you have avoided the trap of designing stability into the system simply for stability's sake.
2. Make sure that the pre-analysis phase has been thoroughly carried out. There is no point in labouring away at a system which is basically poor.
3. Carefully identify all the loops in the model picking out their:

   Polarity
   Gain elements
   Delay elements
   Non-linearities
   Pure integrations
   Input variables and structures

   Draw a separate influence diagram for each loop, and summarize the gains and delays into a loop summary table to show elements which are common to more than one loop.
4. Estimate the magnitude of gain, delay, and delay order, for each loop.
5. Now do the really hard part and try to write down, in one sentence, what the loop is supposed to achieve in the system. This is rather more difficult than it appears and it is essential not to take the easy answer as gospel. Some hard thinking may be needed, but this is the key to the approach. Clearly, if we find a loop which proves, on later study of the simulation output, not to be achieving its aim, then we have gone a long way to finding the area of the model where improvements in system performance are most likely to be found.
6. Sometimes it helps to arrange the equations in the order in which they go round the loop. The DYSMAP documentor is useful in this and later versions of DYSMAP will do it automatically.
7. Arrange the PRINT and PLOT cards so that they show at a glance how the dynamics propagate round a loop. For example, if there is a structure such as that in Fig. 7.8, plot A, B, C and E on a common graph of their own and plot X, G, C, E and F on another (or the same one to save sifting a mass of output, but bear in mind the need for at-a-glance comparison). This doesn't mean that every variable in every loop is plotted, in some cases many times, because that would produce an overwhelming mass of output. What is aimed at are the salient variables, but these are not known to start with. How then, shall we proceed?

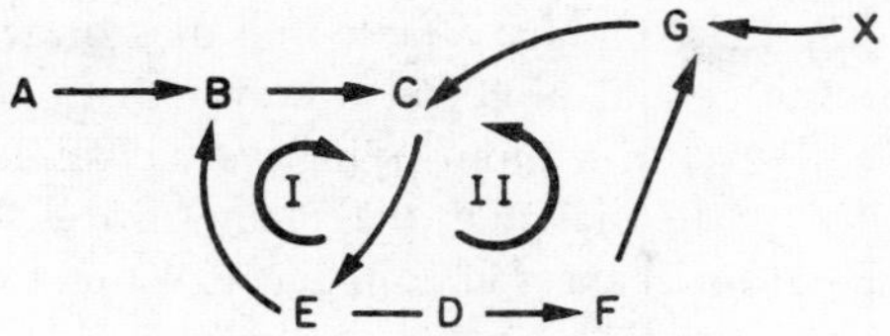

Fig. 7.8

Pick out a few variables which seem to be important; they may lie at loop junctions, be on the receiving end of an input structure, form part of a loop which seems to connect two or more subsystems, or are important components of the system performance measures.

Study of these variables, as in the following steps, will suggest others to be included, and some to be dropped. An iterative approach is called for but DON'T PRINT AND PLOT EVERYTHING.

8. Now do the first run on the model (apart, of course, from the debugging runs and any runs done during the pre-analysis.

   Use a simple input, such as a step, to get the essential feel for the system which is what is needed at this stage.

   From this run you should be able to see how well the system performs, and why. The first is easy: simply the dynamics and final values of the performance measures. The second is more tedious, and depends on the loop plots.

   Going back to the example in Step 7, one might trace how a step in A triggers off changes in B which then propagate round both loops as time passes. Thus, the effect of B and C is eventually transmitted both to B again, and to G, which then exert a joint effect on C. Similarly with X.

   Loop 2 has a delay at D which will cause phase shift and attenuation. However, G must be a level (why?) and this pure integration will cause further phase shift and an integration gain on the net effect of F and X. G will probably therefore re-inforce the dynamics of C to produce a destabilizing effect. All this could be confirmed by studying the dynamic output and this might suggest that the link from G to C might have its gain reduced or delay increased to reduce damping in loop 2.

   All this is speculation, deliberately so because we wish to show the thought process: speculate about how the system behaves; check that it does so by examining the simulation; suggest an area for redesign attention.

This brings us to the point that the three phases mentioned at the end of Section 7.10 are not distinctly separable and, in particular, the second and third phases alternate back and forth.

7.13 Redesign

By now the analyst will have a fairly good idea of how good the system is at doing what it is supposed to. Looking at the loop plots will have shown some of the reasons for this and the loop summary table indicates those factors which affect more than one loop. The central problem, however, is to know what changes to make so as to improve behaviour, without trying every possible change.

The approach is to try to pick out the really critical loop or loops and to attack them, watching out for changes in dominance. Watch for:

1. Variables which are not being controlled e.g. DINV and INV in Section 2.12.
2. Long delays with large amounts in their delay levels e.g. production pipelines or capacity expansion. See if the explicit use of the work in progress in the input to the delay would improve control. This really seems very obvious, except that in practice it often is not.
3. Carefully trace the dynamics of sequences of variables, watching for phase shift and gain and try to find the delays, gain operators or pure integrations which produce them. Increasing delay, reducing gain and adding or removing pure integration will all alter the settling time and phase shift, but decide whether the aim is to increase or reduce the delay in the dominant loop.
4. Remember that effective control often needs extra feedback, so watch for chances to bring in new loops, perhaps allied to the abolition of old loops. Simply removing loops, without replacing them by more effective ones, will almost certainly make matters worse unless the loop to be removed is really badly behaved.
5. Remember that what you are doing is designing a control system rather than, say, a production or financial system even though the control is to be applied in those areas. Do not, therefore, be too obsessed with what is 'always' done in particular types of system, but, on the other hand, be prepared to use your knowledge of that *type* of system to see opportunities for creating new loops, as dictated by the results of the analysis of the controllability of the collection of loops.

The best way of learning all this is by practice and the reader should now work through the loop analysis in Chapter 8, running the program in Appendix B.8, and rereading this chapter as needed.

Chapter 8

Control of a Simple System

8.1 Introduction

In this chapter we shall analyse the structure of a multiloop system in order to design more satisfactory controls for it. We use a simple production and stock system because its components are easily visualized, but the behaviour of this system arises from its *structure* and not from its *type*, i.e. it behaves as it does because of the way its loops are constructed, not because they happen to involve the variables of a production system.

Although dynamics depend on structure, not type, the practicality of changing the structure to improve behaviour depends on the type. The analyst therefore has to master the mental trick of switching between thinking in terms of structure and in terms of type as he proceeds through an analysis.

For the sake of reasonable brevity, not all the steps or graphs are given in detail, and the idea is to raise questions for the student rather than to give all the answers. The student should not take any of the steps on trust, but *must* check each one himself, using the program in Appendix B8.

The actual method of tackling this problem is very simple and involves nothing more than a close examination of what is going on. In this case, the solution to the control problem is rather obvious from the start and we may seem to be making a little ado about not very much. This is because if the approach yields a solution which is obviously correct in a simple case, the user will have that much more confidence in applying it to less transparent problems. In addition, the reader will find the pattern of thought involved easier to learn in a fairly straightforward situation.

The LOOP ANALYSIS method is very close to engineering design in its train of thought and the reader should find that it leads him to think creatively about how the system *might* be operated rather than concentrating on 'optimizing' the way it *is* run.

8.2 The System

(This model has been simplified from components of several real industrial systems, but it is still very close to real life systems, and the parameter values have been chosen to reflect reality. We will show that they are hardly very sensible from a system point of view, but that is also true of real situations. The reader has made progress if he learns to look at the parameters *and* the system, rather than at parameters alone.)

A company has two departments. Distribution hold a stock with which to meet sales, and replenish the stock by placing orders on Manufacturing. Manufacturing adjust their production against the backlog of unfilled orders, delivering the finished goods to Distribution's stock, after a delay. For simplicity in this example, distribution are allowed to have negative stocks and Manufacturing regard the backlog as depleted when the work has started. This is a Manufacturing Man's point of view, but Distribution might argue that the backlog is only depleted when delivery takes place. That would lead to a completely different system and the reader should repeat the analysis of this chapter for that situation. We shall assume that Manufacturing have it their way.

Fig. 8.1 is an influence diagram for the system. full details appear in Appendix B8 but there are two key areas of the model: Factory Order Rate and Indicated Production.

The Factory Order Rate, FOR, is the rate at which Distribution orders goods from Manufacturing. The simplest rule is

$$FOR=ASR \tag{8.1}$$

i.e. a simple replacement policy. If this is used, the links in Fig. 8.1 from DINV and INV to FOR do not exist.

Show that, with this policy, there is no control because there is no overall feedback.

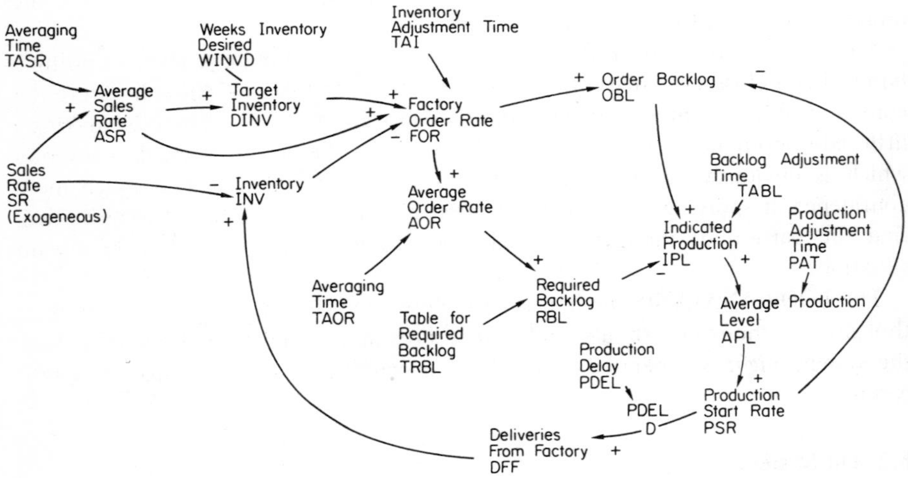

Fig. 8.1 Influence diagram of the basic system

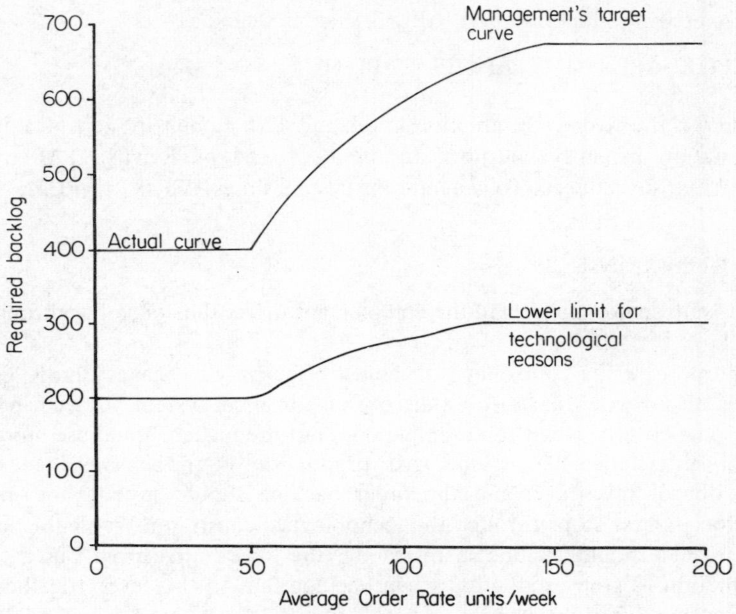

Fig. 8.2 Required backlog policy

A more complicated rule is

$$FOR=ASR+\frac{DINV-INV}{TAI} \qquad 8.2$$

and there are many other options. In each case the system structure may be changed, and the reader should check this. For the present, we assume equation 8.2 to apply and the structure to be as in Fig. 8.1.

Indicated Production, IPL, is affected by the level of backlog desired, RBL,. This is a non-linear function of Average Order Rate, AOR, based on the idea that

$$RBL = f(AOR)$$

as shown in Fig. 8.2.

The lower line on Fig. 8.2 models the need to keep the manufacturing process 'full' because the process technology is very costly to stop if the pipeline is depleted too much (e.g. an oil refinery). The upper curve represents management's wish to keep the production system loaded up for economic and employment stability reasons. The lower limit cannot be changed, except at unacceptable cost, the upper one could.

The Indicated Production Level, IPL, depends on

$$IPL=\frac{OBL-RBL}{TABL} \qquad 8.3$$

The Average Production Level, APL, is then modelled as

APL.K=APL.J+(DT/TAP)(IPL.J.APL.J) 8.4.

where APL is the Average Production Level, and TAP models the technical inability of production instantly to move to the new level of activity. TAP could be increased but not reduced. To generate the system flows, PSR.KL=APL.K

8.3 Model Pre-analysis

There is no environment, and the complement drives the controller through Sales Rate, SR.

The purpose of the controller is to adjust effectively to changes in SR. Precisely what this means would have to be discovered for a real system. All we can do is to assume some criteria, albeit some simple ones but the reader should use more subtle performance criteria. We assume that production is to be smoothed without running out of inventory, and the order backlog should not decline until the production process stops through the technological constraint. We shall ignore this, but the reader should include in his model the lower curve from Fig. 8.2 with a CLIP function to stop production when backlog falls to this level, together with a measure of how often this happens during a run.

Loops

The system's loops are shown in Figs. 8.3, 8.4, 8.5 and 8.6.

Loop A (Fig. 8.3) regulates production to SR. It is negative and contains two pure integrations, at INV and OBL. There is a third-order delay between PSR and DFF, of magnitude six weeks, there are no non-linearities, and there are five inputs

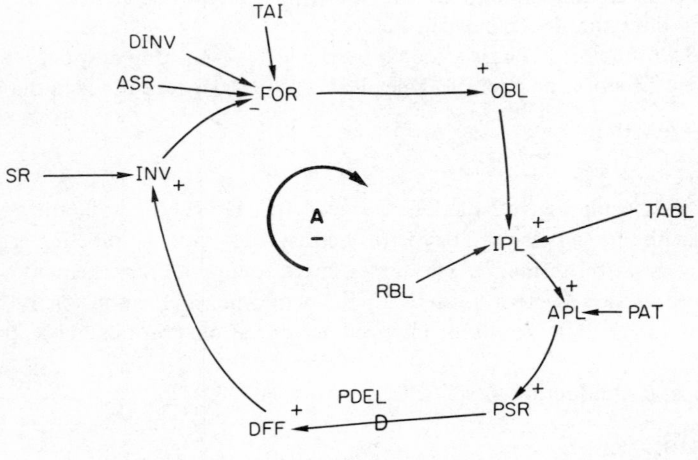

Fig. 8.3 Loop A

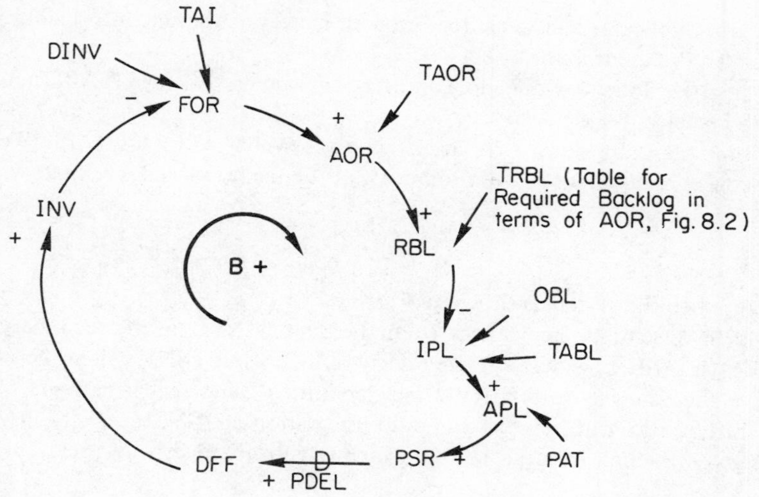

Fig. 8.4 Loop B

SR, ASR, DINV, RBL, and APL and three parameters, TAI, PAT and TABL but, *from the point of view of this loop*, are these gains or delays?

Clearly 1/TAI is a component gain between INV and FOR but because of the set-point, DINV, and the loop to INV, TAI acts as a first-order delay and so do TABL and PAT. The total delay in the loop is therefore PDEL+TAI+TABL+PAT, or 17 weeks, with the initial parameter values.

Loop B (Fig. 8.4) regulates production to control inventory and backlog. It is POSITIVE and is unconformable in a system which is, as a whole, supposed to be goal-seeking. It is, of course, pseudo-positive, because it is paralleled by negative loop A, but it may be capable of reducing the damping of the system. There is one pure integration at INV, four first-order delays with respect to inputs at DINV, AOR, IPL and PSR, and one non-linear gain operator TRBL.

Thus, loop B has a variable gain and a total delay of TAI+TAOR+TABL+PAT+PDEL=21 weeks. Since this delay is longer than loop A's, but B's gain is much larger, we should expect B to contribute to any instability in the system.

The purpose of loop C (Fig. 8.5) is to control backlog by regulating production.

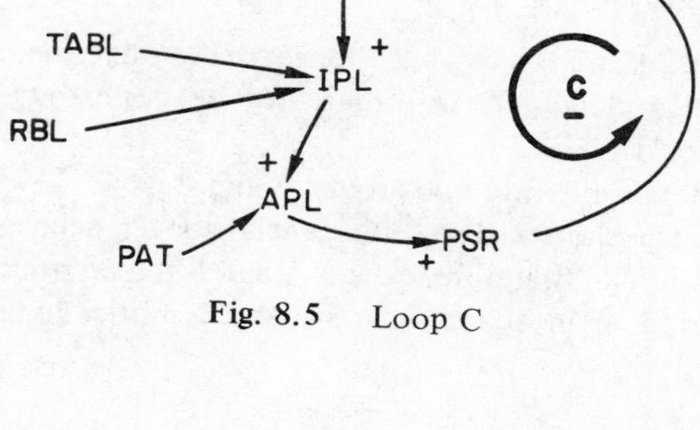

Fig. 8.5 Loop C

It therefore duplicates and reinforces loop B. However, the unconformability of B may weaken the re-inforcing effect.

This negative loop has one pure integration, and two delays, TABL and PAT, totalling 7 weeks.

Not that there appears to be another loop in which APL→IPL→APL. We treat this as a delay component in what follows, but the reader should work out why.

Inputs

The input to the system is shown in Fig. 8.6.

The main instability here arises from the gain WINVD in the secondary input from SR, via ASR, together with the delay TASR. Thus DINV will lag behind SR, will have the gain element WINVD injected into it, and will be affected by the phase shift and gain inherent in the pure integration at INV, especially in view of the – sign on the link from SR to INV which effectively inverts Fig. 6.1.

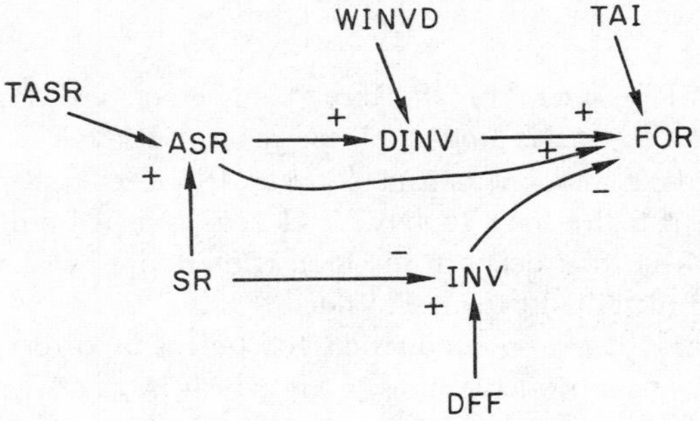

Fig. 8.6 Input

Having located the loops, it is useful to summarize the delays, gains and integrations to show commonalities. Table 8.1 shows the result, laid out for convenience rather than the order in which parameters occur in the loops.

The loop summary table conveys some valuable information about the system, but conclusions drawn from it should be verified by simulation. We shall, however, endeavour to show how the loop analysis guides the simulation as opposed simply to simulating anything that can be thought of.

The salient features are:

1. Two of the loops, A and B have much longer delays than the others, and A and B have high total orders and will therefore be comparatively underdamped. (Review Chapter 6!)
2. The loops fall noticeably into two groups A and B which have long delays and C, which has a much shorter delay. In general longer delays tend to dominate systems and we should, therefore, expect A and B to be critical loops.
3. The natural period of most managed systems is rather long – longer than the

Table 8.1. *Loop Summary Table for Fig. 8.1*

Loop	Type	Delays					Total Delay (Wks)	Total Delay Order	No. of Pure Integs.	Gains
A	–			TABL	PAT	TAI PDEL	17	6	2	
B	+		TAOR	TABL	PAT	TAI PDEL	21	7	1	
C	–			TABL	PAT		7	2	1	TRBL
Input		TASR					4	1		WINVD

*Note: These values are based on the parameters used in the original model and are therefore liable to be changed. Except where stated all the loop summary tables in this chapter are based on the parameter values of the initial model, as listed in Appendix B.8.

longest total delay in the system so the ratio Period: Delay, used in Fig. 6.9, will be fairly large and there will not be much in the way of phase shifts due to these delays – perhaps about ¼ period due to the third-order delay used to model the production process.

4. Loop B is unconformable, as already mentioned.
5. Loop A contains Two pure integrations and, as seen in Fig. 6.15, each of these is capable of causing a phase shift of ½ period. Their effect will be moderated by the other loops, because loop A is NOT an independent entity, so these calculations do no more than add to our 'feel' for the system, but there is an indication of an appreciable phase shift in loop A, which will make deliveries from the factory lag considerably behind Factory Order Rate. This is bound to make the system unstable.

We therefore anticipate, and it is no more than that, that loop B will be a cause of trouble because of its unconformability, and that loop A will also be critical because of its phase shift effects.

This simple analysis suggests that the system is likely to be unstable and a basic run on the model, using the inventory correction rule of equation 8.2 shows that the system explodes (Fig. 8.7). A substantial 40% shock has been used to generate dramatic dynamics, but smaller steps will show the same behaviour mode of explosive oscillation. The period of the oscillation is about 40 weeks, or twice the longest loop delay. The behaviour is seen (more sharply) by examining the tabular output from the run. For example, at week 20 the Average Sales Rate is 136.85,

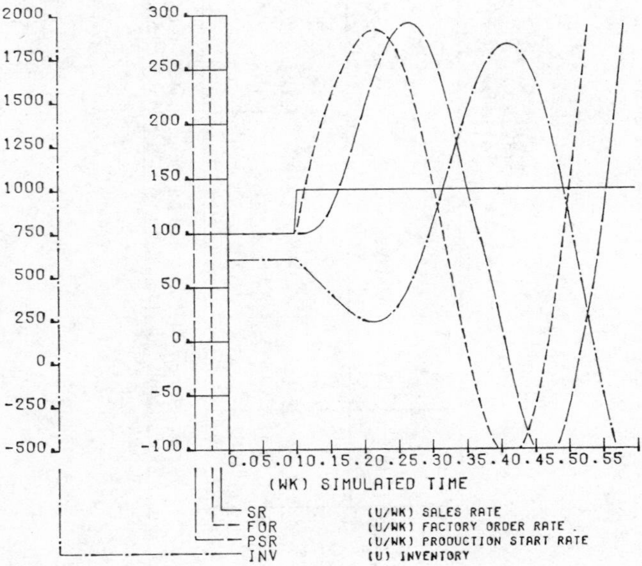

Fig. 8.7 Step response of initial system

Desired Inventory is 821.07 and Inventory is 245.13. Clearly, the inventory-correction term will contribute more than the Average Sales Rate to the resultant Factory Order Rate of 280.83 and, in this way, the instability is generated.

The Tabular Output shows negative order backlog, and even negative work in process (the variable PLA in Appendix B8). We can either complicate the model to prevent this, or ignore it as being merely additional symptoms of what is, in any case, a completely unsatisfactory control system. We shall do the latter as we shall show that the system can be redesigned so as to prevent these happenings. The reader should, however, alter the model, following the suggestions in B8, to see how much difference it makes. Think out the implications of complicating a model to deal with phenomena which are capable of being eliminated by other approaches.

We now see that it is loops A and B which transmit this amplified input through the system and back to Inventory and their instability can be reduced by increasing their delays, or reducing the gain in loop B. A simple approach, such as increasing TAOR, or TABL, or reducing the gain to the lower curve in Fig. 8.2 does not do much, because it only affects loop B and the parallel path through A is not changed. (The student should confirm this by simulation, carefully following the behaviour of more variables than were plotted in Fig. 8.7.)

Notice that increasing TAOR does not affect the natural period but increasing TABL, which affects both loops A and B, noticeably increases the natural period. Why?

By applying the same idea we can see that increasing, say, TAI would increase the delay in both A and B and, by reducing the contribution of the inventory correction terms to FOR, reduce the input gain. Raising TAI to 12 produces behaviour which is slightly damped, but which is also extremely sluggish.

Clearly, we are not going to get very far simply by playing with the parameters, and a more radical attack is called for. Before doing that, we should comment on the connection between Fig. 8.7 and a real system.

It is tempting to assume that no real system could conceivably operate like Fig. 8.7 and this is true in the sense that no real system can explode. However, despite its minor approximations, such as permit negative work in progress, Fig. 8.1 is very like real systems and the parameter values – most of which are monthly averages or adjustment periods – are akin to real life. We must therefore conclude that Fig. 8.7 is simply the diagramatic representation of the panic and collapse situation which is actually found in many companies and is attributed by Manufacturing to market movement and by Distribution to production delay. We can see that it is neither of these, but arises from control interactions.

8.4 Eliminating Gross System Faults

It seems possible that the positive loop B is harmful to the system. We now need to look at the system as a production system to find a way of correcting this fault, which we detected by looking at it as a fairly abstract structure.

The positive loop could be completely removed by making Required Backlog, RBL, constant at a suitable value (provided RBL>300) but if, for example, RBL is

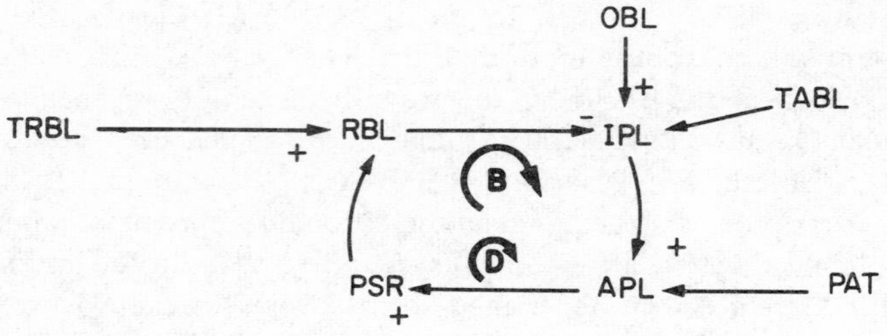

Fig. 8.8 Loops B and D

set constant at its maximum value of 675 the system still explodes, though its period is very slightly reduced. This is because in Fig. 8.7, RBL swings between its upper and lower bounds which means there are periods of lower gain in B, so the system explodes slightly more slowly than when RBL is high and constant and B does not exist to exert any control on the oscillations produced by loop A. Removing control loops without adding better ones is not likely to get us very far.

An alternative approach, which makes sense from a production continuity viewpoint, is to relate RBL to APL (Fig. 8.8). This converts the positive loop into a negative one, thereby eliminating the unconformability and thus also making sense from a system structure point of view. This leads to a new influence diagram and loop summary table, Fig. 8.9 and Table 8.2.

At this point it would be necessary, in a practical case, to check that the curves in Fig. 8.2 which give Required Backlog in terms of Average Order Rate, would still be valid when RBL is to depend on APL, and we shall assume that they do.

The first thing to ensure is that removing the unconformable loop has indeed improved system performance. This is shown in Fig. 8.10 which has the same parameter values as Fig. 8.7.

The improvement in the system's behaviour is not dramatic—there is only a slight reduction in the tendency to explode. This must be due to the parallel path through loop A which must obviously be the next focus of our attentions. However, changing loop B to a negative one has made an improvement and we keep it as such in what follows.

To attack loop A we consult Table 8.2 and observe that TAI is a delay which operates only on that loop and we know that it acts on the inventory discrepancy term in equation 8.2. At this point we can exploit the fact that, as discussed in Chapter 6, gain and delay are not independent and although TAI is a delay in loop A, its reciprocal acts as a gain on the term (DINV–INV) in equation 8.2. When TAI is fairly small, this gain is comparatively large and that, coupled with the long delay elsewhere in loop A, and the phase shifts due to integration, makes it very hard for INV to settle to the eventually constant value of DINV. Increasing TAI to, say, 12, from its original value of 4 should improve matters as shown in Fig. 8.11. The system now has a response which is slightly more damped than that of the original system with the positive loop B, when TAI=12. The negative production backlog and work-in progress referred to in Section 8.3 have all been eliminated.

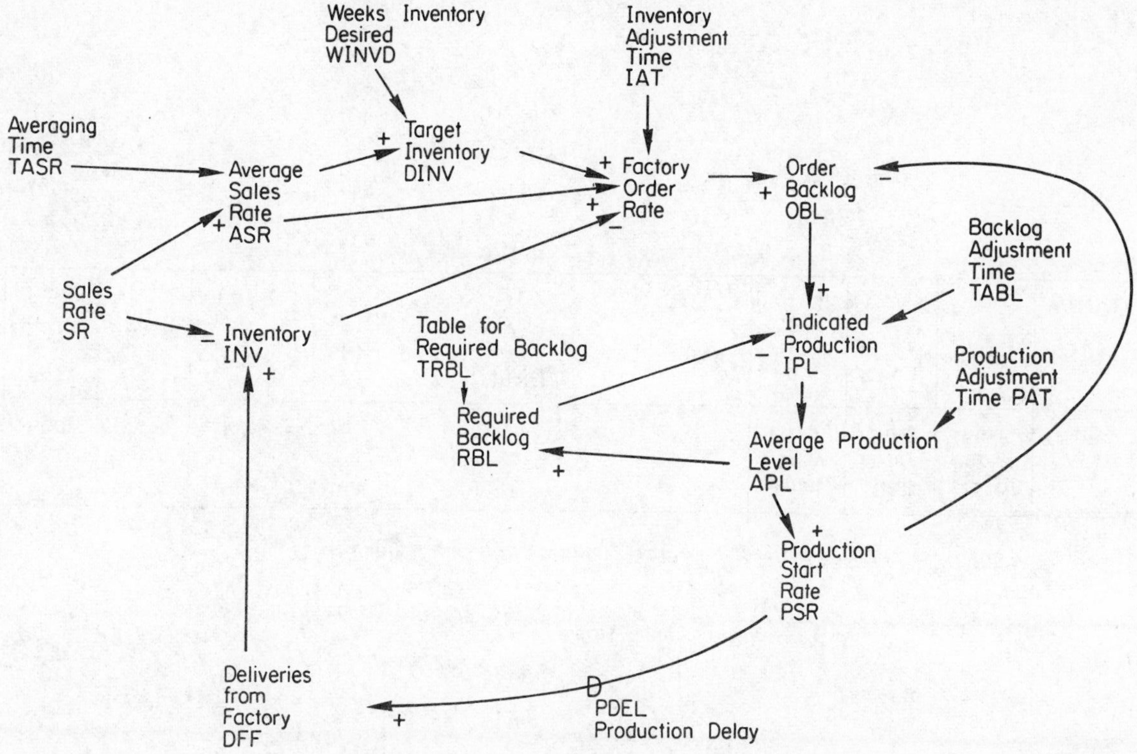

Fig. 8.9 Influence diagram of revised system

Table 8.2. *Loop Summary Table for Revised System*

Loop	Type	Delays	Total Delay (Wks)	Total Delay Order	No. of Pure Integ.	Gains
A	–	TABL PAT PDEL TAI	17	6	2	
B	–	TABL PAT TAP	11	3	0	TRBL
C	–	TABL PAT	7	2	1	
Input		TASR	4	1		WINVD

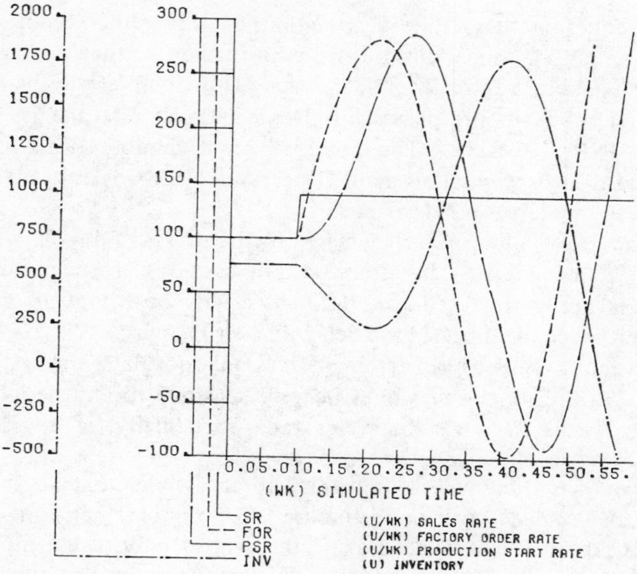

Fig. 8.10 Step response for negative system

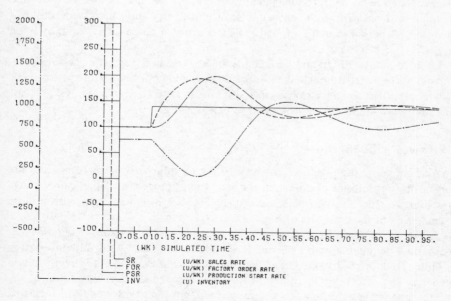

Fig. 8.11 Negative system – longer Loop A delay

However, Production Start Rate lags behind Factory Order Rate, and has a slightly higher amplitude, and Deliveries lag even further, so that the lag between FOR and DFF is about 12 weeks or 1/5 period. This cannot easily be related to Fig. 6.15 as that refers to the phase shift between a net rate and its level, e.g. between (FOR–PSR) and BLOG. The reader will see from more detailed plots that BLOG peaks and troughs where PSR and FOR cross i.e. BLOG is at its maximum or minimum when the net rate is zero (obviously).

Increasing the delay in loop A has increased the natural period to 57 weeks from approximately 38 in Fig. 8.10. This is not, strictly speaking, a true natural period, as the system is not perfectly linear, but it is clearly near enough. Both these periods are about twice the longest loop delay of their systems.

Although an improvement has been made it is not enough, as the system has an extremely long settling time and would be very sensitive to any periodicities of around 50–70 weeks in a variable sales rate, particularly of course annual seasonality.

A little experimentation will demonstrate to the reader that no further real improvement can be made with this structure. For example, increasing TABL, as well as TAI, makes matters worse, leaving the system only just damped. Other changes, such as increasing PAT or reducing TABL make the system explode. Work out why.

In Fig. 8.11 the phase shift between FOR and DFF is 12 weeks but with TAI and TABL both 12 it is 19 weeks. This is produced by the loop delays, and increasing the delays increases the phase shift, as discussed in Chapter 6. The phase shift is important because the longer it is the longer DFF will continue to rise after FOR has started to fall, and the larger, therefore, the fall in FOR and the lower the damping in the system. We might try to reduce this phase shift, increase it until it is so close to 1 period so that DFF and FOR are nearly *in* phase, or bypass it by some means.

Loop A connects FOR and DFF and is the key one in the system's performance. This points out that although unconformability is an important feature which sheds light on the nature and purpose of a system, its detection and removal will not necessarily give dramatic improvements in performance.

8.5 Changing the Structure

We now know the direction in which to try to improve the structure. There are very many possibilities of doing so, and we need to proceed systematically to have any hope of success. We therefore inspect Fig. 8.9 in conjunction with a close examination of the dynamics of Fig. 8.11.

In Fig. 8.11, there is an appreciable phase difference of 5 weeks or about .10 of the natural period, between FOR and PSR which is affected by the pure integration at Order Backlog. Bypassing Order Backlog by linking FOR to IPL directly would moderate the effects of this phase difference, and reduce the sluggishness of the system, but we must do it with a negative loop, not the positive one which existed in Fig. 8.1.

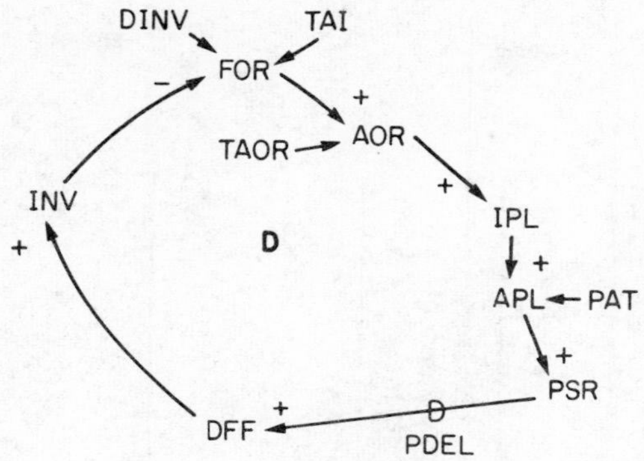

Fig. 8.12 Loop E

One way of doing this would be to put

$$IPL=\frac{OBL-RBL}{TABL}+AOR \qquad 8.5$$

and this, of course, will keep OBL near to RBL. We have thus introduced a new loop, D, which is shown in Fig. 8.12 (Note that TAPL does not affect this loop because AOR affects IPS *directly* and not, as RBL did, by delayed adjustment to a target value).

The loop is negative, is intended to keep IPL close to AOR, and has total delay of TAI+TAOR+PAT+PDEL. The resultant loop summary is Table 8.3 and the dynamics are in Fig. 8.13 with TAI=12. The phase shift between FOR and PSR is now about 6 weeks or .117 of the period–an increase to 117% of its former value in terms of periods but nearly the same absolute value.

The improvement in backlog control is considerable (not shown in Fig. 8.13). For example at week 80 in Fig. 8.11, Required Backlog is 670 and OBL is 1259 whereas in Fig. 8.13 the equivalent values are 665 and 647. This illustrates the need to watch out for variables, in this case OBL, which are supposed to be controlled, finding out whether they are, deciding if it matters, and doing something about it if necessary.

The degree of damping in Fig. 8.13 is virtually the same as in Fig. 8.11 but the natural period has been reduced. In this type of system, which is often exposed to annual cycles, this might be a disadvantage – another instance of the need to synthesize one's thinking about structure and type.

The system is still, however, far from being satisfactory as inventory is not being controlled, damping is poor, natural period may well be awkward for this type of system, and the settling time is still in the order of 2 years. The natural period is about twice the longest delay.

As before, increasing TABL, and TAOR to 12 *reduces* the damping because with other loop delays increased, A loses the dominance it has when TAI=12 and they are 4.

Table 8.3. *Loop Summary Table for Second Revision of Model*

Loop	Type	Delay						Total* Delay (Wks)	Total Delay Order	No. of Pure Integs.	Gain
A	–		TABL	PAT	PDEL	TAI		25	6	2	
B	–		TABL	PAT			TAP	11	3	0	TRBL
C	–		TABL	PAT				7	2	1	
D	–		TAOR	PAT	PDEL			25	6	1	
Input		TASR						4	1		WINVD

*Note: Based on TAI=12

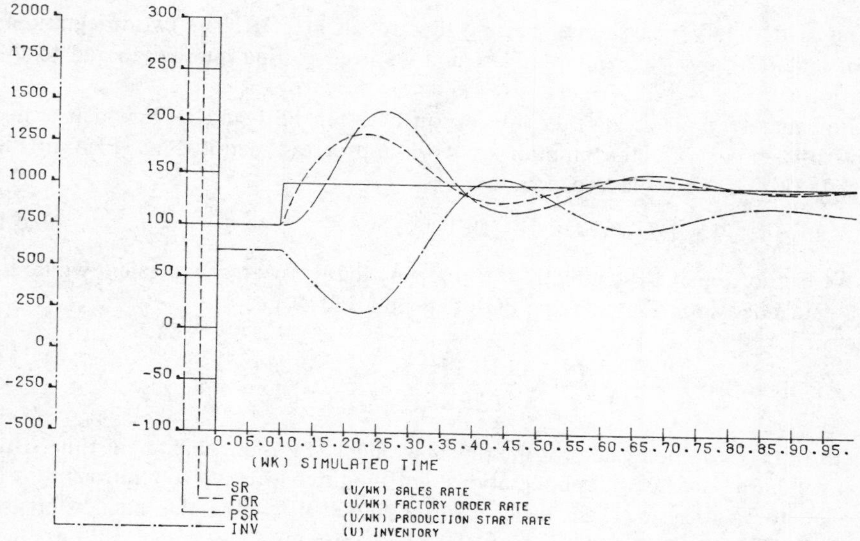

Fig. 8.13 Negative system with new Loop E

We, therefore, return to Fig. 8.9, mentally putting in the new loop D, and recall that the critical loop is still A. If we can bypass the pure integration at INV, with its associated phase shift, and reduce the effect of PDEL, even though we cannot change its magnitude without changing the production process, we may be able to do something.

8.6 Introducing More Loops

So far we have removed the unconformability of the original loop B and replaced it by a negative loop B, and the new loop D, which has improved matters, but by no means sufficiently. We must therefore introduce more control variety into the system, and the key to this is the double pure integration in loop A. This produces a phase shift of ½ period and we must either remove an integration from A, or add one, or create a loop with three pure integrations (which would have a phase shift of ¾ period) which will give better control. This conclusion stems from the system *structure*, as abstractly protrayed in Table 8.3, but the practicality of making the change will depend on the concrete *type* of the system, as in Fig. 8.10 with loop E inserted.

Phase shift due to integration is not easy to handle. For example, the shift due to integration at OBL operates on the difference (FOR–PSR), but these two belong to different loops so that the consequences are not easy to see except by plotting several variables from the loops and carefully tracing the dynamics. Thus we can see that from Fig. 8.14 PSR continues to rise after FOR has peaked, i.e. OBL turns down where the two cross (they are on the same scale). DFF rises after PSR and the effect is that INV peaks, thus causing the trough in FOR. The two integrations, OBL and INV are 180° out of phase and we need to find another to add to the control.

Fortunately there is another integration between PSR and DFF which is not explicitly shown, i.e. the amount of work-in-progress, denoted by PLA in the model. This will have the level equation

$$PLA.K=PLA.J+DT*(FOR.JK-DFF.JK) \qquad 8.6,L$$

We cannot inject PLA directly into loop A, and the nearest equivalent would be to create a new loop E by making FOR depend on PLA, i.e.

$$FOR=ASR+\frac{DINV-INV}{TAI}+\frac{PLD-PLA}{TAPL} \qquad 8.7$$

where PLD is a desired value for the pipeline, and TAPL is an adjustment time. This loop will have only two pure integrations, not the three we seem to require, but it is parallel to A, has a total delay shorter by PDEL plus the net difference (TAI–TAPL) and, since PDEL is modelled as third-order, its order will be 3 less than A's.

We shall need a means of determining PLD and there will be additional loops. There are many ways of fixing PLD, e.g. a fixed number of weeks of Average Sales Rate, ASR, such as

$$PLD=WPLD\times ASR \qquad 8.8$$

and the obvious value of WPLD is DDEL.

An alternative would be to use some measure of the Delivery Delay experienced by Distribution. A version of this is built into the program in Appendix B.8 for the student to try after detecting the loop structure which results. We shall restrict ourselves to equation 8.8 and the resulting system is shown in Fig. 8.14. and Table 8.4, the new loops being shown in Fig. 8.15.

Notice that loops G and H are both pseudo-positive loops. This appears to take us back to square one, but pseudo-positive loops have less effect than might have been expected–hence their name.

The resulting dynamics of this system, with TAI=12, are shown in Fig. 8.16.

The behaviour in Fig. 8.16 is now heavily damped, the natural period is reduced and so is the settling time. The improvement has come about because we have matched loop A (a slow loop with two integrations) with loop E, which also has two PIs and therefore the same phase shift as A, but is much faster. In commonsense terms we have more control because we are taking account of the large amount of goods inside PDEL, as well as INV and OBL.

This is really fairly obvious as it simply amounts to Distribution taking account of goods ordered but not received, which they would do in any case. We are not, however, writing about production–distribution systems but about loop analysis, and we are simply using that example to make sure that the results accord with reason in a simple case.

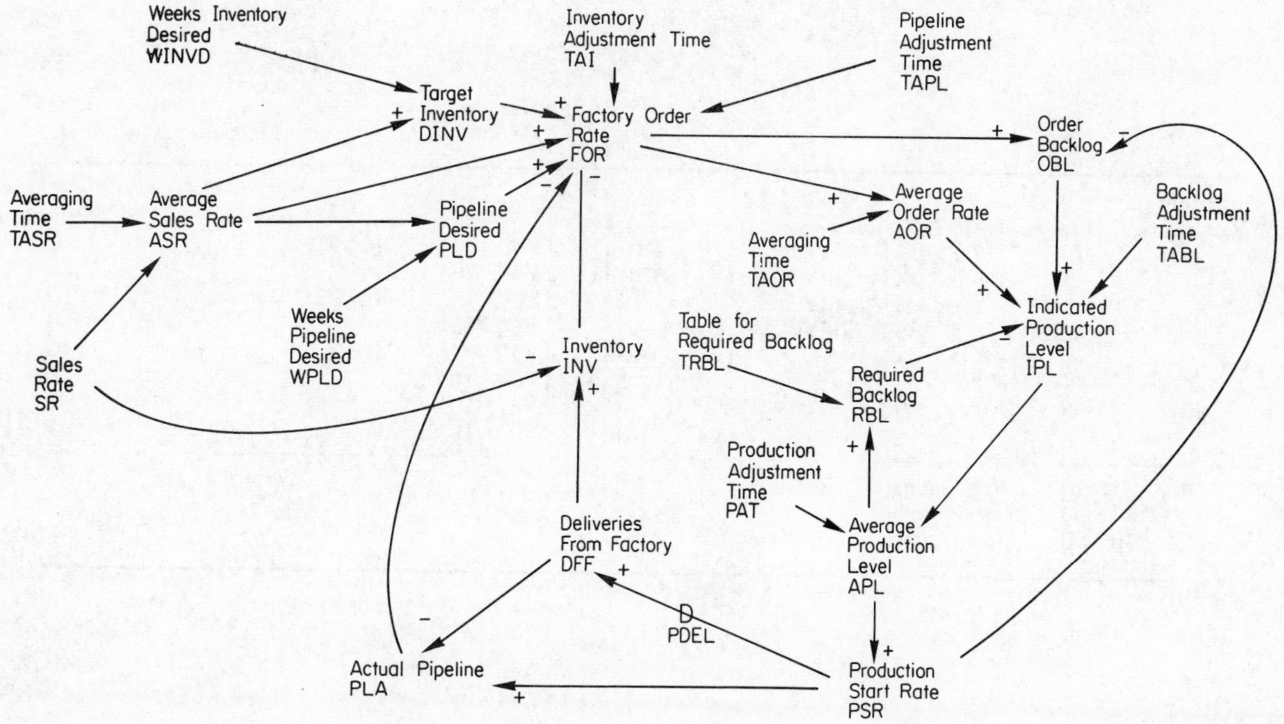

Fig. 8.14 Influence diagram for further revision of system

Table 8.4. *Loop Summary Table for System with Pipeline Control*

Loop	Type	Delays							Total* Delay (Wks)	Total Delay Order	No. of Pure Integs.	Gains
A	–			TABL	PAT	PDEL	TAI		25	6	2	
B	–			TABL	PAT			TAP	11	3	0	TRBL
C	–			TABL	PAT				7	2	1	
D	–		TAOR		PAT	PDEL	TAI		25	6	1	
E	–			TABL	PAT			TAPL	11	3	2	
F	–		TAOR		PAT			TAPL	11	2	1	
G	+			TABL	PAT	PDEL		TAPL	17	6	2	
H	+		TAOR		PAT	PDEL		TAPL	17	6	1	
Input		TASR										WINVD WPLD

*Note: Based on TAI=12

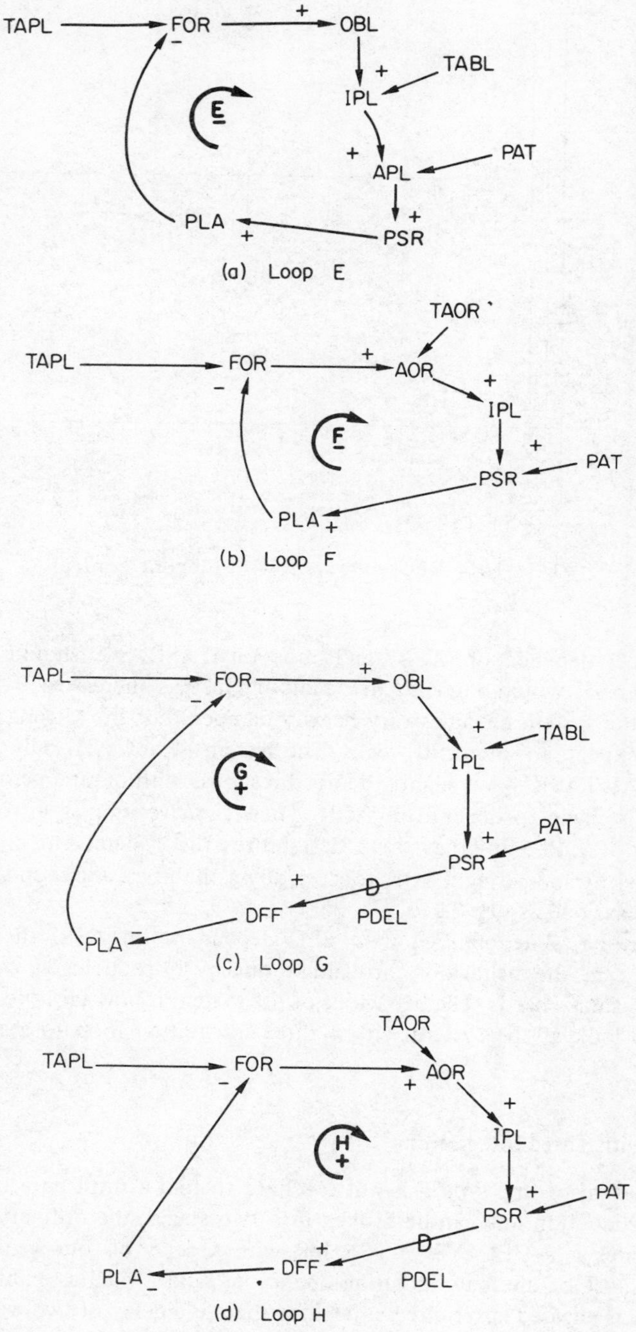

Fig. 8.15 New loops

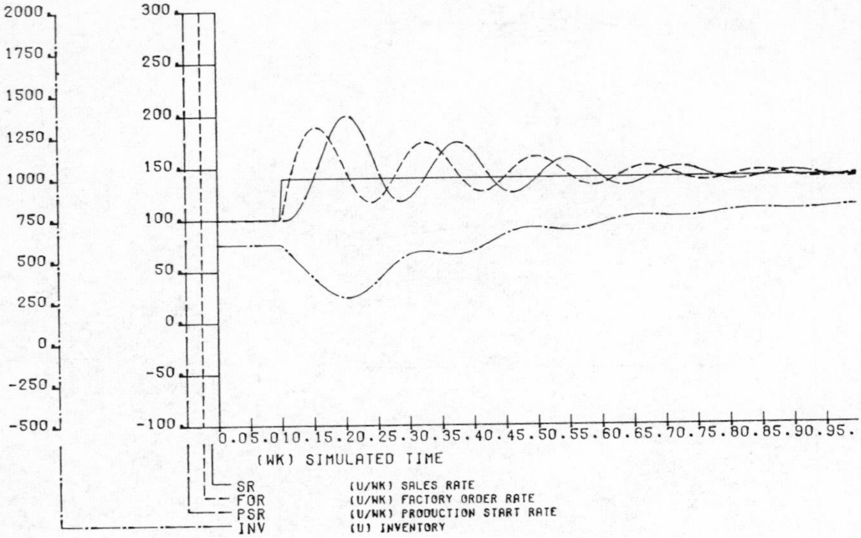

Fig. 8.16 Revised system with pipeline control

Since PLD depends on ASR, and thus on TASR, we should expect that increasing TASR would increase the damping just as increasing TAI did. With TASR=12, the system becomes very heavily damped, but the settling time rises, as we should expect, to about 90 weeks. The system is, however, only fairly robust and with TAI=TASR=4 has slightly damped response with natural period 22 weeks.

If PLD is made to depend on AOR a new *positive* loop is introduced. With TAI=12 and TASR=4 this new loop destabilizes the system, but control can be re-established by increasing TAOR to slow down that loop and loop D, though the result is not as good as Fig. 8.16.

Whatever the value of TAI, with PLD dependent on ASR, there is a large improvement in the behaviour, brought about by introducing new feedback to match the system's needs. The behaviour of the system is now fairly reasonable, but we need to look at the system with a three integration loop to see how it will behave.

8.7 Loop with Three Integrations

For a system of this type it is not too hard to find a third pure integration to use. The production line can be broken into two stages; the ordinary stage which we still model as a DELAY3 of PSR but with a delay of four weeks and which represents most of the manufacturing operations; and a final assembly stage from which work is released as required and where there is a delay of two weeks. This has a total delay of six weeks, as had the original DELAY3 used to model the production flow, but now we regulate both the inflow and the outflow of production as opposed simply to controlling the inflow, as we used to. We reiterate that we were able to see the *possibility* of doing this from the type of the system but we decided that we *needed* to look for that possibility by considering the structure in the light of the observed inadequacies in its dynamics.

We can introduce the third integration by using equations:

```
R PCR.KL=DELAY3(PSR,JK,PDEL)
          PCR      =(U/WK) RATE OF COMPLETION OF BASIC MANUFACTURING STAGE
                    IE RATE OF ENTRY INTO FINAL ASSEMBLY POOL

          PSR      =(U/WK) PRODUCTION START RATE AT FACTORY

          PDEL     =(WK) PRODUCTION PROCESS DELAY


L WIFA.K=WIFA.J+DT*(PCR.JK-DFF.JK)
N WIFA=(COVER+(1-D6)*TAWIFA)*AOR
C DT=0.125
          WIFA     =(U) WORK IN FINAL ASSEMBLY

          DT       =(WK) SOLUTION INTERVAL IN SIMULATION

          PCR      =(U/WK) RATE OF COMPLETION OF BASIC MANUFACTURING STAGE
                    IE RATE OF ENTRY INTO FINAL ASSEMBLY POOL

          DFF      =(U/WK) DELIVERY RATE FROM FACTORY


R DFF.KL=(WIFA.K-DWIFA.K)/TAWIFA+D6*AOR.K
C TAWIFA=2
C D6=1
          DFF      =(U/WK) DELIVERY RATE FROM FACTORY

          WIFA     =(U) WORK IN FINAL ASSEMBLY

          DWIFA    =(U) DESIRED WORK IN FINAL ASSEMBLY

          TAWIFA   =(WK) TIME TO ADJUST WORK IN FINAL ASSEMBLY

          D6       =(1) ZERO-ONE SWITCH TO TEST EFFECTS OF INCLUDING
                    AVERAGE ORDER RATE IN EQUATION FOR DELIVERIES FROM FACTORY

          AOR      =(U/WK) AVERAGE ORDER RATE AT FACTORY


A DWIFA.K=COVER*APL.K
CP COVER=2
          DWIFA    =(U) DESIRED WORK IN FINAL ASSEMBLY

          COVER    =(WK) NUMBER OF WEEKS COVER REQUIRED IN FINAL ASSEMBLY

          APL      =(U/WK) AVERAGE PRODUCTION LEVEL
```

Notice the order of the terms in the equation for DFF and compare with the inventory terms in the equation for FOR. Work out the reason for the difference, and the reason for the form of the initial condition for WIFA as compared to that for INV.

The restructuring of the production sector has created many new loops and we now have a new influence diagram, Fig. 8.17, the loops in Fig. 8.18, and the loop summary table in Table 8.5. The reader will see that a ‘simple’ model has suddenly become complex.

Table 8.5. *Loop Summary Table for System with 3rd Integration*

Loop	Type	Delays							Total* Delay (Wks)	Total Delay Order	No. of Pure Integs	Gains
A	–	TAI		TABL	PAT	PDEL	TAWIFA		25	7	3	
B	–			TABL	PAT				7	2	0	TRBL
C	–			TABL	PAT				7	2	1	
D	–	TAI	TAOR		PAT	PDEL	TAWIFA		25	7	2	
D′	–	TAI	TAOR						16	2	1	
E	–			TABL	PAT			TAPL	11	3	2	
F	–		TAOR		PAT			TAPL	11	3	1	
G	+		TAOR		PAT	PDEL	TAWIFA	TAPL	17	7	3	
G′	+		TAOR					TAPL	8	2	1	
H	+	TAI	TAOR		PAT		TAWIFA		21	4	1	COVER
I	–		TAOR		PAT		TAWIFA	TAPL	13	4	1	COVER
J	+			TABL	PAT	PDEL	TAWIFA	TAPL	17	7	3	
K	+	TAI		TABL	PAT		TAWIFA		21	4	2	COVER
L	–		TAOR	TABL	PAT		TAWIFA	TAPL	17	5	1	COVER
M	–						TAWIFA		2	1	1	
Input								TASR				

*Note: Based on TAI=12

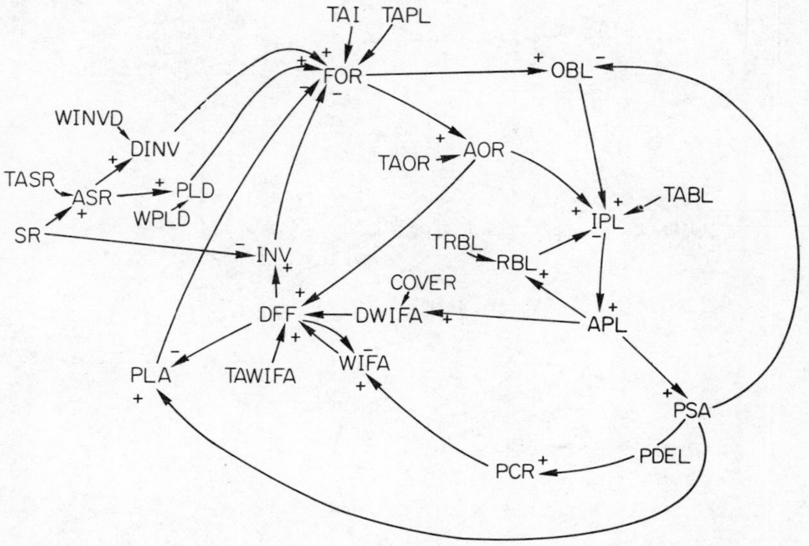

Fig. 8.17 Influence diagram for system with three-integration loop

We now have two loops with three integrations and several pseudo-positive loops have appeared. Some of the loops come and go when parameters are changed from 0 to 1, e.g. D′ and G′ would vanish if AOR was not used in the equation for DFF. Loop M will only act as a delay on DFF.

We introduce the third integration to damp the system and the effects are shown, for TAI=12, in Fig. 8.19. Behaviour has improved a little—natural period is reduced and settling time has decreased. There is still however, a fair amount of overshoot and DFF obstinately refuses to come into phase with FOR. From the damped behaviour of inventory it is clearly not causing the oscillation. PLA does, however, oscillate as does WIFA. This suggests that TAPL and TAWIFA (which has the rather small value of 2 and operates in loops I and J which have only 1 integration but have the gain COVER=2) are important parameters and that increasing them will improve the damping.

The result of increasing TAPL to 12 and keeping TAI at 12 is shown in Fig. 8.20. The change is noticeable, but it would be necessary to decide whether it represents an improvement in the light of the type of the system.

The frequency response for the model in Fig. 8.20 is shown in Fig. 8.21. It is obtained by putting DS=0 and doing a series of reruns with progressively larger values of PERD. In each case, the value of ARAT at week 120 is used for the system gain and it is this, plotted against PERD, which appears in fig. 8.21. (For greater clarity, the graphs in this chapter have been restricted to 100 weeks.)

Fig. 8.21 shows that the system gain, which we have arbitrarily chosen to measure as the ratio of the amplitude of PSR to that of SR, rises steadily. There is some sign of levelling-off at 30 and the natural period is clearly about 40

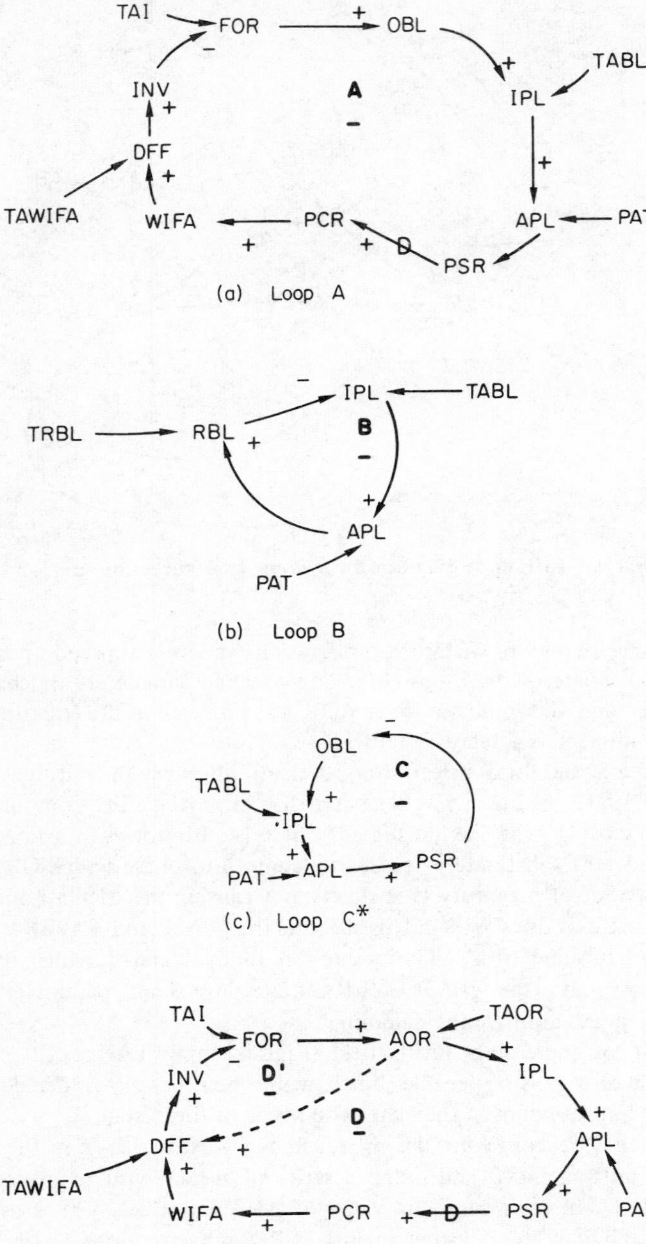

(a) Loop A

(b) Loop B

(c) Loop C*

(d) Loop D and D'

Fig. 8.18 The loops in the model with the extra integration

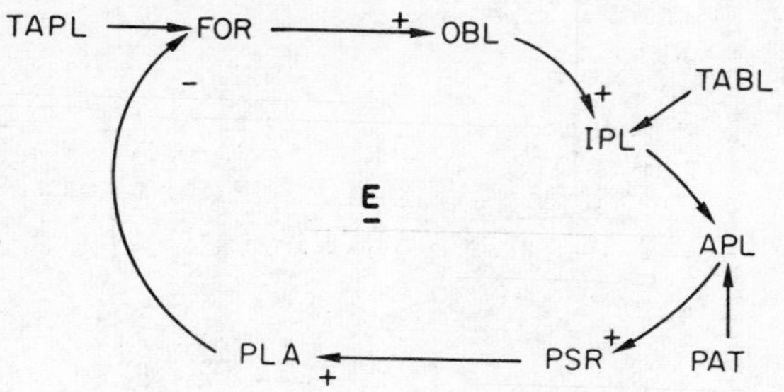

(e) Loop E

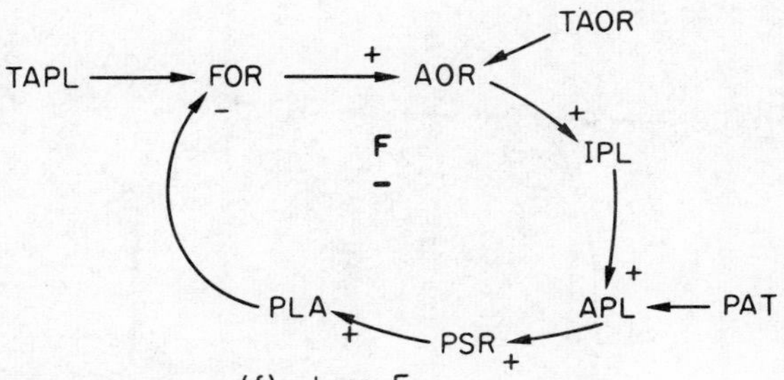

(f) Loop F

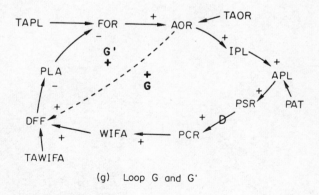

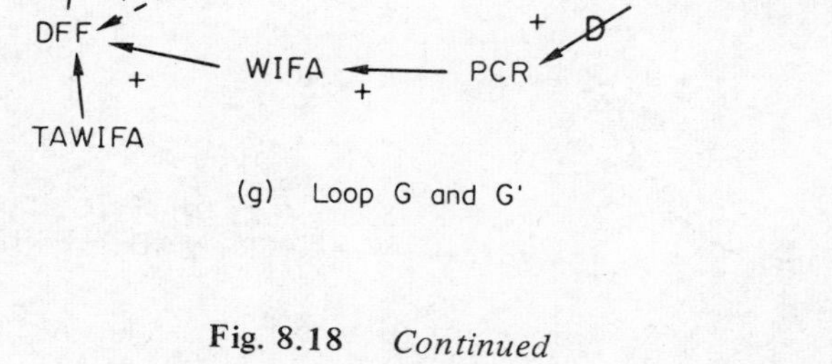

(g) Loop G and G'

Fig. 8.18 *Continued*

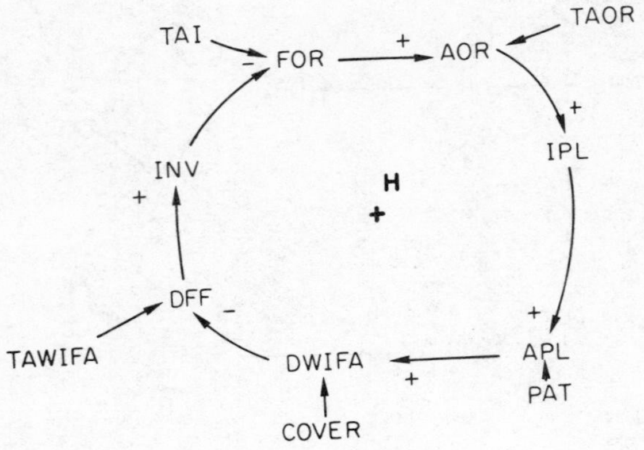

(h) Loop H

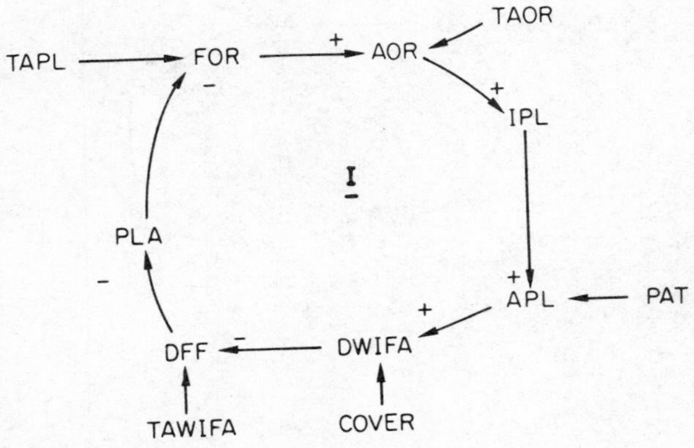

(i) Loop I

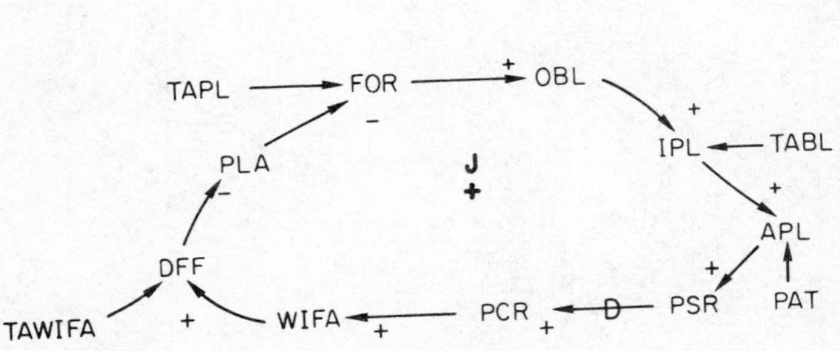

(j) Loop J

Fig. 8.18 *Continued*

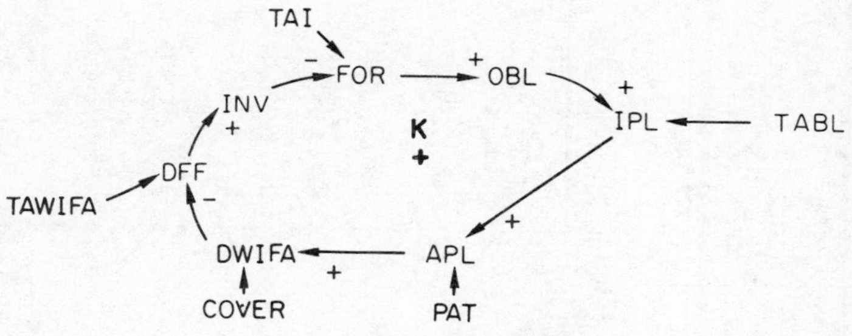

(k) Loop K

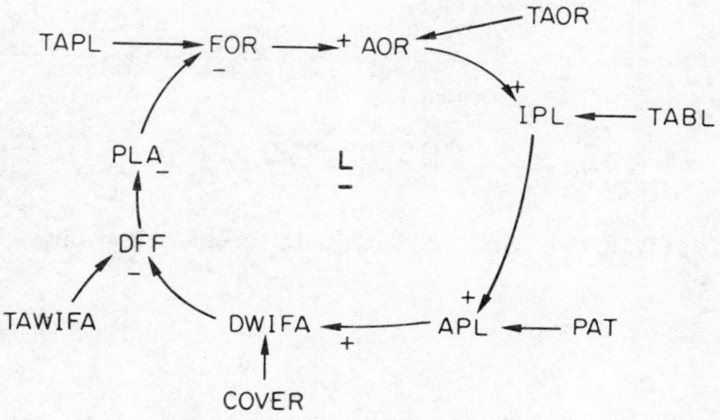

(l) Loop L

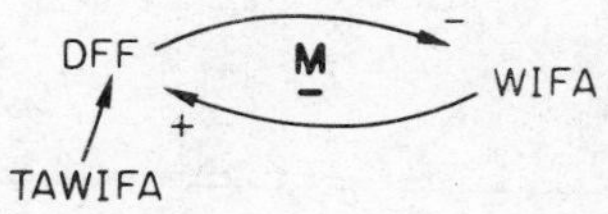

(m) Loop M

Fig. 8.18 *Continued*

weeks. The estimate of 36 derived from Fig. 8.20 is low, partly because of the non-linearities in the system and partly because of the difficulty of reading from the heavily damped response of Fig. 8.20. There is not much point in measuring it exactly because the curve in Fig. 8.21 is rather flat and the system is obviously very sensitive to inputs of about 1 year periodicity, i.e. to annual seasonality, as shown in Fig. 8.22 for a 40 week period.

The reader should try long-period inputs of about 200 weeks to test the effects of the 4-year business cycle, with a suitable choice of LENGTH.

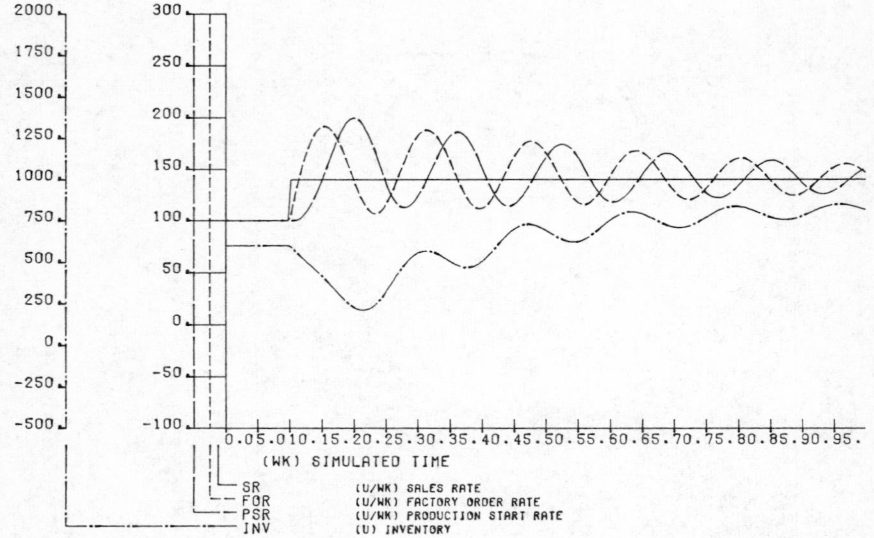

Fig. 8.19 Step response with three integrations

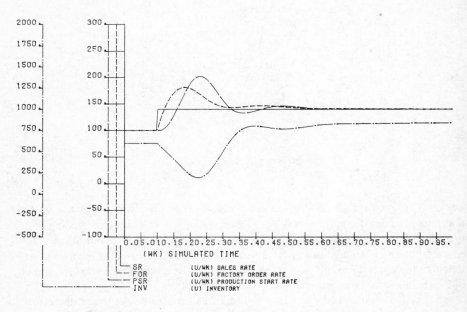

Fig. 8.20 Step response with TAI and TAPL=12

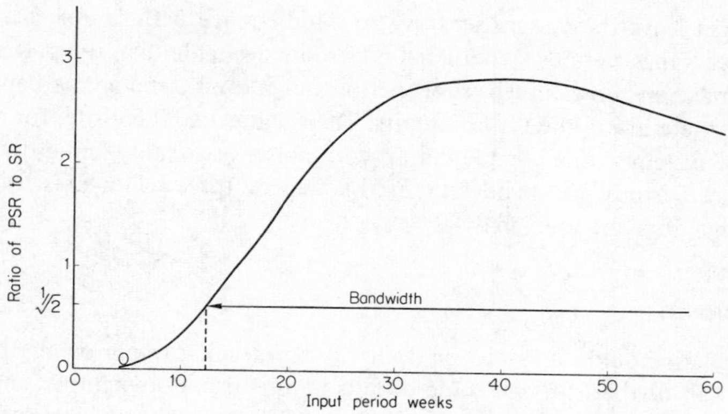

Fig. 8.21 Frequency response for Fig. 8.21

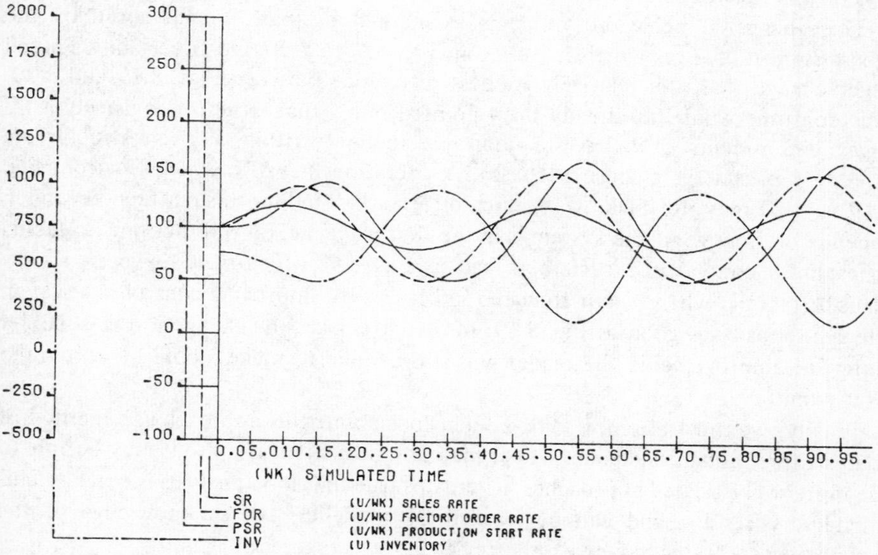

Fig. 8.22 Response to 40-week input period

The bandwidth in Fig. 8.21 is fairly large but the system would deal very well with short-period noise. The reader should now experiment with the system to find how increasing delay and/or reducing gain moves the frequency response to the right to reduce the bandwidth, evaluating the trade-off against settling time. Alternatively one might tolerate a damped step response if it gave rise to the type of peaked frequency response shown in Fig. 2.20 with an attempt to have a lower peak and a lower right-hand tail.

This might leave the system sensitive to middle-range periods and insensitive to low and high range periods. Whether this is more desirable than the pattern of Fig. 8.21 depends on how much short-period noise and semi-annual and annual seasonality there are in the typical inputs. The final test is, therefore, for the reader to drive the system with a complex input with noise, seasonality and growth or step components to study the behaviour of the two or three candidates for the final system design after the loop analysis.

8.8 Conclusions

The reader should by now be well on the track of improving the system behaviour still further, though some pitfalls lie in store. For example, changing D5 to zero makes IPL depend on APL and not AOR. This removes loops D,F,G,H,I, and L, leaving D′ and G′, and creates a new, positive, loop, shown in Fig. 8.23 with a short delay. The effects of this increase in positive loops, and decrease in long-delay negative loops, is a very long period, moderately damped, oscillation.

This change in behaviour mode is as marked as that brought about by the progression of changes which led us from Fig. 8.7 to 8.20 and is certainly a much larger change than the relatively modest difference between Figs. 8.19 and 8.20. This confirms what has already been pointed out – that structure is usually a far larger determinant of behaviour than parameters. Within a given structure, a three-fold parameter change, such as the alteration in TAPL which led from Fig. 8.19 to 8.20 may not make very much difference. Whether this matters very much depends on how good the system is in the first place, i.e. on whether one is seeking to establish control in a system as bad as Fig. 8.7, which was, however, a very plausible animal when it first appeared in Fig. 8.1, or improving control in a system which is already as good as Fig. 8.19. In the latter case, the change to Fig. 8.20, let along the improvements the reader will still be able to make, could be accounted worthwhile.

Finally, we must attempt some conclusions about this approach as a method of analysis for, though the reader whose mathematical talent is up to it may be able to use more sophisticated approaches, it remains true that loop analysis is quick, yields satisfactory results, and can be learned by modellers whose knowledge of the system they are trying to tackle is the key element in a successful study.

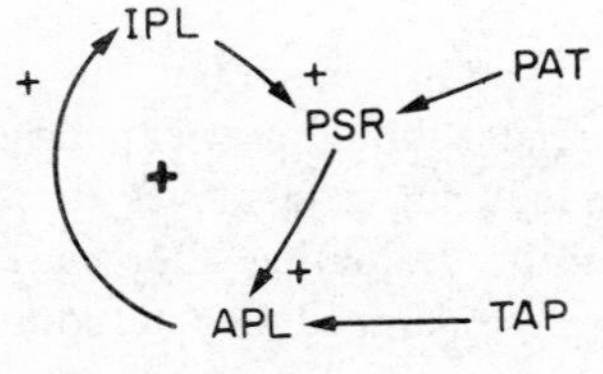

Fig. 8.23

The method is based on simply looking at the system and what it does and trying to explain its behaviour in the light of its loops and some rough rules of thumb about how loops operate. In applying the method there is no substitute for close and careful examination of the simulation printout taking account of more variables than we have had space to show here. The reader will benefit greatly from running the model in Appendix B8, examining the behaviour, and confirming and improving on what has been done here.

The main weakness of the method is that there is little in the way of explicit rules to guide one, except that increasing delay and/or reducing gain are the key to improving stability, where that is a desirable system characteristic. The problem is, of course, to decide which gains and delays to change, and this is done by finding the key loops from the loop summary table and the printout. Later versions of DYSMAP will produce much of this information automatically.

It is naive to suppose that any complex analysis, mathematical or loop-based, is easy, Standard, automatic, formulae only exist for trivial problems and in any work on difficult problems experience is a great help. We have tried to provide some of this experience synthetically by inviting the student to work with this model in detail and to attempt to apply the same approach to the more complicated cases in Chapters 10–12, even though space considerations prevent a fully detailed loop presentation in those chapters.

Chapter 9

Dynamic Processes in the Firm

9.1 Introduction–The Firm as a Dynamic System

Before examining, in Chapters 10–13, system dynamics as a problem-solver, we consider it as an approach to the theory of the business firm. This also allows us to illustrate some submodels of corporate structures, which the student may find useful as frameworks for more detailed work, but these are not *the* correct bases which must always be used. Each situation has to be treated on its own merits.

Our basic view is that the firm disposes of resources, as the means by which it creates performance to generate future resources. Resources could be capital investment to create production capacity to generate cash flow to support capital investment. They could be product development spending to increase sales by improving the product, thereby generating the means for further product development in the future, or investments in productivity to reduce costs and provide the wherewithal for future productivity improvements and so forth. (These ideas develop some of J.W. Forrester's early work. I am indebted to him for suggesting this approach to the theory of the firm.)

Essentially, therefore, the dynamic view of the firm can be expressed as shown in Fig. 9.1 as a simple and appealing arrangement which is fraught with all manner of pitfalls of inadequate definition, oversimplification and lack of empirical verification. We shall, however, attempt some first steps in this minefield because the insights which are offered into the firm's evolutionary processes are a complement to the micro-economic approach.

This idea of the firm as a feedback loop implies that the loop may be negative or positive, and there would be many loops in a more realistic treatment. We shall treat some of them individually first, and then collectively in a simple corporate model later in the chapter.

The mechanisms described do not necessarily exist in all firms, and may not be exactly as described here. In any particular case it is necessary to look fairly closely to find the structures discussed in this chapter, and they will often be modified by the existence of other mechanisms. To practice detecting dynamic processes, reread this chapter in conjunction with the case studies which follow.

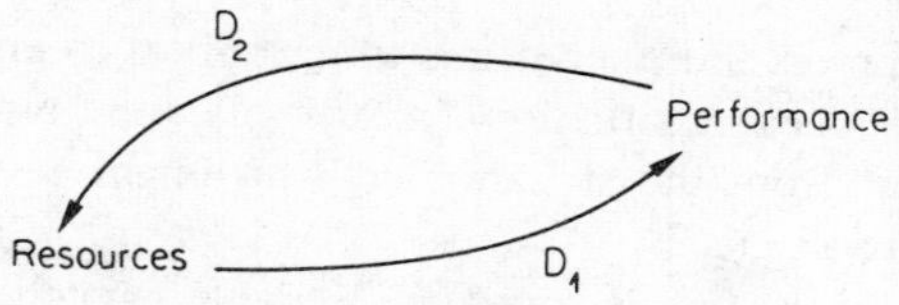

Fig. 9.1 Dynamic view of a firm

9.2 Performance Loops and Delivery Delay

One of the factors which determines the business received by a firm is its performance in dealing with previous business. Other factors will be external shocks, to include general economic movements and actions by competitors, and the firm's promotion of business by advertising, product development etc. We can summarize as in Fig. 9.2 (space limitations necessarily compel a simplified treatment).

Loop A implies a 'performance variable', PV, which the outside environment recognizes and reacts to. The loop may be positive or negative, but for the present we confine ourselves to the latter case, by far the best example of which is delivery delay.

When new orders have to wait before they can be started, this time, together with the manufacturing time, is the delay experienced by customer. If there is a surge of new orders, delay rises and orders fall off because customers won't wait. (There is no doubt about the existence of this process. For example, early in 1974, Jaguar Cars mounted a large and costly advertising campaign based on the slogan 'How right you were to wait'. There is, in fact, some reason to believe that matters may be more complicated than this. In the household goods industry it is widely believed that rising delivery delay *increases* orders as dealers place extra orders to secure a place in the queue. After a time the delay reaches a point where orders start to fall. Recall Section 1.7.)

For example, car dealers usually hold only a demonstration stock of new cars and order a car from the manufacturer to meet the buyer's exact requirements. The

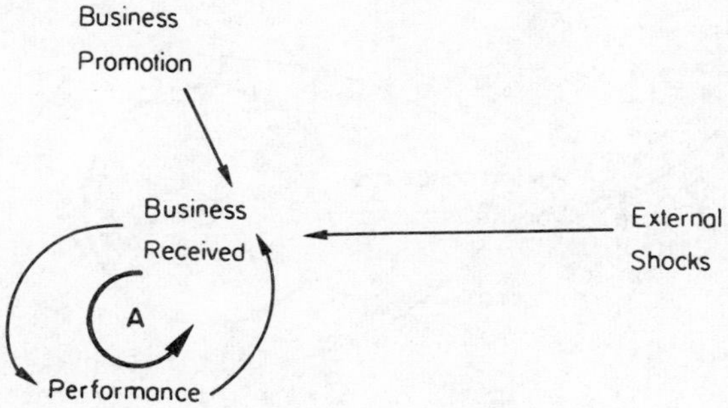

Fig. 9.2

situation is rather common and many firms which appear to manufacture for stock are, in fact in an order-backlog situation. In such cases the orders come, not from the end consumer, but from the stockholder, who might be a retailer or even a department of the company itself.

Clearly, delivery delay is the Performance Variable in such a situation, and the firm's allocation of resources to create production capacity is translated via the PV, into the order load placed on that capacity. Incidentally, all products take some minimum time to manufacture and deliver, due to the technology of manufacture and distribution. Changing the technology would change this minimum, and one might assess a proposed technology on the usual grounds *and* by the effect which the changed minimum delivery delay has on the behaviour of the firm.

There may be another minimum, namely the forward planning period for the buyer. Thus, if an electrical goods dealer orders TV sets for delivery in six weeks time he may well refuse to accept them any earlier. There is then nothing to be gained by the manufacturer offering delivery in less than six weeks. Indeed the manufacturer may suffer if the dealer starts to expect delivery from stock and ceases to order forward.

We need to elaborate this approach to account for two major features—time lags and external influences.

Time lags occur in that if, for example, the firm increases the resources devoted to production, then the PV should improve, but this might not be immediately recognized by the market. The delay can be modelled by using Recognized Performance (RP) which is simply the market's view of what the Performance Variable is (See Fig. 9.3).

The time-lag, D, may be complicated. For example, bad performance may be recognized more quickly than good, it may take longer to prove an improvement in performance following a decline than following an earlier improvement and so on, but these may be unnecessarily detailed refinements.

Demand is affected, not only by the firm's resource allocation, but also by external factors such as the economic climate, and modelling resource allocation and demand requires the idea of LATENT DEMAND. This is the demand which would exist if the performance variable was regarded by the market as

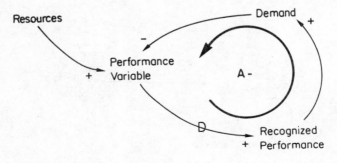

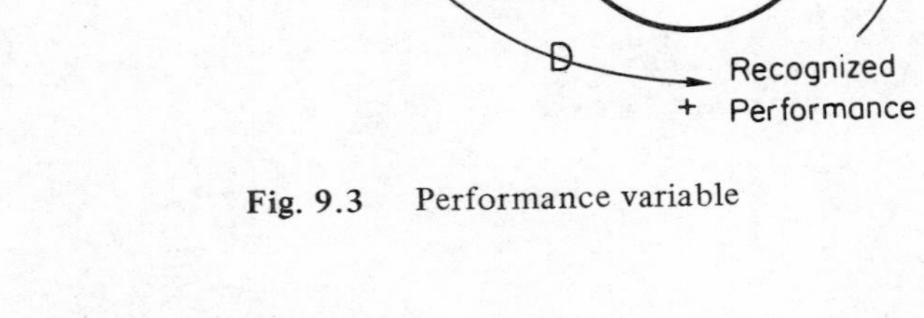

Fig. 9.3 Performance variable

'satisfactory'. The latent demand is influenced by external factors in the complement, such as economic growth, or by competitors in the environment.

We can apply this idea where we can assume that latent demand represents 'base load' and that actual, current, demand may be above or below the latent level. In an actual case one would have to be clear about what happens if, say, actual demand exceeds the latent level. Does it mean a genuine increase, or is it latent demand brought forward from the future which will cause a drop in latent demand at some later time? There are several options and it is essential to make clear, most of all to oneself, just what has been assumed about the market's workings.

A useful way of expressing this whole relationship would be to write:

Actual Demand AD = Performance Factor PF × Latent Demand LD

and the performance factor depends on the recognized performance, Denoting Recognized Performance by RP.

Then

$$AD = PF \times LD$$

and

$$PF = f(RP)$$

The precise form of f would depend on circumstances but for the example where the performance variable is Delivery Delay might be roughly as shown in Fig. 9.4. The effect is that, no matter how good the Recognized Performace becomes, the demand never reaches more than 40% above the normal level and no matter how bad it gets it does not fall below 70% of normal. When RP is normal (i.e. 1.0) then Actual Demand would be the same as Latent Demand because PF=1.0. This is simply an illustration and the actual curve might differ considerably, and might be rather hard to determine, and the concept of internal robustness might have to be applied.

We now have an extended diagram (see Fig. 9.5) for the delivery delay example, into which we can inject some detail for a production system model in which the firm simply produces at capacity.

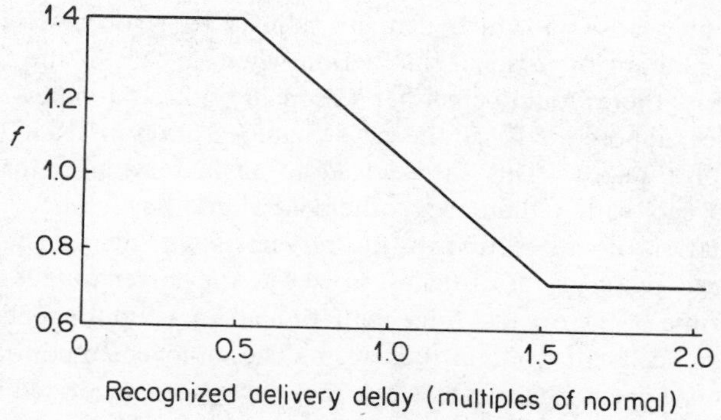

Fig. 9.4

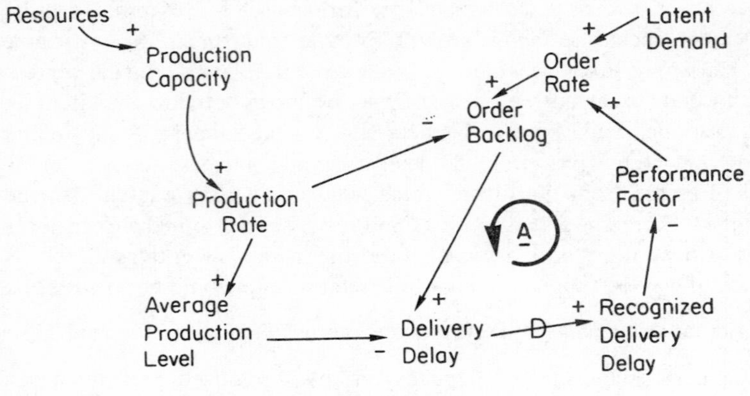

Fig. 9.5

In an actual case one would have to decide what happens to the latent orders not placed—are they lost, accumulated into another pool of latent business, transferred to the competition or what?

In the general case we can see that the action of loop A *is to regulate the market demand to the resources the firm has invested*. Thus, if the firm increases resources the PV will improve and demand will rise, but, if demand rises when resources do not, the PV will worsen and demand will be driven down until it matches resources again, though the adjustment may not be orderly.

If Order Backlog was a determinant of production rate, new loops come into play which would reinforce the effects of loop A, but without changing its consequences.

The existence of the feedback loop, and the fact that it attempts to control demand to resources, has nothing at all to do with conscious choices made by the firm. The loop exists because that is the way the market works. The firm cannot abolish the loop and the best that it can do is to allocate resources and control its actions so as to minimize the loop's harmful effects, or to enhance its beneficial ones.

Oscillation may arise in which demand adjusts to resources but has all the appearance of random or seasonal fluctuation when, in fact, if the feedback loop worked smoothly, there would be neither seasonality nor randomness.

For example, suppose the PV in the car demand case was price, and that this was reduced for some reason. This should lead to an increase in actual demand, as owners replace cars earlier than they otherwise would have done. This leads to a younger population of cars so that, in future years *latent* demand falls off, as the cars do not need replacing. Eventually, however, the larger number of new cars bought in the time of the original price reduction all come up for replacement more or less together, so latent demand rises again. These movements in latent demand might have every appearance of noise but they have been generated by the system itself. Whether in a practical case one would treat the movements in latent demand as being random or system determined would depend on the purpose of that particular study, but it is the kind of consideration which would dictate the choice between the use of statistical methods and those of system dynamics.

9.3 A Simple Model of a Performance Variable

In this illustration, demand depends on delivery delay, a 'normal' delay being 4 weeks. This is expressed by the following table which is arrived at by a study of past records and modified by the Sales Manager's experienced judgement.

Delivery Delay (weeks)	Actual Demand (Units/weeks)
2	800
4	350
6	100
8	50

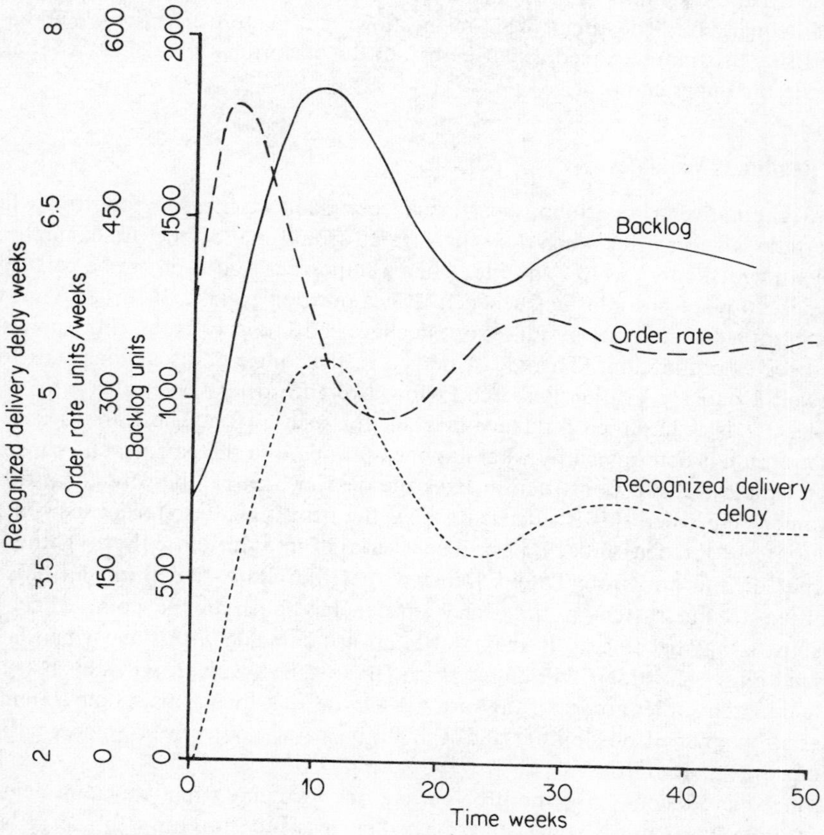

Fig. 9.6 Basic behaviour of simple performance variable model

The market takes some time to recognize changes in the delivery situation, and it is believed that an increasing delay is recognized in about 4 weeks, but when an improvement takes place it requires about 8 weeks before the market really starts to believe in it. Perhaps an advertising campaign could reduce this latter parameter.

The firm has a capacity of 350 units per week and has an order backlog of 4x350=1400 units. One of the firm's major customers suddenly cancels orders for 700 units so the firm can now offer delivery in (1400–700)/350=2 weeks. The resulting behaviour of the system is shown in Fig. 9.6. Before studying Fig. 9.6 and the discussion of the behaviour of the system, the reader should try to write the equations for himself using the third influence diagram in Section 9.2. For simplicity we can ignore the idea of Latent Demand and take Actual Demand as being simply the values in the table.

Fig. 9.6 shows a poorly damped, ultrastable response to the shock, the settling time being about 40 weeks. Clearly the market is very sensitive to delivery delay, which is due to the very steep performance factor curve which we have used. In this case, reducing the delays in performance variable recognition makes little difference to the system's stability or settling time, so the advertising campaign mentioned earlier would have little effect. What might, however, be more effective would be an advertising programme aimed at flattening out the performance variable curve, or at any rate the upper end of it.

9.4 Readiness Variables

An alternative to continuous performance variables is needed for firms which undertake large projects whch are only offered from time to time. An example is a shipyard which only receives an order when a shipowner wants, or can be persuaded to want, to place an order *and* when it is in a position to take the order. Thus the backlog of work increases by pulses but declines continuously as the firm applies its resources to production. (This idea emerged in the course of discussions with John Raiswell, Honorary Visiting Research Fellow at Bradford.)

Fig. 9.7 is a simplified influence diagram for such a firm. The Current Rate of Construction is determined by what has been promised in the past, and by capacity. The Feasible Rate of Construction fixes the delay which can be offered to future customers and when this is sufficiently low the firm is ready to receive new orders *if* any are latent. The Order Rate will be a series of intermittent pulses, signified by the bar on the link from Latent Demand to Order Rate, and they can only be admitted to the system if the Delay offered has fallen to the point where the 'Readiness' has become 1. If these times do not coincide with times when latent demand exists, no orders will be received. The readiness variable will stay at 1 until the next latent order arrives, if the firm can survive that long. This simple statement raises all manner of possibilities and it might be a study objective to model policies to improve survivability.

Other factors come into the problem e.g. the firm may stimulate latent demand by offering lower prices which may mean that it is then too busy to accept later, more profitable orders. If the business cycle affects latent demand, and it usually

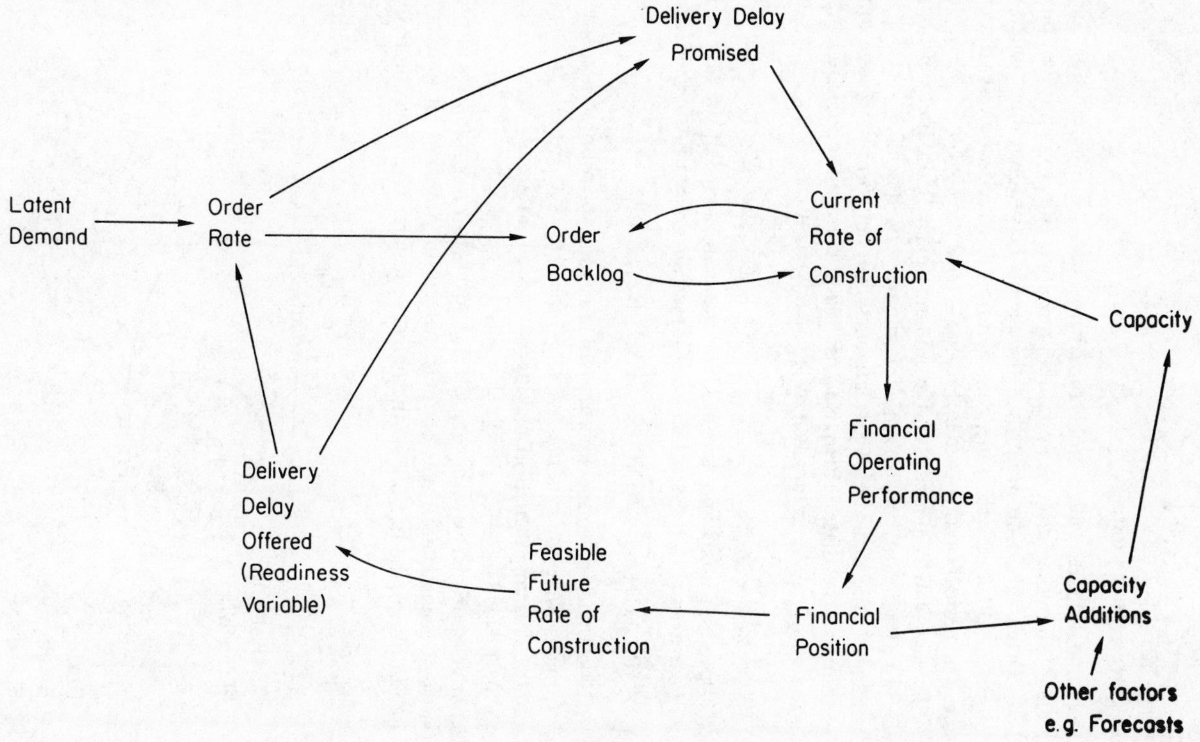

Fig. 9.7 A 'readiness variable' situation

does to a marked extent in shipbuilding, civil engineering and aircraft construction, the firm has an acute problem of timing its readiness to the market's needs. It may even take on so much cut price work on the economic down turn that it misses out on the boom.

9.5 Positive Performance Loops

Having examined loops which hold the firm back, we now consider growth producers. The simplest case is product development to improve product 'acceptability'. Often this is no more than the age of the product line though there are many more complicated situations, consideration of which would require a text on marketing. The basic loop is shown in Fig. 9.8.

In Fig. 9.8, clearly B is positive and would not exist if product development spending was not affected by profits, which would be very unusual.

If products age is the measure of attractiveness we can formulate equations for this loop as follows. The first problem is to calculate Average Product Age, APA. If no products are withdrawn and replaced by new ones, then

APA.K=APA.J+DT 9.1, L

If there are N product items and new products are being added at the rate of RINP products per year with old products being withdrawn *at the same rate* then

APA.K=APA.J+DT*(1–(RINP.JK*APA.J)/N) 9.2, L

where we assume that the products withdrawn are evenly distributed across the age range. This formulation allows us the use of N, the size of the product range, as a control variable. This might help in making the system internally robust (see also *Dynamica*, Vol. 2, No. 3).

An alternative approach is to have a series of levels to carry numbers of product items in the first, second, etc., ages of their lives, filling up the first one with new products and emptying the last one of the oldest products. This needs more lines of

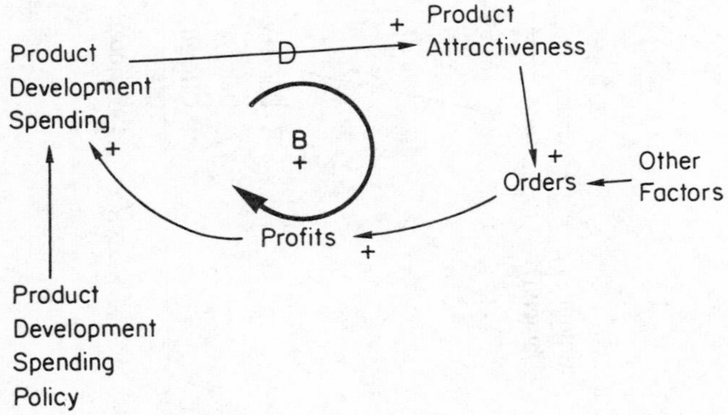

Fig. 9.8

model, and forces us to assume that products have to be withdrawn at a certain age. It could also lead to the model having no products at all. This might well be a realistic approach providing the size of the product range also affected the demand and, of course, the number of items in the line could be a control variable for product developmemt spending rather than, perhaps, simply taking a fixed proportion of profits.

Product age situations are easy to model in the sense that some estimate can be made of product development cost (£ per product) and one then has:

RSNP.KL=PDS.K/PDCOST

where

RSNP	Rate of Starting New Products [Products per year]
PDS	Product Development Spending [£ per year]
PDCOST	Product Development Cost [£ per product]

and

RINP.KL=DELAY*n*(RSNP.JK, PDDEL)*PPC

where

PDDEL	Product Development Delay [Years]
PPC	Probability of Successful Project Completion [1]

and we rely on the nth order delay to model the time uncertainties in the product development process and PPC to model the other uncertainties.

The trouble comes at the other end, when we try to relate Product Attractiveness to Order Rate, and we shall suggest one approach to that later in the chapter. For the moment we remark that this is a perfectly ordinary positive loop and its growth rate would be affected in the usual way by altering its gain and delay.

9.6 Financial Flows

In the earlier chapters we have taken most of our examples from simple production situations, because they are so easy to describe that the explanation could concentrate on the system property being studied. To try to balance our treatment of the firm we need to consider some of the financial loops involved. Dr. Hussien Shehata, then a Ph. D. student at Bradford and Lecturer in Finance at Al-Azhar University Faculty of Commerce, Cairo suggested several of the ideas in this section.

Our general approach will be to trace cash payments flows and to link these to cash balances and the conventions of balance sheets. We start by studying sources of funds, then treat uses of funds and finally link the two by considering a model of financial policy. The treatment is rather general, and somewhat simplified, and will need detailed refinement in application to particular cases. There is a fuller example of financial modelling in Chapter 12 and in the reference by Shehata in Appendix C.

Sources of Funds

Most of the cash flows in a corporate model are fairly easily treated as delayed versions of other flows such as production. For example:

SINCR.KL=DELAY*n*(VDESP.JK,PAYD) 9.3,R

where

SINCR	Sales Income Rate [£/month]
VDESP	Value of Goods Despatched [£/month]
PAYD	Payment Delay [Months]

and

VDESP.KL=DFF.K*PRICE

where

DFF	Deliveries from Factory [Units/months]
PRICE	Price [£/Unit]

Note the use of DFF as a smoothed level or auxiliary to avoid rate-dependent rates, and compare with Section 5.5. It would be easy to treat Direct Cost Rate, DCR, in the same way as a delayed version of the production start rate, PSR and unit cost.

This allows one to treat profit in the managerially correct way, i.e., as an average over a period of time, with:

AINC.K=SMOOTH(SINCR.JK,APERD)
ACOST.K=SMOOTH(DCR.JK,APERD)

where

AINC	Average Income [£/month]
ACOST	Average Costs [£/month]
APERD	Averaging Period [Month]

and

PROFIT.K=AINC.K–ACOST.K

with PROFIT having [£/month]. This allows one to treat APERD as a parameter to study the connection between accounting convention and the system's performance as it reacts to accounting information.

To balance the flows, we also need

CASH.K=CASH.J+DT*(SINCR.JK–DCR.JK) 9.4,L

which provides the firm's cash position in £. Obviously, one could construct a set of equations to forecast cash position for any number of future time points, using, say,

ACP1.K=CASH.K+SFCAST1.K*PRICE–PPLAN1.K*COST

where

ACP1	Anticipated Cash Position at end of 1 month [£]
SFCAST1	Forecast Sales Rate during month 1 [U/month]
PPLAN1	Planned Production Rate during month 1 [U/month]

The apparent dimensional inconsistency from SFCAST1.K*PRICE to the end is due to the implied, but unstated, presence of a multiplier of 1.0 which is the number of months ahead for which ACP1 is being calculated. Obviously, more detailed formulations could be arrived at.

Equation 9.4 is incomplete because it makes no reference to tax payments, and so far we have not mentioned depreciation.

Usually, depreciation is related to production capacity (time depreciation) and/or production (wear depreciation), and it may depend on the historical actual cost of the capital equipment or on the expected future cost of the replacement. The simplest case is historical-price time depreciation. This can be modelled by:

WDV.K=WDV.J+DT*(RAVP.JK–DCFL.J) 9.5,L

where

WDV	Written-down Value [£]
RAVP	Rate of adding to Value of Plant by new capacity Acquisition [£/month]
DCFL	Depreciation Cash Flow [£/month]

and

DCFL.K=DR*WDV.K 9.6,A

where

DR	Depreciation Fraction allowable for tax [1/month]

DCFL could be modelled as a rate and averaged for later use if that was more in accordance with accounting conventions.

We now have,

ATPR.K=(PROFIT.K–DCFL.K)*(1–TAXPC) 9.7,A

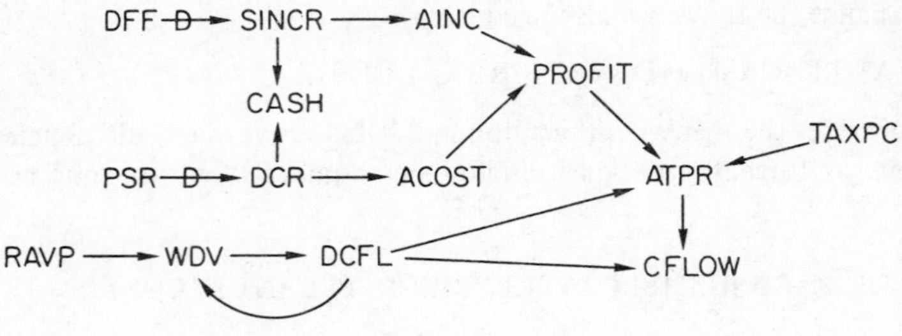

Fig. 9.9

where

ATPR	After Tax Profit Rate [£/month]
TAXPC	Fraction of net profit paid in tax (1)

This approach then provides cash flow, CFLOW [£/month], as:

CFLOW.K=ATPR.K+DCFL.K 9.8,A

ignoring dividend payments. Fig. 9.9 is the influence diagram.

From ATPR and CFLOW we can obtain cumulative retained earnings, CRE, and from this and the Shareholders Capital, SCAP, we get the maximum long-term debt the company could hold

CRE.K=CRE.J+DT*ATPR.J 9.9,L

MXLTD.K=GEAR*(CRE.K+SCAP.K) 9.10,A

where

GEAR	Gearing Ratio
MXLTD	Maximum Permissible Long-Term Debt

We are now in a position to model the firm's ability to generate finance for its growth, replacement of plant, or repayment of debt.

ARS.K=(MXLTD.K–LTD.K)/PHOR+CFLOW.K 9.11

where

ARS	Achievable Rate of Spending [£/month]
LTD	Long Term Debt [£]
PHOR	Planning Horizon [Months]

This models the firm controlling its long-term debt over some period of time, PHOR, which can be as large or small as required, subject to constraints on the availability of borrowed money.

Uses of Funds

The use of funds is more easily modelled than the detailed sources. The important thing is to distinguish clearly between planned and actual spending.

Planned, or target, spending rates are fairly easily calculated. A forecast of demand for PHOR time periods ahead gives target production capacity and the working capital needed to support that level of activity. Comparing these values with their present levels and allowing for plant replacement during PHOR easily provides the amount by which they have to be increased during PHOR, if expansion plans are to be met and it is then a simple matter to find the Required Rate of Spending, RRS.

9.7 The Main Financial Loop

It is fairly evident that the main financial loop in the firm is that involved with capital spending, i.e., the link between the sources and use of funds. The loop must relate the firm's Achievable Rate of Spending, ARS, to its Required Rate of Spending, RRS, and must involve the Actual Spending Rate, ASR. This closes the main financial loops as shown in Fig. 9.10. This diagram requires careful study as there are several loops in it. The reader should identify each one. The diagram implies far more detail than has been shown in the equations. The reader should try to work out the full equations for himself and compare the result with Appendix B.12 when he studies Chapter 12.

A particularly important feature of the diagram is that there are two parallel loop sets going through production capacity and written-down value. There are two gain operators, 'physical wear' and 'depreciation policy', the former usually being the smaller. This offers an explanation, in terms of feedback loops, of the phenomenon of accelerated depreciation and provides us with a dynamic explanation of how it can be used to allow the firm to cope with shocks from the market.

The essential idea is to use ARS and RRS to define the Financing Ratio, FRAT, by

$$FRAT=ARS/RRS \qquad 9.12$$

which reflects the proportion of its spending plans which the company is capable of supporting from its own earnings and its borrowing ability (or its borrowing policy, because there is no particular magic value of GEAR in equation 9.10).

The next stage is to consider the extent to which future spending *plans*, RRS, will be cut back because of any shortage there might be in current financing *expectations*, ARS, i.e., to ask how ASR, the actual rate of spending is to be related to RRS, the planned rate. This is link A in Fig. 9.10. A little thought will show that this is the proper way to model the process. It would be incorrect to relate ASR to ARS directly, in place of the link A to RRS. It is important to know which one to choose because otherwise we should have a completely different loop structure.

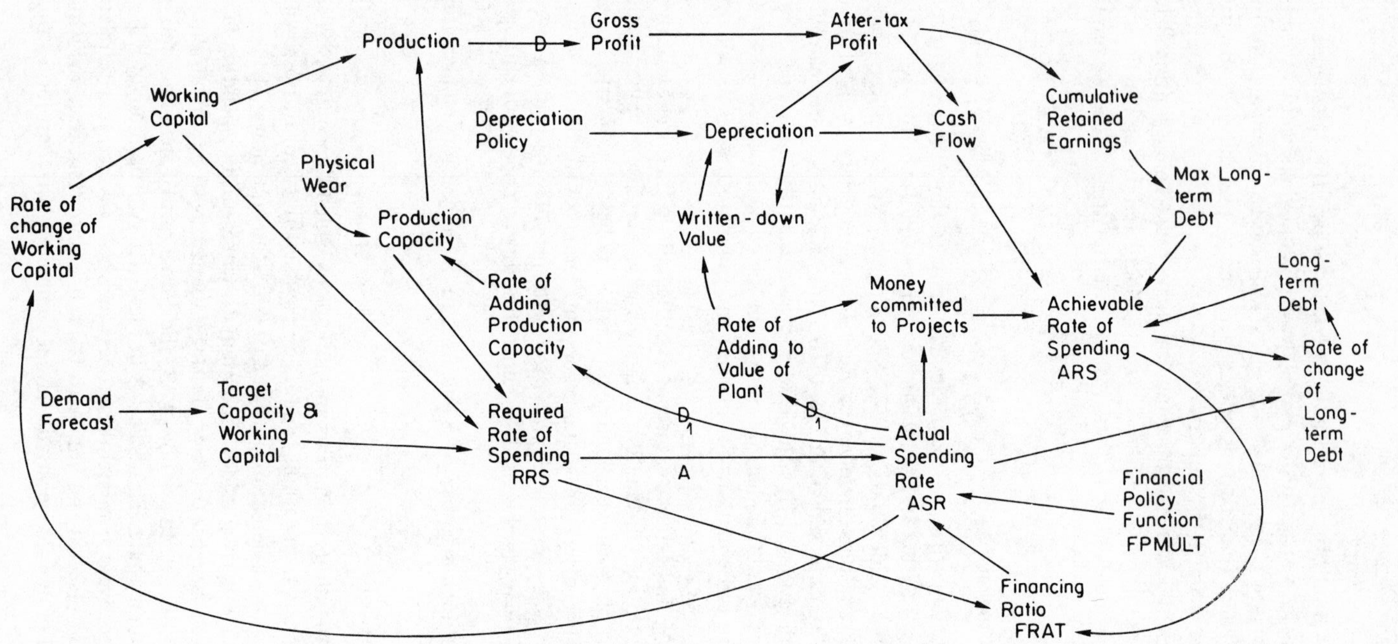

Fig. 9.10 Closing the financial loops

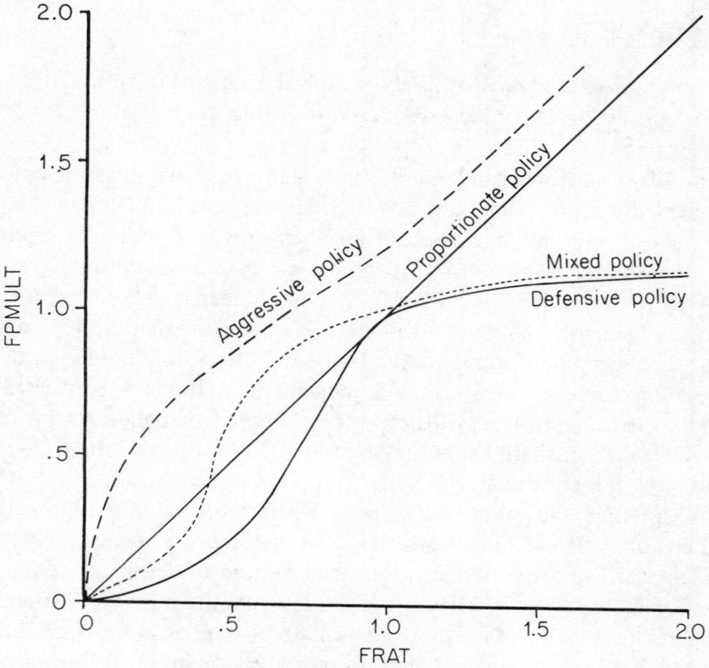

Fig. 9.11 Financial policy curves

There are many ways of treating this, but one is shown in Fig. 9.10. This involves defining FPMULT, a financial policy multiplier, as a function of FRAT, and then putting

$$ASR=FPMULT*RRS \qquad 9.13$$

Four policy curves for FPMULT are shown in Fig. 9.11.

The proportionate policy cuts the coat exactly to the cloth. When the company is rich (FRAT > 1) it spends *more* than demand forecasts call for. This does not matter because the forecasts are bound to be wrong and there are mechanisms which will correct for overshoot (where?).

Obviously the proportionate policy cannot continue for ever. At some point the curve must level off and the surplus be used to repay debt.

The Aggressive and Defensive policies reflect different financial attitudes, and the Mixed Policy a compromise between them. There is no reason at all for the Mixed Policy curve to cross the proportionate line or, indeed, for any of the curves (except the proportionate policy) to be the shape they are. They are purely illustrative and we now have complete freedom to make them whatever shape gave the greatest robustness.

At this point, having introduced the ideas, we leave them to deal with forecasting.

In Chapter 12, we shall have more to say about financial modelling.

9.8 Forecasting in the Firm

In Chapter 5 we studied the purely technical problems of formulating equations to model forecasting. We now consider forecasting as it relates to the firm as a dynamic system.

The problem of forecasting in business and economics has an extensive and sophisticated literature. There are two approaches, based on statistical analysis and on experienced judgement respectively. They are not mutually exclusive but, broadly, the former is used for forecasting the macro-variables of the economy, such as GNP, or total demand, either for economic management or as they relate to the firm. Judgement, or less formal methods, tends to be applied to the micro variables, such as demand for a particular one of the firm's products. In industrial practice, a mixture of statistical and judgement methods is employed. We deal briefly with the characteristics of these approaches before concentrating on another, system-related, rather than technique-orientated, view of forecasting.

Essentially, the statistical approach is to find an equation or set of equations which will 'predict' the value of a variable V, as it will be in h time periods, from past values of V and other variables, and, perhaps, expected values of other independent variables. By contrast, the main feature of the judgement approach is that the forecasters say, in effect, 'we believe, on the basis of all that we know, including its past behaviour, that the value of V in h months time will be x'. Because this kind of statement is hard to reproduce or support, is open to bias, and because the maker of it has little or no defence against the accusation of error of judgement if things go wrong, the method has, perhaps been less favourably treated than it ought to have been, and has tended to be supplanted by statistical approaches with their appearance of objectivity.

Whatever the approach, there are several assumptions behind the use of forecasts:

1. that the way in which the forecast is acted on can be considered as a separate issue.
2. The value of h is usually found by considering past data and finding a combination of a method *and* h for which good 'forecasts' can be made. We write 'forecasts' because this analysis is a pencil-and-paper exercise on old information. The assumption is that the firm, or system, can respond to the forecast within h time periods. This can lead to problems. If, for example, h is 3 months but new capacity takes 18 months to install it is hard to see the relevance of such a short-term forecast to the capacity-acquisition decision.
3. In most applications data are selected because a convincing statistical procedure can be applied to their historical values. This begs the very important question of whether V is a *useful* variable to forecast and effort may be devoted to forecasting variables which are not particularly relevant to the control of the corporate system or which, at any rate, are only part of the problem of control.

Although we can comment on the weaknesses of conventional forecasting it would be quite misleading to imply that it is useless. We can, in fact, use the

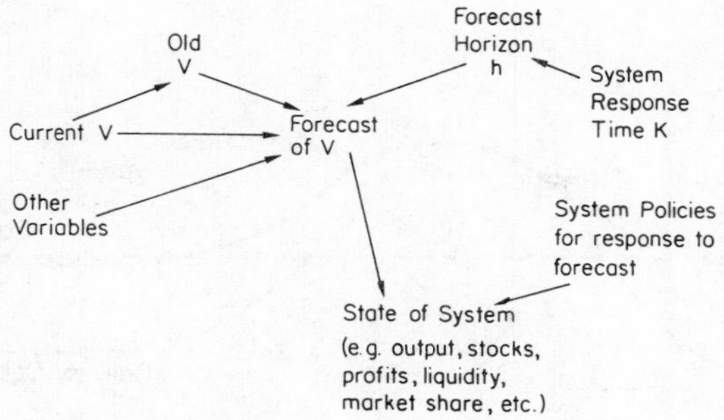

Fig. 9.12

assumption, particularly 1 and 2, to consider the conditions which have to be satisfied for conventional forecasting to be useful in managerial problems. These are that there should be no connection between the way that the system reacts to the forecast and the variables on which it is based, and that the forecast horizon, h, should be approximately the same as the system's response time K. These considerations can be expressed diagrammatically as Fig. 9.12.

If any of the other variables, or V itself, is part of the state of the system we should have feedback loops, since the forecast and the system's response to it were not independent and the condition mentioned would not hold.

The methods available for assessing the accuracy of a forecast are many and varied, but they usually boil down to a comparison between the forecast and the eventual actual value. If the comparison is close the forecast is usually said to be good or accurate.

The literature of forecasting is, however, replete with references to self-justifying and self-defeating forecasts. As an example of the former consider the system from Section 9.3 in which the actual orders placed depend on the general level of demand, and the delivery delay experienced by potential customers.

For a system like that it is easy to see that, if order rate, or forecast order rate, rises but capacity is not increased, then delivery delay will rise and choke off order rate, as shown in Fig. 9.13.

Clearly this would justify a manager who had said that rises in demand were always transient so there was no point in considering a capacity expansion.

With our superior knowledge of feedback loops we are tempted to condemn him but, if increases in capacity are put in hand and, because of long delays the company is not able to take advantage of the situation it may finish up worse off than before, as shown in the next diagram, Fig. 9.14.

In this case, the capacity time lag is 6 time periods and, unless the forecasting horizon is also 6 and the forecasting method can give reasonably *precise* results at that horizon, there is not much point in forecasting.

Even these simple examples illustrate the idea that one must assess a forecast, not by how it compares with the actual, but by how it benefits the system to have the forecast. Another example of this appears in Fig. 9.15.

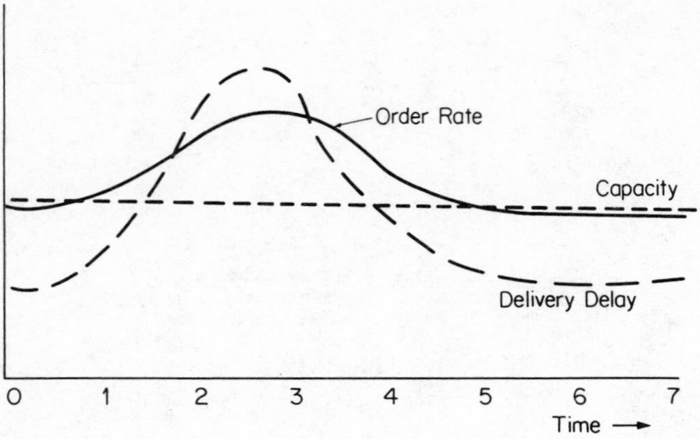

Fig. 9.13

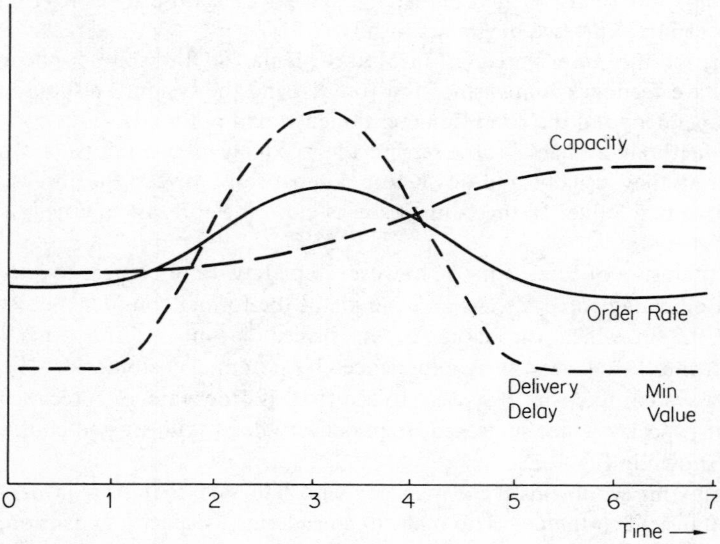

Fig. 9.14 Dynamics with unbalanced forecast horizon and response time

At zero the company has a capacity, demand and sales of 100 units per week. In one years time the demand is going to be 200. They do not know this but they try to forecast it, and two different forecasts are shown. If the forecast is 100 too low, Forecast 1, no change in capacity is indicated so none is made. At the end of a year the sales rate is, as was forecast, 100 and the forecaster is congratulated.

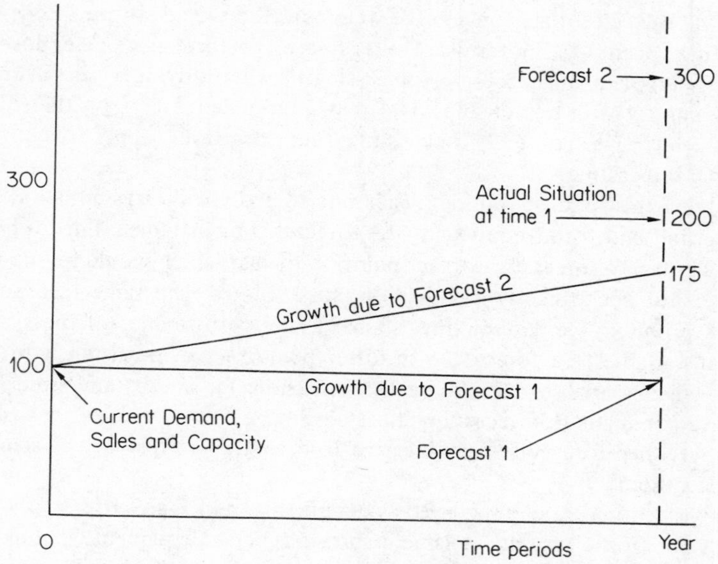

Fig. 9.15 'Accuracy' of forecasts

If the forecaster had been 100 too high, Forecast 2, steps would have been taken to increase capacity. This could not, however, be trebled in a year and the best that can be done is to reach a level of, say, 175 units per week. Compared to 300, this looks rather inaccurate but it is really a *better* forecast for the firm.

In this example, it would not necessarily have been best of all to have got the forecast right and predicted a demand of 200, depending on the company's capacity expansion policy. If this was such that the difference between the apparent need for 200 units per week and the current availability of 100 led to a capacity increase of 50 the result would be worse than the erroneous forecast of 300 triggering more urgent capacity expansion leading to the level of 175. The point is that what matters for the system is the *combination* of the forecast and what is done in response to it, rather than the forecast alone. The example suggests that not all forecasting mistakes are equally bad for the system. In Chapters 10 and 11 we shall encounter cases where wrong forecasts make no difference to performance, and even where they are slightly better for the system than perfect ones.

9.9 A System View of Forecasting

Most definitions of forecasting can usually be paraphrased into a tautology involving 'prediction'. We shall define forecasting as: *the act of giving advance warning in sufficient time for beneficial action to be taken*.

This definition puts the emphasis where, in a business system at least, it ought to be: on the *consequences for the system*, rather than on the techniques. It also brings in the essential concept of reaction time, and it puts assessment of forecasting 'accuracy' in terms of the system performance. One does not, for example, complain of inaccurate forecasting if, when driving a car, a forecast that one is going to hit a tree enables one to avoid it. One merely says that the forecast was valuable. We contend that similar considerations apply to business and economic forecasting.

Applying this kind of thinking leads one to analyse the system so as to specify the precision and form required of the forecast. For instance, it might be possible, by using radar, to forecast the exact point on the car which would be the first to hit the tree, but such precision would be superfluous. The same idea applies to a business problem; we should aim to analyse the performance of the system when different variables are forecast with different degrees of precision, with a view to finding which variables should be forecast, how far ahead, and how robust the system can be made to forecasting uncertainties.

Can we, therefore, relate the ideas of forecasting to those of the structure of a dynamic system?

The basic structure of a managed system is shown in Fig. 9.16.

The value of the forecast variable is shown as being dependent, via links A and B, on either or both of its own prior values or on proxy variables. The forecast is fed into a decision process which affects the system state after a delay.

The system state is, of course, also affected directly by the actual values of the forecast variables and by the actual values of other exogeneous variables which have not been forecast. (There may, of course, be more than one forecast variable, but there are always non-forecast variables.)

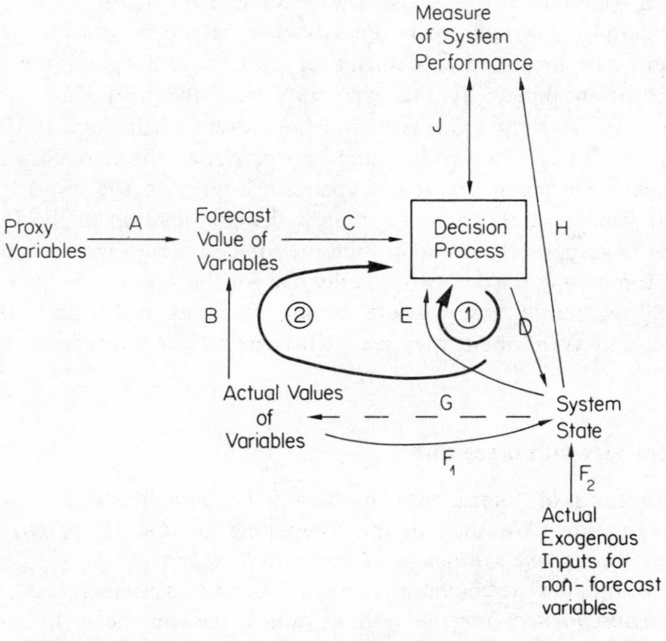

Fig. 9.16 Some ideas of system structure

The system state forms one of the inputs to the decision process and thus creates feedback loop 1, and the state also provides some measure of system performance, values of which may also take part in the decision process.

The simplest question is, therefore, how the forecasting mechanism based on the existence of links A and/or B, C, D, E, F_1 and F_2, H and, perhaps J, will interact with the system's decision rules and delays to generate system performance. This may well involve a study of decision processes to see whether they can correct for forecasting errors.

It is legitimate to speak of errors if the forecast does not react back on to the current values. If, however, a link G exists, then it is almost meaningless to refer to forecast errors in the sense of comparing forecast with actual outcome.

In Chapters 10 and 11 we shall examine real industrial systems which turn out to be very insensitive to large errors in the forecasts.

9.10 An Example of Dynamics and Forecasting

Analysing the dynamics of a system generally involves far more than the forecasting component. We have therefore chosen to describe the performance of a fairly simple system, in order to illustrate the general points made this chapter. The system described here is based on actual situations, but the account has been substantially simplified for economy of space.

The structure of the system is shown in Fig. 9.17. The essential idea is that the firm disposes of resources in the form of production capacity. These resources are loaded by the customer by the placing of orders which create a backlog. The backlog and the capacity create delivery delay which affects demand via the Recognized Delivery Delay.

This process can be represented as in Section 9.2.

The problem for the firm is to decide how to use the two control strategies which are open to it. The short-term controller is to regulate the production and the longer-term action is to adjust production capacity.

For the short-term the firm wishes to maintain a backlog of about 1.5 weeks of production capacity in order to ensure a smooth flow of work and it wishes to keep production up to the level of average order rate. Although there are more sophisticated policies, a fairly effective short-term controller is to calculate the production rate from:

$$\text{PROD} = \text{SORATE} + \frac{\text{BLOG} - \text{DBLOG}}{\text{TABL}}$$

where

PROD	Production Rate [U/WK]
SORATE	Average Order Rate [U/WK]
BLOG	Backlog [U]
TABL	Backlog adjustment period [WK]
DBLOG	Desired Backlog [U]

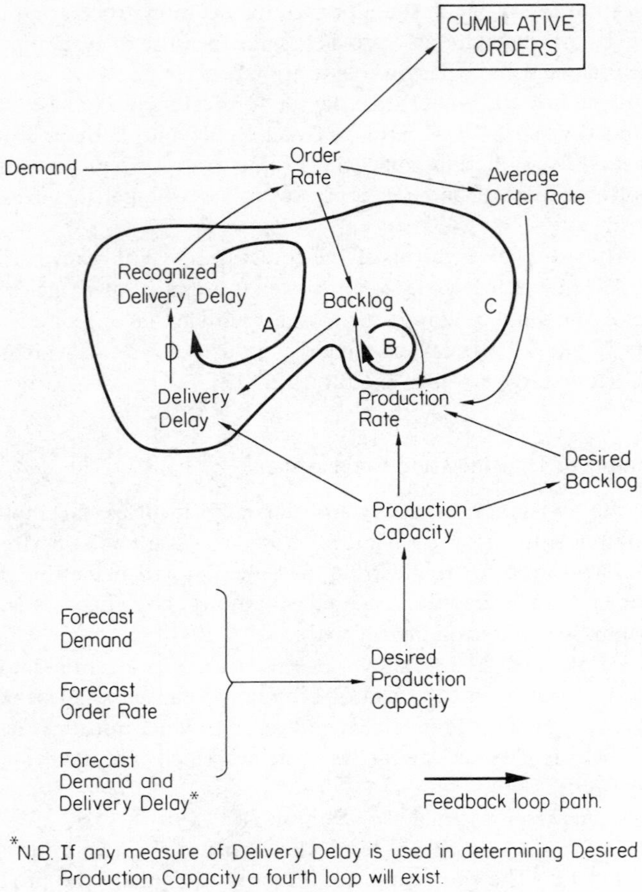

Fig. 9.17 Structure of illustrative forecasting system

Obviously, PROD may not exceed the production capacity, PCAP, and PROD has to fall to zero if the backlog vanishes.

The long-term controller is to regulate PCAP towards a desired value, and the difficulty for the firm is to know what to use in determining this desired value. Clearly it has to use some kind of forecast but of what? Three of the many options are shown on the diagram and the object of the exercise is to see how the performance of the *system* will be influenced by the choice of a variable to forecast and by errors in the forecasting procedure. For our purposes the performance of the system is assessed by considering the cumulative orders taken over a period (there are, of course, more sophisticated possibilities). This performance indicator is relevant to the purpose for which the corporate system exists, but it has little connection with the usual way of judging a forecasting procedure by comparing the forecast value with the actual outcome.

In this case, the performance of the system could be regulated by choosing a suitable value of TABL in the production rate equation, or by using a different formula. This would affect the performance of loop C.

The second point is that we are not engaged in developing a particular forecasting procedure. Rather the aim is to find the best choice of variable(s) to forecast and the boundaries of precision within which the forecasting mechanism must operate. In addition, since we know that errors will be inevitable, it is reasonable to enquire whether, for example, it is better to be optimistic or cautious. Once this has been done, the construction of the forecasting procedure to do the required job is a separate exercise. We could, of course, equally well turn the approach round and say that 'this is the best precision that forecasting method A can achieve, can the system tolerate it?'.

In order to plan capacity, the obvious variable to forecast is Demand. The model shows that, if perfect forecasts of demand could be made, the cumulative orders received, CUMORD, would amount to 35394. We have no means of knowing whether this is good or bad, but we would expect it to be good performance because perfectly accurate forecasting seems like the impossible ideal. We shall refer to this as the *Base Case*.

The Base Case involves forecasts for three time periods ahead, because that is the time needed to adjust capacity. What would happen if we forecast demand, still perfectly, but with forecasting horizons of 1 and 6 time periods respectively. The answers are CUMORD of 35646 and 34799 respectively, or 100.7% and 98.3% of the Base Case. Although the difference is very small it is rather surprising to find performance which is better than what we had assumed to be the 'perfect' case. The reasons lie in the relationship between the capacity–control decision rule and the capacity–acquisition delay.

What would happen if the forecasting mechanism turned out to be always 20% overoptimistic? In fact CUMORD rises to 42119 or 119% of the Base Case performance. The reason for this astonishing difference is not far to seek – the higher the capacity the lower the delivery delay and the higher the orders, and these are what matter, not the forecasting 'accuracy'.

This suggests that we might be better employed in forecasting the actual order rate. This is more difficult since, as we have seen, order rate is partially determined by the system state. The crudest method is to take a recent average of the order rate and call it a forecast, which it is not, but we can make it serve as one. This gives CUMORD of 46640 which is 110.7% of the optimistic demand forecast's performance (which was a forecast, even though a biassed one) and 131.8% of the Base Case!

If we use average order rate and bias it upwards by 20%, CUMORD rises to 60419 or 170.7% of the Base Case.

Finally, we may go the whole way and model a rather rough attempt to control the system by forecasting the effects of delivery delay on demand, i.e. to focus the forecasting on the actual order rate determining mechanism. This gives CUMORD of 60931 which is no less than 172.2% of the 'perfect' result from the Base Case.

We are certainly not entitled to draw any conclusion from these experiments, other than to hope that they have demonstrated the importance and possibility of an analysis of the system's properties as a guide to the design of a forecasting mechanism. It will be an exercise for the reader to attempt to construct a model with the same characteristics.

9.11 Dynamics of Management Information Systems

The basic reason for wanting a Management Information System,MIS, especially a computer-based one, is that much of the information the firm is using is either late, inaccurate, or both. If the lateness or inaccuracy can be eliminated or reduced it is assumed that 'better' decisions will be made. This seems to be a strikingly naive basis on which to invest the large sums needed for developing a sophisticated MIS because, as we have repeatedly seen in the earlier chapters, what matters is the combination of the information *and* the control rules used to act upon it. How, therefore, can we use the SD model to study this problem?

In Chapter 8 we used the formulation that the Indicated Production Level, IPL, depended on the actual Order Backlog, OBL. The model calculates OBL precisely and feeds it instantly, or at least every DT, into the production decision. In effect, we are assuming the presence of a perfect MIS which always produces accurate and current information. There is, of course, no such thing as a *perfect* MIS and in practice OBL would be out of date and probably inaccurate. We could easily represent this by delaying OBL so that the value used in the Production Decision was what the OBL used to be, say 6 weeks ago. We could also introduce errors to represent OBL being consistently underestimated or subject to random errors or whatever. In short, we could define a new quantity, OBLU, the Order Backlog Used and put this into the production equations in place of OBL and also have:

$$\text{OBLU}=f(\text{OBL},\textit{lateness},\textit{errors})$$

where the function can be whatever we wish. The same procedure could be applied to all the other variables such as Inventory INV, Work in Process PLA, Average Sales Rate ASR and so on. Make a list of all such variables in Chapter 8.

We can now run the Chapter 8 models (do it) with this DEGRADED information to see how much difference it makes *with those decision rules*. This opens up a very considerable range of options:

1. Changing the control laws to make the system do better with the degraded information.
2. Increase the amount of degradation to see how wrong the information could be before system performance deteriorates.
3. Reduce the modelled degradation to find out how much the better information improves the performance.

This last idea is the key one. If we can find out which are the critical information channels–probably those in the dominant loop–we can assess the performance criteria for an MIS. We could then specify to the MIS expert that, say, information on Backlog should be no more than two weeks old and accurate to within +5% and –2%. If the MIS people say that this is technically impossible we can then try to find control laws which will be robust with the information they *can* provide.

The implications of this are very striking in terms of the cost/benefit of an MIS, particularly if it can be shown that only some of the information channels need to be upgraded. A further advance would be to use the model to examine the sequence in which information channels should be improved.

Notice that we are discussing MIS from the viewpoint of overall control and have not considered the other side of MIS, namely the need to keep track of individual orders to answer customer queries or to ensure timely delivery of components from external suppliers. This is quite a different matter and SD would not help very much with it. On the other hand, a company might well be unwilling to invest in an MIS if it was only going to be used to answer customer enquiries, particularly if it could be shown that improving information accuracy was not going to lead to improved system performance.

Finally, there is a difference between a study of information degradation and sensitivity analysis of parameters. In the latter case we are saying that, as modellers, we cannot estimate some parameter accurately and we wish to know how important it is. In the former we are using the model as a design tool to predict the consequences of a change which might be implemented.

9.12 A Model of Corporate Growth

We conclude the treatment of the individual corporate mechanisms by drawing them together into a *simple* model of a growing firm. We shall show how such a firm can generate markedly different modes of behaviour and great variations in performance by varying its policies. (The aim of Sections 9.12 and 9.13 is not to give a realistic model of a particular firm, but to demonstrate some ideas. The model should, however, be useful as a basis for some types of corporate planning situations—especially new venture planning. For a contrasting corporate model, see Chapter 12.)

The firm adopts a simple policy of applying a fraction of its cash flow to development of new products and the remainder to capacity acquisition. If the fraction is large enough, new products will be developed and new demand will be created, not all of which need be accepted as orders. It the level of orders exceeds capacity, the extra business is dealt with on overtime or subcontracted to outside firms at a greatly reduced profit. The problem for the firm is therefore to balance its activities so as to generate its own growth.

In this case we are considering a firm which is developing a new product or a new market area. The potential market is assumed to be much larger than the firm's present size, so that we can ignore the limits to growth and concentrate on the generation of growth.

Before discussing the model we shall explain it, though the reader should also examine Appendix B.9.12 and run the model. The model is fairly simple, but it presents a reasonably plausible Board's eye view of a firm, uncluttered by detailed treatment of cash flow, taxation and depreciation.

The Cash Flow, CFLOW, is modelled from the Order Level and the Unit Order Value, less the cost of maintaining and operating the firm's capacity, which we assume to be constant whether the plant is used or not.

```
A CFLOW.K=OV.K*OL.K-CAPAC.K*CMC
C CMC=55
        CFLOW    =(£/WK) CASH FLOW FROM ORDERS

        OV       =(£/U) ORDER VALUE

        OL       =(U/WK) ORDER LEVEL

        CAPAC    =(U/WK) PRODUCTION CAPACITY

        CMC      =(£/WK/(U/WK)) COST OF MAINTAINING CAPACITY
```

Product Development Spending is geared to CFLOW, as described below, but takes some time to be fully effective.

```
R PDSR.KL=MAX(0,PDSFAC.K*CFLOW.K)
        PDSR     =(£/WK) PRODUCT DEVELOPMENT SPENDING RATE

        PDSFAC   =(1) PROPORTION OF CASH FLOW SPENT ON PRODUCT DEVELOPMENT

        CFLOW    =(£/WK) CASH FLOW FROM ORDERS
```

```
L PPDS.K=PPDS.J+(DT/TPPDS)(PDSR.JK-PPDS.J)
N PPDS=MAX(0,PDSFAC*CFLOW)
C DT=0.0625
C TPPDS=4
        PPDS     =(£/WK) PRODUCT DEVELOPMENT SPENDING LEVEL AS PERCEIVED
                  BY CUSTOMERS

        DT       =(WK) SOLUTION INTERVAL

        TPPDS    =(WK) TIME FOR CUSTOMERS TO PERCEIVE PRODUCT DEVELOPMENT
                  SPENDING

        PDSR     =(£/WK) PRODUCT DEVELOPMENT SPENDING RATE
```

PPDS [£/WK] can be translated into equivalent products in the market place by dividing by a factor PSDEC [£/WK/(U/WK)] which can be thought of as the weekly cost of advertising, sales, service, and so on needed to support demand. We can therefore model the equivalent order level which is, as it were, being created and supported by the product development spending.

```
A PPDSE.K=PPDS.K/PDSEC
C PDSEC=1
        PPDSE    =(U/WK) EFFECTS OF PERCEIVED PRODUCT DEVELOPMENT SPENDING
                  IN TERMS OF MARKET DEMAND

        PPDS     =(£/WK) PRODUCT DEVELOPMENT SPENDING LEVEL AS PERCEIVED
                  BY CUSTOMERS

        PDSEC    =(£/WK/(U/WK))  PRODUCT DEVELOPMENT SPENDING COEFFICIENT-
                  AMOUNT OF AVERAGE PRODUCT SPENDING NEEDED
                  TO SUSTAIN 1 UNIT PER WEEK OF DEMAND
```

Order Level, OL, is modelled by considering changes to take place as a result of the effects of product development spending which manifest themselves in PPDSE, the Perceived Product Development Spending Effect (or Equivalent). If, say,

PPDSE is twice the Average Order Level then the market reacts to the firm's product development activities by eventually increasing its orders. The extent of this increase has been found, we shall suppose, to be rather less than the ratio of PPDSE to AOL. For example, if PPDSE/AOL=2 then, after a year, the market would have increased its orders by 70%. This would, of course, increase AOL and by the end of the year, if PDSR was kept at that high level, the ratio would have fallen to 2.0/1.7 = 1.176. This would create a small increase in the second year but eventually AOL would move up to the level dictated by PPDSE. The increase would be slower than this because the model increases AOL continuously. Conversely, if PPDSE fell to zero, orders would eventually die away to match. This is all expressed in the following equations.

```
R RCOL.KL=PDSM.K*AOL.K
        RCOL     =(U/WK/WK) RATE OF CHANGE OF ORDER LEVEL

        PDSM     =(1/WK) PRODUCT DEVELOPMENT SPENDING EFFECT MULTIPLIER

        AOL      =(U/WK) AVERAGE ORDER LEVEL


A PDSM.K=TABHL(TPDSE,PPDSE.K/AOL.K,0,2,0.5)/50
T TPDSE=-0.5/-0.2/0/0.5/0.7
        PDSM     =(1/WK) PRODUCT DEVELOPMENT SPENDING EFFECT MULTIPLIER

        TPDSE    =(1) TABLE FOR PRODUCT DEVELOPMENT SPENDING EFFECTS
                  AS A MULTIPLIER OF AVERAGE ORDER LEVEL

        PPDSE    =(U/WK) EFFECTS OF PERCEIVED PRODUCT DEVELOPMENT SPENDING
                  IN TERMS OF MARKET DEMAND

        AOL      =(U/WK) AVERAGE ORDER LEVEL
```

The factor of 50 in the second equation scales yearly figures to weeks.

TPDSE is a simple model of buyer behaviour and other formulations can be achieved. (Try some!). The graph is shown in Fig. 9.18. Clearly, this curve could never be known with certainty and the reader should apply the concept of internal robustness to it.

The firm bases many of its decisions on the relationship between demand and capacity—the Demand/Capacity Ratio DCR, defined by:

```
A DCR.K=OL.K/CAPAC.K
        DCR      =(1) DEMAND/CAPACITY RATIO

        OL       =(U/WK) ORDER LEVEL

        CAPAC    =(U/WK) PRODUCTION CAPACITY
```

This is averaged to get the Average Demand/Capacity Ratio, ADCR, which is actually used by the firm as the control variable:

```
ADCR.K=ADCR.J+(DT/TADCR)(DCR.J-ADCR.J)
ADCR=1
DT=0.0625
TADCR=4
        ADCR     =(1) AVERAGE DEMAND/CAPACITY RATIO

        DT       =(WK) SOLUTION INTERVAL

        TADCR    =(WK) AVERAGING TIME FOR DEMAND/CAPACITY RATIO

        DCR      =(1) DEMAND/CAPACITY RATIO
```

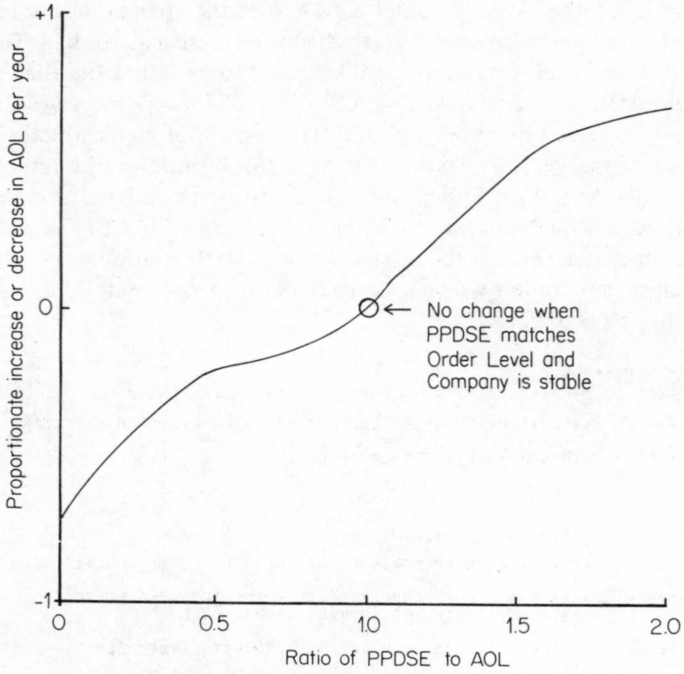

Fig. 9.18 The effects of product development spending on growth of Order Level

The Demand/Capacity Ratio, DCR, is important because it governs the economics of the situation.

```
OV,K=TABHL(TDCR,DCR.K,0,2,0.5)
TDCR=66.7/66.7/66.7/40/30
        OV        =(£/U) ORDER VALUE

        TDCR      =(£/U) TABLE GIVING ORDER VALUE IN TERMS OF DEMAND/CAPACITY
                   RATIO,DCR

        DCR       =(1) DEMAND/CAPACITY RATIO
```

The Order Value can be regarded as what remains after deduction of profit and taxes. This differs from the traditional costing approach but is, perhaps, more relevant to the problem dealt with here of whether the firm can generate its own long-term growth. (Alter the model if you don't agree.)

The table shows that the value of orders drops away if capacity does not match demand as excess orders are met by overtime or subcontracting at greatly reduced value, as show in Fig. 9.19. The actual current value of DCR is used rather than the smoothed value ADCR. The latter is used by the firm as a control variable, but the former represents the present state of affairs and therefore determines the value of processing current orders.

Note that the values in the table are for all orders, not just the marginal ones in excess of capacity. If DCR=2, OV=30, which implies that the marginal orders are

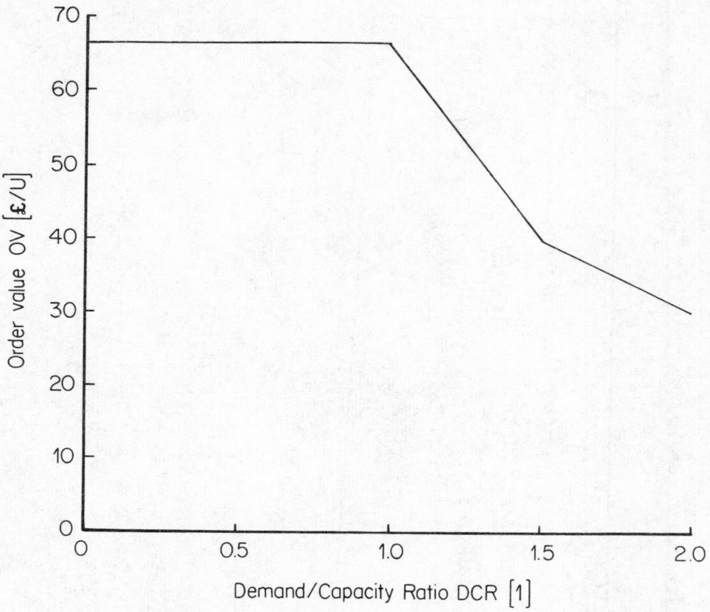

Fig. 9.19 Relationship between Order Level and Demand/Capacity Ratio

being processed at a loss, if orders up to capacity have a unit value of 66.7. Thus if CAPAC=100 and OL=100, DCR=1 and OV=66.7 and OLxOV=6670.

If OL rises to 200, then DCR=2 and OV=30 and

OLxOV=200x30=6000

– an actual fall in revenue. Cash Flow for these two cases would be 1170 and 500, a much greater change.

The salient policy area in the model is that on Product Development Spending which, as we have seen, is a fraction PDSFAC of Cash Flow. The company's current policy is shown in Fig. 9.20. When ADCR falls below 0.8 all the cash flow is spent on Product Development. Above this value, PDSFAC falls–the aim being 15% of cash flow when Demand and Capacity balance. When ADCR>1, PDSFAC is cut back to leave more money for capacity construction.

Capacity starts are governed by the cash flow left over from product development spending.

```
R CSR.KL=MAX(0,((1-PDSFAC.K)*CFLOW.K)/CCOST)
N CSR=(CAPAC/PLC)
C CCOST=170
        CSR      =(U/WK/WK) NEW CAPACITY START RATE

        PDSFAC   =(1) PROPORTION OF CASH FLOW SPENT ON PRODUCT DEVELOPMENT

        CFLOW    =(£/WK) CASH FLOW FROM ORDERS

        CCOST    =(£/U/WK) CAPITAL COST OF CAPACITY
```

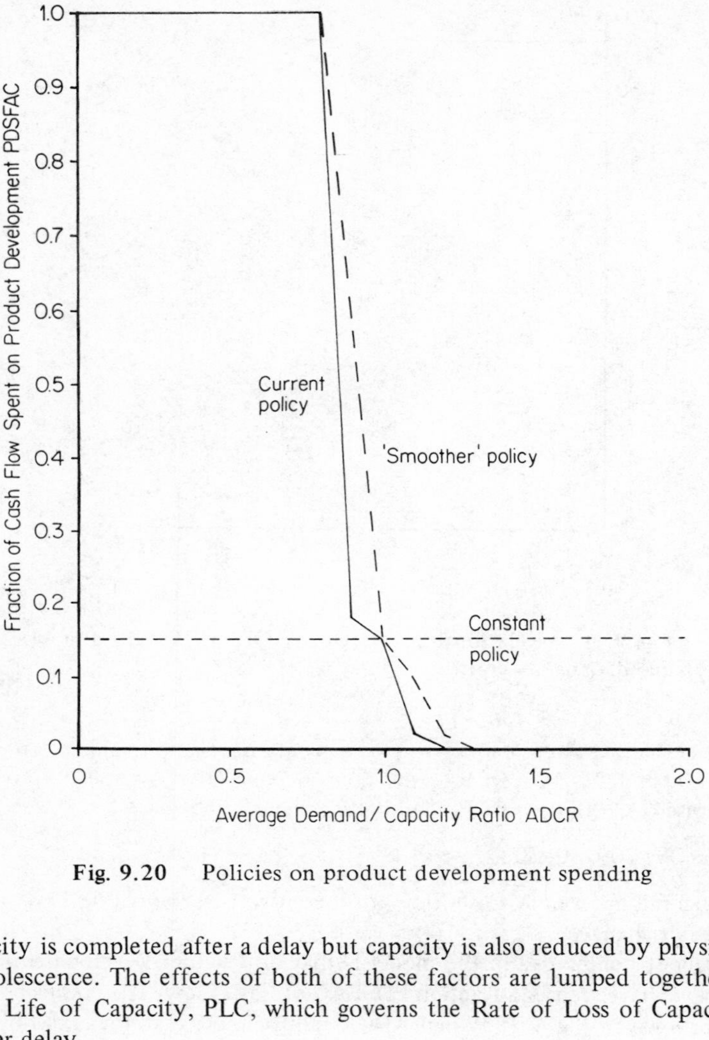

Fig. 9.20 Policies on product development spending

Capacity is completed after a delay but capacity is also reduced by physical wear and obsolescence. The effects of both of these factors are lumped together in the Physical Life of Capacity, PLC, which governs the Rate of Loss of Capacity by a first order delay.

```
R CAR.KL=DELAY3(CSR,JK,CCDEL)
C CCDEL=13
        CAR      =(U/WK/WK) CAPACITY ADDITION RATE

        CSR      =(U/WK/WK) NEW CAPACITY START RATE

        CCDEL    =(WK) CAPACITY CONSTRUCTION DELAY


R RLC.KL=CAPAC.K/PLC
C PLC=20
        RLC      =(U/WK/WK) RATE OF LOSS OF CAPACITY

        CAPAC    =(U/WK) PRODUCTION CAPACITY

        PLC      =(WK) PHYSICAL LIFETIME OF CAPACITY
```

```
L CAPAC.K=CAPAC.J+DT*(CAR.JK-RLC.JK)
N CAPAC=100
C DT=0.0625
        CAPAC   =(U/UK) PRODUCTION CAPACITY

        DT      =(WK) SOLUTION INTERVAL

        CAR     =(U/UK/WK) CAPACITY ADDITION RATE

        RLC     =(U/UK/WK) RATE OF LOSS OF CAPACITY
```

All this can be summed up into an influence diagram, Fig. 9.21 and 9.22. The reader must trace out all the loops and check the loop summary table, Table 9.1.

The reader will see that Table 9.1 is rather different from the loop summary tables in Chapter 8. This is because the model we are dealing with now is structurally much more complex. For example, loops E and F contain parallel subpaths which feed loop components, forward, such as CFLOW in loop E, or backward, such as CAPAC in loop F, so that they re-appear as gain components in their own loops. This happens because the model is highly non-linear in that not only are its policy curves non-linear as in Fig. 9.20 but so are its behavioural relationships, as in Fig. 9.18 and 9.19, and many of the model equations are multiplicative functions of the system variables, as in the equation for CFLOW. This contrasts with the Chapter 8 model which contains only a few non-linearities notably in the policy curve for RBL.

For such a complex non-linear case as this model the full loop summary approach is hardly feasible and a better method is to attempt a qualitative identification of gain and delay rather than an explicit analysis. This will be greatly helped by a careful inspection of the model so that any peculiarities it has can be accomodated or even exploited. In this case we may observe the following points.

1. Most of the parameters of the system cannot be changed. For example, TPPDS, PDSEC, TPDSE and TAOR between them model the process of product development and the market's behavioural response to product development spending. In this case TAOR does not model the controller averaging orders, as it did in Chapter 8, but the *workings of the environment*.
2. Loop B acts as a delay in bringing Order Level to match PDSR. The only parameter which is within the controller is TADCR and the only policy is TPDSF. This is, of course, with this particular structure and the reader should experiment with other ways of controlling the whole thing.
3. The link involving TDCR only operates negatively when DCR exceeds 1. Below that value OV is constant and that link does not exist and therefore loops D and I do not operate. This is shown by the *on the link from DCR to OV.
4. There are two genuinely positive loops, A and E, both of which contain pure integration and the gain factor TPDSF. They will therefore both produce growth providing PDSFAC>0, but when ADCR>1.2 this is not so and growth will cease, and the loop E will not exist as changes in ADCR within that range no longer alter the value of PDSFAC.

Armed with these considerations, we now arrive at Table 9.1 which is only an outline, not a true loop summary. The delays are fairly clear, the only exception being TAOL which is shown in brackets to signify the operation of loop B.

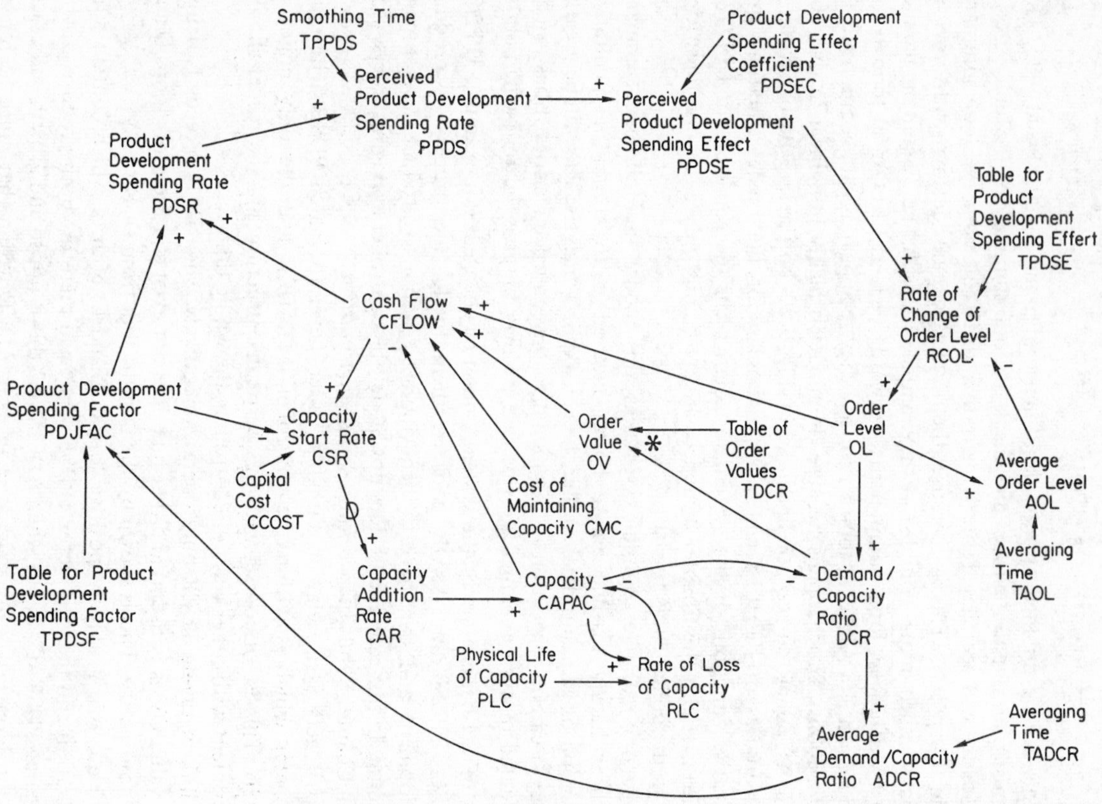

Fig. 9.21 Influence diagram for corporate growth model

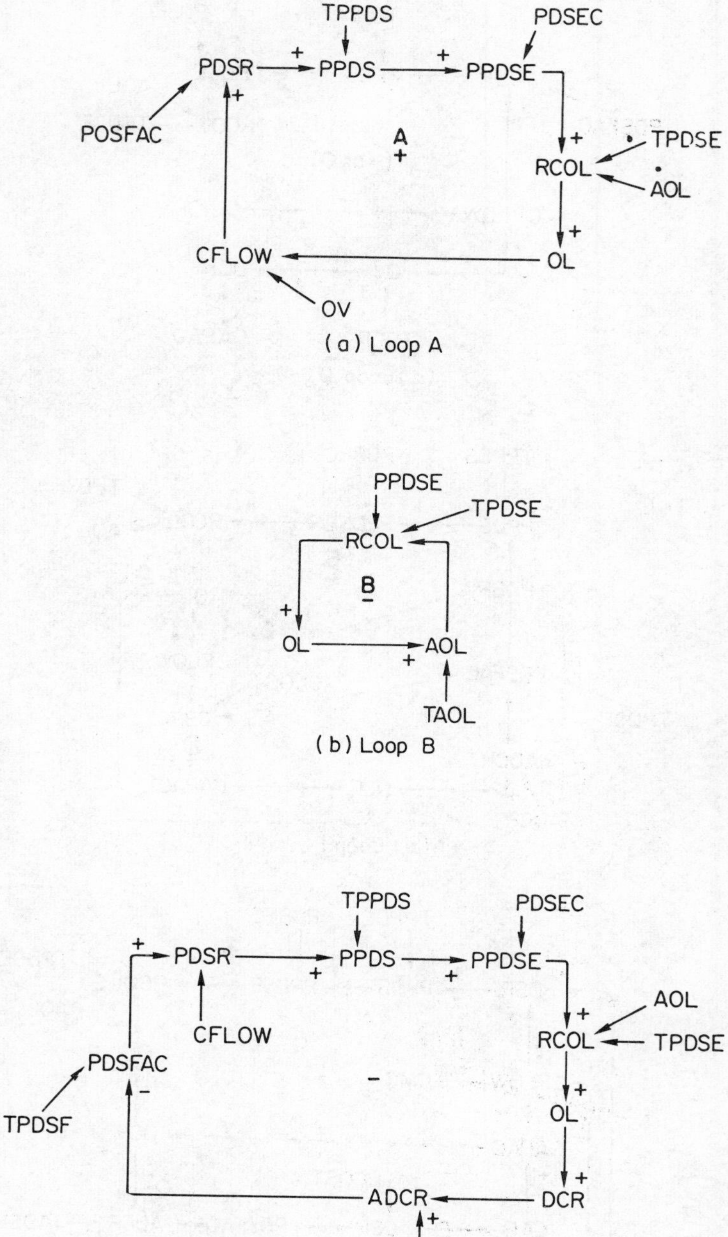

Fig. 9.22 Feedback loops of Fig. 9.21

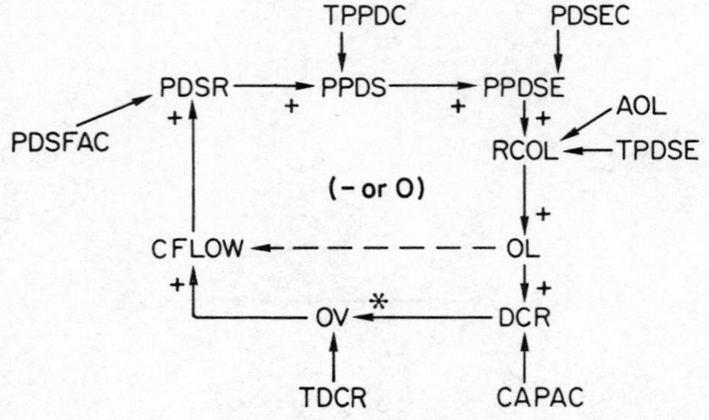

(d) Loop D

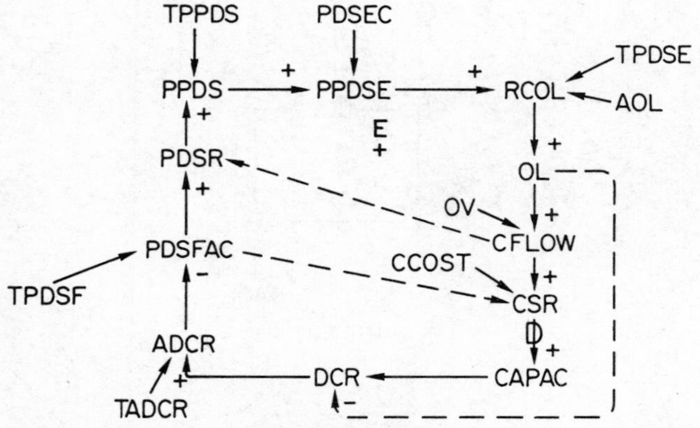

(e) Loop E

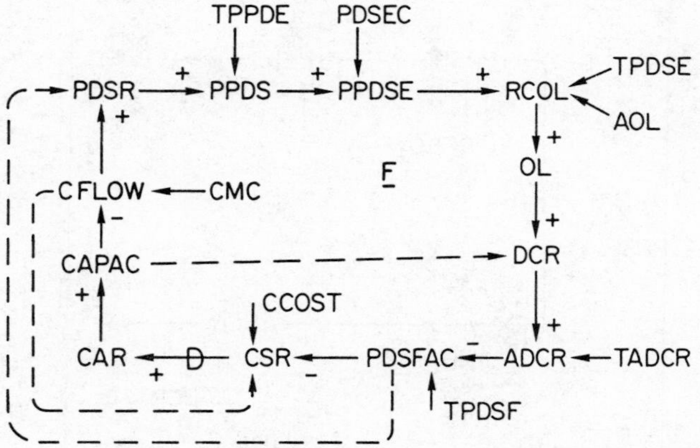

(f) Loop F

Fig. 9.22 *Continued*

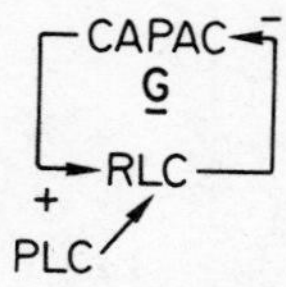

(g) Loop G

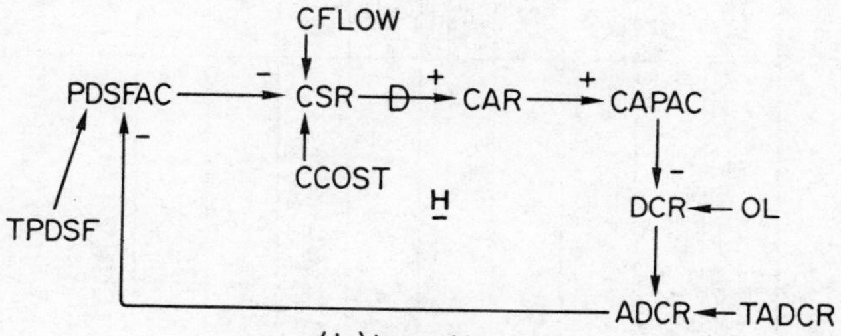

(h) Loop H

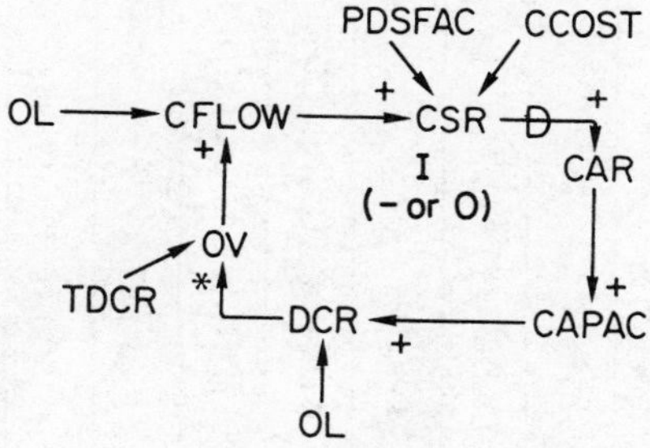

(i) Loop I

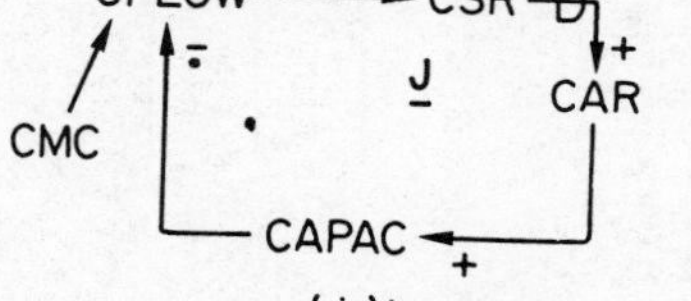

(j) Loop J

Fig. 9.22 *Continued*

Table 9.1. *Loop Summary Table for Fig. 9.22*

Loop	Type	Delays	Gain	Pure Integs.
A	+	TPPDS, (TAOL)	TPDSF, TDCR	OL
B	–	TAOL	TPDSF	OL
C	–	TPPDS, TADCR, (TAOL)	TPDSF	OL
D	–/0	TPPDS, (TAOL)	TPDSF, TDCR	OL
E	+	TPPDS TADCR, (TAOL), CCDEL	TPDSF, TDCR	OL, CAPAC
F	–	TPPDS, TADCR, (TAOL), CCDEL	1–TPDSF, 1/CMC	OL, CAPAC
G	–	PLC		CAPAC
H	–	TADCR	1–TPDSF, 1/CCOST	CAPAC
I	–/0	CCDEL	1–TPDSF, TDCR	CAPAC
J	–	CCDEL	1–TPDSF, 1/CMC, 1/CCOST	CAPAC

The gains are those which are either fixed parameters of the system, policies in the controller, or the economic factor TDCR (the variation of which must clearly be a matter for potential concern). In loops F, H, I and J, the product development spending factor appears as 1–TPDSF because it is the *remainder* of CFLOW which affects CSR. The reader should carefully examine Table 9.1 and Figs. 9.21 and 9.22 bearing in mind our warning that the table in this case is merely a guideline and memory-prodder.

Even with this simple approach we can see that the model can have a variety of behaviour modes. If PDSFAC increases, the gain in loops A and C rises and growth takes place because loop C has a longer delay and cannot hold loop A back. However the eventual rise in demand cuts back OV, reducing the gain in the growth loops and ADCR cuts back PDSFAC, re-inforcing the gain reduction in the growth loops and increasing the gain in the negative loops. Loop F is clearly important because it holds down cash flow. Loop G would ease the position by allowing capacity to fall of its own accord, but it has a very long delay. Clearly the non-linearity of TPDSF will act as a dominance transfer mechanism.

We should therefore expect to see stair-step growth, and possibly overshoot and collapse, as even though there is no explicit upper limit, the negative loops which operate when DCR is low may drag the system down before the positive loops can pull it up again. We should like to see the system produce smooth exponential growth, because that is clearly what it is supposed to do.

This model therefore contains the behaviour modes which we seek, and we are in danger of confusing the dynamics of the system with those of equations which can never be more than a convenient approximation to a system. Thus there is a real risk of getting answers which are misleading. This is a problem which is inherent in dealing with situations of this kind, that is we face the risks of formulating a model which gives the answers we want to get. How, then, shall we proceed?

In a real case, as opposed to one manufactured for a text book, there would be two possibilities. We might back away from the problem and say, in effect, that the overall problem of corporate growth is too difficult to approach, and we can give no help to management. Alternatively, we devote some effort to ensuring that the formulation adopted is, in a real case as opposed to this textbook example, actually a good description of the processes involved and generates the dynamics because of that and not because it has been chosen to create the dynamics we wish to see.

In this section we are attempting to do precisely that, in that we deliberately want to create behaviour modes in order to study some ideas of corporate growth, rather than to model exactly a real situation, so we shall keep this formulation. The purpose of this digression is to point out the need for very careful modelling if one *was* dealing with a real situation of growth generation.

9.13 The Behaviour of the Corporate Growth Model

The dynamics of a trial run on the model are shown in Fig. 9.23. The magnitudes in the model are rather arbitrary and we have deliberately used rather small time delays to magnify the dynamics but the stair-step growth shows up

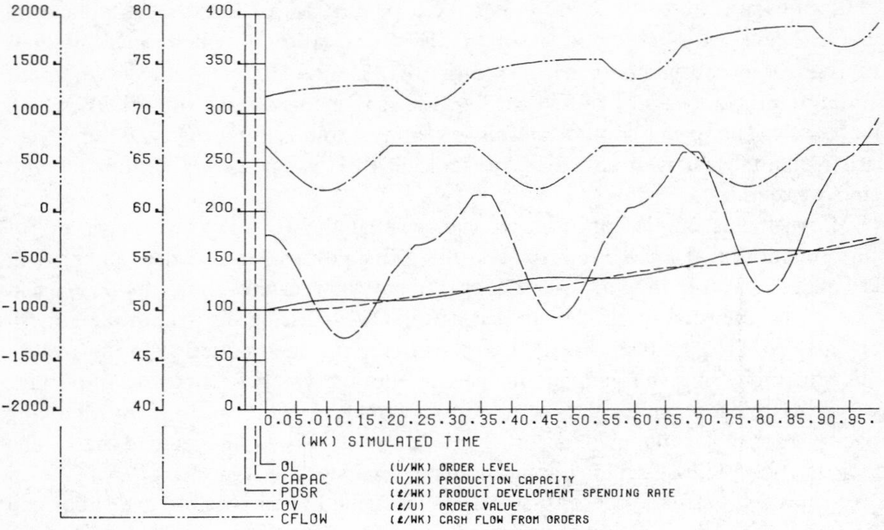

Fig. 9.23 Current product development policy

rather clearly, as do the cycles in product development spending which would, of course, be very unpleasant to live with and impractical in the real world. Note that this does not imply that the *model* is invalid, but that the *system* is inoperable if run the way management intended.

Again, we note that the natural period is approximately 45. In this case the time unit is the week but if we had worked in months (which would probably be more realistic), the natural period would be very close to the 4-year business cycle. In fact, this behaviour has nothing to do with the business cycle, and is entirely system-induced.

Remarkably enough, if plant is used which lasts longer, PLC=40, the result is most unpleasant, as shown in Fig. 9.24). However a rise in the price of capacity to CCOST=190, with the original PLC, has little effect on the behaviour mode. What does this imply about the robustness of the model?

The demand/capacity averaging period is also important, e.g. if TADCR=2 growth is smoother, but if TADCR=12, Fig. 9.25 is the result. The reader must work out what has happened in terms of the loop delays.

The basic behaviour is clearly unsatisfactory and, in seeking improvement, it could be argued that the more realistic course would be to make TPDSF a constant at, say, 0.15 which is the stable value the company would like to adopt. This would have the effect of switching off several loops (which?) and might well be much closer to industrial practice in which research budgets are not constantly oscillating, as they do in Fig. 9.23. The results of such a policy are shown in Fig. 9.26.

Clearly, this is something of a disaster as the smooth growth which does take place does not last, and there are signs of collapse or of very long period

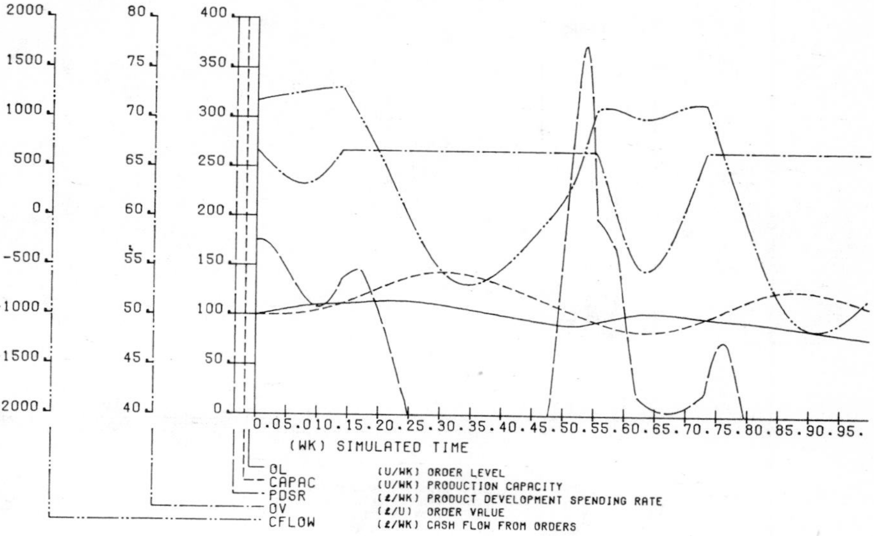

Fig. 9.24 Longer plant life

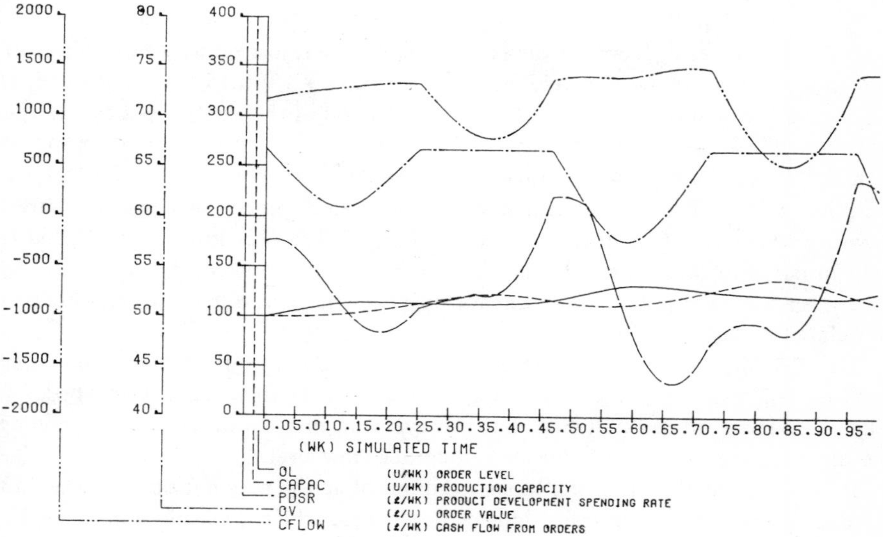

Fig. 9.25 Slower averaging of DCR

oscillations. The reason is that loops C, E, F and H, no longer function so there is very little to regulate the order level away from the unprofitable zone. Clearly we have the classical case of a company experiencing fairly rapid growth and steady expansion in capacity, but a constant cash flow with collapse apparently looming ahead.

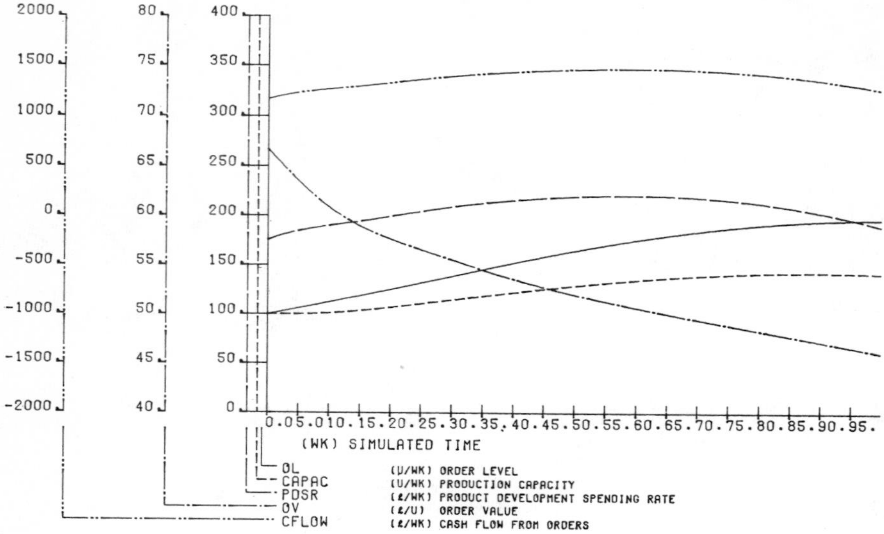

Fig. 9.26 Constant product development spending

There is, apparently, some virtue in a constant TPDSF providing we can stave off the collapse and one way would be to use TPDSF = 1.0/0.15/0.15/0.15/0.15/0. This curve is not shown in Fig. 9.20 but it is a hybrid of the existing policy and the constant which we have just rejected. It provides for capacity expansion if ADCR=0.8, constant product development spending between ADCR=0.9 and 1.2 inclusive, and no product development if ADCR>1.3. Since the table only varies PDSFAC when ADCR is in the ranges 0.8–0.9 and 1.2–1.3, loops C, E, F, and H only function in those intervals and they do not exist elsewhere. We shall refer to this in an 'on–off' policy, which is not exactly what it is but the title is used to stimulate the reader to think.

The performance as shown in Fig. 9.27 is clearly an improvement, the final capacity slightly, and the final order level considerably, exceeding those in Fig. 9.23. There are no stair-steps but the cycles in product development spending are unsatisfactory and order value tends to rather a low level.

It is now clear that allowing loops to go out of action may not be beneficial but we do not want the sharp transitions of dominance which took place in Fig. 9.23. Further, we want to see OL and CAPAC grow more nearly together rather than the high DCR which appears in Fig. 9.27 and which could well produce problems if a longer LENGTH is taken. Indeed Fig. 9.27 may simply be showing the first stages of stair-step growth of much longer period than that of Fig. 9.23.

Another attempt is therefore, to smooth TPDSFas shown in Fig. 9.20, the result being shown in Fig. 9.28. This behaviour does not change, except of course, in magnitude if prices drop to, say, TDCR=66.7/66.7/30/20 showing this is a fairly robust policy.

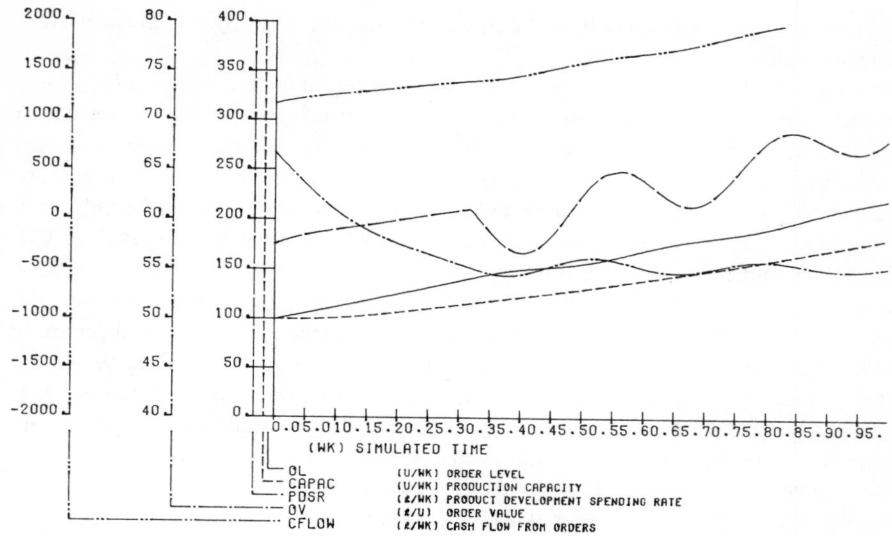

Fig. 9.27 On–off product development spending

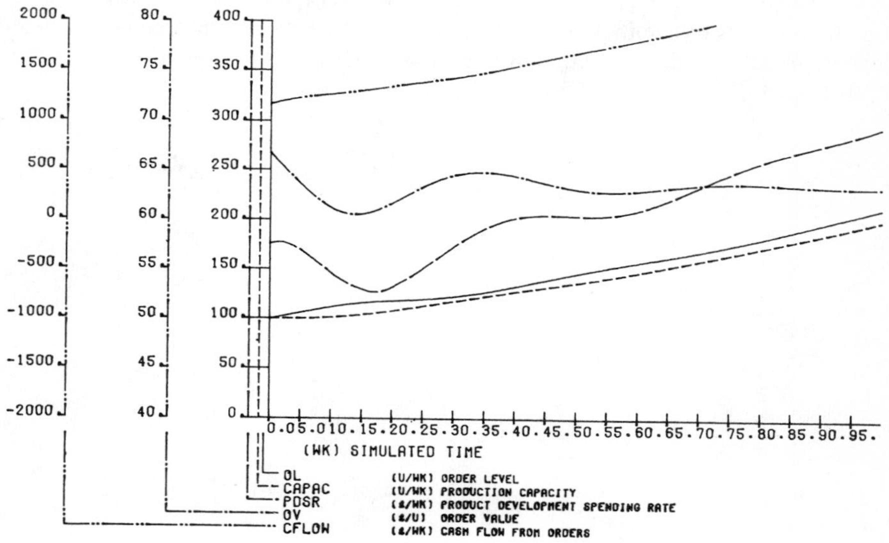

Fig. 9.28 Smoother product development spending

As usual, we have merely pointed the reader in a direction, and further experiment and loop analysis on his part may well be fruitful.

In Chapter 12 we shall encounter a much more detailed and, in fact, fundamentally different, sort of corporate System Dynamics model. What then, is the difference between that model and this and does this model have any value apart from suggesting some ideas of corporate modelling and showing how different behaviour modes arise?

Chapter 12 is an explicit model of the 'what would happen if' type and it should be useful for corporate planning and for seeing how managerial policies would, or would not, achieve the objectives the company has. This chapter's corporate model is much more 'broad-brush' but it could have value in a real situation, with perhaps comparatively little amplification of detail. To be provocative, one might only need to double its size to make it a 'good' (what?) model of General Motors, Shell or ICI. Its value would lie in showing how the whole 'style' of corporate policy could, or could not, produce growth, stagnation, decline or oscillation, and, in short, the extent to which the company produced its own successes and failures and generally made itself easy or hard to manage. Paradoxically, the greatest difficulty in that kind of project, as opposed to the quite different corporate model in Chapter 12, is the need to avoid getting swamped in more and more detail whilst still carrying conviction. The best people to build such a model would be managers, as analysts often seem to feel that they need the security apparently, but misleadingly, provided by lots of data, model detail, and 'optimization'.

Finally, we would stress that this model, and indeed this whole chapter, does not involve any prescriptions for how firms should or should not be run. The book is about dynamic modelling, not business administration, and what we have tried to show here is how dynamic processes arise, how they *might* be modelled (in practice much depends on individual circumstances), and what implications there might be in consideration of the loops involved in those processes.

Notes on Chapters 10 to 12

Having studied the way in which dynamic models are built and analysed, we now turn to the problem of the practical industrial use of the methods. To do this we shall study three case studies, which are described in the next three chapters, and which deal respectively with a consumer-goods company, an oil company, and a metals company. The problem areas are respectively production planning, short-term resource acquisition, and corporate planning.

Each case is based on work done by the author, and others, in real British firms. The case studies have been adapted from the project reports, partly to preserve confidential information, but mainly to put them into a form more suitable for student use. The reader should NOT expect to find out from the case studies:

1. How this or that company is managed.
2. How to build production, or corporate planning, or any other kind of model.
3. The 'correct' modelling method to use in a particular situation.

He should, however, look for:

1. Some guidance on how particular modelling problems have been approached in other areas, becuase this may shed light on some of the snags he may encounter in his own work.
2. Some insight into the way models are formulated and analysed in the practical world, as opposed to the classroom world of the preceding chapters.
3. Some practice in handling large and complicated models, especially, perhaps, in simplifying them to get at the essence of the problem, uncluttered by the detail which is often inserted into models to make them credible to managers.
4. A feel for the kind of problem where system dynamics might be usable, so long as he lets his imagination free and does not restrict himself to the three areas described in these chapters.
5. Some comprehension of the way in which practical SD studies are carried through their various stages.
6. An idea of how complicated situations and models can be reduced to their essentials in preparing reports.

Chapter 10

Case Study – DMC (Export) Ltd.

10.1 Introduction

This chapter is a case study of an actual dynamic problem, involving relationships between production and distribution for an export market. After a brief description of the company and its operations we shall examine the reasons for doing a dynamic analysis, rather than using some other technique. After that we shall discuss the way in which the model was built and verified as being realistic, and then describe the experiments which were done on the model and the reasons for choosing those particular ones.

For the sake of reasonable brevity many of the experiments which were done are not recounted here. The main computer programs for the model are listed in Appendix B and the reader should carry out his own experiments in order to make a fuller analysis of the problem.

The study described in this chapter was a real investigation done in a major British company. For obvious reasons some features of the company and the background information have been disguised, without losing the basic sense of the study. Some of the recommendations made to the real company have been excluded from this account.

10.2 The Company and its Operations

DMC may be regarded as three related companies. First there is a holding company, based in England, which exercises control over all DMC's varied activities. Second there is a subsidiary, DMC (Home) which manufacturers and sells the company's products in Britain, and also manufactures products for sale in all the other European markets. Thirdly there is an export sales company – DMC (Export) – which markets the company's products in the European countries and elsewhere.

DMC manufactures a wide range of high-quality goods. It has an appreciable market share in all of the countries in which it operates but competition is keen both from local companies and from other international groups.

The day-to-day export operations in the major European markets are in the hands of locally-based Sales Companies. Each of these sell to dealers rather than directly to the public. The dealers may be individual shops, small groups, or large chains with branches all over the country in question. The dealers are not restricted to selling only DMC products so that if DMC acquire an unfavourable reputation for, say, delivery then the dealer may very well switch his efforts into selling a competing product. DMC advertise nationally in each country and there is a certain amount of brand loyalty, but availability at the point of sale is accepted as being a key factor. It is generally believed in DMC (Export) that an ultimate purchaser will be prepared to wait only a few days for a DMC product. If it is not available then he will usually buy a competing article and in this sense DMC will have lost a sale.

The Export distribution system is organised on the following basis, apart from a few minor exceptions. The dealers who retail DMC products are supposed to hold stocks equivalent to about four weeks sales. There is no way of making them do this or of knowing if they do it, but DMC believe that generally they do. (DMC's marketing people would hardly say anything else. Is the system sensitive to this belief?) The dealers order their requirements from the local DMC Sales Company, who generally deliver within a few days if they have the goods. The DMC Sales Companies are each supposed to have eight weeks stock, which is held in depots at suitable locations.

At the beginning of each month the various sales companies submit to the Head Office of DMC (Export) their latest forecasts of their sales *to dealers*, and expected stock position for each of the next six months, and calculating from this information an amount called the *Purchase Quantity*. This is the amount to be manufactured and delivered by DMC (Home) during the next month but two, and is the quantity needed to bring the Sales Company's stock up to the required level by the end of the next month but three, assuming all sales forecasts are fulfilled.

DMC (Export)'s Head Office collates all this and passes it on to DMC (Home) as firm orders for manufacture. Once this has been done the orders have to stand, apart from very minor changes on an 'old boy' basis. The advantage for DMC (Export) is that delivery is guaranteed, barring accidents. However the schedule cannot be revised to meet rapid market changes and there are perfectly good reasons why products made for sale in, say, France are unsaleable anywhere else.

10.3 Dynamic Phenomena

The basic dynamic phenomenon is that DMC know that demand, by the public, for one of their products rises and falls during its marketable life. This knowledge is based on factual information such as the return of guarantee cards (DMC believe this to be a very poor indication of actual sales), reports from the larger dealers and the general experience of DMC (Export)'s marketing staff.

When a product is launched, following suitable market preparation, demand rises fairly rapidly from an initial value and then falls off. At this point DMC (Export) usually undertake further sales promotion actions which cause demand to pick up again without quite reaching its earlier peak. After this, demand falls away fairly steadily until the initial level is reached. At this point, in most cases, the product is withdrawn or replaced, remaining stocks are sold off and the product life cycle has ended.

A typical cycle would be:

Time from Introduction (weeks)	*Demand (Units per week)*	
0	75	Product launch
10	85	
20	120	Peak demand
30	95	Repromotion
40	110	
50	105	
60	100	
70	95	
80	90	
90	82.5	
100	75	Product withdrawn

This will be referred to as the *pattern*.

The other dynamic phenomena are:

1. It is known, from experience, that stock at the Sales Companies varies sharply,
2. It happens occasionally that marketing opportunities are missed because the sales company is out of stock.
3. The variations in the purchase quantity are quite wide and make life rather difficult for the manufacturing side of DMC (Home).

The company has never had any reason to keep statistics on these matters but they could be quantified to some extent. This would, however, delay the study and it would be hard to find a reasonably representative period without 'special factors' which would render the data suspect.

These phenomena are all matters of common knowledge to the DMC Management and therefore inevitably influence their decision-making, at least informally. A dynamic model would permit an explicit treatment of these matters.

Since the methods of system dynamics are directly aimed at dynamic phenomena of this sort it was natural to build a dynamic model. The objectives of the investigation were:

1. To study and explain the behaviour of the system with a view to devising improved operating strategies.
2. To determine whether the failures of the system were inherent in it or the result of inadequate operation of it.

DMC were considering revising their whole distribution network in the hope of getting better control over their dealers and of reducing distribution costs. However, they also wished to know whether the alternative distribution systems would run into the dynamic problems which occurred in the existing system, and how the proposed systems should be controlled. To study this was a subsidiary aim of the study, using system dynamics as a design tool.

10.4 The Models

Ultimately four models were constructed to represent variations on two principal themes. The models were:

IA The present system.

IB The present system with dealers stocks eliminated.

IIA A hypothetical system in which there are no dealers stocks, the sales company holds stock at a Company Distribution Centre and the manufacturing is regulated at the factory in order to maintain a stock from which DMC (Export) can draw.

IIB Model IIA proved to be fundamentally unstable so this model consists of Model IIA with improved control policies applied to the sales company and the factory.

10.5 Test Inputs

Since the models represent the way in which DMC's system performs, or might be made to perform, they have to be driven in the way that the company is 'driven' by the consumer. This is done by driving the model by the pattern of demand mentioned earlier.

Although the pattern is assumed to be 'typical' it is not known by the system. The analyst is in the superior position of knowing what the driving force is going to be, but the system does not know this. It therefore does what any sensible system would do in the circumstances and forecasts. The forecast chosen is the general level of demand i.e. the height of the pattern at some future time.

There are, however, other possibilities such as predicting the occurrence of the second peak in the pattern, or predicting x months in advance the time at which the product will reach the end of its life cycle, thus allowing production to be stopped and stocks to be run down. What should x be? Is it connected to the system's settling time?

Forecasting is, however, modelled in a way which is intended to represent two quite separate operations – the making of the forecast and management's willingness to believe it. The forecast itself is denoted by, for example, AFCAST1 referring to the next period. This is modelled by a weighting of the estimated demand pattern for one PERD ahead and the current value of actual sales, ACSALES. AFCAST1 is then fed into a smoothed level, FCAST1, the smoothing time of which represents management adjusting from the old FCAST1 to the new information represented by AFCAST1. Similar processes apply to the other imminent periods and to the long-term demand forecast. The reader may wish to experiment with the parameters or even to rewrite this part of the model.

The pattern may not be the same as that used here. It might be sharper, with its peaks occurring at different times. The pattern might, in fact, start all over again, either because of market forces or because DMC promote the product again. The pattern might suddenly collapse, due to competitive actions, with a possible revival due to counter-action by DMC. There will certainly be noise in the pattern and in internal information channels. This may suggest some additional tests to which the recommendations might be subjected.

Actual Sales will be the same as market demand, provided the system has stocks available at the right time. The system's purpose is, therefore, to ensure that demand is met, and it has to be flexible enough to adapt to variations in demand. We can, therefore, judge performance on several criteria.

1. The gap, if any between demand and actual sales measured cumulatively over a 100 week period and called Lost Sales.
2. The total amount of stock in the system.
3. The dynamic behaviour of the system when faced with the steady decline in sales over the last 60 weeks of the pattern.
4. The degree of fluctuation imposed on the production system by the inability of the inventory system fully to absorb the fluctuations in demand.

10.6 Model IA. The Present System

By a process of discussion with DMC (Export) a model was arrived at which shows the basic structure of the system in the form of an influence diagram Figure 10.1. The diagram calls for a little study in conjunction with the description of the salient variables given below. However, it will be seen at once that the system contains feedback loops. For example Dealer Order Rate acts through Company Sales to Dealers and Deliveries Stock to affect Dealers Stock which in turn helps to determine Dealer Order Rate. This loop operates such that a fall in Dealer Stock triggers a rise in Dealer Order Rate which acts to restore Dealer Stock to its proper value. The speed of this reaction is influenced by the physical delay in movement of goods between Company Sales to Dealers and Deliveries to Dealers Stock and also by factors such as the Dealer Stock Adjustment time which reflects the willingness or ability of the dealers to adjust their stocks to new situations.

There is no reason to suppose that the dealers will select their policy variables to meet DMC's corporate objectives, and this conflict between the two interacting subsystems may lead to serious imbalances. One way for DMC to overcome this situation is to choose their own policy variables in such a way as to make DMC's sectors of the system 'robust' in the sense that they will then be able to cope adequately with whatever shocks are imposed from the dealers sector. The magnitude and form of these shocks will depend on the way in which the dealer organisation amplifies or attenuates the shocks created by DMC's manipulation of ultimate consumer demand.

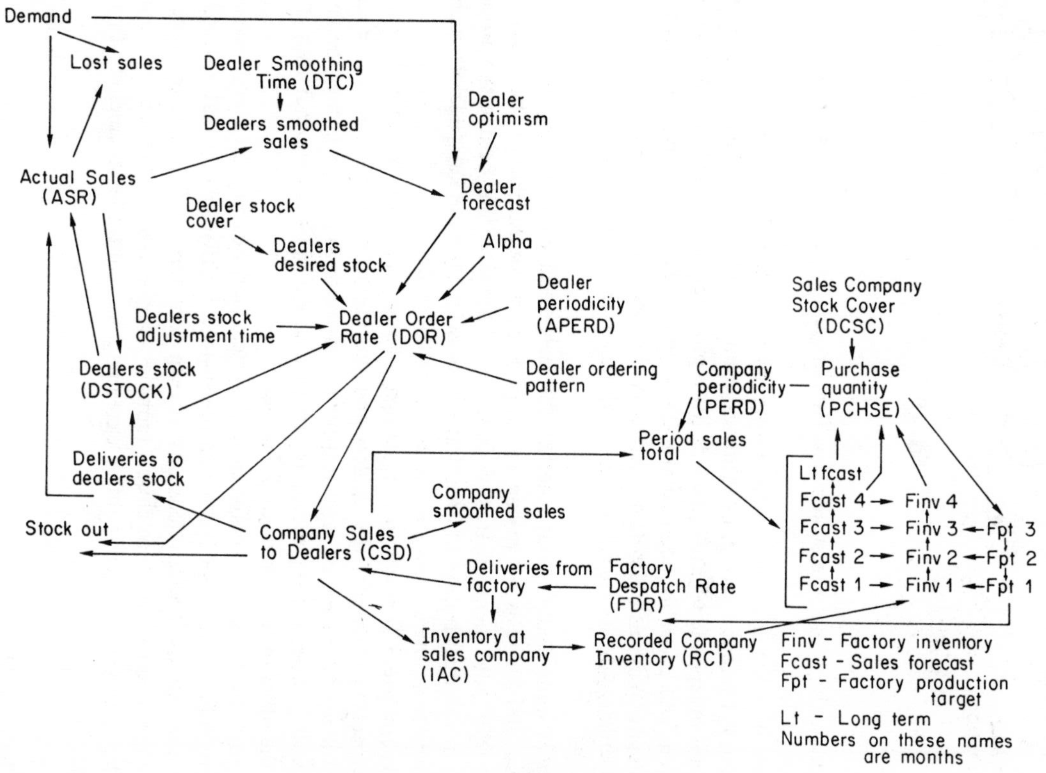

Fig. 10.1 Model I influence diagram

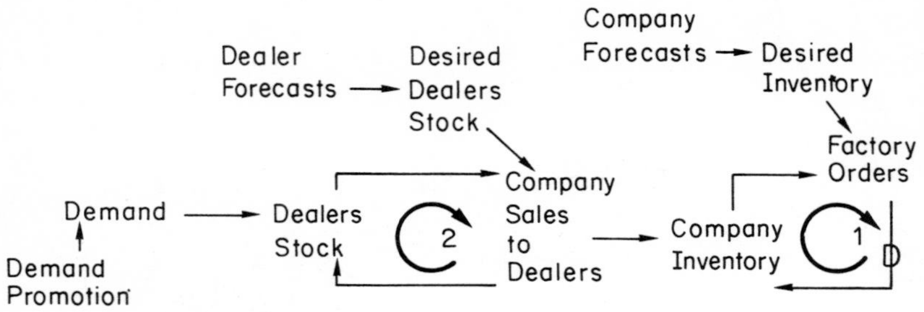

Fig. 10.2

Basically, therefore, the system has two unconnected control structures, shown in Figure 10.2.

System 1 is the controller, with a long delay. System 2 is the environment (i.e. the dealers). There is a very poor feedback connection between them, and, since the dealers are independent firms, there is little prospect of creating one. Any noise in the system will probably be in DEMAND i.e. in the way the complement drives the environment. Will this noise be amplified by the environment and then passed on to the controller, and, if so, what can be done, *in the controller*, to ameliorate the situation?

There is some doubt about whether the dealers really follow the rules laid down for them and there is no reason to suppose they would change their ways just because DMC told them to. It is, in fact, part of the reasoning behind the proposed changes in the distribution system that stocks (which are the key to sales) might be taken away from the dealers so as to be better controlled.

In general the S D problem is to design a better controller which will be robust with respect to whatever the dealers and the public between them do to it. In addition it has to be robust with respect to assumptions about whether the dealers follow DMC's stockholding requirements or not. We shall assume that they do, so testing the other options is up to the reader.

We note the similarity between this system and Section 2.12 and therefore expect to find severe amplification in the system. Finally observe that, through the market promotions, DMC is acting as its own complement. This implies some very interesting modelling work to close a loop between Market Promotion and variables from the controller, possibly after extending the controller to include financial variables.

10.7 The Basic Model

The model works in the following way. The demand pattern (DEMAND) is imposed on the system externally. If dealers stock (DSTOCK) is sufficiently large, all the demand is met. If dealers stock is lower than demand then only that portion which can be met is met and lost sales (LSALE) accumulate. If dealers stock is practically nil only the flow of deliveries to dealers stock is available to meet demand and the link with Actual Sales (ASR) comes into operation.

The actual sales rate is used by the dealer to calculate the dealers smoothed sales (DSSR) which his average sales over a period of time called the dealer time constant (DTC).

The dealer is also assumed to forecast demand (DFCST). This takes the form:

Dealer Forecast = Factor x Actual Demand

If Factor = 1 the dealer makes perfect forecasts of demand while if Factor is greater than 1.0 he is consistently optimistic and so on.

The dealer determines his order rate such that it provides for his sales level together with a stock correction factor. The dealers desired stock level is taken to be his estimate of sales level as calculated earlier multiplied by the number of weeks stock he is supposed to hold. This ordering pattern is reduced in the early part of the dealers accounting period (denoted by APERD in the model), and increased towards the end to reflect the dealer behaviour of trying to bunch orders at the end of the month to get extra time to pay (see Section 4.12).

At the sales company level the company sales to dealers (CSD) is determined by dealer order rate in the same way as demand determines actual sales subject to the stock level. At the sales company this leads to another performance indicator called Stock Out (STKOUT). This is the cumulative total of sales to dealers lost by the sales company's inability to meet demand. This formulation models the risk to DMC from the dealers who sell other products as well as DMC's and will, to some extent, switch selling effort to those other products if the DMC sales company is unable to meet their demands.

The joint effect of sales to dealers and deliveries from the factory is that the Inventory at the sales company (IAC) is recorded once every month and is then called Recorded Company Inventory (RCI). At present the month is the decision-making period both for the company. As there may be advantages in changing this the two variables APERD and PERD have been used to indicate the dealers and the company's decision-making periodicities respectively. The effects of changing these periodicities will be described below.

The company's sales to the dealers (CSD) are used to calculate actual sales month by month (or PERD by PERD to be exact) and this value (denoted by ACSALES) is used to revise five forecasts which are used in the factory production planning function. These are called FCAST1 to FCAST4 and LTFCST for the forecasts of sales for each of the coming four PERDs and the long term monthly level. These values are then used to calculate a quantity called purchase (PCHSE) which is to be received by the start of the fourth month hence and manufactured and despatched in the month before. These amounts become the factory production target for the coming three periods and are denoted by FPT1–FPT3. (Factory Despatch Target would be a better terminology.) This procedure is based on a desired eight weeks stock at the sales company. The complex pattern of linkages between these variables in the diagram reflects the rather simple calculation which is performed.

Although both the dealer and the company sector of the model contain feedback loops there is very little in the way of linkage between the two sets of loops. This is almost certainly the root of much of the trouble with this control system and performance could be improved by linking the two sets of feedback loops in a properly designed master loop. One way of doing this would be for DMC (Export) to have a formal policy relating their market promotion activity to variables in the company sector such as Purchase Quantity and/or Recorded Company Inventory. If the reader visualizes a line running from either of these variables to Demand he will see that the control systems of the company and the dealers have been encased in a master control loop. This would almost certainly improve performance providing the control actions were properly integrated with the system.

During the discussion which led to the building of the model several variables were identified as being likely to be of interest:

DCSL Desired Company Stock Level. This is the number of weeks stock the sales company is supposed to hold.

PERD The decision-making frequency in the company.

APERD The same frequency at the dealer level.
Initially both PERD and APERD are 1 month or 4.3 weeks.

DTC Dealer Time Constant. The period over which the dealer averages his sales for ordering purpose.

Some computer runs were as follows.

Initial Run

PERD=4.3 APERD=4.3 DCSL=8 DTC=4.3

This is the basic system with accurate forecasting, monthly decision-making and 8 weeks stock cover at the sales company level.

The behaviour is shown in Fig. 10.3, for the pattern input. The fluctuations are more severe in production than in demand, showing that the system amplifies the loads placed on it. There are large swings in inventory but there are no lost sales. One disturbing feature is the very poor ability of the system to adjust to the decline in demand from week 40 onwards. The system has a major time-lag of about 25–30 weeks from the rise in demand in weeks 30–40 to the rise in inventory level from weeks 55–70. Steps to reduce this time constant would probably be beneficial.

This basic run was discussed with the company and agreed by them as being a satisfactory model of the system. It was impossible to 'test' the model against recorded data because they were not available. Because of the way in which the study had been framed to meet the needs of DMC management it is highly debatable whether testing the model against data would have had any meaning (Why?)

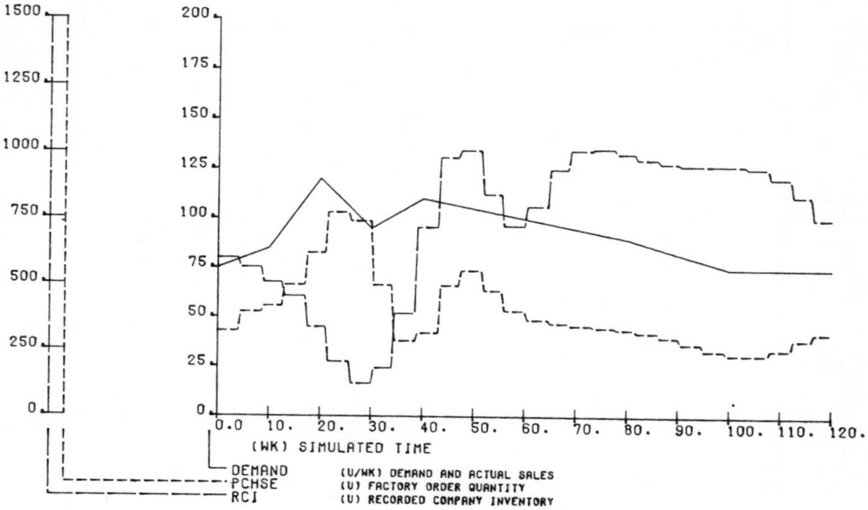

Fig. 10.3 Model IA pattern input

Reduced Stock Levels

PERD=4.3 APERD=4.3 DCSL=4 DTC=4.3

In order to assess the amount of inventory in the system it was subjected to the same demand patterns as before but with the Company Stock Level (DCSL) reduced from 8 weeks to 4 weeks cover.

The response to the pattern input is shown in Figure 10.4. The interesting feature is that there are no lost sales, in fact the amount of inventory in the system could probably be drastically reduced without detrimental effect. Since the value of the total inventory held by DMC runs into millions of pounds the savings should be appreciable.

Several other possibilities were tested and may be reported briefly.

Since forecasting is part of the model a run was made in which the dealer consistently overestimated demand by 20% and DMC made the same overestimate of initial demand. Apart from some minor variations in purchases in the early stages of the demand pattern there was no appreciable difference between this run and that of Fig. 10.3. The system is insensitive to errors in forecasting because its control mechanisms automatically correct for them.

DMC's accounting and decision-making were based on the calendar month. It had been argued in the company that control would be improved by making decisions every two weeks instead of monthly. This was tried in the model and the results proved that the oscillations generated by the dealers were such that the system behaviour was made worse. This was due to the dealer's periodicity no longer matching the company's. If the company decision-making was put on a

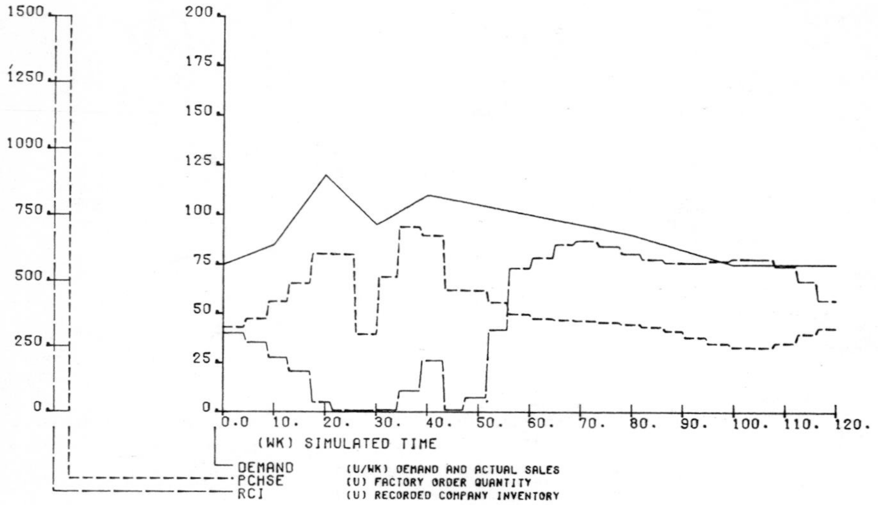

Fig. 10.4 Model IA reduced stock levels

fortnightly basis and the dealers were also induced to accept a fortnightly accounting period another run showed that the system behaviour was materially improved. The curves were smoother than those of Fig. 10.3 and the variations in inventory are less severe.

10.8 The Present System without Dealers Stocks

DMC had questioned whether there was any real need for the dealers to hold stocks, apart from a few items for demonstration and immediate sale. It was reasoned that DMC could deliver from its depots in the Sales Country to any dealer in the country within a day or so.

This proposition was tested in model IB in which the dealers stock was eliminated from the model together with the equations representing those aspects of dealer behaviour which dealt with stockholding. The results are shown in Fig. 10.5. This shows a marked improvement in behaviour and demonstrates the way in which the dealers destabilize DMC. There would be little direct saving to DMC from the reduction in inventory caused by eliminating the Dealers Stock, but the side effects from smoother operation might be significant. This would need to be balanced against increased distribution costs.

Model IB was also tried with Sales Company Stocks cut from the present value of eight weeks to four weeks. The computer output is shown in Fig. 10.6. The system copes slightly less well with the peak demand and some lost sales arise—161 units out of 9574. The total stock in this system is, however, only four weeks compared with eight weeks at the Sales Company and four weeks at the dealer in the present system.

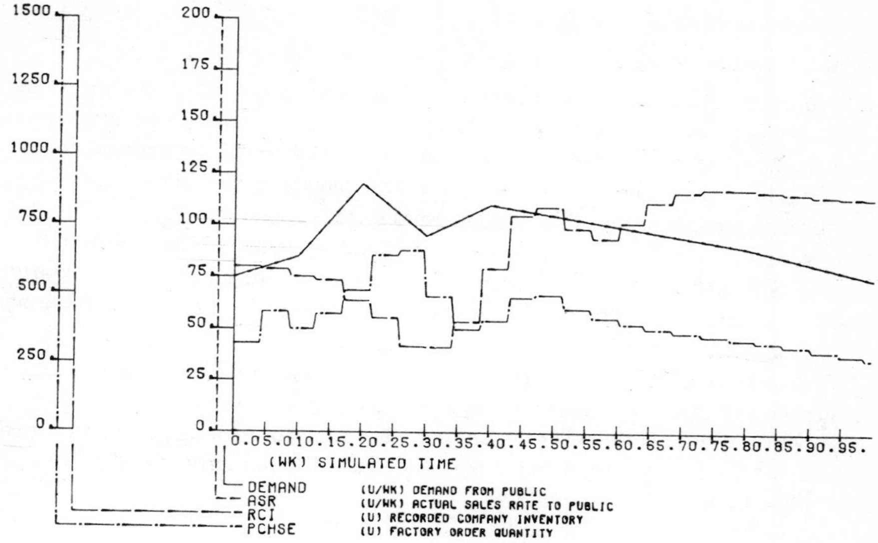

Fig. 10.5 Model IB pattern input

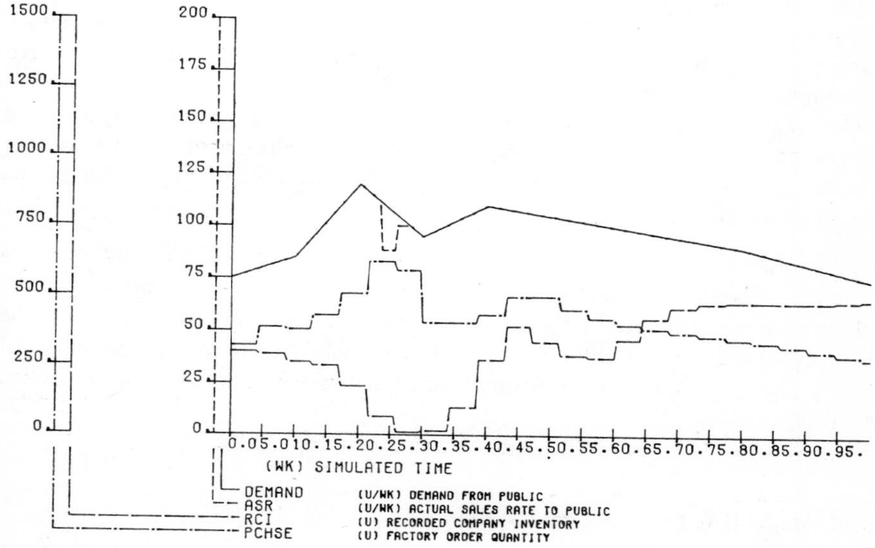

Fig. 10.6 Model IB reduced stock levels

10.9 Some Conclusions from Model I

A number of points emerge from a study of models IA and B.

1. The present system should work perfectly well in the sense that there are no lost sales. It is known that occasionally the system does not work properly so either the model is wrong (though DMC management accept it), or the demand pattern is sometimes more severe than had been anticipated, or the failures are due to incorrect operation rather than fundamental system unsuitability.
2. The system is not very much affected by forecasting errors but even with perfect forecasts it does not cope very well with the decline in demand during the latter part of the pattern. It would probably pay DMC to bias their forecasts downwards as the product life cycle proceeds.
3. The dealers stocks serve no very useful purpose and probably makes matters worse. In any case there is far too much inventory in the system.
4. The whole system is far too sluggish and this is due to the long period of notice given to the factory. Reducing this would make life harder at the factory end but might be worthwhile.

10.10 Model II: An Alternative Distribution System

For some time DMC had been considering changing the whole distribution organization to one in which the principal features would be

1. The factory would hold a stock of finished goods and would gear its manufacturing programme to keeping that stock at a required level, representing a given number of weeks of demand.
2. The Sales Companies in the European countries would hold stocks at Company Distribution Centres in strategic locations. They would draw from the factory inventory in such a way as to keep these centres stocked up to a required number of weeks demand.
3. The dealers would hold very small stocks for demonstration purposes and immediate sales. They would draw from the Company Distribution Centre at need, and DMC would guarantee delivery within a day or so, subject to the availability of inventory.

The purpose of Model II was to study the controllability of this system, not its economics. The model was built because it was suspected that this proposed system had serious disadvantages in that it would be more vulnerable to lost sales if the Company Distribution Centre ran out of stock while there was a possibility of running into very high total inventory levels unless the system could be controlled effectively.

The principal features of the system are shown in Fig. 10.7.

10.11 Model IIA: Basic Proposed System

The initial version of Model II covered the system as it was proposed to operate it. The essential idea of the proposed system is to decouple the market and the

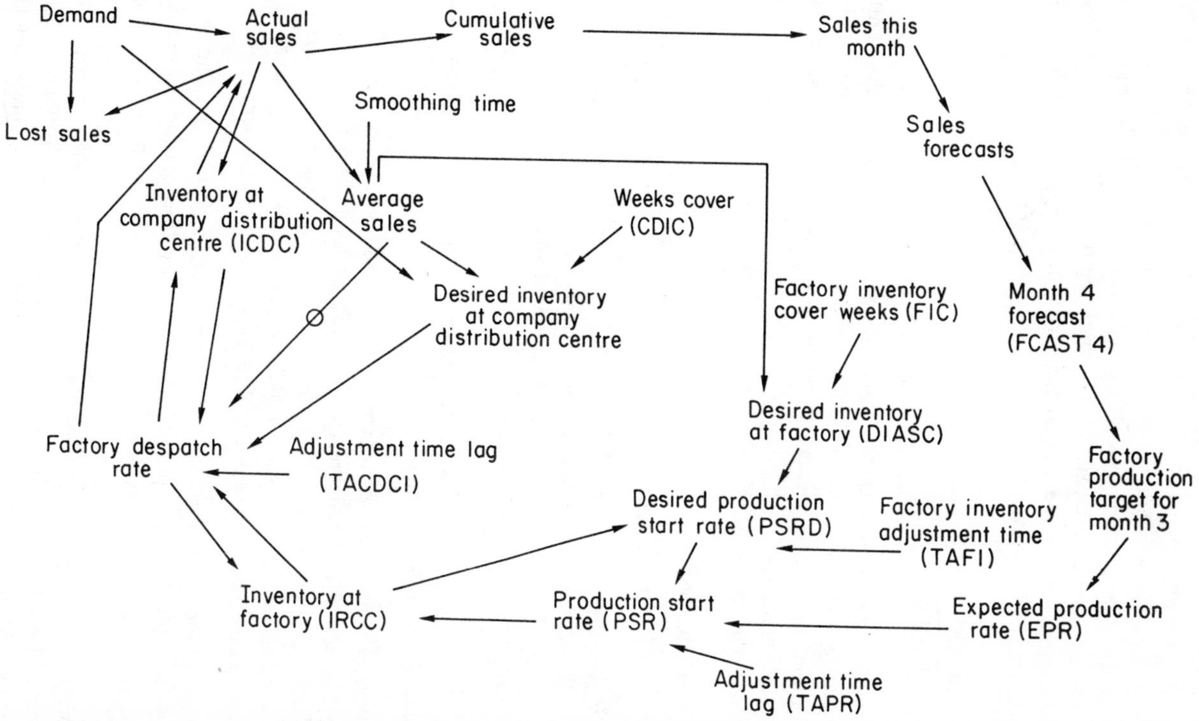

Fig. 10.7 Model II influence diagram. (Note: See section 10.2 for explanation of the circle on one of the links.)

factory. The dealer holds no stock but orders from inventory held at a Company Distribution Centre (ICDC). The factory manufactures not to a purchase order, but to maintain a factory inventory (IASC).

The Model is driven by the same test input of demand as Model IA and IB and lost sales arise in the same way and for the same reasons.

Inventory at the sales company (ICDC) is replenished by an order rule based on ICDC and a desired inventory DIACDC. The ordering rule is that, in any week, the rate at which goods are to be despatched from the factory to the company distribution centre (FCR) is given by:

$$\text{FDR} = \frac{\text{DIACDC} - \text{ICDC}}{\text{TACDCI}}$$

In this equation the term TACDCI stands for the time to adjust the company distribution centre inventory and models the speed at which the sales company attempts to make up deficiencies in its inventory. The value of TACDCI is an important model parameter.

The desired inventory, DIACDC, is a multiple of average sales i.e.

$$\text{DIACDC} = \text{CDIC} \times \text{SSR}$$

where

SSR	Average Sales Rate
CDIC	Weeks cover to be held at the distribution centre Initially CDIC=8 to parallel the present situation in the sales companies.

An equation such as that for FDR represents a control system which is inherently unstable in the sense that:

1. Actual inventory can never be brought to the desired level.
2. Whenever the actual inventory reaches the desired level the input to the system (FDR in this case) ceases and the actual inventory falls again. Until the system reaches a stable state the control policy imposes fluctuating loads on the production system unless the value of TACDCI is carefully chosen.

This is true even when demand on the system is perfectly uniform. The choice of TACDCI which leads to stable production loads may thus conflict seriously with that which leads to a satisfactory level of stock for dealing with sudden changes in demand.

The factory has a long-term production target fixed by a sales forecast for four months ahead. This determines an Expected Production Rate (EPR) for which the factory prepares. At the time of production, however, the factory inventory (IASC) has changed because of ordering by the sales company and this gives rise to a desired production start rate (PSRD). The gap between PSRD and EPR can only be partially closed at that late date, thus providing a simple model of factory response.

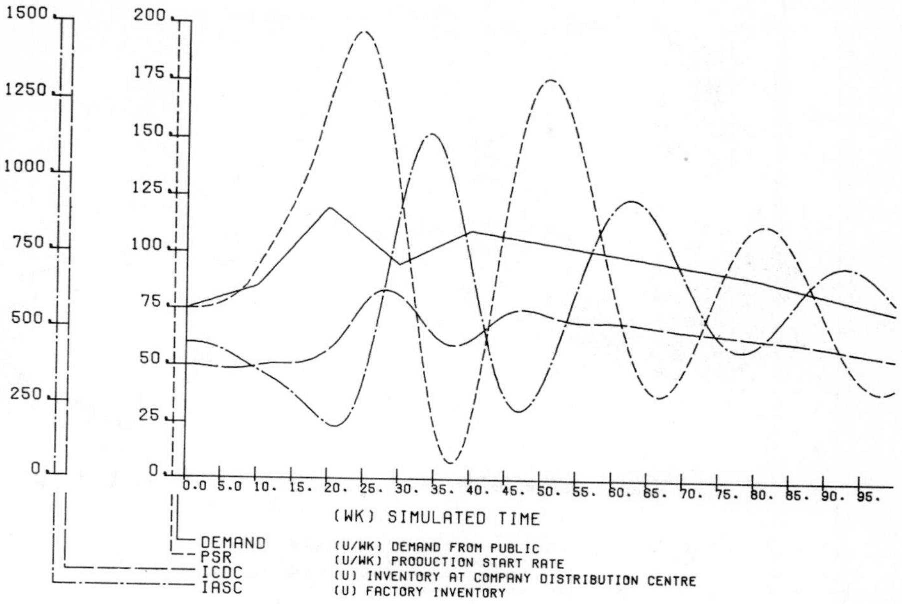

Fig. 10.8 Model IIA accurate forecasts

The basic equations are that Production Start Rate, PSR, is a level, related to PSRD by

PSR.K=PSR.J+(DT/TAPR)(PSRD.J–PSR.J)

Where the time delay, TAPR, is 4.3 weeks to model the factory gradually adjusting to changing targets.

This formulation of PSR is convenient as it provides facilities for later modelling of raw material orders in terms of PSR as a level. (Recall Chapter 5). Note that production is now a continuous variable, rather than the sampled quantity of models IA and IB, though EPR is still discontinuous.

The connection between factory production and the factory inventory, IASC, is the policy by which PSRD is adjusted in an attempt to regulate inventory. This is based on a Production Start Rate Desired (PSRD) and depends on the inventory at the factory (IASC–Inventory Awaiting Shipment to Sales Company), the desired inventory (DIASC), and the Expected Production Rate (EPR) as shown by

$$\text{PSRD} = \frac{\text{DIASC}-\text{IASC}}{\text{TAFI}} + \text{EPR}$$

where TAFI is a time constant to reflect DMC's eagerness, or otherwise, to see factory inventory at the desired level.

At the factory, the Desired Inventory, DIASC, is a multiple of Average Sales (SSR) such that

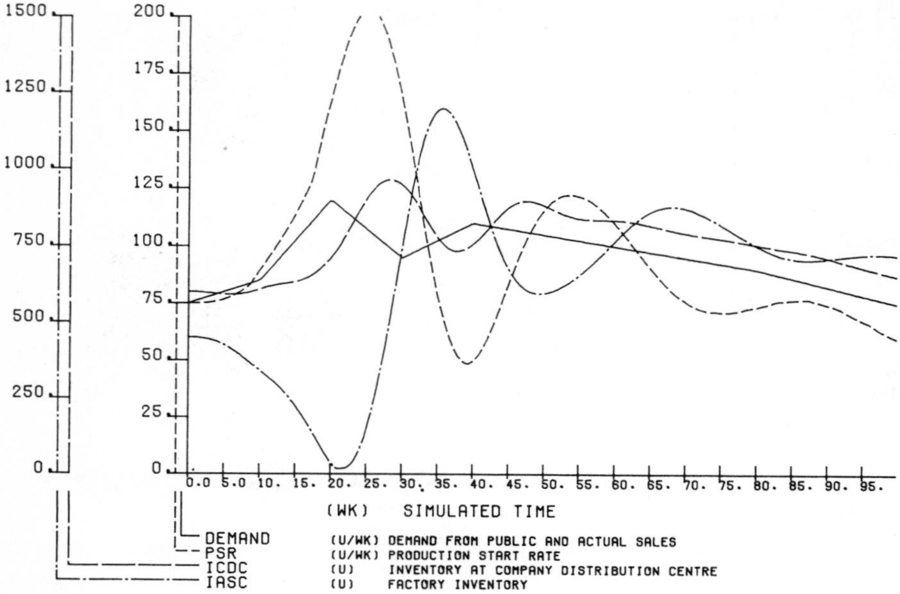

Fig. 10.9 Model IIB pattern input

DIASC=FIC*SSR

Where FIC is Factory Inventory Cover (Weeks).

If we take the value DMC proposed to use of FIC=6 and CDIC=8, the system would have 14 weeks of desired stock, which is even larger than Model I. The reader should show that the stable condition for the system is actually to have

(CDIC–TACDCI)+FIC

weeks of stock cover, where

CDIC	Desired Distribution Centre Inventory (WK)
TACDCI	Time to Adjust Company Distribution Centre Inventory (WK)
FIC	Factory Inventory Cover (WK)

This comes to 11 weeks, so total inventory would still be very high in the stable state.

Note that CDIC and TACDCI have been bracketed together to show that, if the latter exceeds the former, the stable state of the Company Distribution Centre will be out of stock by an amount determined by the difference between TACDCI and CDIC, and the Average Sales, SSR. Thus the Sales Companies could be out of stock while the Factory had stock, and the system would not correct the error. For this reason, Model IIA will be a poor control system unless it is carefully designed.

All the models in series II contain the same provision for assessing the effects of forecasting errors as Model IB.

Several computer runs were made on this model.

Initial Run

CDIC	= 8	TACDCI	= 3	CFAC	= 1.0
TAPR	= 4.3	TAFI	= 4	FIC	= 6
CALPHA	= 0.9				

The value of CALPHA reflects the importance DMC were going to attach to true forecasts as opposed to Actual Sales in their forecasting.

The behaviour with the pattern input is depicted in Fig. 10.8. This shows that the system would be capable of dealing with the demand pattern, but only in the sense of there being no lost sales. The production pattern is however, ridiculously unstable. This arises from the fact that TACDCI is smaller than TAFI so that the sales companies can control their inventory very well but transfer the problem back to the production system, where matters are made even worse by the relatively small value of TAPR reflecting the production system's attempts to keep the factory inventory in line with targets by being willing to adjust production comparatively quickly.

This run demonstrates fairly clearly the common managerial situation in which a proposed system is installed on the basis of costing information, but without having considered questions of controllability. In this case the system appeared to be cheaper to operate because DMC had not realized that it would affect the factory in any particular way. In practice it would, of course, be extremely expensive in total and, in fact, probably impossible to put into effect.

It might, of course, be possible to improve matters considerably by redesigning the production sector, or choosing other parameter values and the reader should make a loop analysis on the lines of Chapter 8.

Leaving that task to the reader, we shall assume that DMC were so unimpressed by the controllability of Model IIA that they wished to see if an alteration in the control structure could be devised which would enable the hypothetical system to be operated more effectively. This was Model IIB.

10.12 Model IIB Improved Control of the Proposed System

This model is essentially the same as Model IIA except that steps have been taken to try to stabilize the control system. This consists of adding SSR—average sales rate—to the equation determining the Factory Despatch Rate (FDR). This is shown in Fig. 10.7 by the link with a circle drawn on it.

The result of this change in the control equations is a system which will attain the desired stock levels in the steady state but which will therefore contain large amounts of stock. Since the stock correction factors present in Model IIA also depend on average sales rate, including SSR again will tend to reinforce fluctuations. This is shown by the simulation experiments.

Initial Run

CDIC=8	TACDCI=3	CFAC=1.0
TAPR=4.3	TAFI=4	FIC=6

The program for Model IIB is not given in Appendix B.10, it being left as an exercise for the reader to amend Appendix B.10 for Model IIA to obtain Model IIB.

The pattern response is shown in Fig. 10.9.

Attempting to smooth the Factory Despatch Rate–the rate at which goods are withdrawn from factory inventory to replenish that at the sales companies' distribution centres–has made controllability even worse than it was in Model IIA.

Recall, however, Section 2.12 in which changes to one parameter converted the system from explosive to stable oscillation. The reader should try to determine what can be done to improve Model II. We shall attempt to get better control of the existing system to avoid the wholesale reorganization of DMC which would be implicit in the implementation of a controllable Model II, if such a thing exists.

10.13 Improving the Present System

The preceding sections have dealt with the analysis of the present system, and have shown some comparisons between the system as it exists and some alternative configurations for the production distribution system. These are, however, only some of the possibilities and we shall conclude this case study by analysing the possibilities for improving the system as it stands by using altered decision rules. The reasoning is that there is not likely to be much point in considering a wholesale reorganization of DMC's production and distribution, if much the same result can be achieved by fairly simple changes in Head Office procedures, using the existing information from the field offices.

We have already seen, from Fig. 10.3, that the system amplifies considerably the loads placed on it by the demand pattern. For example, the ratio of the highest demand (120 units per week) is 160% of the lowest (75 units per week), but the highest production rate at the factory (870 units per month at week 26) is 580% of the lowest (150 units per month at week 34). (PCHSE does decline a little further at the end of the run, but we are analysing the dynamics due to the early variations in the demand pattern.) Thus the amplitude variation in demand is magnified 580/160 or 3.6 times in the process of translating it into factory production targets. (It must be emphasized that DMC is a real company which, by no stretch of the imagination, could be regarded as backward or inefficient in its management methods. This phenomenon of overamplification of shocks is quite, in fact, common in business practice.)

The situation for the controller is quite easily modelled in terms of its feedback loops by using the influence diagram in Fig. 10.10, which is a simplified version of the key processes of Fig. 10.1.

The negative feedback loop from Actual Stock to Expected Stock Gap to Factory Production and, eventually, back to Actual Stock, is clearly seen. The gain in this loop depends on the Expected Stock Gap and this, in its turn, is increased by the rise in Required Stock re-inforced by the fall in Actual Stock when Demand rises. It is fairly evident that the Expected Stock Gap will rise more sharply than Demand when the latter rises, and fall more steeply when it falls. The situation is

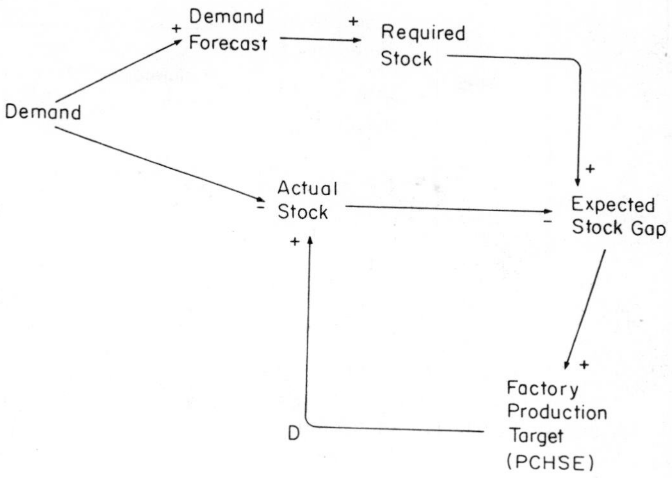

Fig. 10.10

not eased by the comparatively long delay before production ordered by DMC (Export) actually reaches them. This delay is three months but this is less than the time needed for the demand pattern to go through some fairly sharp swings.

One way of diminishing the in-built amplification of the system is to reduce the input gain of the negative feedback loop by taking the stock-correction factor over a longer period. In the present system, any Expected Stock Gap, is eliminated in one month by making this month's purchase quantity equal to the Expected Stock Gap. Thus, in the present system

$$\text{PCHSE}=\text{ESG}$$

where ESG is Expected Stock Gap, and

$$\text{ESG}=\frac{\text{DCSL}}{2\text{xPERD}}(\text{FCAST4}+\text{LTFCST})-\text{FINV4}$$

The first term on the right hand side is the stock required to provide DCSL weeks of cover against the expected level of demand, and the second term is the expected inventory at the end of the fourth month if nothing is ordered for despatch from the factory during the third month and hence receipt during the fourth month.

As we have already argued, this formulation, acting through the feedback loops shown in the simplified diagram Fig. 10.10, corrects the situation too sharply, and thus gives rise to the amplification phenomenon. We could replace this decision rule by one which equated the Purchase Quantity, PCHSE, to the long-term demand forecast, but this would take no account of stock. We, therefore, use a production policy based on the influence diagram shown in Fig. 10.11.

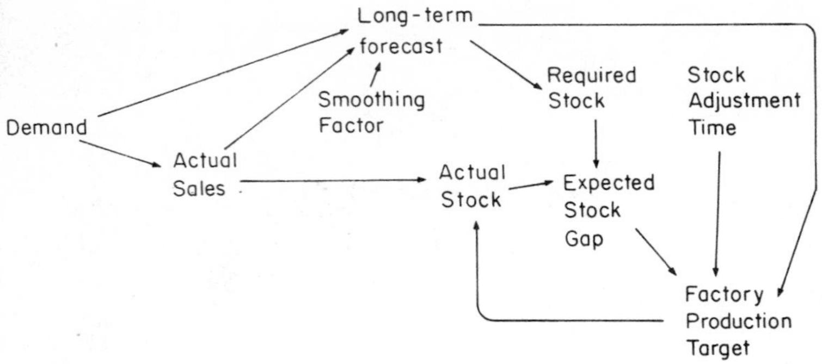

Fig. 10.11

If the stock adjustment time and the smoothing factor are chosen properly, this policy will weaken the effect of the feedback loop linking Actual Stock with the Factory Production Target, without completely removing its ability to ensure that the system keeps to a reasonable stock position. The basic equation will be

$$\text{PCHSE} = \text{LTAFCST} + \frac{(\text{DCSL/PER}) \times \text{LTAFCST} - \text{FINV3}}{\text{TASCI}}$$

where

LTAFCST	Forecast demand approx. 6 months ahead [Units/month]
DCSL	Required stock cover [week]
PERD	Basic period [4/3 weeks or 1 month]
FINV3	Expected Inventory 3 months hence
TASCI	Adjustment time for Stock Position [weeks]

The second term on the right-hand side of this equation ensures stock correction by using an adjusted target stock of

$$\frac{\text{DCSL}}{\text{PERD}} \times \text{LTAFCST}$$

and correcting this over TASCI weeks.

A system based on this kind of policy ought to have better performance than the existing system, mainly because of the longer period over which stock discrepancies are to be adjusted. It may, however, be more vulnerable to forecasting errors (not necessarily, because of the control exercised by the stock-correction factor and because, appearances notwithstanding, this policy is no more dependent on the ability to forecast correctly than is the existing policy).

Basic Performance Run

The basic performance of this suggested policy is shown in Fig. 10.12, which should be compared with Fig. 10.3. The behaviour is strikingly smoother, as the

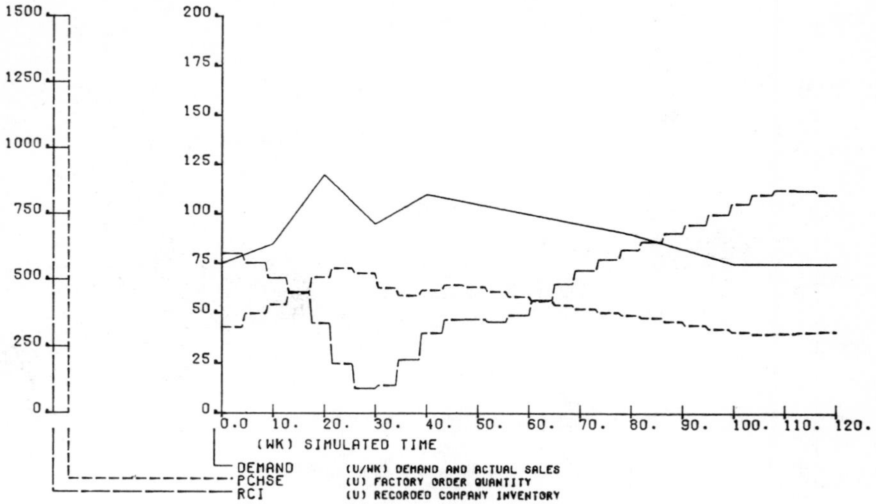

Fig. 10.12 Proposed system – accurate forecasts

variation in the Purchase Quantity shows The maximum value of Purchase is approximately 1.5 times the minimum value, instead of about 5.8 times as in Fig. 10.3. The Inventory Position is slightly less attractive at first glance but Recorded Company Inventory is amply supported by the dealers' stock, however, and Lost Sales arise. In fact, the general level of total stocks in Fig. 10.12 is much less than it is in Fig. 10.3, even though the same target stock levels are used in both runs.

Clearly, the inventory behaviour towards the end of the run is not satisfactory and further improvement should be attempted by the reader. It may be that the problem is now to decide when to shut down production, and run down stocks to the end of the life cycle. This carries the risk of having to start up again, or forego demand and the reader should experiment, using better performance measures and broadening the model to consider DMC's policy on new product introductions.

The proposed rule also stood up to being run with a 20% forecasting bias. The output was similar to Fig. 10.12, the main exception being that the general level of Recorded Company Inventory was rather higher. This system might, therefore be more sensitive to forecast errors than Model IA. Consider!

Damping of Forecasts

Apart from consistent biasses there is another important aspect of forecasting– damping of forecast values by mixing in present values. This statement reflects the fact that there are two forecasts one can make of events at some future point in time. One is the Actual Forecast and the other is a Damped Forecast, in which the actual forecast is modified by incorporating some weighting by present values. This

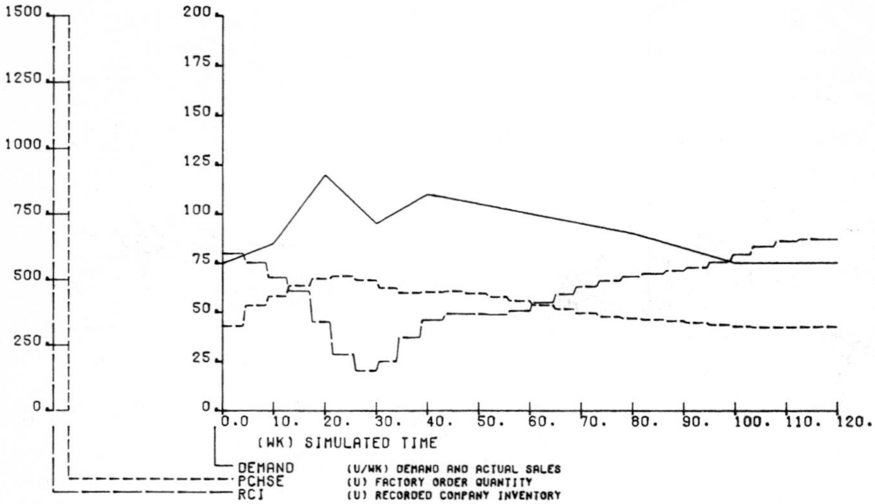

Fig. 10.13 Proposed system – undamped forecasts

process can be represented by

$$\text{Forecast as Used} = (1 - \text{CALPHA}) \times \text{Present Sales} + (\text{CALPHA}) \times \text{Forecast as Made}$$

This is a model of a fairly common process, which amounts to reforecasting of forecasts.

The model was tested with two values of CALPHA. The initial run, from which Fig. 10.12 is derived has CALPHA=0.3, but when the model was run with CALPHA=0.7 (i.e. a fairly light weight given to present sales) the performance was even smoother. This is shown in Fig. 10.13.

Sharper Patterns

The pattern of demand which was used throughout the earlier work had been suggested by management as being ‘reasonably typical’. How would the proposed production policy perform in the face of a demand pattern much sharper than the one so far used? This is shown in Fig. 10.14. There is, in fact, very little difference from Fig. 10.13, except that stocks are rather lower. The system still has no Lost Sales, even when the nominal stock cover was cut from 8 weeks to 6 weeks. Since this reduction made very little difference to production and stock figures, and, since stock values are far below the nominal in any case, there seems little point in pursuing that angle of attack.

How would Model IA stand up to this pattern?

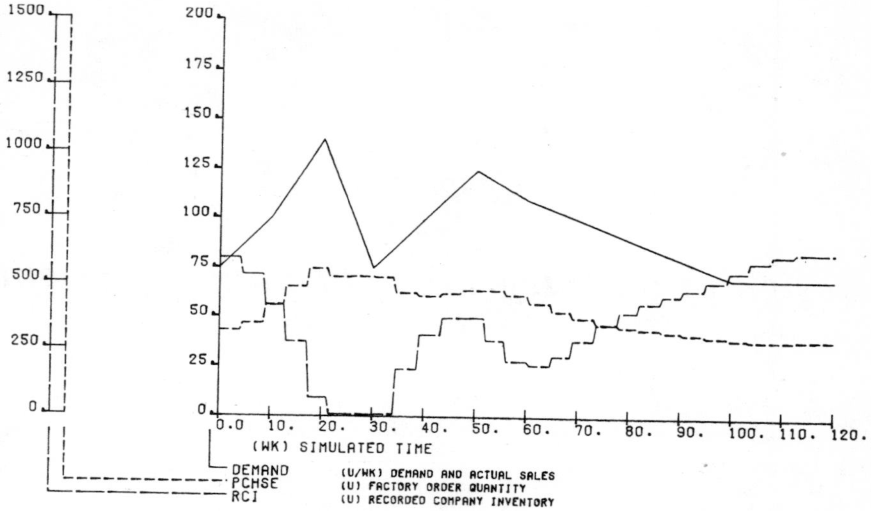

Fig. 10.14 Proposed system – sharper demand

10.14 Conclusion

The full details of the original and revised model equations are given in Appendix B.10. However, without going into them in detail, it seems clear that the present system can be markedly improved, without a wholesale reorganization.

The new planning procedure would be as follows:

1. The field salesmen continue their present practice of making a forecast of demand for a period about 4 or 5 months hence. This is exactly as at present.
2. DMC (Export) Management will use this to produce a damped forecast based on the Field Sales Forecast and Last Month's Sales. Thus:

LTAFCST=(1–CALPHA) × LAST MONTH'S SALES+CALPHA×FIELD SALES FORECAST

with CALPHA=0.7

This damped forecast, LTAFCST, will then be used in the equation for calculating the purchase quantity, given in Section 10.13.

Such a system should have the following advantages:

1. A general reduction in stock levels.
2. A considerable smoothing of fluctuations in purchase quantity, and hence in production demands on the factory.
3. Being very similar, organisationally, to the present system it should be very easy to implement.
4. The tests carried out, not all of which are reported here, show that the proposed system should be robust in the sense that it always has much the same performance regardless of errors, or fluctuations in its environment

Chapter 11

Design of an Integrated Oil Supply System

11.1 Introduction

This chapter is based on work done by the author, and others, for an oil company. Since the purpose is to provide the reader with an exercise in dynamic analysis, rather than simply to recount an historical investigation, important parts have been altered and rephrased to disguise what the real firm does.

11.2 Background Situation

The major oil companies operate in various parts of the world and have to ship their raw material to North Western Europe (NWE) for further processing and for eventual sale as finished product. It is with this aspect of their operations, the so-called Supply System, that this case is concerned. We do not deal with exploration, oil field development, refinery construction, or marketing, and conclusions regarding, say, the importance of forecasting to the supply system should not be applied uncritically to those other areas.

The data used in this chapter and in Appendix B11 are based on what they used to be in 1970 in a real company, and the policies in the model are versions of what a large oil company might have done in the days of relatively low oil prices, and practically complete freedom from control by the producing countries. The prices and costs have been completely distorted so that there is no point in the reader wasting his efforts on trying to infer from this case study how any particular oil company he thinks he recognizes is run. He should concentrate on the exercise in dynamic analysis, regarding the oil industry background of the case as simply additional colour.

In practice, oil companies employ various well-developed procedures to deal with short-term movements of raw material and product within the market area and, of course, to make the detailed day-to-day arrangements on ship movements necessary to get the raw material to the right part of North West Europe. In this case, however, the firm had had no previous experience in the use of system dynamics and wished to see if the methods were likely to be of use as an addition to their available problem-solving methods.

One of the problems which might be used as a testing-ground for evaluating system dynamics could be stated somewhat as follows:

'We already have a very effective process for managing the shipping, storage and refining of oil and products (collectively called the supply system) which we have carefully *evolved* over many years. Is it, however, possible to use system dynamics to evaluate alternative ways of controlling such a system in a "typical" company?'

This chapter therefore describes a system design study in which a basic model of a 'company' supply system was built, partly from discussions with experienced executives and partly from common sense, which was then analysed to determine its basic dynamic characteristics, and for which various design alternatives could be investigated. It should be noted that this chapter is written from the point of view of designing an oil supply system for a 'company' operating in the early 1970s. Although this means that the results have no relevance to the operations of an actual oil company in the changed world of the mid 1970s it should add to the number of problems which the student will be able to pose in order to test his own ability as a system designer.

11.3 Sources of Shipping

(It will be realized that we make several abbreviating assumptions for the sake of presenting the case study in a book. In a practical model for a real company these assumptions might not be realistic and more detailed modelling could be done. The reader should attempt to identify such areas where the 'book' model might need alteration to make it 'real'.)

For simplicity, the geographically separate oil-producing areas can be grouped into long-haul and short-haul sources. The voyage times for these two categories are 67 days and 25 days respectively and the proportion of short-haul oil to total imports will be called the Short-haul Ratio. Clearly the difference between the respective voyage times means that a small change in the Short-haul Ratio would have a large effect on the number of ships required to transport a given total flow of oil.

A typical oil company owns and operates a number of oil tankers which however, provide only part of the required shipping. The balance is made up of vessels chartered from their owners. Charters are of two types:

1. Spot-Charters (S-C).
2. Time-Charters (T-C).

Spot-Charters (S-C)

The ship is taken for one voyage only and, after completing it, may be chartered by some other company for a voyage to some other part of the world. Owners keep their available ships at sea in the vicinity of the oil fields so that a spot-chartered ship can 'present' for loading within 24 hours of the charter being placed.

Chartering takes place through a highly-developed tanker market. The price paid for the vessel is expressed in World Scale units, with WS100 representing, we assume £4 per ton of crude oil transported from Bahrein to Rotterdam. This is based on a ship of stated size steaming at a stipulated speed (a 'Notional' tanker) and for any actual ship or any other voyage the actual cash payment is readily calculated.

Like most commodity markets, the tanker market is very volatile, and prices can range from WS30 to WS300 or more. If charter prices became *very* low, ships may be laid up and become unavailable at short notice.

Time-Charters (T-C)

The ship is chartered for an agreed period of time, usually 3–5 years and is, in effect, at the complete disposal of the charterer for that period. Ships are sometimes chartered for 15–20 years but these can be regarded as part of a company's own fleet.

The price paid for T-C ships is fixed by the charter market and is expressed in £/year for the ship. Knowing the capability of the vessel, this can readily be converted to the same basis as spot-charter prices and World Scale Equivalent, WSE, is used in this chapter. WSE generally rises and falls with WS, but by no means as sharply. However a high WSE price, though lower per crude ton transported than the WS price at the same time, has to be paid for far longer, during which time both WS and WSE may fall considerably.

Time chartering can be done in two ways. In immediate time chartering the vessel presents for service within about a month. In distant time-chartering the charter contract is signed now but the vessel presents at an agreed date in the future, possibly several months ahead. In each case the price paid throughout the charter is that which ruled on the day the charter was placed.

Because there is usually an upsurge in demand for oil during the winter, it is probable that quite large tonnages of shipping may be required to present in the winter, during a period called the Presentation Band. Distant chartering is used to ensure security of supply of good quality shipping from reliable sources.

11.4 The Dynamics of the Design Problem

So far we have described briefly what we shall take to be the custom and practice of ship chartering. We must now consider the dynamics of the problem in order to see what the design of the system has to be able to deal with.

Clearly money could be saved by chartering fewer ships, but this might lead to loss of sales from inadequate throughput. Optimization would trade-off lost sales against tanker savings to find some minimum-cost solution. In effect, market-share considerations dominated the problem and a tanker-chartering system design which produced measurable lost sales could not be regarded with much enthusiasm.

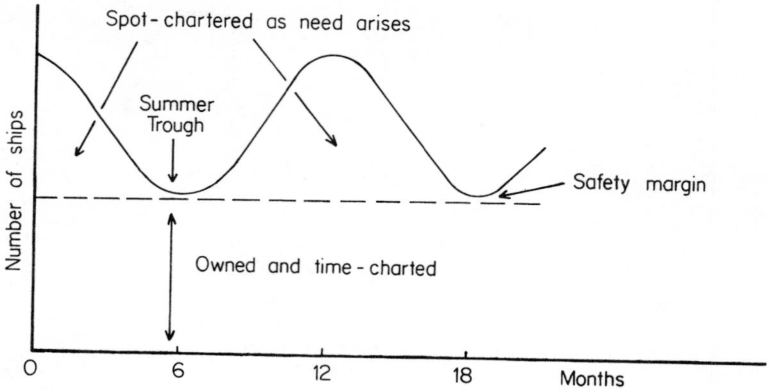

Fig. 11.1 A simple view of the shipping problem

The key to the dynamics is that, in aggregate, the demand for refined product shows a marked seasonality with the peak at mid-winter. Individual products have different patterns which can to some extent be dealt with by short-term planning procedures. There remains a residual seasonality in total product offtake which means that crude movements must also be seasonal, though the amplitude may not be the same as that of product offtake (it could be larger, see Section 2.12).

The simplest way of providing shipping for a seasonal demand is to have just enough owned and time-chartered shipping to meet the minimum at the midsummer trough, with a small margin against overchartering, and to spot-charter enough shipping, as and when needed, to fill the winter peak. This is shown in Fig. 11.1.

In this system it is not necessary to worry much about the amplitude of seasonality and whether it will vary from year to year (which it very definitely does). Many large oil companies operate a system which is a sophisticated version of Fig. 11.1.

If the seasonal amplitude is reasonably constant, an alternative approach, shown in Fig. 11.2 and described below, can be considered. As it stands, it might not be feasible for an actual company. However, in the remainder of this chapter we shall assume that it is and that the amplitude of seasonality is constant. The reader now has the problem of adapting the policies for the case of Fig. 11.2 to make them work for a variable seasonality and comparing this with a system based on Fig. 11.1, following the general lines shown in this chapter and drawing on material from earlier chapters. The equations in Appendix B.11 are for the situation in Fig. 11.2 and the reader should encounter no particular difficulty in formulating those for Fig. 11.1.

It must be understood that Fig. 11.2 represents a forecast, at one point in time, of the next 18 months, and not 18 months as they pass.

If no further time-chartering is done, then, other things being equal, the existing time-charter (T-C) fleet would decay as ships come to the end of their charters.

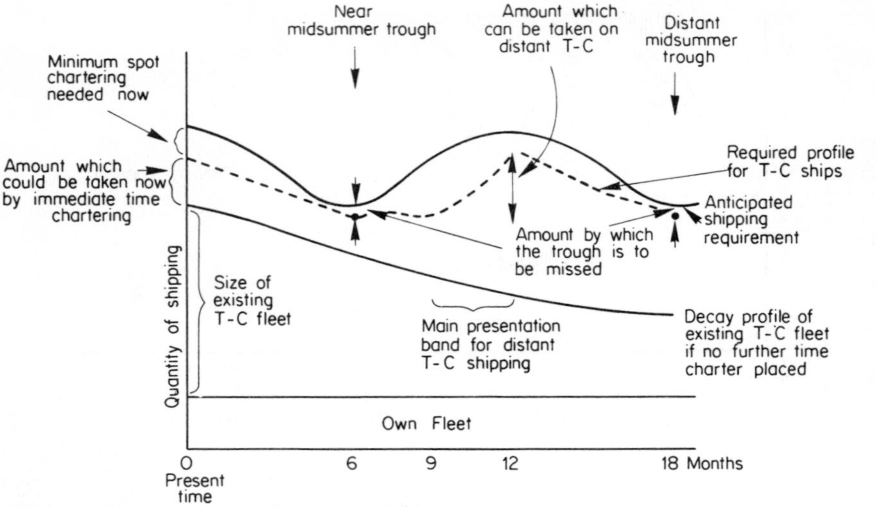

Fig. 11.2 An alternative approach to the shipping problems

Ideally, extra time-charters would be placed so that the owned and T-C total would miss the troughs in shipping demand by a small amount, called the Spot-Charter Fraction (SCF), in order to avoid over chartering if at all possible. The area between the dashed line and the shipping requirement curve then has to be spot-chartered whatever the price, or stocks have to be run down.

In the model the spot-charter fraction is taken as a simple constant, 5% being chosen for a numerical value. A real company might use a much more subtle formulation and some other average value. The student should experiment here.

Fig. 11.2 exagerates for clarity, but it can be seen that the near midsummer trough and the SCF fix the maximum T-C tonnage at that time and, by projecting this backwards, the gap to be met by immediate T-C is found. Anything above this has to be made up by S-C vessels. This understates the case, as immediate T-C vessels do not present for a month, so still more S-C is needed to hold the fort until the T-C ships beome available. Alternatively, stocks have to take up the slack.

Looking to the distant midsummer, one can anticipate the surge of T-C shipping needed in the autumn and winter to provide the increased carrying capacity for the winter peak. Thus the quantity of shipping which has to become available during the presentation band can be found, assuming that the near-midsummer T-C is properly done and, as shown in Fig. 11.2, the required profile for T-C ships passes through the black spot at month 6. Distant charters up to this amount can be placed at any time early in the year when suitable ships come on the market. (Owners will enter into new charters for distant presentation whenever it suits them to do so and they think the price is right. In practice they usually wait until the ship is within a few months of completing her present charter as prices are very volatile and they have to ensure that the ship will not be prevented from presenting by unexpected maintenance, in order to avoid heavy contract penalties.)

The 'company' could, therefore, take both immediate and distant charters in the first half of the year. In the second half they need take only immediate charters to top up the presentation band as their forecasts about the 'distant' midsummer, which, by that time of year, has become the 'near' midsummer, become clearer. There could therefore, be a process of telescoping and expanding forecasting.

Many of the complexities of the situation cannot be shown in Fig. 11.2. Basically, the Anticipated Shipping Requirement is rather elastic and is affected by several factors. In discussing these factors it should be borne in mind that we are referring to the pre-1973 situation. Since then, the general changes in the world environment have markedly affected the way in which things might have to be done.

1. FORECASTING Traditionally, product offtake can be predicted very accurately for a month or so ahead, but the oil industry as a whole knows from bitter experience that 18 months is another matter. A mild winter or a poor summer has a very large effect, as do changes in economic growth, fuel policy etc. How then, do these factors affect a Supply System and how can it be protected against them?

2. PRODUCTION AND STOCKS The shipping required will obviously be affected by the way in which the refineries are run and the stocks are managed. A possible policy would be to build stocks in advance of seasonal offtake and operate the refineries in a seasonal manner. How should this best be done?

The Supply System has to cope with seasonality in the sales of final product, or 'product offtake'. There are, of course, many hundreds of refined products, but in this account we aggregate them into a single product, the offtake of which peaks in December and troughs in June. The peak:trough ratio is modelled as being constant at 1.25. The offtake is met from refined stocks which are planned to peak at the end of October and trough at the end of April.

Refinery activity fills product stocks and depletes crude stocks which are, in turn, replenished by imports of crude oil. The company is modelled as having a certain amount of freedom in the amounts of oil it brings from short-haul and long-haul sources provided that, in the long term, it meets its contractual obligations about the proportions of its oil taken from the various producer countries.

3. PRICE EXPECTATIONS Tanker prices are markedly volatile and pre-1973, affected the system in two ways.

Long-haul oil was usually cheaper than short-haul, but if tanker prices were very high it payed to increase the short-haul ratio (i.e. the proportion from short-haul sources) but this had a marked effect on shipping requirements.

In parallel with this are the direct effects on chartering. If shipping prices are expected to fall does it make sense to defer present chartering plans?

The problem, therefore, is to design a loop structure which will provide a satisfactory control system when faced with these various shocks and dynamic factors.

There are many other questions which can be raised about this system and the purpose of this case study is to stimulate the reader to think of them himself rather than to present all the answers.

11.5 Shipping Requirements, Surpluses and Deficits

Before proceeding with more detail it is necessary to understand some of the factors affecting shipping.

Let RSCP2 be required shipping capacity at the start of month 2 [Tons of crude/month], and CCP2 be carrying capacity of ships known to be available at that time [Tons of crude/month].

Then, if

$$\text{RSCP2} > \text{CCP2}$$

there is a shipping deficit and, conversely, there would be a surplus.

The size of a ship is measured in Deadweight Tons, DWT, which is basically the amount of cargo which can be loaded into the ship at any one time. The speed at which a ship can sail is obviously important and, the shipping industry works on a standard definition of a 'notional tanker', which is 19450 DWT and sails at such a speed that it can carry 99580 tons of crude *per year* from Bahrein to Rotterdam via the Cape i.e. from a long-haul source.

From a short-haul source the same ship could carry:

$$99580 \times \frac{\text{LHVT}}{\text{SHVT}} \text{ tons per year}$$

if LHVT and SHVT are respectively the long-haul and short-haul round-trip voyage times in days (ignoring additional days in port on the short-haul voyages).

Now if C_1 and C_2 are the monthly tonnages of oil to be brought from long and short-haul sources the shipping required, SR, in DWT will be

$$\text{SR} = \frac{19450 \times}{99580 \times 12}\left(C_1 + C_2 \times \frac{\text{SHVT}}{\text{LHVT}}\right)$$

If

$$\frac{C_2}{C_1 + C_2} = \text{SHR, the short-haul ratio,}$$

then

$$\text{SHR} = \frac{19450}{99580 \times 12}(C_1 + C_2)\left((1 - \text{SHR}) + \text{SHR} \times \frac{\text{SHVT}}{\text{LHVT}}\right)$$

The last term on the right is called the Short-haul Weighting Factor, SHWF, and the factor (19450 x 12)/99580 is called the Shipping Constant, CONS. (Much of this terminology has to be invented by the project team. Recall the dimensional analysis in Chapter 5.)

In a surplus situation one can stop any further chartering and/or choose a SHR which will use up the available shipping time on the long-haul routes. The only alternatives are to allow ships to stand idle or to attempt to charter them to some other user. The system design specification is assumed to rule out these options.

11.6 Measures of System Performance

The simplest measure for the performance of a supply system is the Cumulative Expenditure on All Shipping, CUMEXP, over a 10-year period. The length is dictated by the 3–5 year time-charter duration and the need to allow time for dynamics to work their way through the system. The settling time for the model turns out to be about 3 years, and this also affects the choice of LENGTH.

There are, however, many other aspects to system performance which are judged qualitatively in this model:

1. The stability of the rate of spending on shipping, TSER, because of its implications for cash flow, and its total value.
2. The whole purpose of the system of stocks and shipping is to operate smoothly in the face of seasonality in offtake. The Crude Oil Arrival Rate, COAR, should ideally be flat despite the seasonality in Product Offtake, PO. If therefore, we observe that COAR has an amplitude as large as, or larger than, PO we should conclude that that set of policies was fundamentally incapable of doing what it was supposed to do.
3. The stability of the refinery production rate, CRPT, because of the high capital cost of surplus refinery capacity, to cope with seasonal peaks of production.
4. The ability of the system to cope with forecasting errors.
5. The extent to which stock plans are achieved.
6. The risk of lost sales due to lack of refined stocks.
7. The risk of stopping the refineries due to lack of crude stock.

11.7 The Initial Design

The basic design evolved from a series of discussions over several months. An essential feature of the process was the production of several Working Memoranda. These are informal documents which describe what the analyst thinks he has found out about the problem so far. The earlier ones especially are intended to be demolished by the criticism of the client, and they therefore provide an invaluable means of ensuring that the design requirements have been properly understood.

The essential features are shown in Fig. 11.3, in which CIF Demand means crude oil transported to NWE and sold as crude to other refiners. Fig. 11.3 encapsulates a great deal of detail, but the basic design stands out fairly clearly.

The system as a whole is driven by the demand for oil in NWE which derives from the growth of GNP in the region. This ignores any changes in the fuel mix in NWE and assumes that market share is constant. To relax either of these

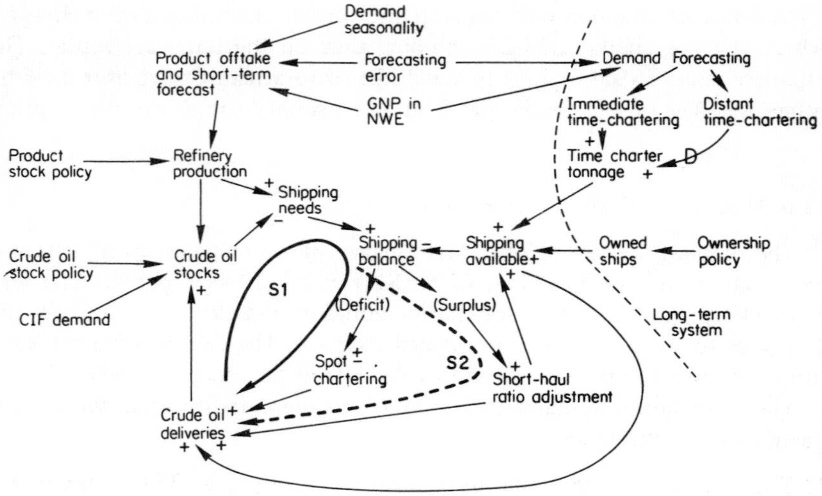

Fig. 11.3 The structure of the initial design

assumptions would be fairly easy so one could study the response of the design to changes in either of these factors.

In the model a GNP time series is injected exogenously and the ability of the system to cope with changes in GNP growth rates can be studied. Forecasting Error is treated by an Error and Bias Ratio, using the approach of Chapter 5, so that we can examine the consequences of forecasting attitude and error for the system.

On the left-hand side of the diagram, the drive is transmitted through the refinery sector, so if that sector's gain can be attenuated we should be able to insulate the model structure from seasonalities, and vice-versa. (Recall Section 2.12.)

The central area of the diagram shows the main control mechanisms on shipping. Essentially there are two, S1 and S2, both negative. For the deficit situation S1 provides for the normal spot-chartering activity and S2 does not function at all. Whenever a surplus arises, S1 is switched off, and S2 comes into play until the surplus has passed.

Collectively, S1 and S2 constitute a short-term control in the sense that they provide a response to external shocks which comes into play very quickly but which does not last very long, as spot-chartered ships are only effective for half their voyage time. The long-term is dealt with by the controls on the right-hand side of the diagram. The salient feature is that the long-term system simply drives the short-term one but, even though immediate time-chartering of ships is a fairly prompt response, there is no feedback from the main control loops to the long-term input. Thus all the current shocks to the model arrive via refinery production and crude oil stocks, and are met exclusively by S1 and S2.

A much more detailed explanation of the model appears in Appendix B.11.

11.8 Tanker Prices in the Model

Prices in the tanker market are notoriously unstable, as shown for a recent 5-year period in Fig. 11.4, the resemblance to the peaked oscillation mode of Fig. 2.11 being fairly noticeable.

The period is fairly typical of recent history except that it excludes the 1967 and 1973 Middle East wars, which had a very severe effect on spot rates. Generally, it is apparently cheaper to use T-C ships, though a T-C ship chartered towards the end of the third year would have been fairly expensive as WSE150 would have to be paid during the life of the charter, during most of which even spot rates were below WS100. It is fairly evident, however, that if the emphasis in chartering can be moved further towards time-chartering, money will be saved in the long run. The problem is to devise a control system which will bring this about, and which will cope with the loss of flexibility inherent in such a policy. First we must consider the extent to which tanker prices are exogenous to the actual supply system, i.e. whether they belong in the environment or the complement.

There are differing opinions on the extent to which tanker chartering by any one company affects prices in the tanker markets. Clearly, in special circumstances such as the sudden closure of the Suez Canal in 1967 the market is very unstable because all the users of shipping are in the market. For the general situation, however, the picture is far less clear cut. It was agreed that any company would affect charter prices for a few days by its chartering activities. This would not matter as they

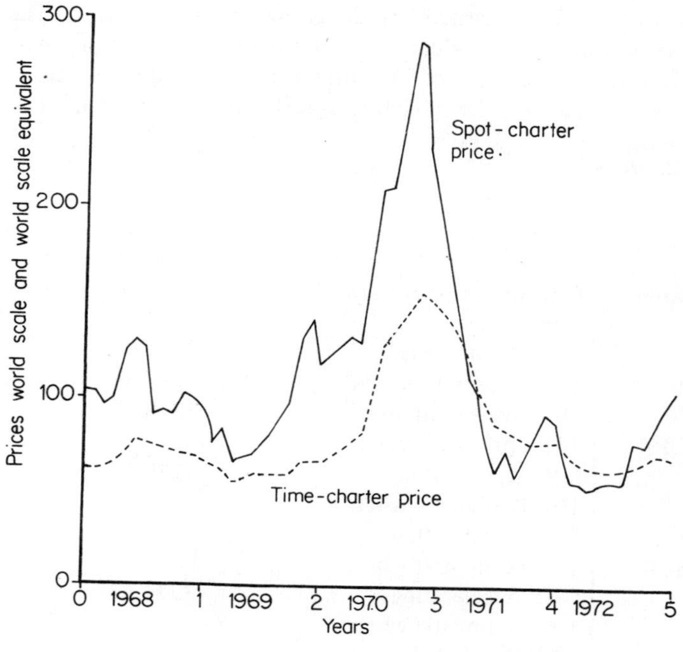

Fig. 11.4 Spot and time-charter prices

would not usually re-enter the market so soon. In the long-term it seemed likely that there was no identifiable effect because of the possibility of owners bringing out ships which has been laid up, switching ships from other trades, and even building new ships. An attempt was made to build a simple model of tanker price generation but this proved to be inadequate and, in view of the purpose of the project it was decided to treat tanker prices as arising in the complement, and not specifically to model the tanker market.

In the event, it was decided to use the price series from Fig. 11.4, folding it over to get a 10-year series. Whilst these are not, of course, actual prices in any sense, they do expose the system to the kind of severe shocks it would have to deal with. The reader may wish to try modelling the environment and he should certainly test the model against the severe drop in demand and tanker prices which has taken place since 1973.

Modelling the tanker market in this way implies that any conclusions about the design will have to be divided into two parts: those independent of tanker prices, and those affected by them. In practice it was felt that the price data were generally typical of price movements in the markets, even in wartime. Thus any cost differences indicated by the model for different designs of control system could not be relied upon to be precise values, but would certainly be correct as to relative magnitude.

The model for the initial system design employs forecasts of GNP, particularly for the planning of distant time-chartering. Any forecast is liable to error and we therefore need to test how sensitive the design will be to such errors. The simplest way of tackling this is to evaluate the effect on total shipping expenditure, CUMEXP, over the 10-year period for different error situations. Accordingly, a series of runs were made on the model using different patterns of forecasting error on GNP Growth.

The results were:

Run Name	*Forecast Error Pattern*	*Cumulative Shipping Expenditure* (£M)
PERF	Perfect Forecasts	3275.96
PESSI	10% Underestimation	3310.31
OPTIM	10% Overestimation	3218.12
EXAGER	10% optimistic when GNP Accelerating 10% Pessimistic when GNP Decelerating	3269.55
CONSER	10% Pessimistic when GNP Accelerating 10% Optimistic when GNP Decelerating	3244.63

It will be realized that the artificial tanker price series, coupled with the use of a fairly arbitrary series for GNP growth, mean that these values for total shipping expenditure are, although correct in their relation to one another, so heavily disguised that it is impossible to link them to any particular real company.

The results show that the basic design of the system is good in the sense that the total amount spent on *shipping* is remarkably insensitive to forecast errors. Whether this would be true for a more sophisticated measure of system performance is something the student should test.

It also appears that it pays to be wrong in the sense that over forecasting is slightly cheaper than under. To find out why, run the model and examine the dynamics.

It must be recalled that the forecasts referred to are of the general level of oil demand, not of seasonality, which we are assuming to be constant. Do the foregoing conclusions apply if this assumption is relaxed?

For the remainder of this chapter, the run labelled EXAGER will be used as a base case in evaluating changes in the system design and comparisons will be made in terms of CUMEXP. This should suggest some checks which the reader may wish to carry out.

Some output for the base case is shown in Fig. 11.5a and b. The model produce many pages of graphs but for reasons of brevity the essential features have been condensed into two. For the sake of clarity, the graphs run for 96 months, but LENGTH in the model is 120 months and the quoted values for CUMEXP are for that period. In what follows it should be made clear that we are referring to the performance of the model and not to what the real company does.

The first point is that the Total Shipping Expenditure Rate, TSER, is very unstable. It rises particularly sharply at two points where periods of heavy spot-chartering coincide with high prices, triggered off by expectations of high prices.

More serious, perhaps, is the apparent seasonality in the Crude Oil Arrival Rate, COAR. Despite the fact that the system is supposed to smooth out a lot of the seasonality in demand, the amplitude of COAR is in fact larger than that of Product Offtake. The seasonality in COAR would probably be even larger if shortage of refinery capacity did not inhibit some of the seasonality in the Current Refinery Throughput, CRPT.

Finally, the stock policies are failing to achieve their purposes as product stocks fall progressively shorter of their targets, though control of crude stock is good.

In summary, the initial control system design cannot be said to be exceptionally good or bad. It generally fails to do what it is supposed to do, except in the sense of ensuring that enough oil gets through to prevent lost sales arising, but it does not produce any disasters. In particular, performance cannot be improved by improving the 'accuracy' of forecasting.

A further run with step changes in the oil demand level showed that the model's settling time was in the order of three years.

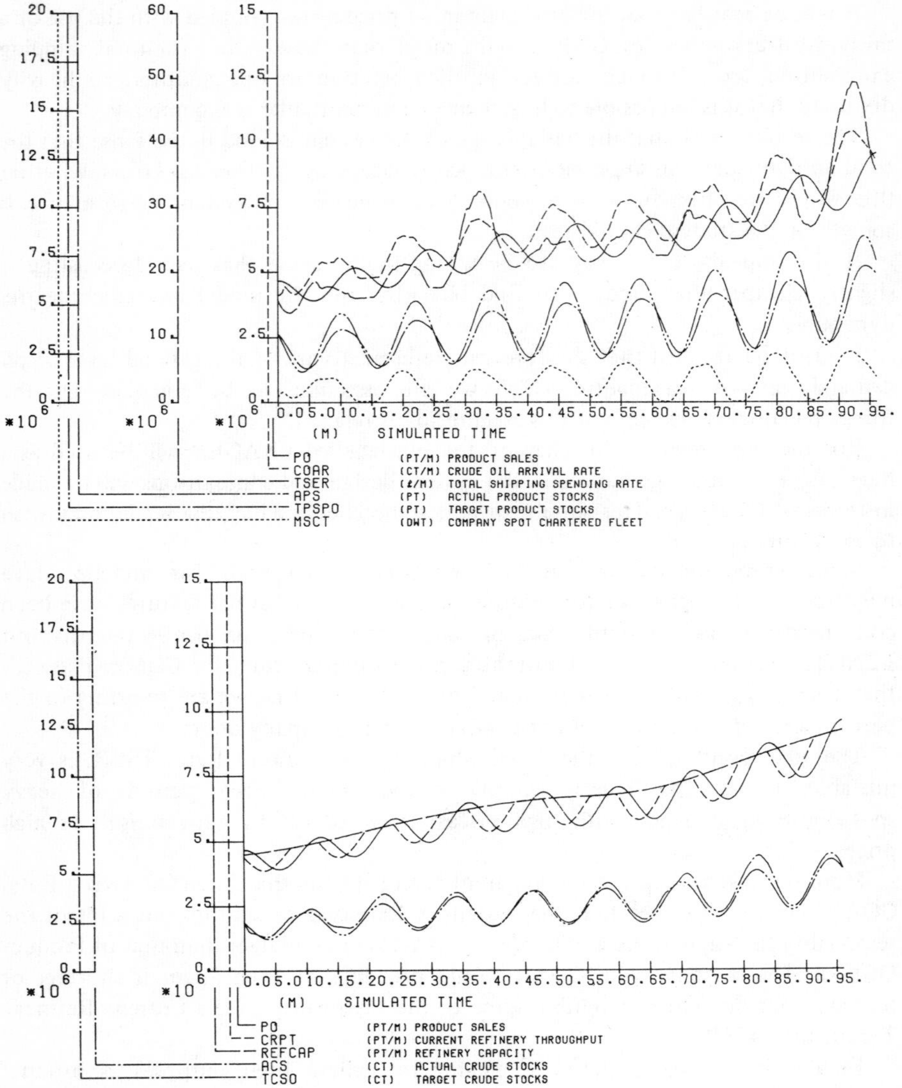

Fig. 11.5 Tne base case (EXAGER)

11.9 Improving the Design

In searching for improvements it would be easy to embark on a vast programme of computer simulation with little guidance as to what should be simulated next, and even less hope of ever seeing the wood for the computer printout. It is therefore useful to examine the feedback structure of the first design in more detail, as shown in Fig. 11.6 and Fig. 11.7.

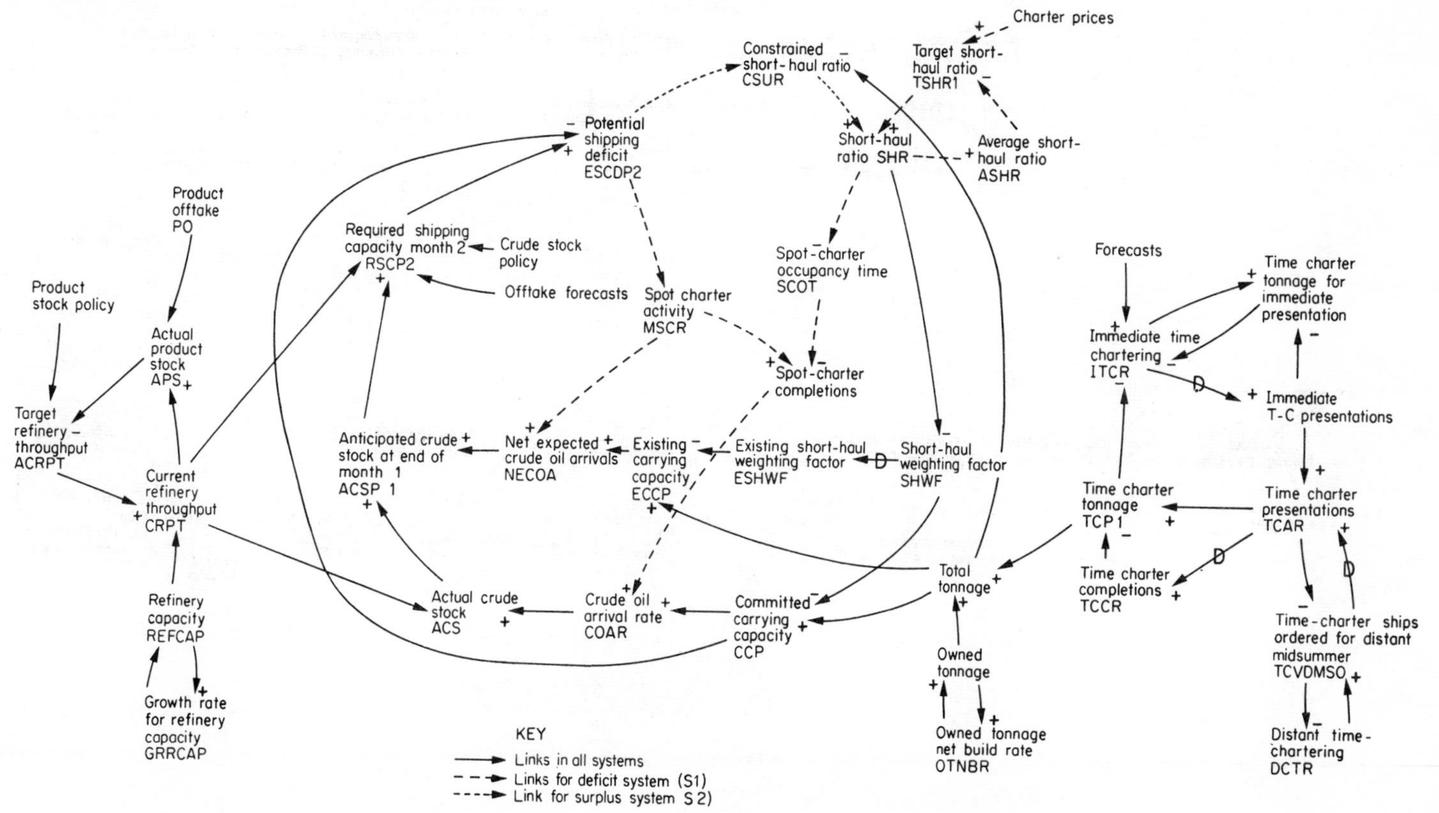

Fig. 11.6 Principal feedback and loops and system drives

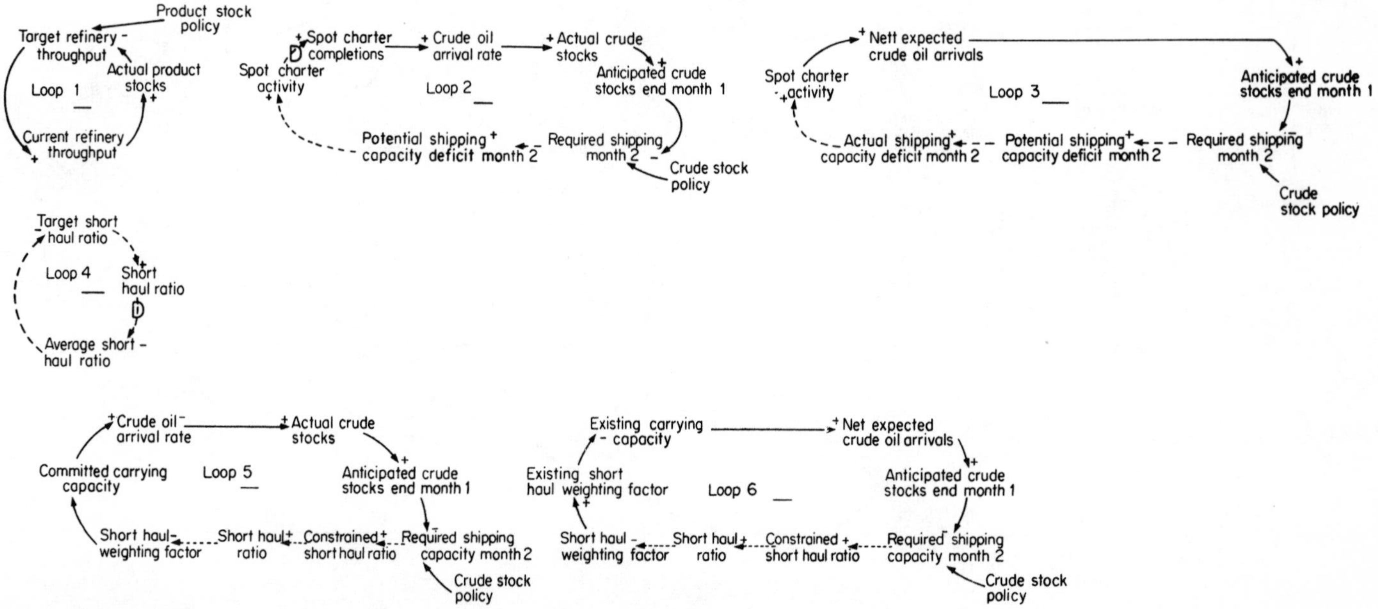

Product stock policy
Target refinery throughput
Actual product stocks
Loop 1
Current refinery throughput
Spot charter completions
Crude oil arrival rate
Actual crude stocks
Spot charter activity
Anticipated crude stocks end month 1
Loop 2
Potential shipping capacity deficit month 2
Required shipping month 2
Crude stock policy
Nett expected crude oil arrivals
Spot charter activity
Anticipated crude stocks end month 1
Loop 3
Actual shipping capacity deficit month 2
Potential shipping capacity deficit month 2
Required shipping month 2
Crude stock policy
Target short haul ratio
Loop 4
Short haul ratio
Average short haul ratio
Crude oil arrival rate
Actual crude stocks
Committed carrying capacity
Loop 5
Anticipated crude stocks end month 1
Short haul weighting factor
Short haul ratio
Constrained short haul ratio
Required shipping capacity month 2
Crude stock policy
Existing carrying capacity
Net expected crude oil arrivals
Existing short haul weighting factor
Loop 6
Anticipated crude stocks end month 1
Short haul weighting factor
Short haul ratio
Constrained short haul ratio
Required shipping capacity month 2
Crude stock policy

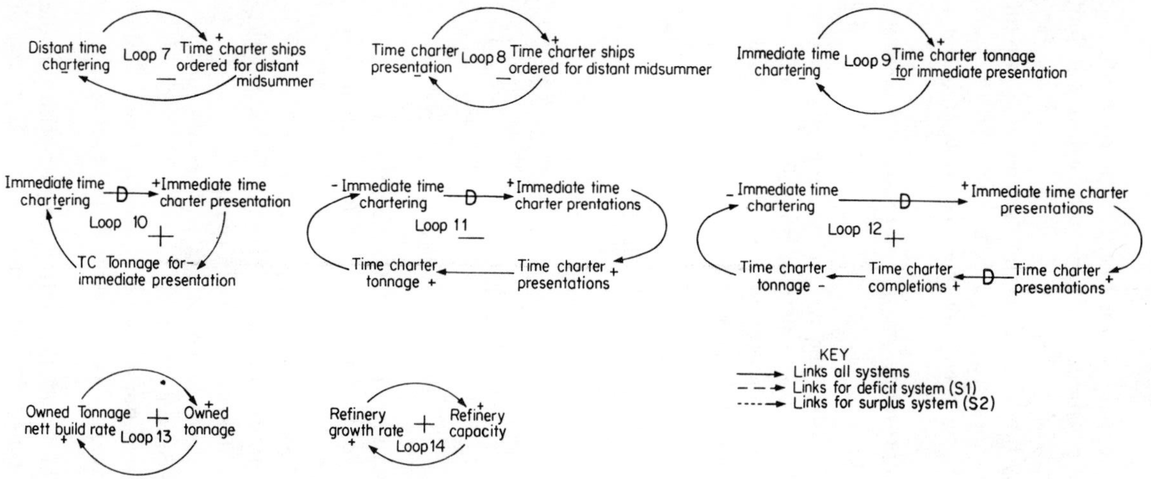

Fig. 11.7 The individual loops of the supply system (no constraints active).

The reader should be sure that he has a good grasp of the general layout of the system. The key points are:

1. The Constrained Short-haul Ratio is that which would use up all the available shipping by transferring as much as possible to long-haul routes so as eventually to wipe out the surplus.
2. The dotted lines show how the feedback pattern is completely changed by switching from S1 to S2.
3. The system is driven through the refinery sector but there is no control of refining by the system (except via a constraint that low crude stocks will limit refinery throughput).
4. Unless the refinery sector is carefully tuned, it will amplify seasonalities into the main system so that, apart from planned stocks not being achieved, the stock planning process may even make things worse rather than making them better as it is supposed to.
5. The long-term system affects the short-term, but not vice-versa and there is a very poor feedback structure.

The crude and product stock policies affect the system in various places. Both policies involve target stocks following a seasonal pattern troughing at a min-ops level and peaking at max-ops. The trough times are at the end of April and are denoted by MINPT and MINCT for product and crude respectively.

The basic equations are:

$$\text{ACRPT} = \textit{offtake} + \left(\frac{\text{TPSP–APS}}{\text{ATPS}}\right) \qquad 11.1$$

and

$$\text{RSCP2} = \text{CIF } \textit{Sales} + \left(\frac{\text{TCS2–ACSP1}}{\text{TACS}}\right) + \text{PRTP2*RRF} \qquad 11.2$$

where

ACRPT	Target Refinery Throughout, [Tons of Product/Month]
TPSP2	Target Product Stock Position in Month 2 [Tons of Product]
APS	Actual Product Stocks [Tons of Product]
ATPS	Adjustment Time for Product Stocks [Months]
RSCP2	Required Shipping Capacity in Month 2 [Tons of Crude/ Month]
TCS2	Target Crude Stock in Month 2 [Tons of Crude]
ACSP1	Anticipated Crude Stocks at End of Month 1 [Tons of Crude]
TACS	Time to Adjust Crude Stocks [Months]
PRTP2	Planned Refinery Throughput in Month 2 [Tons of Product/ Month]
RRF	Refinery Recovery Factor [Tons of Crude/Ton of Product]

The bracketed terms in equations 11.1 and 11.2 introduce gain (through the target stocks) and first-order delay (through the adjustment times). They are, of course, proportional controllers, but the offtake and refinery throughput introduce a measure of feedforward which ought to reduce the potential instability of these controllers.

Loops 2 and 3 have the shortest total delay, and lengthening the delay by increasing ATPS made them so slow that large amounts of lost sales arose to insufficient chartering and low stocks.

The value of max-ops depended on physical capacity and that of min-ops on the need to keep the logistic pipeline (literally) full. Reducing the gaps had a great effect on smoothing the system so that refinery throughput and shipping expenditure were, for all practical purposes, deseasonalized. This meant that at least 200,000 tpa of refinery capacity was saved and operating costs in the refineries would certainly be reduced. The reader may care to ponder the implications of this, apart from the fairly noticeable financial benefits. (It must again be stressed that this does not imply that the real company is run like this.)

Further runs on the model showed that the stock policies could be retimed, and that the values of MINPT and MINCT could be altered.

Having improved the design somewhat by adjusting gain and delay in the stock area it was natural to look for similar improvements in the shipping area. The options are, however, more limited and one can only alter the adjustment times for immediate time-chartering and for spot-chartering.

Increasing the former made loops 9 and 11 so slow that extra spot-chartering had to be done and costs went up and increasing the spot-charter adjustment time led to lost sales due to slowness in loops 2 and 3. (See Appendix B.11 for the relevant equations.)

Having then briefly looked at the possibilities for fine-tuning of the existing system, and summarized a great deal of work in a short space, we turn to the opportunities which system dynamics methodology suggests and facilitates for changing the design so as to improve its performance and controllability.

11.10 Structural Changes

We have already remarked that the refinery sector and the long-term chartering system have no feedback connection from the main body of the model. This is very obvious from the diagrams and it is one of the advantages of this approach that it makes such features so clearly apparent. There are, in fact, two main areas in which improvements to the control structure can be seen to be possible: the crude control and the shipping control.

The Crude Control

Stripped to its essentials, the refinery production and crude stock sector has the structure shown in Fig. 11.8. Actual Crude Stock feeds into the rest of the system and eventually back to the Crude Oil Arrival Rate COAR. The link from ACS to CRPT only acts as a constraint and reflects the inability of a refinery to operate fully without sufficient crude oil.

The situation is even worse than this when viewed in more detail. If ACS is low,

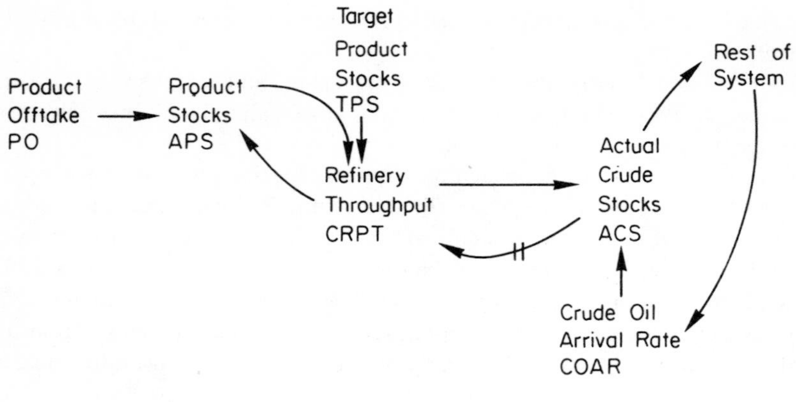

Fig. 11.8

ACSP1, the Anticipated Crude Stock in 1 months time, is low and this forces a curtailment in the Planned Refinery Throughput for the second month hence, PRTP2. This is one of the determinants of the spot-charter rate because of the effect which it has on the Required Shipping Capacity during month 2, RSCP2. This creates two pseudo-positive loops which could drive refinery throughput to extinction if they were not short-circuited by negative loops and by other loops shown as 2, 3, 5, 6, and 7 in Fig. 11.7. This is shown in detail in Fig. 11.9.

The detailed equations for the system and a fuller explanation of the variables appears in Appendix B, Section 11. For the moment the important thing is, not what the symbols in the diagram mean, but the implications for the design of a controller.

1. This is a classic example of an input being open to amplification, and indeed Fig. 11.5 shows that the Crude Oil Arrival Rate has an amplitude at least as large

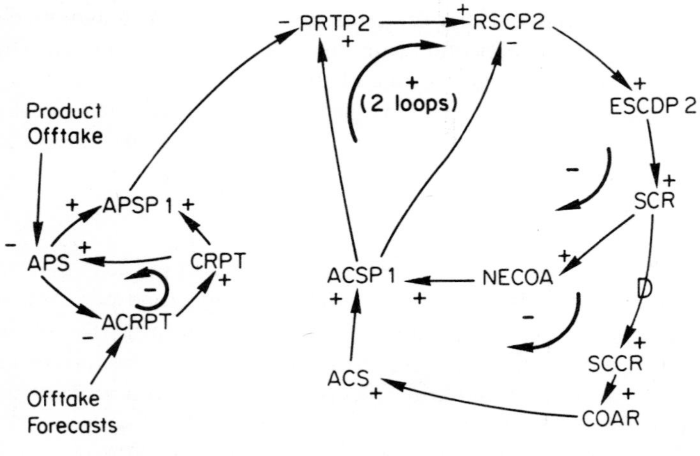

Fig. 11.9

as the Product Offtake which drives this part of the system. At the very least we can say that the various stocks in the system are not fulfilling their function of insulating it from demand seasonality.

2. The only control on stocks of crude is via the fairly slow loops which charter ships and there is no control of crude by regulating the rate at which it is used.

The diagram shows very obviously that linking either ACS or ACSP1 to ACRPT will create two new negative loops to act as system controllers. We therefore introduce a link from ACS to ACRPT such that:

$$\text{ACRPT}=\text{OFP1}+\frac{\text{TPSP1}-\text{APS}}{\text{ATPS}}+\frac{\text{ACS}-\text{TCS1}}{\text{ATCS}*\text{RRF}} \qquad 11.3$$

where ACRPT	Target Refinery Throughput for coming month [Tons of Product per month]
OFP1	Forecast of Offtake for coming month [Tons of Product per month]
TPSP1	Target Product Stock Position at end of coming month [Tons of Product]
APS	Actual Product Stocks [Tons of Product]
ATPS	Adjustment Time for Product Stocks [Months]
ACS	Actual Crude Stocks [Tons of Crude]
TCS1	Target Crude Stock Position at end of coming month [Tons of Crude]
ATCS	Adjustment Time for Crude Stocks [Months]
RRF	Refinery Recovery Factor [Tons of Crude per ton of Product]

This ensures that the new link from ACS to ACRPT will be positive thereby creating a new *negative* loop. A little carelessness at this stage with the third term on the RHS of equation 11.3 would create havoc in the model. We want the new loop to be fairly quick acting, so ATCS is set to 1 month.

This improves performance considerably, leading to large cash savings and generally smoother operation. It still leaves the control of the system entirely to proportional action, allied to some measure of feedforward from the offtake forecast. The serious reader should run the model in Appendix B, or a simplified version of it, and experiment with more refined controllers.

The Shipping Controller

We now turn to the other area in which an examination of the control structure reveals the design to be less than completely satisfactory – ship chartering. (At the risk of repetition, this is not necessarily how the actual Company is operated.)

The main problem is in the area of immediate time-chartering which, as explained earlier, is the chartering of ships which become available within about one month but are otherwise perfectly ordinary time-charters of 3–5 years duration. The essential feedback structure of this part of the model is shown in Fig. 11.10.

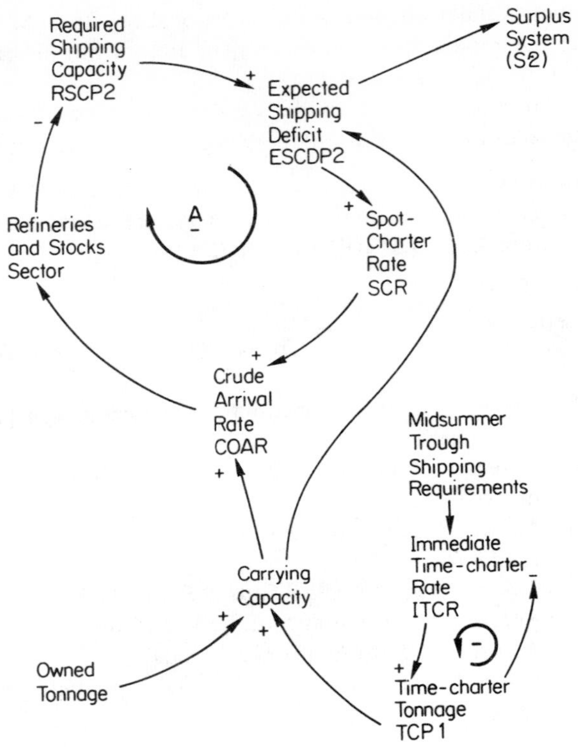

Fig. 11.10

Two things stand out from Fig. 11.10:

1. When there is a shipping surplus, spot-chartering stops and control passes to the S2 system to use up the available shipping. Now there is not much difference between spot-chartering and immediate time-chartering, but the latter can continue even in S2.
2. More fundamentally, all the control in the system is exerted through loop-system A and there is no control over immediate T-C apart from the forecast requirement for the coming summer trough.

It seems reasonable to expect that, if time-chartering proceeds without control from loop-system A, then it could drive A out of control, i.e. to force S2 into operation more frequently than need be the case. For the base case, EXAGER, S2 does not operate very often but one might reasonably expect to get either smoother, or cheaper, performance from the system by linking ITCR to ESCDP2 and splitting ESCDP2 so that some fraction, PSCD, would be spot-chartered and (1 – PSCD) would be met by immediate time-chartering. It is also logical to stop immediate T-C when the system is in surplus (S2) but one would expect the effects of this on system performance to be masked by the other changes in the system.

In practice, of course, a smoother system will almost certainly be a cheaper one because there will be fewer shipping peaks to be met by relatively expensive spot-chartering. One might therefore look for the trade-off between smoothness and cheapness, accepting that cost, as measured by cumulative expenditure on shipping, will be rather a poor measure to use in discriminating between the two situations.

Before discussing these two design objectives in more detail we shall deal with two minor matters affecting shipping.

Price-expectations and Shipping

If spot-charter prices are high it may be cheaper to use the more expensive short-haul oil and, in the model, the Short-haul Ratio which one would like to have, TSHR, rises and falls as expected price rises and falls. The short-haul ratio actually used, SHR, is affected by various constraints, as discussed in the model equations in Appendix B.11, but obviously the Expected Spot-charter Price, ESCPR, affects the short-haul ratio.

ESCPR also affects spot-chartering directly in that, if the price is expected to rise, more chartering is done by the model than is strictly called for by the shipping deficit. The effects of this on loop-system A are shown in Fig. 11.11.

Clearly, variations in ESCPR have a multiplicatively-variable effect on gain in loop-system A. Since charter prices are very volatile it is clear that this gain effect is likely to be quite marked.

A more sophisticated treatment is to recognise that the cost of shipping affects the cost of oil landed in NWE. This raises the question of whether the design should allow for the option of foregoing business when the cost of landed crude is so high as to make it unprofitable.

Expected charter prices were forecast using a method due to Zannetos (*The*

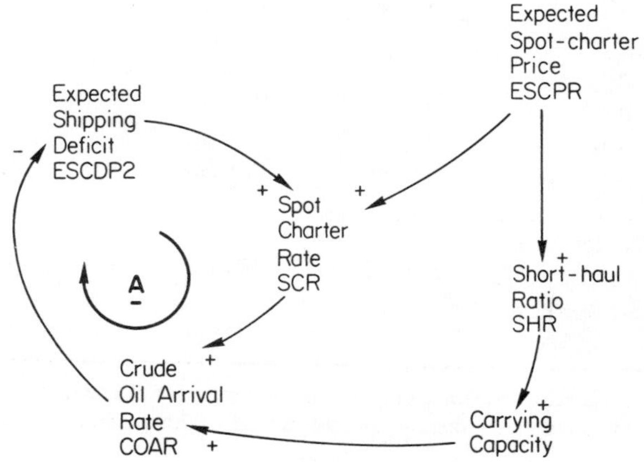

Fig. 11.11

Table 11.1. *Summary of Salient System Changes*

Parameter Changes	*Base System* (EXAGER)	*Least Cost System* (MXC1)	*Smoothest System* (CRCON2)
Min-ops./Max-ops. Product Stocks range	2.7 m.t.	1.6 m.t.	1.5 m.t.
Time of year for min.	1st May	1st March	1st April
Adjustment time of product stocks. (months).	3	1	1
Min-ops/Max-ops. crude stocks range.	1.2 m.t.	1.0 m.t.	0
Time of year for min. crude stocks.	1st May	1st March	1st March
Adjustment time of crude stocks. (months).	1	1	1
Structural Changes			
Proportion of deficit spot chartered.	100%	0	50%
Block on Immediate Time-chartering if surplus shipping.	Not Present	Present	Present
Response of Target Short-haul Ratio to expected price (max change from policy figure).	25%	0	0
Response of Spot-chartering to Expected Price (% change from indicated amount).	20%	0	0
Crude Control on Refinery Throughput	Not Present	Present	Present
Performance (over 10 year period)			
CUMEXP (£M)(fictitious values)	3269,55	3071.12	3170.9
Average Annual Saving (£M) on Shipping		19.8	9.9

Note: the numerical values in this table are roughly indicative of the magnitudes which one might encounter in a large oil company but are subject to the qualifications already made.

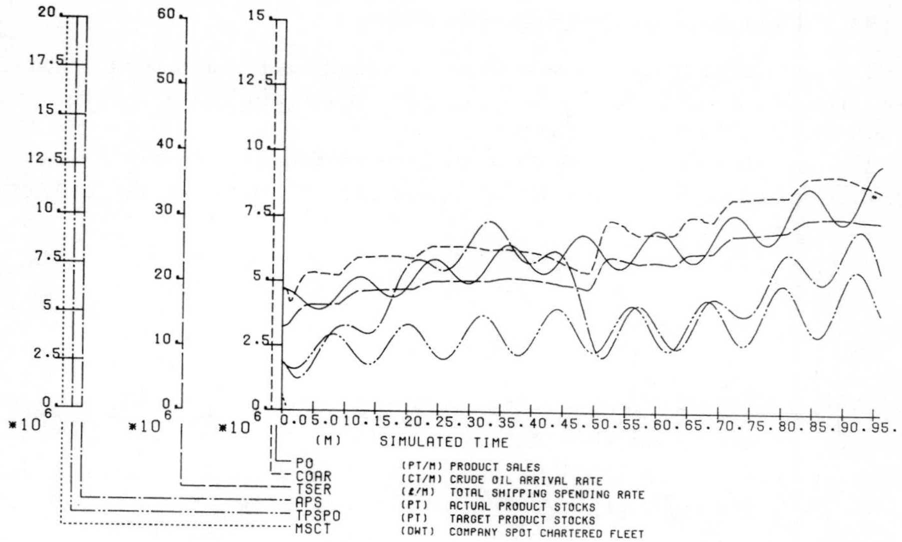

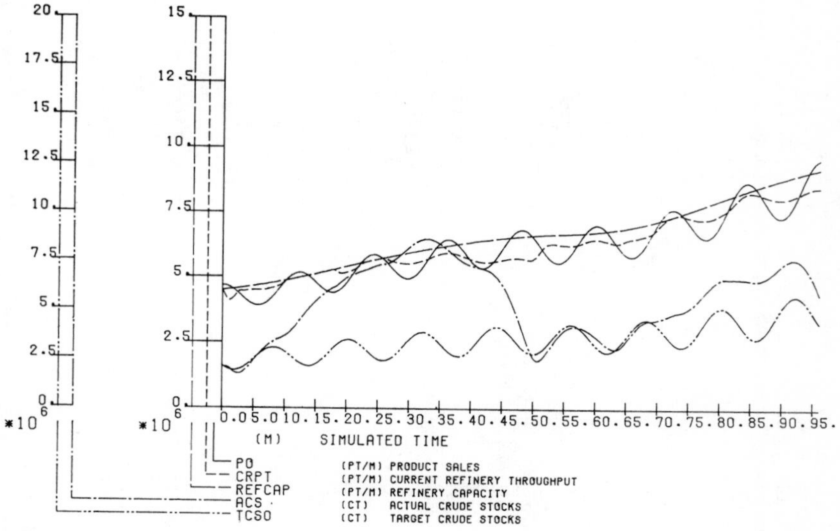

Fig. 11.12 The cheapest system (MXC1)

Theory of Oil Tankship Rates, MIT Press, 1969) but, removing the effect of charter prices on loop-system A improved the model's smoothness of response. This was another case of the rather remarkable response of this model to forecasts. In this instance, accuracy or otherwise of the price forecasts was irrelevant, as their use made performance demonstrably less smooth.

11.11 The Smoothest and the Cheapest Systems

The two computer runs which generate the smoothest and the cheapest systems are summarized in Table 11.1 and computer output for them is shown in Fig. 11.12a and b, and Fig. 11.13a and b.

The main point is that the cheapest system, run MXC1, does not do very well at controlling stocks. During the periods of optimism in forecasting, the levels of

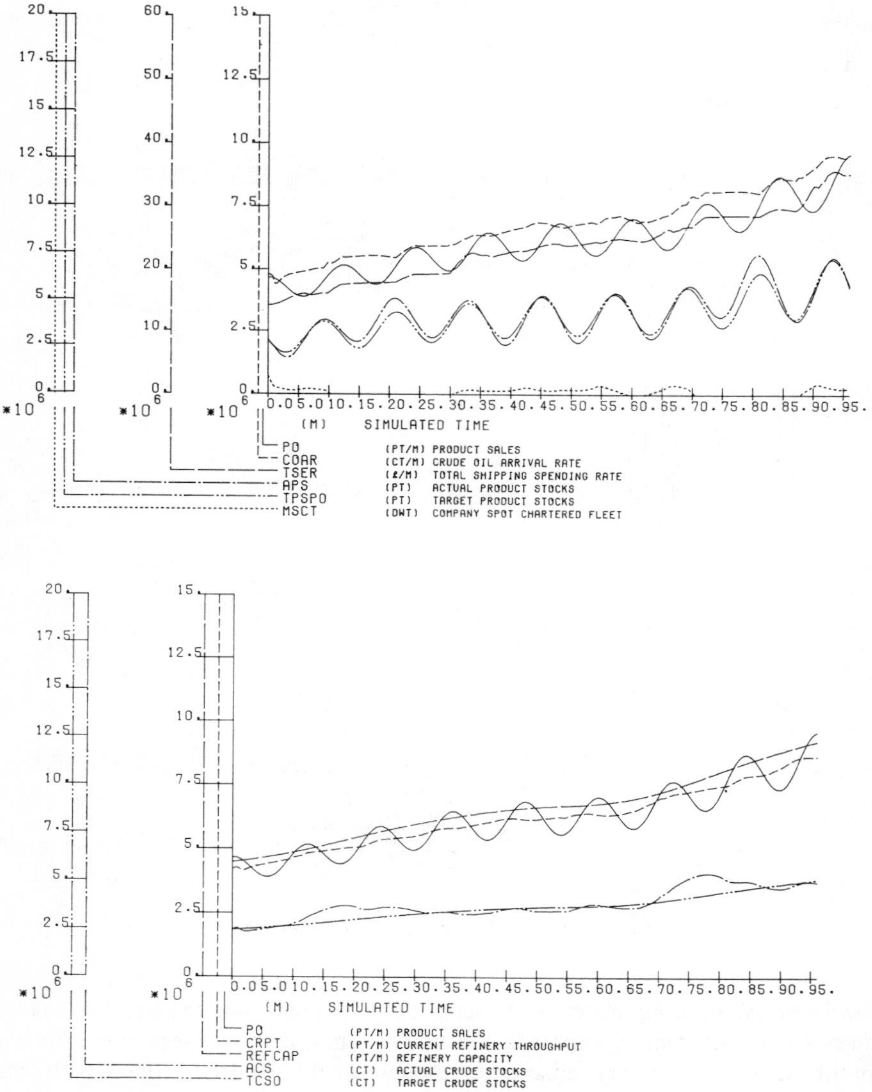

Fig. 11.13 The smoothest system (CRCON2)

crude and product stocks far exceed what the 'Company' can hold. This would involve renting storage space which would probably more than wipe out the shipping savings. (This may suggest some questions relating to forecasting accuracy, and measures of system performance!)

It can, however, be seen that the smoothest system, CRCON2, is extremely smooth, *despite* the forecasting errors. Stock plans are almost perfectly achieved, cash flow is stabilized, and refinery throughput is deseasonalized (which has major implications for costs and capital spending).

The reader should note that these results are not the last word that can be said about the system, as they have been chosen to stimulate thought, rather than to give definitive answers. The reader should also note that the oil industry generally knows that the smooth system is the economical one to run. Work with the model should start at this point.

Chapter 12

United Metals Ltd.

Some of the features of this model have been drawn from published information about several real companies in the metal, and other, industries but 'United' is NOT a pseudonym for a real firm, and there is no connection with any particular company.

12.1 Introduction

This chapter applies SD to the study of the strategic future of a firm. The main detail of the modelling is in the financial area but the essence is to study the interconnection of the firm's activities as a whole. In this way we show the application of SD to corporate planning and business policy but, in doing so, we present some difficult ideas and the chapter calls for careful study.

The model incorporates a rather fuzzy environment sector, which prompts some interesting questions of robustness and sensitivity analysis, and it may be too simple to carry conviction with management. The serious student should therefore decide whether he needs to refine some of the concepts in the model before doing a policy design for the system. (The model in Appendix B. 12 has been written with this in mind, and is not necessarily how I would model the situation.)

Paradoxically, whilst the model lacks enough detail to be a good model of *a* firm (in what sense?), it may well be too detailed to be a model of firms in general. An exercise of equal value would be to produce a simpler model – not by simplifying this one but by starting again from the problem description.

12.2 The Industry and the Company

United Metals is a subsidiary of a parent company which expects United to fund its own future. For our purposes, United comprises two divisions: Semifabrication, and Manufacturing. Respectively, they produce particially completed goods such as tubes, and final products, ranging from tubular garden furniture to vehicle bodies, which can be sold to end customers. The starting point for semifabrication is an ingot of metal produced from ore in a converter. United's parent also owns ingot producing companies which, as they are located overseas, are operated independently from United. These producers cannot meet all United's metal needs.

The variety of products is rather large and United's principal direct competitors are the other major integrated metal companies, who also produce ingot, semi-fabricated, and final-use products. There are, however, a large number of independents, especially in the manufacturing end of the production chain, but also in ingot production.

United has, for historical reasons, an established demand for ingot from other semifabricators who actually compete with United's own semifabrication division. This is met by buying ingot on the market and reselling it – a profitable business. The profit flow from this metal broking business is not included in United's profit flow equation as it is regarded as a separate business. The only effect it has in the model is to alter the average cost of metal used by United and thus, indirectly affect the profitability of United's semifabricating and manufacturing operations.

United's parent company have a number of strategic options open to them. For example, they could encourage United to grow by expanding into manufacturing, or growing as a semifabricator. On the other hand, they could decide to reduce their involvement in the highly competitive metal industry and gradually run down United, using it as a money machine to generate cash to put into other business areas. Alternatively, they might expand the ingot producing company to give United access to a large supply of cheaper metal. This project would be costed and financed in its own right by United's parent, but they wish to see what synergistic effects there might be if United were, at the same time, expanding either downstream or laterally.

The Semifabricator Division takes metal from the ingot producers, and bought metal, and produces a wide range of standard shapes, mostly bars and tubes. This output is sold to companies which transform it into products of final use. United's Manufacturing Division make such products themselves, in competition with companies which buy their bars and tubes from United's Semifabricator Division.

The manufacturing sector of the industry consists of a large number of companies, some of which are subsidiaries of United or the competitors, the rest being independents who are, however, totally unconnected with the ingot producing independents. The independents range from Fords, to very small firms supplying a local, or very specialized demand. The independents buy ingot to semifabricate themselves, or semifabricated metal which they process. The metal flows in the industry are shown in Fig. 12.1 and may be described as complicated.

United have roughly 25% of the UK market for semifabricated metal and a much smaller share of the *metal* used in final products. They also sell semifabricated metal to the competitors manufacturing side, or buy it from the competitors. This is either a purely short-term arrangement or because of an inability to supply particular qualities. It is a small-scale trade and we ignore it.

United's finance comes from three sources; the original share capital, retained earnings, and long-term borrowing from the parent holding company. This latter source is supposed to be restricted to a maximum of 60% of the original share capital plus cumulative retained profits.

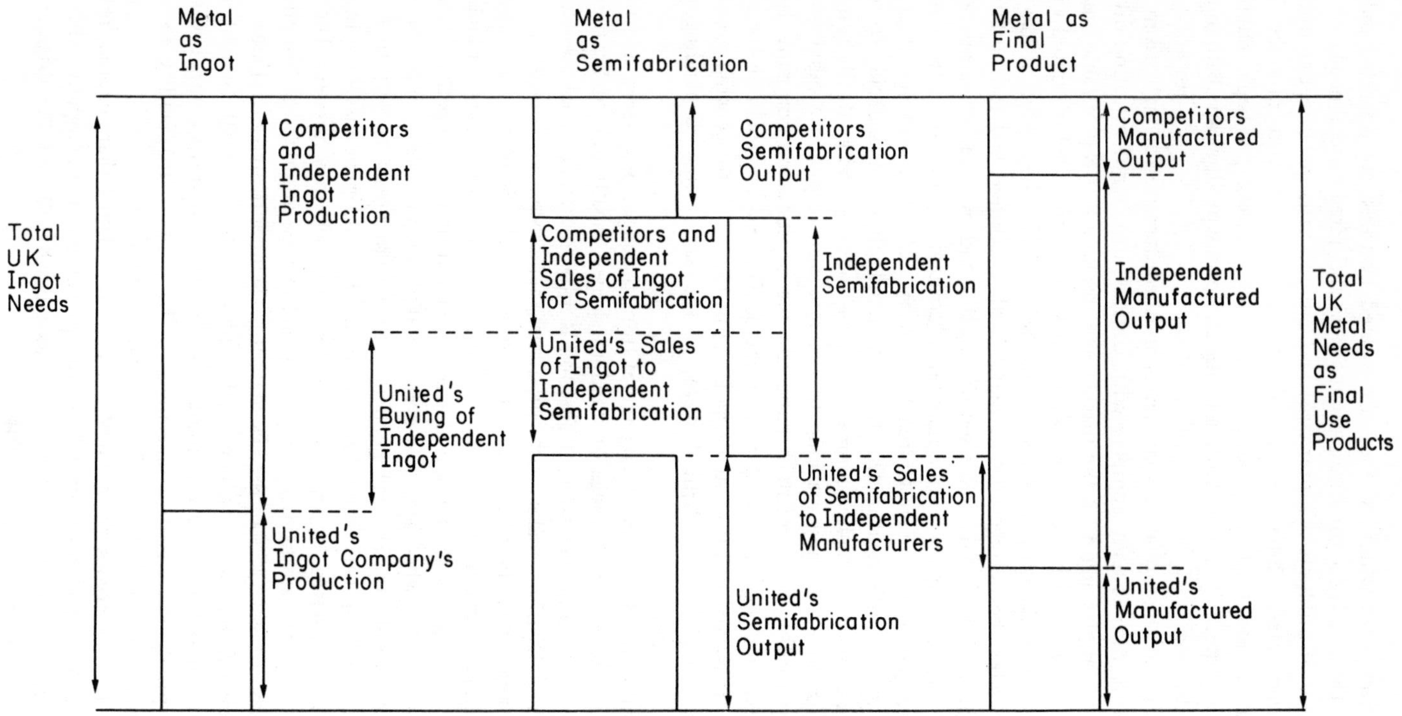

Fig. 12.1 Schematic metal flows in the industry

12.3 The Purpose and the Model

The object of this case study is to show how an SD model can be constructed of this situation, and to see what light it sheds on the issues involved. In particular, we wish to show how United's policies interact to produce performance over a substantial period of time, and how robust that performance will be. We therefore do not need a complicated and detailed model of production or of capacity, because we are mainly interested in the ability of the system to generate money to meet its spending needs. Our model must therefore contain those loops which produce money, and those which absorb it and in particular it must contain the mechanism by which the firm attempts to harmonize the two – its financial policy.

We are assuming that we do not need to model in detail how the money is spent. For example, if we show that the strategic option of downstream integration into manufacturing is attractive we are not, in this chapter, going to model the choice of whether the integration should be in windmills or washing machines, or by expanding United's own companies as opposed to buying others. This is not to say that these choices are unimportant but that an SD model would not necessarily be the best tool for that analysis.

In practice the value of detail of this sort would depend very much on the type of industry and how far it was from demand saturation and therefore, how the firm might review its objectives. That is, however, very different from the dynamic problem we posed at the outset, which reminds one of the need to suit the model to the problem.

It will probably help the reader to understand the model if he constructs his own influence diagram from the information here and in Appendix B.12. For this reason we do not give an influence diagram in this chapter (but see Fig. 9.7).

12.4 Cash Generation in the Model

Cash comes from three sources in the model: Depreciation, Profits, and Growth of Long Term Debt.

Depreciation Cash Flow, DCFL (£/month), is simply a fraction of the written down value of the plant, which is thereby further depreciated.

Working Capital, WCAP, is a composite of accounts receivable and payable and inventory and it is assumed that the level of production which can be sustained is proportional to the Working Capital available. This ignores the possibility of short-term financial action on liquidity management because it is assumed that this has no real affect on long-term dynamics.

Profit is treated on a simple basis of (price–direct cost) and the profit flow derives from the production flow. Target production is determined on the basis of controlling the backlog of orders to a reasonable level and this, therefore, determines the amount of working capital which would be needed to support such a level of activity. Actual working capital, WCAP, then becomes an important variable as it may not be adequate, and production would therefore have to be cut back and backlog would increase. The model assumes that backlog has no effect on demand, though it is believed in United that there would be an effect.

Profits come from production throughput, TPUT, which is basically measured as semifabricated metal. A fraction UPC of TPUT goes on to Manufacturing and thereby earns a second slice of operating profit. The fraction UPC, the United Percent, is a key variable in corporate planning because it is the fraction of semifabricated output which United is going to keep to itself for future processing by the Manufacturing side. Changes in UPC would therefore model various policies toward integration etc.

This uncomplicated formulation of the production flow only acknowledges the presence of two divisions by two delays of different magnitude between production and profit flow.

Profit flow is also affected by the average cost of ingot. For simplicity it is assumed that United's company sell them metal for £150/ton while United has to pay £250 on the open market. (Which suggests some interesting dynamic problems or a deranged costing system.) United's products are priced on an average metal cost of £210 and departures from this affect profits. The profit flow is also affected by interest payments on long term debt and by receipts from short-term investments. Short-term investments are made from any cash i.e. unallocated money the company has, or money earmarked for new projects but not actually fully spent, half of which is assumed to be available for short-term loans.

```
R PR.KL=DELAY3(TPUT.JK,SFD)*(1-UPC.K)*GMSF.K
X   +DELAY3(TPUT.JK,MFD)*UPC.K*GMMG.K-MSS.K+(210-UMC.K)*LTPUT.K
X +(INTPC/1200)*(MCNP.K/2+CASH.K-LTDEBT.K)
C SFD=1
C MFD=3
C INTPC=15
        PR        =(£/M) RATE OF RECEIPT OF PROFIT FROM PRODUCTION

        TPUT      =(T/M) THROUGHPUT IN MANUFACTURED-METAL SECTOR

        SFD       =(M) SEMI-FABRICATION PRODUCTION DELAY

        UPC       =(1) 'UNITED PERCENT' - ONE OF THE KEY VARIABLES IN THE MODEL.
                   THE FRACTION OF SEMI-FAB PRODUCTION WHICH IS RETAINED BY
                   UNITED FOR ITS OWN MANUFACTURING ACTIVITIES

        GMSF      =(£/T) GROSS MARGIN ON METAL SOLD AS SEMI-FABRICATED

        MFD       =(M) PRODUCTION DELAY IN MANUFACTURING

        GMMG      =(£/T) GROSS MARGIN ON METAL SOLD AS MANUFACTURES (INCLUDES GMSF)

        MSS       =(£/M) ACTUAL MARKET SHARE SPENDING

        UMC       =(£/T) AVERAGE COST PER TON OF INGOT METAL

        LTPUT     =(T/M) AVERAGE LEVEL OF SEMI-FAB THROUGHPUT

        INTPC     =(%/Y) INTEREST RATE ON CASH RESOURCES

        MCNP      =(£) MONEY COMMITTED TO NEW CAPACITY ADDITION PROJECTS

        CASH      =(£) CASH RESERVES IN COMPANY

        LTDEBT    =(£) LONG-TERM DEBT OF COMPANY

A GMSF.K=IGMSF*(1+STEP(SFPFAC,PRSTM))
C IGMSF=117
C SFPFAC=0
C PRSTM=60
        GMSF      =(£/T) GROSS MARGIN ON METAL SOLD AS SEMI-FABRICATED

        IGMSF     =(£/T) BASE VALUE FOR GROSS MARGIN IN SEMI-FAB

        SFPFAC    =(1) FRACTION CHANGE IN GROSS MARGIN IN SEMI-FAB

        PRSTM     =(WK) TIME AT WHICH PRICE RISES COME INTO EFFECT
```

```
A GMMG.K=IGMMG*(1+STEP(MFPFAC,PRSTM))
C IGMMG=212
C MFPFAC=0
C PRSTM=60
          GMMG     =(£/T) GROSS MARGIN ON METAL SOLD AS MANUFACTURES (INCLUDES GMSF)

          IGMMG    =(£/T) BASE VALUE FOR GROSS MARGIN IN MANUFACTURING

          MFPFAC   =(1) FRACTION CHANGE IN GROSS MARGIN IN
                    MANUFACTURING OPERATIONS

          PRSTM    =(WK) TIME AT WHICH PRICE RISES COME INTO EFFECT
```

A three-month exponential smooth of PR gives Average Profit Level which gives After Tax Profit Level on the simplifying assumption that a fixed percentage of pre tax profits goes to taxes and dividend payments.

N.B. None of these assumptions is forced upon us because not making them would be too difficult a problem, it is a matter of the purpose and scope of the original study and space factors in a book. The reader is strongly encouraged to alter the model to meet whatever assumptions he would have made in similar circumstances. The numerical values in this and other equations have no particular relation to real firms.

```
A ATPL.K=((100-TAXDPC)/100)*(APL.K-DCFL.K)
C TAXDPC=50
          ATPL     =(£/M) AFTER TAX PROFIT LEVEL.

          TAXDPC   =(%) PROPORTION OF PROFIT FLOW LOST IN TAXATION AND DIVIDEND
                    PAYMENTS

          APL      =(£/M) AVERAGE PROFIT LEVEL

          DCFL     =(£/M) DEPRECIATION CASH FLOW
```

At any one time the firm has an actual Long Term Debt, LTDEBT, and a maximum value which LTDEBT could move towards MXLTD, over a planning horizon, PHOR. Corporate Policy limits MXLTD to a fixed amount of the total of the original investment plus cumulative retained profits to date and expected additional profit during the planning horizon (using rather a simple forecasting formula).

```
MXLTD.K=(CRP.K+SCAP.K+ATPL.K*PHOR)*(GEAR/100)
PHOR=12
GEAR=60
        MXLTD    =(£) MAXIMUM PERMISSIBLE LONG-TERM DEBT

        CRP      =(£) CUMULATIVE RETAINED PROFITS

        SCAP     =(£) SHAREHOLDERS CAPITAL EMPLOYED

        ATPL     =(£/M) AFTER TAX PROFIT LEVEL.

        PHOR     =(M) PLANNING HORIZON IN MODEL

        GEAR     =(%) PROPORTION OF ACTUAL ASSETS WHICH CAN BE TAKEN AS LONG-TERM
                  DEBT
```

The firm can therefore assess its spending capacity in terms of its ability to increase its indebtedness, its expectations of future earnings during the planning period, and its currently available cash. This can be summed up by:

```
ACGR.K=ATPL.K+DCFL.K
        ACGR     =(£/M) ACTUAL CASH GENERATION RATE

        ATPL     =(£/M) AFTER TAX PROFIT LEVEL.

        DCFL     =(£/M) DEPRECIATION CASH FLOW
```

```
A ECG.K=ACGR.K*PHOR
C PHOR=12
        ECG     =(£) EXPECTED CASH GENERATION OVER PLANNING HORIZON

        ACGR    =(£/M) ACTUAL CASH GENERATION RATE

        PHOR    =(M) PLANNING HORIZON IN MODEL
```

```
A ARS.K=(MXLTD.K-LTDEBT.K+ECG.K+CASH.K)/PHOR
C PHOR=12
        ARS     =(£/M) ACHIEVABLE RATE OF SPENDING ON CAPACITY AND WORKING
                 CAPITAL ADDITIONS

        MXLTD   =(£) MAXIMUM PERMISSIBLE LONG-TERM DEBT

        LTDEBT  =(£) LONG-TERM DEBT OF COMPANY

        ECG     =(£) EXPECTED CASH GENERATION OVER PLANNING HORIZON

        CASH    =(£) CASH RESERVES IN COMPANY

        PHOR    =(M) PLANNING HORIZON IN MODEL
```

12.5 Cash Spending Plans in the Model

Indicated capacity in semifabrication and manufacturing depends, in turn, on a demand forecast (modelled by our standard method), a target market share in semifabrication, and a target for UPC, the fraction of semifabricated output to be manufactured by United's own companies.

```
A ISFCAP.K=LTDFC.K*PMS.K
        ISFCAP  =(T/M) INDICATED SEMI-FAB CAPACITY BASED ON DEMAND FORECASTS

        LTDFC   =(T/M) DEMAND FORECAST FOR END OF PLANNING HORIZON

        PMS     =(1) PLANNED MARKET SHARE OF SEMI-FAB MARKET
```

```
A IMFCAP.K=PPCU.K*ISFCAP.K
        IMFCAP  =(T/M) INDICATED MANUFACTURING CAPACITY BASED ON INDICATED
                 SEMI-FAB CAPACITY AND PLANNED PERCENT UNITED

        PPCU    =(1) PLANNED VALUE OF PERCENT OF SEMI-FAB OUTPUT TO BE
                 RETAINED FOR MANUFACTURE BY UNITED COMPANIES

        ISFCAP  =(T/M) INDICATED SEMI-FAB CAPACITY BASED ON DEMAND FORECASTS
```

A simple proportional control law gives the Target Rate of Ordering semifabrication capacity, allowing for the expected rate of replacement of capacity as it wears out.

```
A TROSFC.K=MAX(0,(ISFCAP.K-SFCAP.K*(1-DNRSF*(WR/1200)*CDEL)-SFCOR.K)
X /TASFC)
        TROSFC  =(T/M/M) TARGET RATE OF ORDERING SEMI-FABRICATED METAL
                 PRODUCTION CAPACITY

        ISFCAP  =(T/M) INDICATED SEMI-FAB CAPACITY BASED ON DEMAND FORECASTS

        SFCAP   =(T/M) SEMI-FINISHED CAPACITY

        DNRSF   =(1) ZERO-ONE SWITCH TO TEST EFFECTS OF NOT REPLACING
                 WORN-OUT SEMI-FAB CAPACITY

        WR      =(%/Y) RATE OF PHYSICAL WEAR OF PRODUCTION PLANT

        CDEL    =(M) CONSTRUCTION DELAY IN CAPACITY ADDITION

        SFCOR   =(T/M) SEMI-FAB CAPACITY ON ORDER BUT NOT RECEIVED

        TASFC   =(M) TIME TO ADJUST SEMI-FAB CAPACITY
```

There is a similar expression to model expansion aims in Manufacturing, and a simple calculation gives the working capital needed to sustain that level of activity.

The net effect of all this is the Required Rate of Spending, i.e. the rate at which United would like to be spending money now in order to be ready for the expected demand by the end of the Planning Horizon. The spending rate will be reviewed as time passes but it represents present plans and therefore drives the demand side of the financial system.

```
A RRS.K=TIWC.K+TROSFC.K*CCSFC+TROMFC.K*CCMFC
C CCSFC=5580
          RRS      =(£/M) REQUIRED RATE OF SPENDING ON CAPITAL PROJECTS

          TIWC     =(£/M) TARGET RATE OF GROWTH IN WORKING CAPITAL

          TROSFC   =(T/M/M) TARGET RATE OF ORDERING SEMI-FABRICATED METAL
                    PRODUCTION CAPACITY

          CCSFC    =(£/T/M) CAPITAL COST OF SEMI-FAB CAPACITY

          TROMFC   =(T/M/M) TARGET RATE OF ORDERING MANUFACTURED-METAL
                    PRODUCTION CAPACITY

          CCMFC    =(£/T/M) CAPITAL COST OF MANUFACTURED-METAL CAPACITY
```

12.6 Financial Policy

We now have the model generating an Achievable Rate of spending, ARS, and a Required Rate, RRS and, in general the two will be unequal. Financial policy in the model is the means by which they are to be reconciled. This is an interpretation of 'policy' which is consistent with the way in which we have used the word earlier in this book but it differs from the usual loose usage in which it often means either the objectives to be attained ('our policy is full employment') or the means of attainment ('by a policy of investment in new equipment'). At the risk of repetition we remind the reader that, in this book, 'policy' means 'a rule for regulating a flow in the hope of achieving a target'. Thus in Chapter 2, the simple production system was controlled by a policy of adjusting the production so as to close the gap between Inventory and Desired Inventory in a certain number of weeks.

In this case, 'financial policy' is more difficult, but we can approach it by keeping the idea of regulating the stream of cash spending. We start by identifying the Financing Ratio, FRAT. (This, and other terms, may not be encountered in finance textbooks, but see Helfert, *Techniques of Financial Analysis*, Irwin, 1967, page 5 for an interesting example of financial flows.)

```
A FRAT.K=ARS.K/(RRS.K+CLIP(0,1,RRS.K,0.1))
          FRAT     =(1) RATIO OF ACHIEVABLE RATE OF SPENDING AND REQUIRED RATE

          ARS      =(£/M) ACHIEVABLE RATE OF SPENDING ON CAPACITY AND WORKING
                    CAPITAL ADDITIONS

          RRS      =(£/M) REQUIRED RATE OF SPENDING ON CAPITAL PROJECTS
```

This simply measures the extent to which the expected means, ARS, match up to projected ends, RRS and it is the ratio of a forecast and a target. It is used to model United saying, in effect, 'this is what we *expect* to have, and this is what we *would like* to do' – there being neither certainty nor compulsion in such a situation.

The next step is the spending actually undertaken and one way of approaching this is to say that a percentage of required spending will actually be authorized now, in the light of current expected availability of money.

```
A FPMULT.K=TABHL(TFPMLT,FRAT.K,0,2,0.5)
TP TFPMLT=0/0.8/1.0/1.1/1.1
        FPMULT  =(1) FRACTION OF PLANNED EXPENDITURE ACTUALLY INCURRED
                     THIS IS THE KEY VARIABLE IN THE FINANCIAL POLICY
                     SECTOR OF THE MODEL

        TFPMLT  =(1) TABLE VALUE FOR FINANCIAL POLICY MULTIPLIER FPMULT

        FRAT    =(1) RATIO OF ACHIEVABLE RATE OF SPENDING AND REQUIRED RATE
```

It is not required by this formulation nor, indeed, expressed in that table function, that the reduction be proportionate. That particular table still authorizes 80% of projects when only 50% of the money can be foreseen. This runs down cash and leads to poorer control of long-term debt, thereby possibly storing up trouble for the future. It amounts to United going ahead and hoping the money will be found when needed.

This approach has its drawbacks but it does allow us to explore questions such as whether the firm should follow a growth-orientated investment policy such as this, or whether it should be more conservative, and under what circumstances. It leads us into a crisis intervention situation by allowing us to ask when and how the firm should take emergency action if, say, it is being aggressive and making optimistic forecasts and the market turns down. The list of possible questions is practically endless, e.g. what relative attention should be paid to cash, debt and return on investment – are any new control structures needed? The reader should think of some of these questions and then try to judge just how useful this model would be for them. This should lead him to adapt the model in various ways.

The financial policy multiplier method also models the case of United authorizing more spending more than has been asked for. This is a common enough situation in real life in which managers simply bring forward additional capital projects once they realize that the firm is fairly rich. However, any other situation is easily put in, of course.

The modelling of the Rate of Spending Actually Authorized, RSAA, starts from Working Capital. The model allows for increases in working capital to be less severely cut back, or more agressively pushed forward, by a factor WCPF – the Working Capital Preference Factor – as this will test various attitudes to this aspect of finance. Note that this does not imply that management actually have such a factor. It is rather a modeller's device for representing a particular, and observable, managerial behaviour pattern.

```
A ARWC.K=WCMULT.K*TIWC.K
        ARWC    =(£/M) ADDITION RATE TO WORKING CAPITAL

        WCMULT  =(1) MULTIPLIER TO REGULATE THE GROWTH OF WORKING
                     CAPITAL AS COMPARED TO ITS TARGET GROWTH

        TIWC    =(£/M) TARGET RATE OF GROWTH IN WORKING CAPITAL
```

```
A WCMULT.K=FPMULT.K*WCPF
        WCMULT   =(1) MULTIPLIER TO REGULATE THE GROWTH OF WORKING
                  CAPITAL AS COMPARED TO ITS TARGET GROWTH

        FPMULT   =(1) FRACTION OF PLANNED EXPENDITURE ACTUALLY INCURRED
                  THIS IS THE KEY VARIABLE IN THE FINANCIAL POLICY
                  SECTOR OF THE MODEL

        WCPF     =(1) MULTIPLIER TO REFLECT RELATIVE PRIORITY GIVEN TO
                  GROWTH IN WORKING CAPITAL AS COMPARED TO GROWTH IN
                  INVESTMENT IN PRODUCTION FACILITIES
```

This approach treats Working Capital as simply the net of inventories, payables and receivables and assumes that these always remain in the same proportion to each other and to throughput. Does the fact that, in real life, this is not the case matter in the context of the time scale of this model?

If the company is declining and Working Capital becomes excessive, some of it is siphoned off and used to increase Cash or reduce Long-Term Debt.

Actual Spending on plant is easily found.

```
A ARCMTNP.K=MAX(0,FPMULT.K*(RRS.K-TIWC.K))
        ARCMTNP =(£/M) ACTUAL RATE OF COMMITTING MONEY TO NEW PROJECTS

        FPMULT  =(1) FRACTION OF PLANNED EXPENDITURE ACTUALLY INCURRED
                 THIS IS THE KEY VARIABLE IN THE FINANCIAL POLICY
                 SECTOR OF THE MODEL

        RRS     =(£/M) REQUIRED RATE OF SPENDING ON CAPITAL PROJECTS

        TIWC    =(£/M) TARGET RATE OF GROWTH IN WORKING CAPITAL
```

Hence

A RSAA.K=ARWC.K+ARCMTNP.K

RSAA=(£/M) RATE OF SPENDING ACTUALLY AUTHORISED

in which ARCMTNP feeds a level, MCNP, which is money authorized but not necessarily spent and half of which is assumed to be available for short-term investment on the money market.

Finally, to balance cash flows in the system, we have:

```
A FCSR.K=ACGR.K+(CASH.K/PHOR)
C PHOR=12
        FCSR     =($/M) FEASIBLE CASH SPENDING RATE

        ACGR     =(£/M) ACTUAL CASH GENERATION RATE

        CASH     =(£) CASH RESERVES IN COMPANY

        PHOR     =(M) PLANNING HORIZON IN MODEL


R RSC.KL=MIN(RSAA.K,FCSR.K)
        RSC      =(£/M) RATE OF SPENDING CASH

        RSAA     =(£/M) RATE OF SPENDING ACTUALLY AUTHORISED

        FCSR     =($/M) FEASIBLE CASH SPENDING RATE
```

The difference between the spending authorized, RSAA, and that available, FCSR, drives repayment or borrowing in the Long-term Debt.

```
A CDIFF.K=RSAA.K-FCSR.K
        CDIFF    =(£/M) DIFFERENCE BETWEEN SPENDING AND CASH

        RSAA     =(£/M) RATE OF SPENDING ACTUALLY AUTHORISED

        FCSR     =($/M) FEASIBLE CASH SPENDING RATE


R RALTD.KL=MAX(CDIFF.K,0)
        RALTD    =(£/M) RATE OF ADDING TO LONG-TERM DEBT

        CDIFF    =(£/M) DIFFERENCE BETWEEN SPENDING AND CASH


R RRLTD.KL=MIN(CDIFF.K,0)*CLIP(-1,0,LTDEBT.K,0)
        RRLTD    =(£/M) RATE OF REPAYMENT OF LONG TERM DEBT

        CDIFF    =(£/M) DIFFERENCE BETWEEN SPENDING AND CASH

        LTDEBT   =(£) LONG-TERM DEBT OF COMPANY
```

It will be clear that, as we suggested earlier, the model is too simple and could stand considerable expansion to incorporate areas which a financial manager would regard as important. (I am indebted to Mr. A. Foster who reviewed an early draft of this chapter.)

In particular, although the model outputs balance sheet items it does not take detailed account of:

1. Balance sheet ratios
2. Cash flow
3. Earnings cover for interest charges
4. Short-term liquidity ratio and the proportion of long-term and short-term finance. See H. Shehata, *The Financial Aspects of Growth – A System Dynamics Study*, Ph. D. Thesis, University of Bradford, 1976.
5. Taxation is treated very simply
6. It assumes that United's parent company will cheerfully supply Long-term debt at 15%, regardless of the ROI on that debt.

A valuable exercise for the reader will be to model these aspects (which makes a good term's course in modelling practice, following an introductory course in System Dynamics). It will be found that none of these aspects is very difficult to model in the sense of calculating the required quantities – the problem lies in discovering how they are used in corporate policies. According to circumstances this can be approached by the use of the literature or, preferably, by a case exercise in an actual firm.

12.7 Market Share in the Model

United indulge fairly heavily in spending to obtain market share by R and D, advertising etc.

Market share is modelled as a level, because the propensity to buy United's products would not suddenly go away if United stopped spending. There is, however, a decay of product obsolescence, customer forgetfulness and competitive action.

```
L UMS.K=UMS.J+DT*(RGMS.JK-RDMS.JK)
N UMS=0.25
        UMS      =(1) UNITED'S MARKET SHARE OF MANUFACTURED-METAL
                  DEMAND

        RGMS     =(1/M) RATE OF GROWTH OF MARKET SHARE

        RDMS     =(1/M) RATE OF DECLINE OF MARKET SHARE


R RDMS.KL=UMS.K/ML
C ML=36
        RDMS     =(1/M) RATE OF DECLINE OF MARKET SHARE

        UMS      =(1) UNITED'S MARKET SHARE OF MANUFACTURED-METAL
                  DEMAND

        ML       =(M) LIFETIME OF A GIVEN MARKET SHARE POSITION
                  (FIRST-ORDER DELAY)
```

Growth of market share depends on Average Market Share Spending over the recent past and that market share (the willingness of customers to buy United's products) will decay, as discussed above, if it is not propped up by new spending. In the following equations market share spending of £50,000 per month generates just enough to balance the decay in a market share of 25%.

```
R RGMS.KL=TABHL(TMSGR,AMSS.K,0,2E05,5E04)
T TMSGR=0/.00694/.008/.0085/.009
        RGMS     =(1/M) RATE OF GROWTH OF MARKET SHARE

        TMSGR    =(1/M) TABLE GIVING MARKET SHARE GROWTH RATE IN TERMS OF
                  AVERAGE MARKET SHARE SPENDING,SCALED TO MATCH
                  NORMAL MARKET SHARE DECAY WHEN MARKET SHARE
                  IS 25% AND SPENDING AVERAGES £50000/M

        AMSS     =(£/M) AVERAGE MARKET SHARE SPENDING
```

Target market share spending is regulated by United's observations of the difference between market share and its planned value (Compare this with the alternative approaches to a similar problem in the corporate growth model in chapter 9.)

```
A MSE.K=PMS.K-RMS.K
        MSE      =(1) MARKET SHARE ERROR BETWEEN PLANNED AND ACTUAL

        PMS      =(1) PLANNED MARKET SHARE OF SEMI-FAB MARKET

        RMS      =(1) REAL MARKET SHARE


A TMSS.K=TABHL(TTMSS,MSE.K,-0.1,.1,0.05)
T TTMSS=5E04/5E04/5E04/10E04/20E04
        TMSS     =(£/M) TARGET MARKET SHARE SPENDING

        TTMSS    =(£/M) TABLE OF TARGET MARKET SHARE SPENDING IN TERMS OF
                  MARKET SHARE ERROR

        MSE      =(1) MARKET SHARE ERROR BETWEEN PLANNED AND ACTUAL
```

Market Share is measured at the semifabrication stage and it is assumed that the manufacturing companies can always sell all they produce. This may not be unrealistic because the manufacturing end of the market is fragmented into hundreds of small firms often serving a local, or technically limited, market in which there is little competition. If we allow United credit for the skill to pick good companies to take over in this area it may be reasonable to assume that the 'swings and roundabouts' balance from the perspective of United's board. In semifabrication very large and highly competitive production units dominate the market and the assumption would not even be plausible. The reader should alter this part of the model if he wishes.

Note that this formulation of Market Share Spending as an error-corrector produces *negative* feedback and is therefore incapable of being a growth-generator. Compare with Chapter 9 and revise this model if you prefer.

```
A MSS.K=TMSS.K*MIN(1,FPMULT.K*MSSPF)
C MSSPF=1.5
        MSS      =(£/M) ACTUAL MARKET SHARE SPENDING

        TMSS     =(£/M) TARGET MARKET SHARE SPENDING

        FPMULT   =(1) FRACTION OF PLANNED EXPENDITURE ACTUALLY INCURRED
                  THIS IS THE KEY VARIABLE IN THE FINANCIAL POLICY
                  SECTOR OF THE MODEL

        MSSPF    =(1) MARKET SHARE SPENDING PREFERENCE FACTOR.
                  FACTOR TO REFLECT EMPHASIS GIVEN TO NOT REDUCING
                  MARKET SHARE SPENDING IF AT ALL POSSIBLE
```

Actual Market Share Spending is then affected by financial resources, modified by a factor to reflect the importance United attach to marketing effort. Initially this is fairly high.

12.8 The System Performance

The model can be run under a range of conditions, and the reader should be able to conduct a variety of experiments with it. We shall use it more as a conventional *what-would-happen-if* type of simulation rather than doing loop analysis and system redesign because, as will emerge, the faults in this system are a lack of dynamics rather than too much of them. This 'straight' simulation shows up another facet of SD modelling but the reader should attempt the conventional SD approach of a loop analysis, aiming to engender the growth dynamics seen in Chapter 9.

For initial conditions we take it that the Converter capacity in United's parent's ingot producing company is limited so that no more than 6000 t/m of ingot are available but the reader will be able to vary this by altering the PULSE function in the equation for Converter Capacity, CCAP, in line 202 of Appendix B.12. This allows him to study the third strategic option for United's parent in Section 12.2.

Market Demand for the industry as a whole is a simple step from 400000 tpa to 480000 at month 60 and we study the 10-year dynamics. Initially, United have a 25% market share so semifabrication capacity is 8330 tpm and 30% of this, the United Percent, UPC, is kept for processing by the manufacturing end so manufacturing capacity is 2500 tpm of metal. Initially the company has £990000 in cash from the sale of an interest in an office block. The problem is how will United shape its future under different financial policies and growth objectives?

The 'traditional' financial policy is to regulate spending plans by the Financial Policy Multiplier shown in Section 12.6, i.e. a 'flatter than proportional' policy. This policy is modified in that working capital expansion and market share spending are not cut back as severly as capital spending – see lines 136–137, and 223–224 respectively, and work out how this is modelled.

FRAT % of spending needs actually available	0	50	100	150	200
FPMULT % of spending needs actually authorized	0	80	100	110	110

Initially, the long-term aims of United are to maintain its market share of any market growth and to keep UPC at 30%, i.e. a no real growth situation.

Demand is forecast with an appreciable optimistic bias, using our standard method. (In case the reader needs prodding, we suggest studying forecast sensitivity and shorter-term dynamics, altering the model to bring in short-term liquidity controls, and testing alternative market spending effect assumptions.)

The dynamics of this base case, run STEP1, are shown in Fig. 12.2 and are frankly unexciting. In such a case we need to supplement the plotted dynamics by

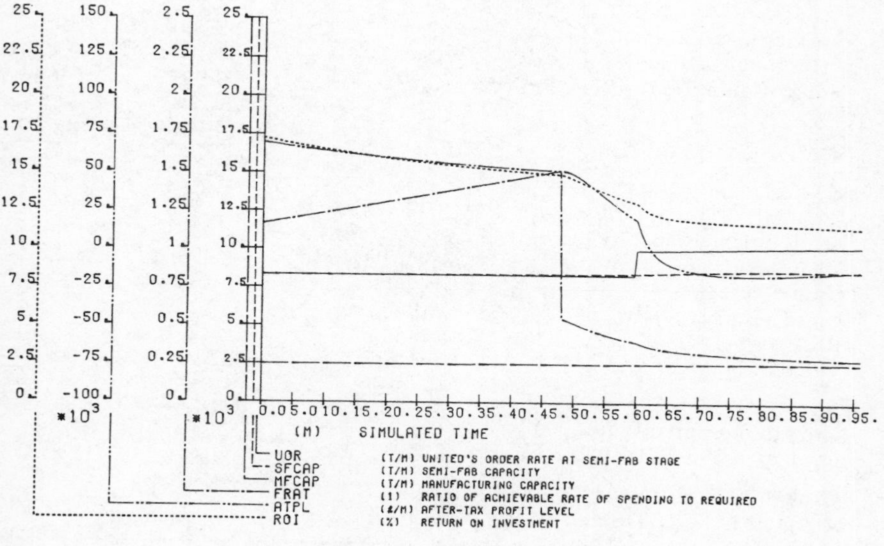

Fig. 12.2 Initial step response

values abstracted from the tabular output and, for the sake of brevity, we use the initial and final values of selected variables as in Table 8.1. The reader should verify definitions before studying the table, altering the model and rerunning it if he prefers other definitions. (For reasons of space, the graphs cover 8 years, but the final values on the tables are at 10 years). Note that the pre-step dynamics are *not* indicative of a fault in the model, but, because the parameters are correct, are part of the inherently unsatisfactory dynamics of this firm.

Some fairly obvious conclusions can be drawn from Fig. 12.2, i.e. that United are not the world's best investment. The company can live with the initial demand – the financing ratio grows fairly steadily, showing that they can support replacement of plant and slight growth which stems from their optimistic forecasts of demand. United are gradually repaying their Long-term Debt, by retaining earnings and this leads to a fall in ROI.

The onset of the step in industry demand at month 60 is foreseen at 48 and this

Table 12.1. *Summary of Run STEP 1*

		TIME	
		0 £x10^6	120 £x10^6
Balance Sheet Items			
Long-term Debt	LTDEBT	28.09	33.41
Cumulative Retained Profit	CRP	15.61	18.17
Shareholders Capital	SCAP	31.21	31.21
Capital Employed	CAPEMP	74.91	82.79
Cash	CASH	.99	.24
Working Capital	WCAP	32.70	39.75
Written-down Value	WDV	31.23	33.54
Money Committed to New Projects	MCNP	9.99	9.26
Check		74.91	82.79
Other Variables			
Cumulative Pretax Gross Profit	CUPROF (£x10^6)	0	104.95
Semifabricate Capacity	SFCAP (t/m)	8333	8452
Manufacturing Capacity	MFCAP (t/m)	2499	2535
United Percent	UPC (1)	.30	.30
United Market Share	UMS (1)	.250	.258
Real Market Share	RMS (1)	.250	.212
Value of Plant	VP (£x10^6)	31.23	48.17
Changes			
Growth in Capital Employed less increase in LTDEBT	(£x10^6)		2.56
Growth in Real Value* less increase in LTDEBT	(£x10^6)		17.19

*Note: 'Real Value' is calculated using Value of Plant in place of Written-down Value.

generates large spending needs and causes the discontinuity in FRAT. After that, ATPL goes negative, and SFCAP cannot keep up and eventually falls when United can no longer finance replacements. (From month 45 the ROI is lower than the interest payable on debt.)

Since SFCAP does not keep pace with orders received, backlog builds up until it is 6.6 times the desired value. In the model this business is lost, because only production generates a profit flow.

Table 12.1 confirms the story. There is growth, but far less than that in industry demand. Net Capital employed is increased by £2.56×10^6 and there is an increase in real value of £17.19×10^6 caused by the plant physically wearing out more slowly than it is depreciated.

The United Market Share, UMS, is the proportion of the industry's customers who would like to buy United's semifabricate and this grows from .25 to .258 under the influence of the relatively aggressive marketing policy. However, the Real Market Share, RMS, which is the proportion of the demand actually met by United's production, *falls* from .250 to .212 due to the inability to finance expansion.

In conclusion then, we can say that, on present form, United could survive for a long time in the present market, though the falling ROI is a danger sign. A rise in market size could be very bad for the firm.

The policy of preferential treatment for Working Capital and Market Share Spending leads to excessive Working Capital, and Market Share ahead of the ability to produce. Removal of this preferential treatment improves performance in several respects.

12.9 Synergistic Effects of Finance

United's parent company are considering a capital restructuring which would involve reducing United's capital employed, to a point where they would have an overdraft. The model used to assess the effects of this.

The dynamics of the run, called NOCASH, are very much the same as Fig. 12.2 so there is no drastic alteration in the behaviour pattern. The other effects are summarized in Table 12.2.

The surprising feature about Table 12.2 is that United seem to do slightly better with £1.5 million less capital employed. It would probably be better to say that they do slightly less badly – the reason is that, being short of money they build up less plant and the after tax profit level becomes less negative. (The tables do not show that the Cumulative Retained Profit is being steadily eroded to meet losses.)

This may imply that United should be folded up and used as a money machine.

12.10 Conservative Financial Policy

The two earlier runs suggest that when FRAT>1, and the United is generating money, the relatively aggressive financial policy leads to the firm overreaching

Table 12.2. *Summary of Run NOCASH*

		TIME	
		0 £x10^6	120 £x10^6
Balance Sheet Items			
Long-term Debt	LTDEBT	27.52	32.92
Cumulative Retained Profit	CRP	14.65	17.25
Shareholders Capital	SCAP	31.21	31.21
Capital Employed	CAPEMP	73.38	81.38
Cash	CASH	.54	.65
Working Capital	WCAP	32.70	38.91
Written-down Value	WDV	31.23	32.73
Money Committed to New Projects	MCNP	9.99	9.09
Check		73.38	81.38
Other Variables			
Cumulative Pretax Gross Profit	CUPROF (£x10^6)	0	103.75
Semifabricate Capacity	SFCAP (t/m)	8333	8274
Manufacturing Capacity	MFCAP (t/m)	2499	2482
United Percent	UPC (1)	.30	.30
United Market Share	UMS (1)	.25	.256
Real Market Share	RMS (1)	.25	.207
Value of Plant	VP (£x10^6)	31.23	47.10
Changes			
Growth in Capital Employed less increase in LTDEBT	(£x10^6)		2.70
Growth in Real Value less increase in LTDEBT	(£x10^6)		16.97

itself. A more conservative policy would be:

Financing Ratio FRAT %	0	50	100	150	200
Financial Policy Multiplier FPMULT %	0	60	90	100	100

with the same growth objectives, and preferential treatment of working capital and market share spending, and the same initial finances as run STEP1. This is run CONSER.

The dynamic mode is not shown as it is the same as Fig. 12.2. The financial performance is summarized in Table 12.3.

Table 12.3. Summary of Run CONSER

		TIME 0 £x10^6	TIME 120 £x10^6
Balance Sheet Items			
Long-term Debt	LTDEBT	28.09	32.02
Cumulative Retained Profit	CRP	15.61	19.21
Shareholders Capital	SCAP	31.21	31.21
Capital Employed	CAPEMP	74.91	82.44
Cash	CASH	.99	.54
Working Capital	WCAP	32.70	39.32
Written-down Value	WDV	31.23	33.18
Money Committed to New Projects	MCNP	2.99	9.40
Check		74.91	82.44
Other Variables			
Cumulative Pretax Gross Profit	CUPROF (£x10^6)	0	105.52
Semifabricate Capacity	SFCAP (t/m)	8333	8408
Manufacturing Capacity	MFCAP (t/m)	2499	2522
United Percent	UPC (1)	.30	.30
United Market Share	UMS (1)	.250	.258
Real Market Share	RMS (1)	.250	.209
Value of Plant	VP (£x10^6)	31.23	47.56
Changes			
Growth in Capital Employed less increase in LTDEBT	(£x10^6)		3.60
Growth in Real Value less increase in LTDEBT	(£x10^6)		17.98

In this situation, conservative policy does better, simply because the company is in no real position to do anything else.

12.11 Downstream Integration

Since the manufactured-metal end of the industry is generally held to be more profitable it might make sense for United to attempt to grow in that area. Run DIVERSIFY tests this, with the same policies and finance as STEP 1 but the addition that there is a target value of UPC 25% higher than the initial value. The dynamics are shown in Fig. 12.3 and summarized in Table 12.4.

Although the dynamic *mode* in Fig. 12.3 has not really altered, the improvement in the value of ATPL is most important as at least it never becomes negative (ROI is calculated on a pretax basis and rather overstates return, as it excludes that half of MCNP which is not available for short-term lending in the money market).

The summary table shows an improvement over previous cases, noting that SFCAP has actually decreased. The real market share has decreased even more than

Table 12.4. *Summary of Run DIVERSIFY*

		TIME	
		0	120
		£x10^6	£x10^6
Balance Sheet Items			
Long-term Debt	LTDEBT	28.09	34.29
Cumulative Retained Profit	CRP	15.61	19.05
Shareholders Capital	SCAP	31.21	31.21
Capital Employed	CAPEMP	74.91	84.55
Cash	CASH	.99	.95
Working Capital	WCAP	32.70	40.64
Written-down Value	WDV	31.23	33.14
Money Committed to New Projects	MCNP	9.99	9.82
Check		74.91	84.55
Other Variables			
Cumulative Pretax Gross Profit	CUPROF (£x10^6)	0	106.09
Semifabricate Capacity	SFCAP (t/m)	8333	8237
Manufacturing Capacity	MFCAP (t/m)	2499	2992
United Percent	UPC (1)	.300	.363
United Market Share	UMS (1)	.250	.256
Real Market Share	RMS (1)	.250	.206
Value of Plant	VP (£x10^6)	31.23	47.60
Changes			
Growth in Capital Employed less increase in LTDEBT	(£x10^6)		3.44
Growth in Real Value less increase in LTDEBT	(£x10^6)		17.90

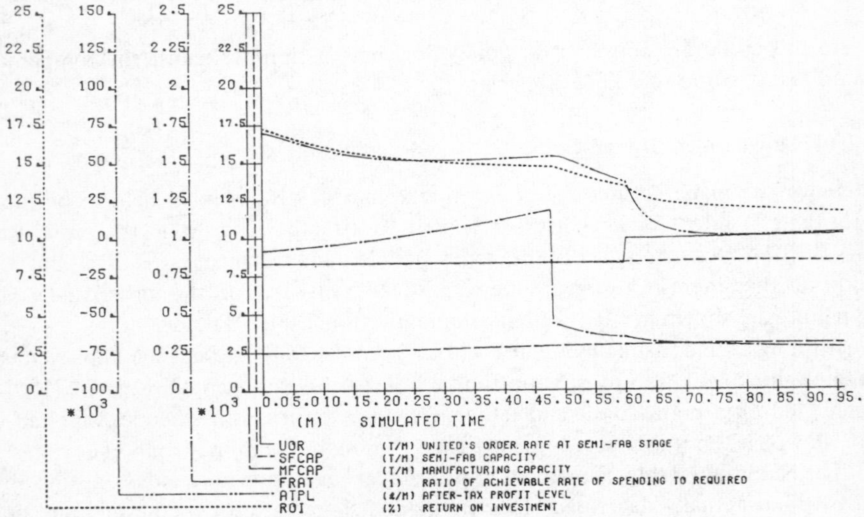

Fig. 12.3 Step response with diversification

in Table 12.1 but the comparison is more complex. In Table 12.1 the market share is .212 of the semifabricated demand plus 0.3 x 212 of the manufactured demand. This .0636 of the manufactured demand is, of course, included in the .212 as shown in Fig. 12.1. In Table 12.4 the figures are .206 plus .363 x .206 i.e. .0748. The reader should construct an appropriate weighting index for this.

Even this run is not really healthy as the semifabrication capacity is falling due to lack of finance and the decay is inevitable.

The alternative strategy of running down manufacturing and building up semifabrication has extremely unfortunate results as the return on semifabrication operations is so poor that the firm finished up worse than it started.

The other policy of running down semifabrication by not replacing plant and putting everything into manufacturing should also be tested.

12.12 Running Down United

One option for United's parent is to run down the firm to provide cash for other ventures as compared to the simpler alternative of selling United. This is tested in

Table 12.5. *Summary of Run GETTING OUT*

		TIME	
		0 £x10⁶	120 £x10⁶
Balance Sheet Items			
Long-term Debt	LTDEBT	28.09	4.68
Cumulative Retained Profit	CRP	15.61	30.76
Shareholders Capital	SCAP	31.21	31.21
Capital Employed	CAPEMP	74.91	66.65
Cash	CASH	.99	0
Working Capital	WCAP	32.70	30.24
Written-down Value	WDV	31.23	28.89
Money Committed to New Projects	MCNP	9.99	7.52
Check		74.91	66.65
Other Variables			
Cumulative Pretax Gross Profit	CUPROF ($£x10^6$)	0	112.67
Semifabricate Capacity	SFCAP (t/m)	8333	1731
Manufacturing Capacity	MFCAP (t/m)	2499	519
United Percent	UPC (1)	.30	.30
United Market Share	UMS (1)	.250	.133
Real Market Share	RMS (1)	.250	.145
Value of Plant	VP ($£x10^6$)	31.23	40.72
Changes			
Growth in Capital Employed including decrease in LTDEBT	($£x10^6$)		15.15
Growth in Real Value including decrease in LTDEBT	($£x10^6$)		26.98

run GETTING OUT which has parameters as in STEP 1 with the exception that the market share aimed for is always 70% of the present value, with UPC held at 30%.

The dynamics are simply a series of downward slopes and the important features are expressed in Table 12.5.

This table shows the effectiveness of United as a money machine and would aid in the choice between selling out now and running the firm down.

12.13 Price Increases

The real trouble with United is that it simply does not make enough money at what it does. We therefore run with a 20% increase in the gross margin, at the same time as the step, with everything else as in STEP1. The dynamics are still not basically altered, except that ATPL, of course, improves rather sharply. The financial performance also improves, as shown in Table 12.6.

It now becomes an exercise to find out how soon before the step the price increase has to be to enable the market share target to be met.

Table 12.6. *Summary of Run 20% INCREASE*

		TIME	
		0 £x10^6	120 £x10^6
Balance Sheet Items			
Long-term Debt	LTDEBT	28.09	38.10
Cumulative Retained Profit	CRP	15.61	24.15
Shareholders Capital	SCAP	31.21	31.21
Capital Employed	CAPEMP	74.91	93.46
Cash	CASH	.99	0
Working Capital	WCAP	32.70	41.82
Written-down Value	WDV	31.23	37.62
Money Committed to New Projects	MCNP	9.99	14.02
Check		74.91	93.46
Other Variables			
Cumulative Pretax Gross Profit	CUPROF (£x10^6)	0	120.1
Semifabricate Capacity	SFCAP (t/m)	8333	9234
Manufacturing Capacity	MFCAP (t/m)	2499	2770
United Percent	UPC (1)	.30	.30
United Market Share	UMS (1)	.250	.271
Real Market Share	RMS (1)	.250	.231
Value of Plant	VP (£x10^6)	31.23	52.86
Changes			
Growth in Capital Employed less increase in LTDEBT	(£x10^6)		8.54
Growth in Real Value less increase in LTDEBT	(£x10^6)		23.78

12.14 Conclusion

The reader will, no doubt, wish to study the stability of the system in response to, say, the 4-year business cycle, a ramp, a downward step and so on. He should also work out for himself just what the differences are between this approach and the more conventional balance sheet producing open loop simulation model. They are, in fact, considerable, and centre on the way in which the loops in this model regulate what happens and tie together its policies. For example, how do the loops connect the preferential treatment accorded to working capital additions and market share spending with the short and long term dynamics?

Chapter 13

System Dynamics in Practice

13.1 Introduction

So far in the book we have expended considerable effort on studying some of the theory of dynamic modelling, looked at the techniques for building and analysing models and examined some practical case studies. We have not, however, treated the crucial question of how one gets started on a dynamic modelling exercise in practice and the steps which need to be taken to carry such a study to a successful conclusion. To do so is the function of this chapter.

Clearly we cannot offer a set of standard rules which will always ensure a successful applied study. The best we can hope for is to distil some general guidelines from some years of experience and to offer these to the intelligent reader to apply with commonsense in his own situation. Some of what we shall say is generally applicable to any problem-solving work in business and industry and none of it is particularly original. In fact, the reader would do well to compare this chapter with the comparable material in almost any good management science text.

In some parts of this chapter the reader may feel that 'any good management scientist should have known that'. He is quite right, but a casual acquaintance with much of practical management science work will reveal that these rules are often forgotten, and this is our justification for taking space to reiterate them.

13.2 Essential Rules

There are four fundamental rules in practical system dynamics:

1. Keep a sense of proportion.
2. Remember the purpose.
3. Pay your way.
4. Keep computers in perspective.

Keep a Sense of Proportion

System Dynamics is a stimulating and exciting field and its techniques enable one to tackle problems of great complexity, with multiple objectives and at a high level of managerial responsibility. The approach also has some powerful implications as a tool for the theory of the firm. At the same time it is NOT the only pebble on the beach and it is not the most appropriate approach to all problems at all times. It is, therefore, essential for the practitioner to develop a 'nose' for dynamic problems and to bear in mind that he is there to solve problems as they arise (or even before) and not to do system dynamics all the time.

Remember the Purpose

Nobody does system dynamics for fun; it is too difficult and there are more amusing ways of entertaining oneself. However, it is very easy to get so involved in a project that you forget why you started on it in the first place. It is therefore vital to develop the mental discipline of always asking oneself 'what does this piece of the project contribute to the overall achievement of the purpose we started out with?'

It is very difficult indeed to do this as fully as it should be done, far more so than, say, analysing the feedback loops in a model. It is hardly ever done properly and it is easy to fob oneself off with superficial answers such as 'I am debugging the program in order to get the model working properly'. Of course you are, but why did you write it in the first place? Perhaps the influence diagram would have been enough satisfy the purpose. Perhaps the assumptions you are so busy programming are so vague that they will not really take you any further forward than you could have got by careful thinking about the influence diagram. In fact, they may even take you backwards by obscuring a clear conclusion (which might be that the project should be abandoned) under a pile of computer printout.

Pay Your Way

Practical projects have time scales, after which the answer will be too late to be of any use. They also have cost/benefit ratios which dictate how much effort it is worth putting into it. Usually the benefits are not easily seen in advance, or even afterwards.

One should therefore proceed with caution and by stages. At the start of each stage ask whether it is going to contribute enough to the project, over and above what has already been gained, to make it worthwhile. As a rough guide it is worth considering three stages in a project and looking at the answers one *usually* gets from comparing the likely benefits with the expected costs. We assume that the criteria to be discussed below for identifying a dynamic problem have been applied and have shown that a dynamic model is the right one to build in this case.

1. Building the Influence Diagram. This nearly always contributes a very large benefit compared to its cost, especially if the firm has not used system dynamics before. We have usually found that it opens up a whole new perspective.

2. Simulation Modelling. We find that when this stage is appraised, it usually shows one of two answers; either a negative one, that the simulation model will probably not show very much that we do not already know from careful thinking about the influence diagram and the loops it contains; alternatively, the answer may be weakly positive, we cannot see any good reason for not building it and there are areas which the influence diagram will not answer, but we cannot be sure that simulation will really help until we have done it.
3. Full Loop Analysis and program of simulation experiments. This is the expensive phase of the work and the answer to whether it is worth the effort is usually decisively 'no' or clearly 'yes'. The latter usually applies when we are confident that we have a good, dependable model with few fuzzy areas, where we wish to do fairly detailed policy design and where we can see a reasonable prospect of getting our suggestions implemented if they are sensible and beneficial. This is not the same as saying that we can see that we will get useful answers – merely that we get on well with management.

In practice, of course, the steps are not as clear cut as we have made them and the modeller who is following the method suggested in Chapter 3 will find many small steps rather than three large ones.

In short we are saying that it is necessary to make sure that the project earns its keep and this sometimes conflicts with the analysts' natural desire for technical excellence. When this dilemma arises perhaps the best guideline is the apocryphal definition of management science as 'the art of giving bad answers to questions to which, otherwise, worse answers would be given'.

Keep Computers in Perspective

Simulation models in DYSMAP are generally easy to write and seem to be cheap to be run. In fact they aren't cheap at all because they produce a lot of information about the system which should be absorbed and digested before proceeding. It is very tempting to keep on making runs and reruns, without really understanding what has been learned, in the hope that some striking improvement will be found. It almost certainly will not and the way to make real progress is to recall that the feedback loop is the basic unit of analysis and to make sure that a set of output has been understood in terms of what it shows about the system's loops before the next run and reruns are made.

13.3 Criteria for Project Selection

Management science is a difficult discipline, because there are nearly always more problems than resources to tackle them, and it is often by no means easy to know which is the best way to approach a particular problem. We therefore discuss below a set of guidelines which should aid in identifying whether or not a particular managerial problem is likely to be a suitable candidate for dynamic modelling.

It is not suggested that the criteria are a set of formal tests which each problem must pass before qualifying – they are simply pointers for the experienced management scientist. Nor should it be thought that any of the criteria have to be expressed in strictly numerical form. Managerial opinion is just as valid as time-series, which are usually incomplete and out of date in any case. We assume, of course, that the analyst is not so naive as to believe everything he is told by managers, and that he will not ignore interpersonal and political factors when applying these criteria.

We believe that the following 14 criteria are important, but they are not the only ones, nor should they be applied uncritically.

1. Is there any dynamic behaviour? This requires a careful cross-check of dynamic behaviour, relying both on statistical time series and on peoples' opinions and experience. Look for related behaviour (e.g. between inventories and production flows), leads and lags and, in particular, time periods (monthly or annual dynamics). Pay particular attention to inputs to the controller from the complement and/or environment, looking for seasonalities, noise and possible step changes. Bear in mind the need to look for dynamics which may occur in the future as well as those which have happened in the past. Watch out for feedforward of inputs, as when planned stocks are related to expected seasonalities.

   When investigating dynamics it is useful to try to get some preliminary feel about why people mention them and what they tend to do about them. See below.
2. Do the dynamics matter, and why? Dynamic behaviour is not necessarily either good or bad, of itself. For example, dynamics in inventory may well indicate that the stock is doing what it is supposed to, and insulating the controller from the environment. This can be checked by asking whether the amplitude of production dynamics seems reasonable in relation to input and inventory dynamics. This should show whether or not the controller is making things worse, as in Chapter 10.

   Equally, growth in capacity, or dynamics in product introduction, may simply show that the controller is flexible and responsive to the environment. If, however, we can answer the question 'so what?', in relation to observed dynamics, we have come a good way to identifying if the problem is worth studying and to getting a good feel for the boundaries of the system we shall need to examine.
3. Are there any loops? A dynamic model has to contain loops, so it seems reasonable to see if there are any identifiable loops to model. The best way of approaching this is to use the interview notes from steps 1 and 2, and the list extension method of Chapter 3, to build up a simple influence diagram. This can be discussed with managers, and experience shows that this nearly always brings out large numbers of loops. Tact is needed to avoid being swamped in detail. It is important for managers to have at least some idea of why loops are being looked for, otherwise it seems meaningless to them.
4. Are there any alternative system structures or control policies? If the answer to

this is 'no', the study is pointless because if nothing can be changed the system is either perfect or paralysed. Usually there are many options for change and the problem is to reduce them to manageable number rather than to find them. We need to know more than that they exist, we also need at least a rough idea of what they are so as to get more feel for the most probable system boundary. This does not mean that there is no need to model those features which cannot be changed, because they may contribute important dynamics, but clearly the model must include those areas where change can be effected.

5. Can it be done? This is rather a weak criterion at this stage in an investigation. At best it can receive only a rough answer which, if positive, will need clarification by the application of, particularly, criteria 6–10. Basically, therefore, we seek an intuitive feel for whether system dynamics is likely to be the best approach, or whether we should be looking for a probabilistic approach from decision theory, game theory, replacement, bidding or whatever, or perhaps a more deterministic approach from linear or dynamic programming. There are many complementary approaches in management science, and the analyst ahould not get too closely wedded to any of them if he is to be a problem solver rather than a technique peddler.
   The main things to look for are the likely snags and problems, both technical and behavioural, and to ask not so much 'can it be done?' as 'how should it be done, if at all?'
6. What data and information are available? This is a matter of making some kind of inventory of sources of data and information and comparing it to what is likely to be needed. At the early stages of the project this will be hardly more than an inspired guess, but it should still be attempted. Recall that much of the most useful information will be qualitative rather than strictly quantitative. It is also generally true that if data are not available there will be little prospect of being able to collect them from scratch in time to be of any use to the project, if this is going to involve setting up and operating a special-purpose recording system, as opposed to digging the data out of existing files.
7. Can we define the variables? It is essential to pay careful attention to the problem of defining what is meant by each of the variables, or the people who have to be convinced of the virtue of the recommendations will either not understand them, or not believe them, and the project will be futile.
   This seems easy enough if one is modelling a controller-only system where there are variables such as production rate, workforce or cash flow. Difficulties arise, however, if the modeller's definition of, say, cash flow does not accord with the accountant's. Should one conform to his practice, or educate him into accepting yours?
   The problems become more severe with variables such as 'demand' and 'market share'. By 'demand' do we mean that of the buying public, or that of intermediaries such as dealers (Chapter 10)? Do we mean total industrial demand or that for a particular firm? What happens if the demand is not met; does it vanish or accumulate, and, if so, for how long? Does 'market share' mean the public tendency to buy one firm's goods, or the proportion of our firm's sales to the industry total?

This raises the related question of whether one should include variables such as Lost Sales (as in Chapter 10). By definition, such variables cannot be measured, which is a validation problem, but they do have real meaning for the firm as indicators of system performance. Curiously enough, experience indicates that management scientists who have not worked on dynamic models are very resistant to including such variables, whereas managers regard them as credible and important.

At the extreme of the scale of difficulty of definition one has system parts such as those shown in Fig. 13.1.

Fig. 13.1

This means something, presumably, or companies wouldn't spend money in this way. On the other hand, while one may intuitively recognize 'product quality', defining and measuring it is a different matter. One approach is to define an arbitrary quality scale, with say, 5 representing 'normal quality in relation to competition' and then to use a table function.

One's intuitive reaction to this is that it would be practically meaningless, but matters are not quite so simple, as was argued in Chapter 9. However, this kind of situation is one in which there is an appreciable risk of losing managerial confidence by over simplifying and thereby becoming contentious. This does not mean that one should always avoid such areas, because to do so would be to withdraw from problems which may be of great managerial importance. However, great care *is* needed.

8. Where are the dangers of oversimplification? A common cause of failure in management science work in general, no less and no more than in system dynamics in particular, is the oversimplification of the model, either because reality is too complicated to model, or because a realistic model is too expensive or too difficult to analyse. Programming in DYSMAP is so easy that it is rarely too difficult to program a formulation, the problem is that the managerial question, or the system interactions, may be very subtle and hard to formulate. It is an advantage of the Working Memorandum technique described below that it forces one to explain the formulation in time to get comment about it, and this improves the chances of getting a successful (i.e. acceptable) model.
9. What level of aggregation is needed? The other risk of oversimplification is that insufficient detail has been modelled e.g. that production has been treated as being of a single undifferentiated product whereas the firm really manufactures several hundred separate products. A similar problem could arise in modelling projection of future cash flows of the firm projects month by month, and the model contains annual figures. Deciding what level of aggregation is needed, without overcomplicating the model, is difficult but it affects both the realism

of the result and the cost of the project, though this is usually not seriously affected as DYSMAP programming is so easy.

There are serious risks that an aggregated model will produce results which are completely different from a disaggregated equivalent. There are mathematical approaches to aggregation (e.g. J. A. Sharp, *A study of some problems of system dynamics methodology*, Ph.D. Thesis, University of Bradford, 1974).

An alternative, and less sophisticated approach, is to use the list extension method in Chapter 3 to build a simple, aggregated, model. This can then be progressively disaggregated until no appreciable changes in the dynamics of the *performance indicators* is observed. The value of DT needs to be continually monitored in the compromise between run times and numerical stability. This approach embodies part of what is meant by using the model to guide its own evolution. The rest of the meaning involves using feedback loop analysis to identify the critical model areas.

Another version of this method, which has been used in research into the World Models of Forrester and Meadows, is to duplicate the model. This requires some changes in variable names and some linkages between the resulting two models, but the end product is an easy way of creating a model in which there are two competing sectors from a single, highly aggregated model. This approach has its simplicity to commend it, but its implied assumptions have to be carefully checked.

10. What facilities are needed? This is largely a matter of the practical conduct of the study. Can we get enough computer time? What manpower can be deployed and when? When will the manpower be needed? Have we the required staff skills and, if not, how do we get them?

    It is not essential to have a special simulation language available in order to do dynamic analysis. A perfectly good study can be done using FORTRAN, though a graph-plotting package, using either the line printer or a CALCOMP plotter, saves a good deal of effort. In practice, however, a special language is a great convenience. For advanced work there are particular advantages in DYSMAP, which produces a FORTRAN equivalent of the DYSMAP statements which can then be operated on as a FORTRAN program with standard graph-plotting subroutines.

    Generally, computing costs themselves are a very small fraction of total study costs. Do not forget, however, that computing produces output which has to be analysed. This is expensive unless the loop structure has been understood, and is used to direct the simulation experiments, so that the output simply confirms predictions made on the basis of loop analysis. If this is done properly the output analysis cost is reduced, because the analyst, as it were, knows what he is looking for.

11. What training will be required? It is essential that the managers for whom the study is being done should understand the outline of the methodology involved and what it can, and can not, be expected to do. It is, therefore, worth the effort of trying to get them to attend a short training session. Similarly, training may be needed for some of the analytical team. Some outline training programmes are discussed in Appendix D.

12. How much can we afford?
13. How long have we got? These two questions are obviously interrelated, and are important to the project manager for all sorts of rather obvious reasons. They are rather hard to discuss in detail because there are so many different situations and degrees of skill, but some rough guidelines can be given, based on the three case studies of Chapters 10–12. In each case, the company had had no previous experience with system dynamics but had been fairly heavily involved in other management science work.
    A very rough comparative table is as follows:

Duration (Man-days)

	Chapter		
	10	11	12
Formulate basic model	10	10	6
Program and debug	10	10	5
First simulation experiments	3	10	6
Refinement of model	2	20	10
Analysis of model and formulation of recommendations	10	20	20
TOTAL	35	70	47

These times do not include writing up the project notes into case study form or the necessary alterations to the model to make them suitable for this book. They do not include managerial time spent on the project i.e. they are simply the duration for a reasonable skilled and hard-working analyst.

14. How about implementation? The object of the exercise, in a practical situation, is to suggest alternative ways of operating the system which some manager will like so much that he will be prepared to go to the time, trouble, and expense of changing what he does to what he does to what has been suggested. Put as baldly as this, it is a fairly tall order, but the study is not a success unless it is achieved. It is very unlikely that the success will be attained unless the problem of implementation is recognized from the outset and steps are taken to solve it, from the very beginning of the project. At the beginning, the analyst does not know what the recommendations are going to be, but he can still make

progress by bearing in mind a number of points and continually updating them as the study proceeds. The importance of good personal relationships, training, and communication are obvious, but there are additional points:

a. Identify the decision-maker. This may be a group and may not be the obvious person.
b. Make sure that you have a model which he thinks is credible i.e. which includes the details which he thinks are important.
c. Find out his pet theories about the system and the way in which it could be improved, and include them in the study. You will find this saves a lot of work because he probably knows the system far better than you can ever do. You do not have to prove his theories, indeed you may very well finish up demolishing them (tactfully, of course), but if you have not even considered them, he will reject your study as incomplete, and he will be right. (Why?)
d. Find out the *political* constraints on *policy*, which are imposed on your decision-maker by others, as well as the *physical and economic* constraints on the system.
e. At the risk of repeating ourselves, make sure about training him to understand what you are doing.
f. Identify the problems he will have in carrying out your recommendations, and apply the earlier points to those problems. Who will have to approve any expenditure involved? Will any staff have to be redeployed or retrained?

   This assumes that the manager is not himself affected by the changes proposed. Where that is the case, then even more care is needed. For example, in one practical case of the redesign of a production planning system, the decision-maker was correctly identified as the Group Production Controller, rather than the Financial Director who actually commissioned the study. It was found that higher management assessed the Production Controller's job performance by his ability to minimize changes in production rate. It was, however, concluded from the study that the system would operate more effectively if backlog was controlled and it was recommended that the Production Controller could more usefully be judged by his ability to use production as a backlog regulator.
g. Identify the political opposition to any change, and apply points 1–6 to them.

Summary on the Criteria

The first step in building a dynamic model is to decide what it is for. A clear statement of the purpose of the model will be a great help in the subsequent work by keeping the attention of the analyst, and of the managers for whom he is ultimately working, firmly fixed on the essentials of the system. If this is not done it is very easy to keep adding more and more detail to the model without increasing its usefulness.

We therefore start off by asking a series of questions which will help in deciding the model's purpose.

1. What are the dynamic phenomena of interest?
2. What is there about these phenomena which causes concern?
3. What is the policy-question which management are asking about this system? In other words, which of the policies in the system have been called in question?
4. What, therefore, is the model supposed to do? Who will use it?
5. How soon are the results of the analysis needed for them to be of any use?
6. How much effort is it worth investing in the model?
7. How are we to judge whether 'better' system performance has been achieved?

The variety of system problems is so very large that it is far from easy to put forward general guide lines which will help in deciding the purpose in any particular instance. Despite the variety of system problems and the impossibility of giving generalized answers, the reader should be able to obtain a good appreciation of the way in which these questions help the analysis by careful reading of the text in the light of this own experience.

13.4 The Working Memorandum Method

As we have argued, one of the essential features in conducting a management science investigation is communication between the analyst and the managers who will have to be persuaded to accept recommendations and who will have to live with the consequences. Obviously, there is no substitute for good personal relations between manager and analyst or for the day-to-day contact which has to take place in any case. However, there are useful supplements to the personal contacts.

Practical experience of consulting in management science in general, and system dynamics in particular, has shown the value of the preparation of informal written descriptions of the work and their presentation to management at appropriate intervals during the project.

The purpose of these accounts is to say, in effect, 'this is what we think we have found out about your problem, is it right?' The whole tone is to seek confirmation, clarification or supplementation of facts, objectives and assumptions as the project proceeds. This helps to ensure a common basis of knowledge and judgment between managers and analysts and, in particular, that the object of the study is held clearly in mind all the time, or is revised, and mutually understood to have been revised, if changes in circumstances indicate that to be a proper thing to do.

It doesn't matter what the descriptions are called, except that 'Report' conveys an impression of finality and conclusiveness. Even 'Interim Report' implies fairly firm conclusion about part of the project which can be implemented while the rest goes ahead. We shall call them Working Memoranda (WM).

The aim of a WM is to check that what the analyst thinks he knows is correct and relevant to the study, and to determine whether the purpose of the study needs to be revised in the light of its own progress. It is, therefore, highly likely that the earlier memoranda will be scrapped and superseded by later work. If this happens the WMs are fulfilling their function. In practical terms the analyst has to be sure that managers know what the WMs are for and do not lose confidence in the project (or the analyst) because the earlier WMs require substantial correction. Against this, of course, it is far better to scrap the first WM than to have the final report rejected because of some early, and uncorrected, misapprehension about the nature of the system. It is no defence, in such a situation, to tell managers that they should have mentioned some aspect of the system which has been ignored. In practice they don't know what has been included or omitted from a model until they see it in print; verbal communication is NOT sufficient.

Although the earlier WMs may be superseded, the later ones become more and more definite and they may then serve two additional purposes. In the first place, they help to keep up managerial interest in the project during the protracted stage of system analysis and model experiment. Secondly, they help to write the final formal report, which can include fairly large extracts straight from the later memoranda. This makes the labour of writing and discussing the WMs seem much more worthwhile. In academic system dynamics it is a practice much to be recommended to Ph.D students.

It is not possible to lay down what the WMs should cover or when they should be produced, but the following is a rough outline of typical practice.

WM Number	*Covering*
1.	Reasons for selecting the project; aims and timetable.
2.	Basic description of the system; influence diagram and outline equations; emphasis on model boundary.
3.	Simple runs on the model; any implications from the dynamics; model validation where possible; plans for programme of simulation experiments.
4.	First set of simulation experiments and loop analysis; comments and implications; plans for second phase of experiments suggested by results of first phase.
5.	Second set of simulation experiments and loop analysis; implications of analysis and preliminary recommendations; weaknesses and limitations in the model.

Each of these WM should be fairly easily written since the aim is not to produce a superbly polished piece of literature, but to get managerial participation. It is suggested that the model, and output from it, should be put forward at a very early stage without waiting to achieve the nth degree of technical perfection in it. Experience indicates that managers find it easier to comment if they have the full collection of influence diagram, equations, and, especially, graphical output.

13.5 Conclusion

The reader who has got this far deserves to be congratulated, but can he now say, briefly, what he has learned? It will prove to be a useful exercise if the reader can write short answers to the following questions.

1. What is the essential purpose of a system dynamics study?
2. What distinguishes SD methodology from other management science techniques?
3. Why is this distinguishing feature important?
4. What kind of problem is SD most suited for?
5. What kind of problem should SD not be used for?
6. Can I apply it in my own firm?

Appendix A

Mathematical Models of Some Simple Systems

A.1 Introduction

Most of the treatment of systems in this book is non-mathematical, in the sense that, although it is numerical and founded in control theory, little use has been made of the formal rigour of the extensive mathematics of control theory. This has been done for two reasons.

In the first place, System Dynamics is an approach, and a set of techniques, for studying the dynamics of very large, non-linear, managed systems, i.e. systems in which policy-choices, rather than hardware design, are the objective of the study. For such systems, mathematical control theory is often inappropriate, if only because the mathematics becomes too difficult, except under unrealistic simplification.

Secondly, we have argued, in the earlier chapters, that a requisite for success in dealing with such systems in knowledge of, and insight into, the workings and structure of the system. This knowledge is usually the property of people who have not had an extensive training in control theory, and who do not wish to have to undergo one before they can deal with their system's dynamic problems.

For these reasons we have presented simulation techniques, with some admixture of classical control theory. In some ways this is unsatisfactory, because we are debarred from using some of the results on stability and optimal control which can be found in modern, state–space control theory. This is not, however, too serious, as stability may be by no means desirable in a socio-economic system, and both stability and optimality are usually rather hard to define in an operationally useful way.

These arguments do not, however, imply that mathematical treatment is never useful, and in this appendix we present mathematical analyses of some simple systems. The purpose is to show, by example, how the system equations can be recast as differential equations. This may help the reader to see the relevance of some of the modern control theory literature, as referred to in the Bibliography in Appendix C.

For the sake of ease of explanation, all the examples are of specific corporate systems—usually involving production activities. The reader whose background is not in business may wish to rephrase the examples so as to make them relate to his own field.

All the examples are treated as being linear and continuous. The reader must note that it is the non-linear and/or discontinuous nature of most real systems which makes their dynamics complicated and interesting. It is often useful to reduce such a system to linear, continuous and low-order form, in order to be able to study its dynamics, especially its stability characteristics, when that attribute can be defined and is desirable. However, the results of such an analysis have to be transferred back to the real system with great caution. Since this kind of treatment should not be attempted until the analyst has had considerable experience with real systems, and is thoroughly familiar with control theory, it is not treated in this introductory text.

A knowledge of Laplace Transform methods is required for reading this appendix.

A.2 A First-order Negative Loop

The simplest example is a pure integration of a production rate, p, into an inventory I (see Fig. A.1). There is a desired value of inventory, D, and production is set so that p is a fraction $1/\tau_1$ of the discrepancy between D and I.

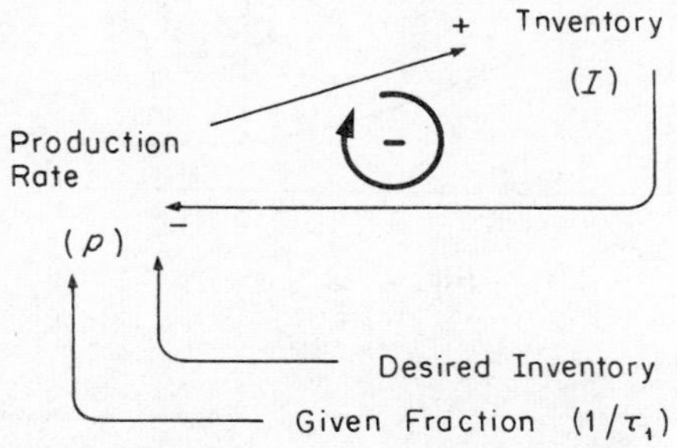

Fig. A.1

Clearly

$$p = \frac{D - I}{\tau_1}$$

and

$$p = \dot{I}$$

so

$$\tau_1 \dot{I} + I = D$$

With

$$I(0) = 0 \quad \text{and} \quad D(0) = 0$$

the Laplace transform is

$$\tau_1 I(s) + I(s) = D(s)$$

or

$$\frac{I}{D}(s) = \frac{1}{\tau_1 s + 1}$$

The transfer function has a single pole at $s = -1/\tau_1$ and the zero frequency gain is 1.

The response to a unit impulse in D is given by:

$$I(t) = \frac{1}{\tau_1} L^{-1} \left[\frac{1}{\frac{1}{\tau_1} + s} \right]$$

$$= \frac{1}{\tau_1} e^{-t/\tau_1}$$

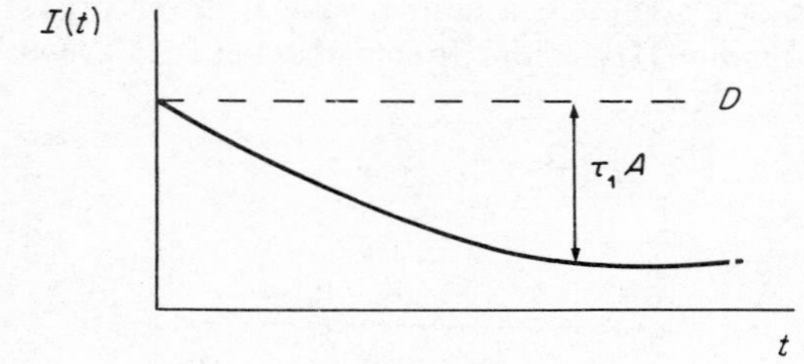

Fig. A.2a

This is shown in Fig. A.2a.

The response to a step input in D corresponds to the behaviour of the system when D takes on a new finite value, A units larger than its old one.

$$I(t) = \frac{A}{\tau_1} L^{-1} \left[\frac{1}{s(s + 1/\tau_1)} \right]$$

$$= A(1 - e^{-t/\tau_1})$$

The response when $t = \tau_1$ is

$$I(\tau_1) = A(1 - e^{-1}) = 0.632A$$

The time, T, needed for inventory to reach 95% of its desired value is found from

$$I(T) = 0.95A = A(1 - e^{-T/\tau_1})$$

For this to hold

$$e^{-T/\tau_1} = 0.05$$

or $T=3\tau_1$ approximately. (See Fig. A.2b.)

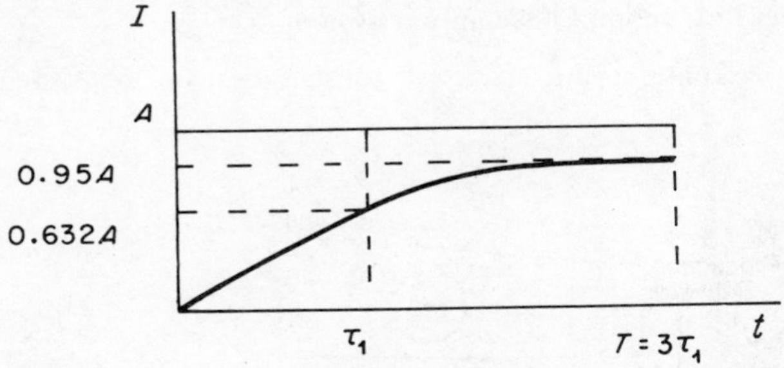

Fig. A.2b

The parameter τ_1 is thus a time constant and has dimensions of [TIME].

The time needed for I to reach to within 2% of A is approximately $4\tau_1$.

The ramp response is of some interest because it shows the ability of this system to respond to a steadily-rising desired inventory (for instance in a growing company).

For a UNIT ramp,

$$D(s) = \frac{1}{s^2}$$

and

$$I(t) = \frac{1}{\tau_1} L^{-1} \left[\frac{1}{s^2 (s + 1/\tau_1)} \right]$$

whence

$$I(t) = t - \tau_1 + \tau_1 \cdot e^{-t/\tau_1}$$

(see Fig. A.3.)

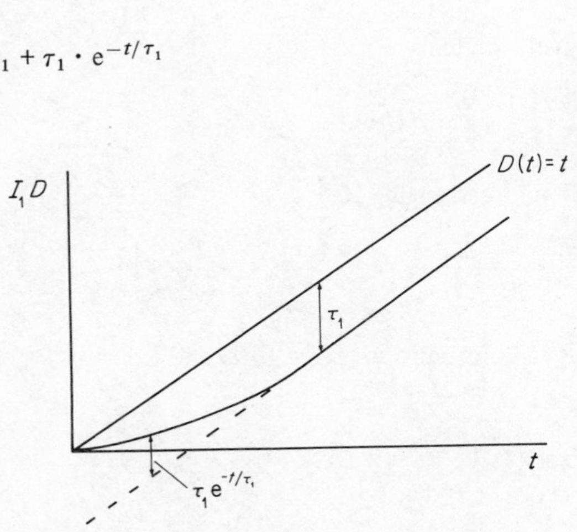

Fig. A.3

In this system actual inventory would always be τ_1 behind the desired level. Since we considered a unit ramp in which $S(t)=t$ the actual inventory shortage would be τ_1 weeks of production.

A.3 A First-order Production Consumption System

If we have to consider the effect of consumption on the system, shown in Fig. A.4.

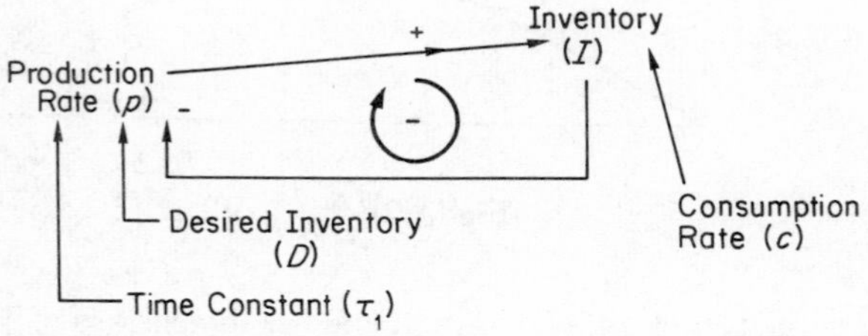

Fig. A.4

Take D to be a parameter, $I(0) = D$ and form the equations

$$\dot{I} = p - c$$

$$p = \frac{D - I}{\tau_1}$$

Then

$$\dot{I} + \frac{I}{\tau_1} = \frac{D}{\tau_1} - c$$

The Laplace transform yields

$$sI(s) - I(0) + \frac{I(s)}{\tau_1} = \frac{D}{\tau_1 s} - C(s)$$

Putting $I(0) = D$ and rearranging

$$I(t) = L^{-1}\left[\frac{D}{\tau_1}\cdot\frac{1}{s(s+1/\tau_1)} - \frac{C(s)}{s+1/\tau_1} + \frac{D}{s+1/\tau_1}\right]$$

For a step input in consumption, $C(t)=Au(t)$ and $C(s)=A/s$ where $u(t)$ is the unit function:

$$I(t) = (D - \tau_1 A)(1 - \mathrm{e}^{-t/\tau_1}) + D\mathrm{e}^{-t/\tau_1}$$

See Fig. A.5.

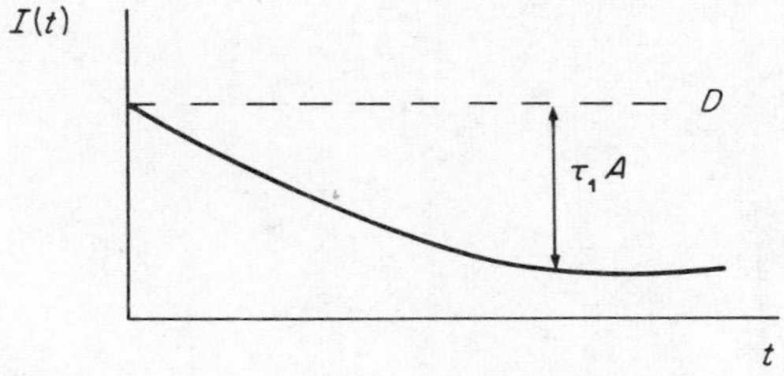

Fig. A.5

Since $\tau_1 A$ may be larger than D there could, in theory, be negative inventory. Note that inventory will always remain below its desired value since this is the only mechanism the system has for ordering production.

A.4 A Second-order Delay

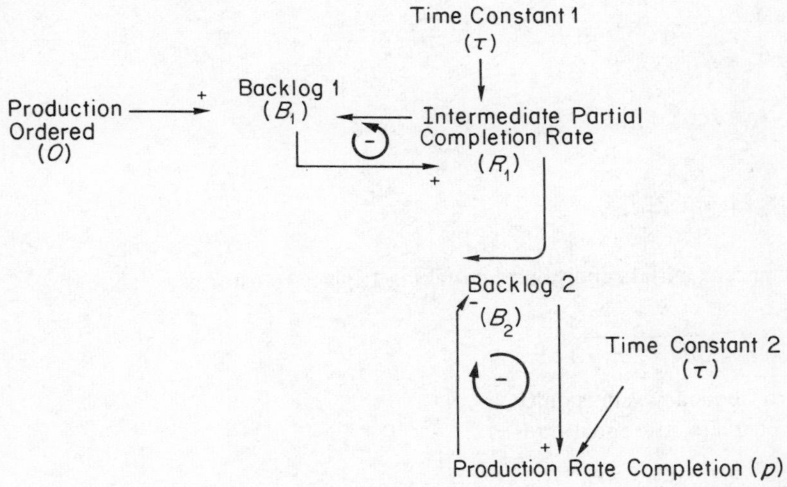

Fig. A.6

Next we consider a simple case in which there is a second-order lag between one rate and another e.g. from the production started to production completed. (See Fig. A.6.)

For example, o is an imposed signal, such as a production start rate, and p is the output, or production completion rate, in response to it.

B_1 and B_2 represent two production stages and it is assumed that the output rate from each of these is $1/\tau$ of the backlog in each case. The total delay is 2τ.

Clearly

$$\dot{B}_1 = o - R_1$$

and

$$\dot{B}_2 = R_1 - p$$

with

$$p = \frac{B_2}{\tau} R_1 = \frac{B_1}{\tau}$$

This leads to

$$\dot{B}_1 + \dot{B}_2 = o - p$$

and

$$\dot{B}_2 = \tau\dot{p}$$

It follows that

$$R_1 = \tau\dot{p} + p$$

and, since

$$\tau\dot{R}_1 = \dot{B}_1$$

$$\dot{B}_1 = \tau(\tau\ddot{p} + \dot{p})$$

Hence

$$\tau^2\ddot{p} + 2\tau\dot{p} + p = o$$

Assuming zero initial conditions, and taking Laplace transforms

$$\frac{P}{O}(s) = \frac{1}{\tau^2 s^2 + 2\tau s + 1}$$

The zero frequency gain is unity.

The characteristic equation is

$$\tau^2 s^2 + 2\tau \cdot s + 1 = 0$$

or

$$s^2 + \frac{2}{\tau} \cdot s + \frac{1}{\tau^2} = 0$$

Comparing with the standard form of the second-order characteristic equation

$$s^2 + 2\zeta\omega_n s + \omega_n{}^2 = 0$$

$$\omega_n = \frac{1}{\tau}$$

and

$$\zeta = 1$$

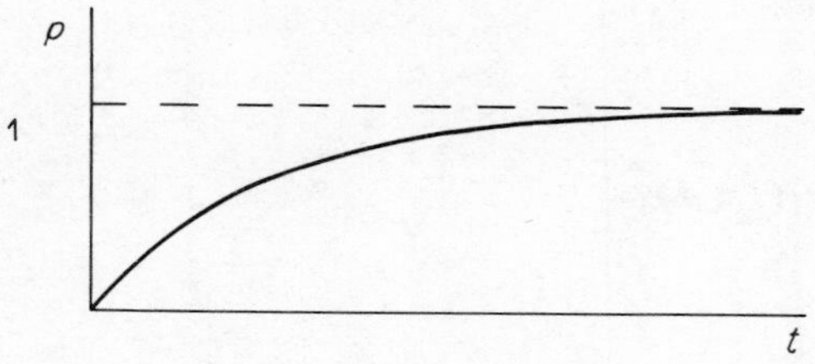

Fig. A.7

The net effect will be as shown in Fig. A.7. It is rather cumbersome to derive formulae for the 95% response time of the system. Generally, the smaller the value of τ_1 the more rapidly will the system respond, and it is impossible for it to overshoot.

A.5 A Second-order Multi-loop System

Consider a system in which production start rate d is linked to an exogeneously determined consumption rate c (see Fig. A.8).

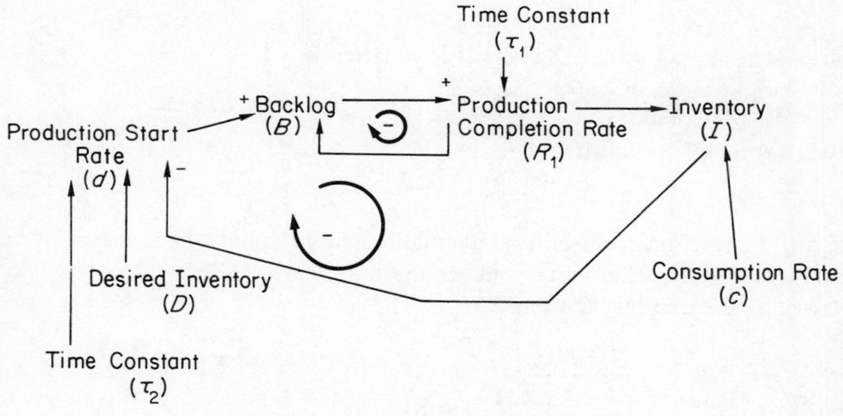

Fig. A.8

B represents the lag in the production process and I is inventory which is controlled to a desired value D.

The basic equations are

$$R_1 = B/\tau_1 \qquad \dot{B} = d - R_1$$

$$\dot{I} = R_1 - c \qquad d = \frac{D - I}{\tau_2}$$

From these

$$\dot{I} = d - \dot{B} - c$$

and

$$\dot{R}_1 = \ddot{I} + \dot{d}$$

$$\dot{B} = \tau_1 \dot{R}_1 = \tau_1(\ddot{I} + d)$$

Eliminating $\dot{B}$

$$\tau_1(\ddot{I} + \dot{d}) = d + \dot{I} - c$$

Now

$$I = D - \tau_2 d$$

so

$$\dot{I} = -\tau_2 \dot{d}$$

$$\ddot{I} = -\tau_2 \ddot{d}$$

so

$$\tau_1 \tau_2 \ddot{d} + \tau_2 \dot{d} + d = c + \tau_1 \dot{c}$$

Laplace transformation with zero initial conditions leads to

$$\frac{D}{C}(s) = \frac{1 + \tau_1 s}{\tau_1 \tau_2 s^2 + \tau_2 s + 1}$$

The static gain is 1.0. The characteristic equation is

$$\tau_1 \tau_2 s^2 + \tau_2 s + 1 = 0$$

and this has complex roots if

$$\tau_2 < 4\tau_1$$

Trying to respond to inventory fluctuations more rapidly than the plant can operate leads to an oscillatory response from the system.

Rewriting the transfer function as

$$\frac{D}{C}(s) = \frac{1}{\tau_1 \tau_2} \cdot \frac{1 + \tau_1 s}{(s - P_1)(s - P_2)}$$

where P_1, P_2 are the roots of

$$s^2 + \frac{1}{\tau_1} s + \frac{1}{\tau_1 \tau_2} = 0$$

the step response can be found to be

$$d(t) = \frac{1}{\tau_1 \tau_2} \cdot L^{-1} \left[\frac{1 + \tau_1 s}{s(s - P_1)(s - P_2)}\right]$$

From this

$$d(t) = \frac{1}{\tau_1 \tau_2} \left(\frac{1}{P_1 P_2} + \frac{\tau_1 P_1 + 1}{P_1(P_1 - P_2)} e^{tP_1} - \frac{\tau_1 P_2 + 1}{P_2(P_1 - P_2)} e^{tP_2} \right)$$

Now

$$P = \frac{-1/\tau_1 \pm \sqrt{\dfrac{1}{P_1^{\,2}} - \dfrac{4}{\tau_1 \tau_2}}}{2}$$

so

$$P = \frac{-1 \pm \sqrt{\dfrac{\tau_2 - 4\tau_1}{\tau_2}}}{2\tau_1}$$

If $\tau_2 < 4\tau_1$, the roots will be complex conjugate and there will be an oscillatory response. Otherwise the response will be of the form shown in Fig. A.9.

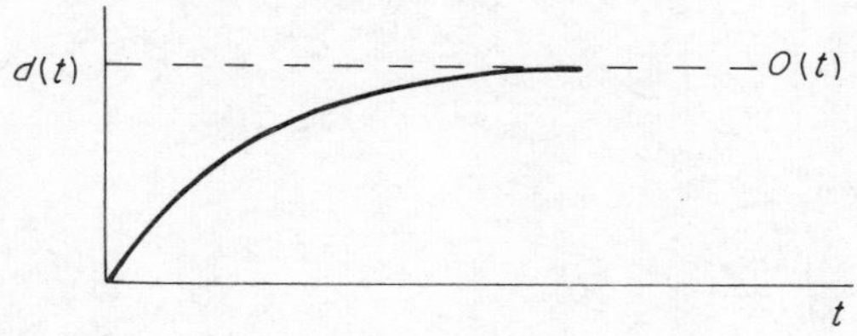

Fig. A.9

As in Section A.4, we may compare the characteristic equation for the system with the standard second-order form.

We find

$$\omega_n^{\,2} = \frac{1}{\tau_1 \tau_2}$$

and

$$2\zeta\omega_n = \frac{1}{\tau_1}$$

Hence

$$\zeta = \sqrt{\frac{\tau_1}{4\tau_2}}$$

Clearly

$$\zeta < 1$$

when

$$\tau_1 < 4\tau_2$$

as before.

Damping characteristics for this system are shown in Fig. A.10 and the reader should simulate the system and compare with the results he obtained for the system of Section A.2.

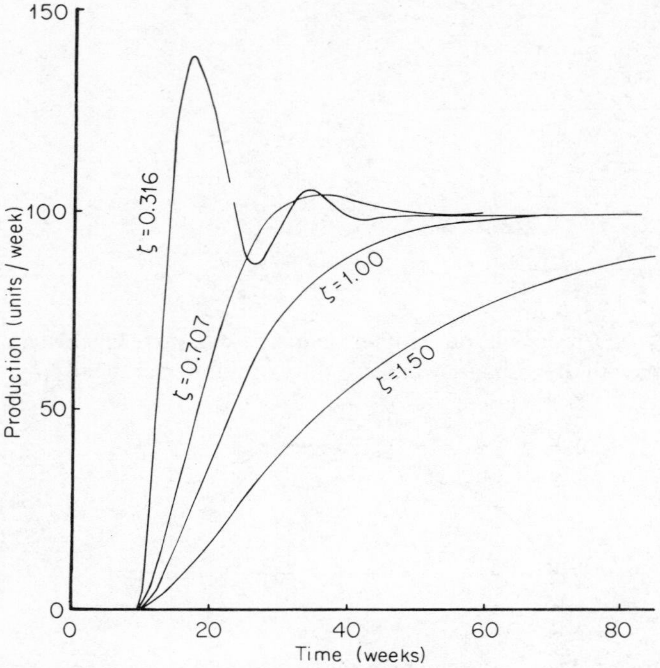

Fig. A.10 Response of a second-order production system to step input of 100 units/week at $t = 10$ weeks

The characteristic equation can have equal roots i.e. $p_1 = p_2 = p^*$, and, for that case, the step response will be

$$d(t) = \frac{1}{\tau_1 \tau_2} L^{-1} \left[\frac{1 + \tau_1 s}{s(s - p^*)^2} \right]$$

This leads to

$$d(t) = 1 e^{-t/2\tau_1} - \frac{1}{4\tau_1} t e^{-t/2\tau_1}$$

because, when $p_1 = p_2 = p^*$, $4\tau_1 = \tau_2$ and therefore $p^* = -1/2\tau_1$. This is the same as $\zeta = 1$ and is the critically-damped case in Fig. A.10.

A.6 A Third-order Production Model

Orders are placed to be equal to average consumption, plus a stock-adjustment factor (see Fig. A.11).

The basic system equations are

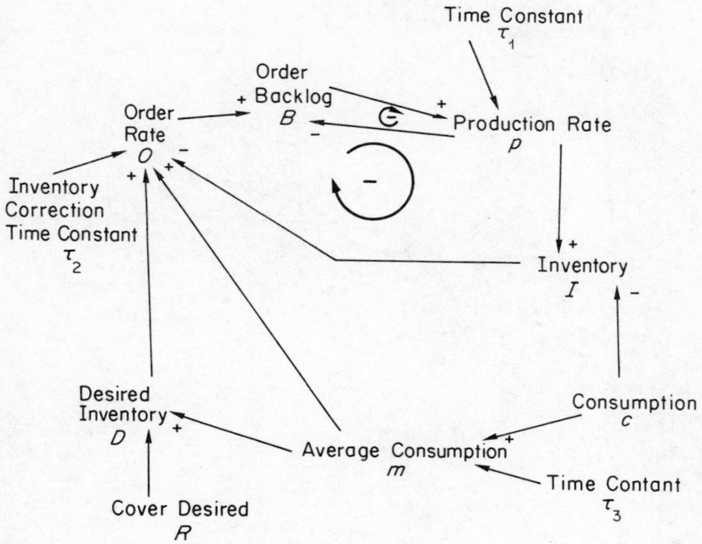

Fig. A.11

$$D = Rm \qquad \dot{m} = \frac{c - m}{\tau_3}$$

$$O = \frac{D - I}{\tau_2} + m$$

$$p = \frac{B}{\tau_1}$$

$$\dot{B} = o - p$$

$$\dot{I} = p - c$$

Solving for production in terms of consumption we have

$$D = I + \tau_2 O - \tau_2 m$$

$$= I + \tau_2 O - \tau_2 \frac{D}{R}$$

Hence

$$D\left(1 + \frac{\tau_2}{R}\right) = I + \tau_2 O$$

and

$$\dot{D}\left(1 + \frac{\tau_2}{R}\right) = \dot{I} + \tau_2 \dot{O}$$

Now,

$$O = \dot{B} + p$$
$$= \tau_1 \dot{p} + p$$

and

$$\dot{I} = p - c$$

so

$$\dot{D}\left(1 + \frac{\tau_2}{R}\right) = p - c + \tau_2(\tau_1 \ddot{p} + \dot{p})$$

and

$$\ddot{D}\left(1 + \frac{\tau_2}{R}\right) = \dot{p} - \dot{c} + \tau_2(\tau_1 \dddot{p} + \ddot{p})$$

Also

$$\tau_3 \dot{m} = c - m$$

and

$$\tau_3 \ddot{m} = \dot{c} - \dot{m}$$

thus

$$\tau_3 \frac{\ddot{D}}{R} = \dot{c} - \frac{\dot{D}}{R}$$

We now substitute in this equation for $\ddot{D}$ and $\dot{D}$ and obtain

$$\frac{\tau_3}{R} \frac{(\dot{p} - \dot{c} + \tau_1 \tau_2 \dddot{p} + \tau_2 \ddot{p})}{1 + \dfrac{\tau_2}{R}} = \dot{c} - \frac{1}{R}\left(\frac{p - c + \tau_1 \tau_2 \ddot{p} + \tau_2 \dot{p}}{1 + \dfrac{\tau_2}{R}}\right)$$

Re-arranging, and taking Laplace transforms with zero initial conditions, leads to the system transfer function

$$\frac{P}{C}(s) = \frac{1 + (R + \tau_2 + \tau_3)s}{1 + (\tau_2 + \tau_3)s + \tau_2(\tau_3 + \tau_1)s^2 + \tau_1 \tau_2 \tau_3 s^3}$$

The system has a zero frequency gain of 1.0.

This system can also be examined in terms of its state–space matrix.

We define the three state variables

$$x_1 = B, \quad x_2 = m \quad \text{and} \quad x_3 = I$$

and regroup the basic system equations so as to involve only these three variables and c. This leads to

$$\dot{B} = \frac{(R+\tau_2)m - I}{\tau_2} - \frac{B}{\tau_1}$$

$$\dot{m} = \frac{c - m}{\tau_3}$$

$$\dot{I} = \frac{B}{\tau_1} - c$$

which can be expressed in state–space matrix form as

$$\dot{x} = \begin{bmatrix} \frac{-1}{\tau_1} & \frac{R+\tau_2}{\tau_2} & -\frac{1}{\tau_2} \\ 0 & -\frac{1}{\tau_3} & 0 \\ \frac{1}{\tau_1} & 0 & 0 \end{bmatrix} x + \begin{bmatrix} 0 \\ \frac{1}{\tau_3} \\ -1 \end{bmatrix} C$$

The stability of the system can be investigated either by finding the eigenvalues of the third-order matrix or by finding the roots of the denominator of the transfer function. This example has been arranged so that the eigenvalues are fairly easy to find but in general they are rather tedious, even for known values of the parameters. Finding the system modes in terms of general parameters is usually extremely cumbersome.

A.7 A System with Positive Feedback

Fig. A.12 is based on a model due to Forrester. In this diagram,

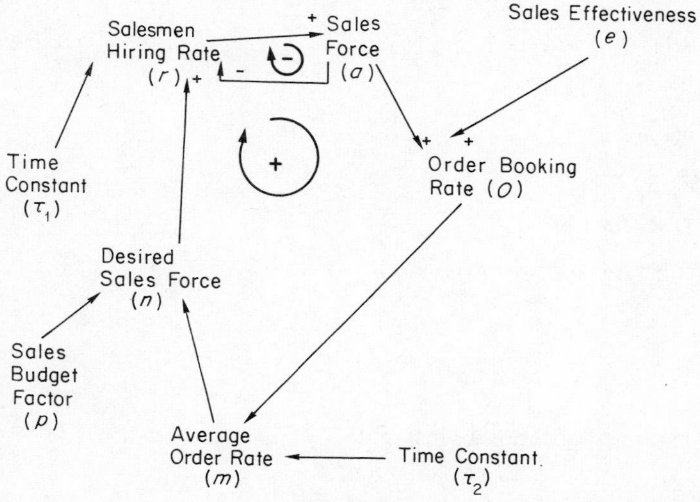

Fig. A.12

r	salesman hiring rate	(men/month)
a	sales force	(men)
e	sales effectiveness	(£/month/man)
O	order booking rate	(£/month)
m	average order booking rate	(£/month)
p	sales budget factor	(£/month/man)
n	desired sales force	(men)
$\tau_1 \tau_2$	time constants	(months)

The equations of the system are

$$\dot{a} = r \qquad r = \frac{n - a}{\tau_1} \qquad n = \frac{m}{p}$$

$$\dot{m} = \frac{O - m}{\tau_2}$$

and

$$O = ae$$

This second-order system can be solved to give any variable as a function of its own derivative and the parameters.

We shall do this for a. Since, from the system equations,

$$m = np$$

and

$$n = \tau_1 r + a$$

We have

$$m = p(\tau_1 r + a) = p(\tau_1 \dot{a} + a)$$

and

$$\dot{m} = p(\tau_1 \ddot{a} + \dot{a})$$

Now

$$\begin{aligned} ae = O &= \tau_2 \dot{m} + m \\ &= \tau_2 p(\tau_1 \ddot{a} + \dot{a}) + p(\tau_1 \dot{a} + a) \end{aligned}$$

or

$$\frac{ae}{p} = \tau_1 \tau_2 \ddot{a} + (\tau_1 + \tau_2)\dot{a} + a$$

Taking Laplace transforms we may put

$$A(s)\left(\tau_1 \tau_2 s^2 + (\tau_1 + \tau_2)s + \left(1 - \frac{e}{p}\right)\right) = f(A(\phi), s)$$

The roots of the characteristic equation are

$$s = \frac{-(\tau_1 + \tau_2) \pm \sqrt{(\tau_1 + \tau_2)^2 - 4\tau_1\tau_2\left(1 - \dfrac{e}{p}\right)}}{2\tau_1\tau_2}$$

The time solution for a will be in the form

$$a(t) = \mathrm{e}^{t/p_1} + \mathrm{e}^{t/p_2}$$

where p_1 and p_2 are the roots of the characteristic equation. Clearly if the larger of p_1 and p_2 is positive the system will exhibit exponential growth.

The condition for this is

$$\sqrt{(\tau_1 + \tau_2)^2 - 4\tau_1\tau_2\left(1 - \frac{e}{p}\right)} > (\tau_1 + \tau_2)$$

or

$$4\tau_1\tau_2\left(1 - \frac{e}{p}\right) < 0$$

whence the equivalent condition for a positive root

$$e > p$$

may be deduced.

This amounts to saying that, if salesmen are more effective than the budget gives them credit for being, the system will continue to grow once it has started. If $p = e$, the budget is right, but the system as it stands will have no growth potential, and if $p < e$ the system will decline. This is shown diagrammatically in Fig. A.13.

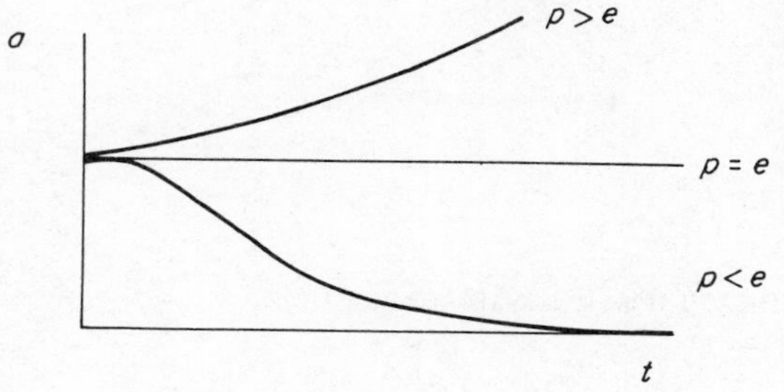

Fig. A.13

This model is considerably oversimplified, as it pays no attention to cash flow, the relationship between e, p and salesmen's actual earnings (which would probably be a very important growth generator), and a host of other factors. The reason is, of course, that such a model would be very intractable mathematically.

A.8 A Fifth-order Production Inventory System

In order to show how complex and laborious this kind of mathematical modelling can become we close this appendix by working through the analysis of a system which is, in fact, still very simple. The system is the same as that shown in Section 2.12.

The influence diagram is shown in Fig. A.14.

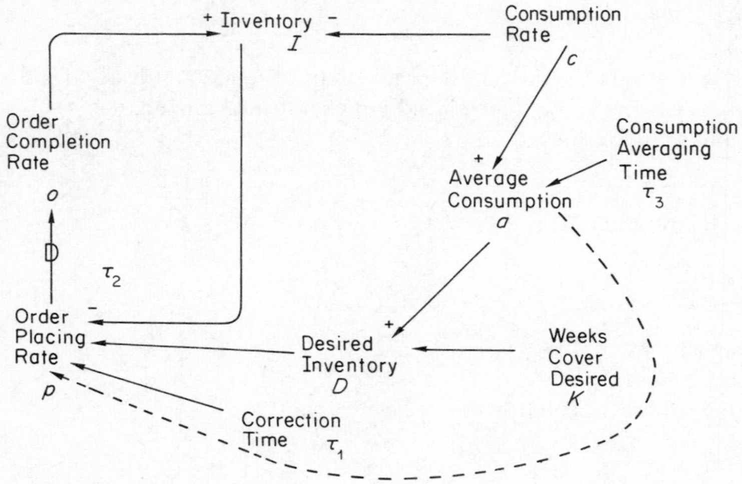

Fig. A.14

The delay is of third order, with magnitude τ_2.

There are many control strategies for this system, two of which are to make

$$p = \frac{D - I}{\tau_1}$$

or

$$p = \frac{D - I}{\tau_1} + a$$

With the first of these rules, the system equations are

$$\dot{I} = o - c$$

$$\dot{a} = \left(\frac{c - a}{\tau_3}\right)$$

$$D = Ka$$

$$p = \frac{D - I}{\tau_1}$$

The representation of the third-order delay in differential equation form by writing out the equations for its three intermediate levels, L_1, L_2 and L_3 and two dummy rates R_1 and R_2, leads to

$$\dot{L}_1 = p - R_1$$

$$\dot{L}_2 = R_1 - R_2$$

$$\dot{L}_3 = R_2 - o$$

$$R_1 = \frac{L_1}{(\tau_2/3)}$$

$$R_2 = \frac{L_2}{(\tau_2/3)}$$

$$o = \frac{L_3}{(\tau_2/3)}$$

To solve for o in terms of c and the system parameters we find

$$\dot{p} = \frac{\dot{D} - \dot{I}}{\tau_1}$$

$$= \frac{K\dot{a} - o + c}{\tau_1} \qquad \text{A.1}$$

Substituting for $\dot{a}$ in equation A.1 gives

$$\tau_1 \dot{p} = \frac{K}{\tau_3}(c - a) - o + c$$

whence

$$\tau_1 \ddot{p} = \frac{K}{\tau_3}(\dot{c} - \dot{a}) - \dot{o} + \dot{c}$$

Re-arranging equation A.1 shows that

$$\dot{a} = \frac{\tau_1 \dot{p} + o - c}{K}$$

and this may be substituted in the equation for $\ddot{p}$ to eliminate $\dot{a}$ whence

$$\tau_1 \tau_3 \ddot{p} + \tau_1 \dot{p} = (K + \tau_3)\dot{c} - o + c - \tau_3 \dot{o} \qquad \text{A.2}$$

This equation can now be re-arranged to give o in terms of c. This involves eliminating p by using the equations for the delay to relate p and o. The delay equations provide

$$p = \left(\frac{\tau_2}{3}\right)^3 \dddot{o} + 3\left(\frac{\tau_2}{3}\right)^2 \ddot{o} + \tau_2 \dot{o} + o$$

Repeated differentiation and substitution into equation A.2 gives

$$\tau_1\tau_3\left(\frac{\tau_2}{3}\right)^3 o^{\mathrm{v}} + \tau_1\left(\frac{\tau_2}{3}\right)^2\left(3\tau_3 + \frac{\tau_2}{3}\right) o^{\mathrm{iv}} + \tau_1\left(\tau_2\tau_3 + 3\left(\frac{\tau_2}{3}\right)^2\right)\dddot{o}$$

$$+ \tau_1(\tau_2 + \tau_3)\ddot{o} + (\tau_1 + \tau_3)\dot{o} + o = (K + \tau_3)\dot{c} + c \qquad \text{A.3}$$

This is the basic equation for the dynamics of the system. The reader should recall that it is based on a particular control policy. Deriving the corresponding equations for all the other policies which could be conceived of would be a monumental task.

Examining the stability of equation A.3 by means of general formulae for all the τ would be very laborious. The easier approach is to assume values for, say, τ_2 and τ_3 and then solve the value of τ_1 at which the system is just oscillatory. Assuming $\tau_2 = \tau_3 = 4$, and taking Laplace transforms with zero initial conditions, leads to the characteristic equation

$$9.5\tau_1 s^5 + 23.7\tau_1 s^4 + 21.3\tau_1 s^3 + 8\tau_1 s^2 + (\tau_1 + 4)s + 1 = 0$$

The Routh Array (see, for example, Gille, Pelegrin and De Caulne, *Feedback Control Systems*, McGraw-Hill, p. 163) for this function is:

s^5	$9.5\tau_1$	$21.3\tau_1$	$\tau_1 + 4$
s^4	$23.7\tau_1$	$8\tau_1$	1
s^3	$18.1\tau_1$	$0.6\tau_1 + 4$	0
s^2	$7.21\tau_1 - 5.24$	1	
s^1	$4.33\tau_1^2 + 7.6\tau_1 - 20.96$	0	

The condition for stability is the value of τ_1 which makes the first entry in the last row zero, i.e. $\tau_1 = 1.49$. At this value, the system is just oscillatory.

To conclude the example, we look at the third condition, which is equivalent to

$$B^2CDE - ABD^2E - B^2E^2 + 2AB^2E < B^2C^2 - ABCD$$

Substitution for *A, B, C* and *D* to find a general formula relating τ_1, τ_2 and τ_3 is impossibly complicated. The best that can be done (and it would still only be valid for one particular decision rule) is to put in numerical values of say, $\tau_2=4$, $\tau_3=4$, and find the critical value of τ_1. This comes down to finding the value of τ_1 for which

$$-13917.76\tau_1^3 + 160897.74\tau_1^2 - 398728.86\tau_1 + 8545.12 = 0$$

The only practical way of doing this is by numerical methods. The answer is $\tau_1 = 3.558703$ (very nearly, an exact calculation requires more than 8 significant figures). When the system is simulated it is seen that those values of the inventory adjustment time (a value very close to 1 month) does lead to sustained oscillations.

This whole effort takes over a day of algebra and calculations, even for this very simple, linear, system.

Appendix B

Principal Computer Programs Used in the Book

This appendix presents listings of some of the more important DYSMAP programs used in the book. The purpose is to provide the reader with illustrations of how a model is transformed into a simulation program, and to demonstrate some of the techniques of programming.

Most of the programs are divided up into sections preceded by text comments. In some cases the whole program is given; in others, to save space, only the most interesting parts, or those which differ from a previous, similar, program are listed. In each case the reader will have to refer to the appropriate chapter text in order to get the full benefit of the sample listings.

The sections in this appendix are numbered according to the section of the book in which the listed program was first or most prominently used. Thus, Section B.10.7 lists the program used in Chapter 10, Section 7.

B.2.12 Program for Example on System Behaviour

The basic equations for the system are relatively straightforward. Note the use of the dummy variables D1 and D2 which respectively control the input and the order rule to be used. This approach to the RERUN facility allows a great deal of experimentation in a single turn round.

It is slightly wasteful of computer time as both inputs, for example, are calculated but only one is used. Note the argument in the sine function and the CLIP, both of which are needed to bring the sine in at TIME=10 as a check for false dynamics.

```
* PROGRAM FOR SECTION 2.12
L INV.K=INV.J+DT*(CONRATE.JK-CONS.JK)
N INV=(COVER-(1-D2)*TAI)*AVCON
N ORATE=AVCON
R CONRATE.KL=DELAY3(ORATE.JK,PDEL)
R ORATE.KL=((DINV.K-INV.K)/TAI)+D2*AVCON.K
A DINV.K=COVER*AVCON.K
L AVCON.K=AVCON.J+(DT/TAC)(CONS.JK-AVCON.J)
N AVCON=100
R CONS.KL=100+STEP(20,10)*D1+(1-D1)*SIN(6.283*(TIME.K-10)/PERD)
X *20*CLIP(1,0,TIME.K,10)
```

These are the initial parameter values:

```
C PERD=25
C D1=0
C TAC=4
C COVER=10
C PDEL=4
C TAI=6
C D2=0
```

This calculates the maximum amplitude for ORATE and is purely for the purpose of plotting the system gain curves easily. The CLIP at TIME=40 ensures that transients have died away and the PULSE function transfers the result to the previous old maximum value, OMAX, when a change is called for.

```
A MAXAMP.K=MAX(ORATE.KL,OMAX.K)*CLIP(1,0,TIME.K,40)
L OMAX.K=OMAX.J+DT*PULSE((MAXAMP.J-OMAX.J)/DT,DT,DT)
N OMAX=0
```

This calculates the minimum amplitude. Note the alternative formulation using a dummy rate, RCOMIN, instead of the PULSE. The CLIP on RCOMIN makes sure that no changes take place until transients have died out and is needed here rather than in the equation for MINAMP because its use in that equation would make MINAMP=0 and drive OMIN to zero if RCOMIN were not clipped.

```
A MINAMP.K=MIN(ORATE.KL,OMIN.K)
R RCOMIN.KL=((MINAMP.K-OMIN.K)/DT)*CLIP(1,0,TIME.K,40)
L OMIN.K=OMIN.J+DT*RCOMIN.JK
N OMIN=100
```

This equation calculates the amplitude ratio, given that the input amplitude is constant at 40.

```
S ARAT.K=(MAXAMP.K-MINAMP.K)/40
```

Note the value of DT to come to the continuous case. In most RERUNS on the sinusoidal input, PLTPER=0 to reduce the bulk of outputs, as all that is needed is the terminal value of ARAT. The bulk can be reduced still further by making PRTPER into an auxiliary and using STEP functions, e.g. if

PRTPER·K=2 + STEP (LENGTH–2,CONST)

then, with CONST=LENGTH, the printing will be at a frequency of 2 and with CONST=0 in a rerun, only the values at TIME=LENGTH will be output. This kind of approach can be used more generally, e.g.

PRTPER.K=1 + STEP(9,10)–STEP(9,LENGTH–10)

will print out at each of the first 10 time points, each of the last 10, and every 10th intervening value. DT has to be an exact binary fraction for this to work exactly.

```
C DT=0.0625
C LENGTH=120
C PRTPER=2
C PLTPER=1
PRINT 1)COMRATE
PRINT 2)ORATE
PRINT 3)CONS,AVCON
PRINT 4)INV,DINV
PRINT 5)MAXAMP
PRINT 6)MINAMP
PRINT 7)ARAT
PLOT ORATE=O,CONS=C,COMRATE=D(-100,300)/INV=I(0,2000)
RUN 1
```

The variables are defined as follows:

```
D ARAT=(1) AMPLITUDE RATIO OF ORATE AND SR
D AVCON=(U/WK) AVERAGE CONSUMPTION
D COMRATE=(U/WK) PRODUCTION COMPLETION RATE
D CONS=(U/WK) EXOGENEOUS CONSUMPTION RATE
D COVER=(WK) NUMBER OF WEEKS INVENTORY COVER REQUIRED
D D1=(1) SWITCH TO TEST STEP OR SINE
D D2=(1) SWITCH TO TEST EFFECTS OF AVCON IN ORATE EQUATION
D DINV=(U) DESIRED INVENTORY
D DT=(WK) SOLUTION INTERVAL
D INV=(U) INVENTORY
D LENGTH=(WK) RUN LENGTH
D MAXAMP=(U/WK) MAXIMUM ATTAINED BY ORATE
D MINAMP=(U/WK) MINIMUM AMPLTITUDE OF ORATE
D OMAX=(U/WK) PREVIOUS VALUE OF MAXAMP
D OMIN=(U/WK) PREVIOUS MINIMUM OF MINAMP
D ORATE=(U/WK) PRODUCTION ORDERING RATE
D PDEL=(WK) PRODUCTION PIPELINE DELAY
D PERD=(WK) PERIOD FOR SINE WAVE
D PLTPER=(WK) PLOTTING INTERVAL
D PRTPER=(WK) PRINTING INTERVAL
D RCOMIN=(U/WK/WK) RATE OF CHANGE OF OMIN
D TAC=(WK) CONSUMPTION AVERAGING TIME
D TAI=(WK) TIME TO ADJUST INVENTORY
D TIME=(WK) SIMULATED TIME DURING RUN
+
```

B.6 Program for Loop Sensitivity

This little program is included here to show how one can generate a large amount of comparative information from one run and three reruns.

We require graphs for each of two gains and each of two delays, so each section of the program is in quadruplicate. The sections should be fairly self-explanatory.

Only some of the reruns are included, but note the first rerun (i.e. RUN 2). With integration the system would go into explosive oscillations, unless gains were *considerably* reduced to maintain a damped response. Try to make a non-integral loop explode.

```
DOC
* SIMPLE LOOP SYSTEM BEHAVIOUR
NOTE
NOTE MODEL FOR CHAPTER 6 DIAGRAMS
NOTE PROGRAM NAME CHAP6-DYN
NOTE
R IN1.KL=D2*G*AO1.K+EX.K+(1-D2)*G*INT1.K
R IN2.KL=D2*G*AO2.K+EX.K+(1-D2)*G*INT2.K
R IN1A.KL=D2*GA*AO1A.K+EX.K+(1-D2)*GA*INT1A.K
R IN2A.KL=D2*GA*AO2A.K+EX.K+(1-D2)*GA*INT2A.K
A EX.K=PULSE(2000,10,LENGTH)
C D2=0
NOTE
NOTE  DELAYED OUTPUTS
NOTE
R OUT.KL=DELAY3(IN1.JK,DEL)
R OUTA.KL=DELAY3(IN1A.JK,DEL)
R DOUT1.KL=DELAY3(IN2.JK,DEL/3)
R DOUT2.KL=DELAY3(DOUT1.JK,DEL/3)
R DOUT3.KL=DELAY3(DOUT2.JK,DEL/3)
R DOUT1A.KL=DELAY3(IN2A.JK,DEL/3)
R DOUT2A.KL=DELAY3(DOUT1A.JK,DEL/3)
R DOUT3A.KL=DELAY3(DOUT2A.JK,DEL/3)
C DEL=10
N IN1=0
N IN1A=0
N IN2=0
N DOUT1=0
N DOUT2=0
N IN2A=0
N DOUT1A=0
```

```
N DOUT2A=0
NOTE
NOTE  EQUATIONS FOR EFFECTS OF INTEGRATION
L INT1.K=INT1.J+DT*OUT.JK
N INT1=0
L INT2.K=INT2.J+DT*DOUT3.JK
N INT2=0
L INT1A.K=INT1A.J+DT*OUTA.JK
N INT1A=0
L INT2A.K=INT2A.J+DT*DOUT3A.JK
N INT2A=0
NOTE
NOTE  PARAMETER VALUES
NOTE
C G=-0:5
C GA=-1.25
NOTE
NOTE  SMOOTHING EQUATIONS FOR LOOP OUTPUT
NOTE
L A01.K=A01.J+(DT/TAO)(OUT.JK-A01.J)
L A02.K=A02.J+(DT/TAO)(DOUT3.JK-A02.J)
L A01A.K=A01A.J+(DT/TAO)(OUTA.JK-A01A.J)
L A02A.K=A02A.J+(DT/TAO)(DOUT3A.JK-A02A.J)
N A01=0
N A02=0
N A01A=0
N A02A=0
C TAO=1
NOTE
NOTE  CONTROL SECTOR
NOTE
C DT=0.0625
C LENGTH=55
C PLTPER=1
PLOT OUT=0,OUTA=A,DOUT3=B,DOUT3A=C(-20,20)
RUN FIG 6.17 IMPULSE RESPONSE WITHOUT INTEGRATION
D A01=(U/WK) AVERAGE OUTPUT FOR 3RD ORDER LOWER GAIN
D A01A=(U/WK) AVERAGE OUTPUT FOR 3RD ORDER HIGHER GAIN
D A02=(U/WK) AVERAGE OUTPUT FOR 9TH ORDER LOWER GAIN
D A02A=(U/WK) AVERAGE OUTPUT FOR 9TH ORDER HIGHER GAIN
D D2=(1) ZERO-ONE SWITCH FOR TESTING EFFECTS OF PURE INTEGRATION
D DEL=(WK) TOTAL DELAY
D DOUT1=(U/WK) 1ST INTERMEDIATE FLOW FOR 9TH ORDER LOWER GAIN
D DOUT1A=(U/WK) 1ST INTERMEDIATE FLOW FOR 9TH ORDER HIGHER GAIN
D DOUT2=(U/WK) 2ND INTERMEDIATE FLOW FOR 9TH ORDER LOWER GAIN
D DOUT2A=(U/WK) 2ND INTERMEDIATE FLOW FOR 9TH ORDER HIGHER GAIN
D DOUT3=(U/WK) OUTPUT FOR 9TH ORDER LOWER GAIN
D DOUT3A=(U/WK) OUTPUT FOR 9TH ODER LOWER GAIN
D DT=(WK) SOLUTION INTERVAL
D EX=(U/WK) EXOGENEOUS INPUT
D G=(1) LOWER GAIN
D GA=(1) HIGHER GAIN
D IN1=(U/WK) INPUT TO 3RD ORDER LOWER GAIN SYSTEM
D IN1A=(U/WK) INPUT TO 3RD ORDER HIGHER GAIN SYSTEM
D IN2=(U/WK) INPUT TO 9TH ORDER LOWER GAIN SYSTEM
D IN2A=(U/WK) INPUT TO 9TH ORDER HIGHER GAIN SYSTEM
D INT1=(U) INTEGRATION OF OUTPUT FROM 3RD ORDER LOWER GAIN
D INT1A=(U) INTEGRATION OF OUTPUT FROM 3RD ORDER HIGHER GAIN
D INT2=(U) INTEGRATION OF OUTPUT FROM 9TH ORDER LOWER GAIN
D INT2A=(U) INTEGRATION OF OUTPUT FROM 9TH ORDER HIGHER GAIN
D LENGTH=(WK) SIMULATED  PERIOD
D OUT=(U/WK) OUTPUT FROM 3RD ORDER LOWER GAIN
D OUTA=(U/WK) OUTPUT FROM 3RD ORDER HIGHER GAIN
D PLTPER=(WK) PLOTTING INTERVAL
D TAO=(WK) AVERAGING TIME FOR OUTPUTS
D TIME=(WK) SIMULATED TIME
C DEL=5
RUN  FIG 6.18 SYSTEM OF FIG 6.17 WITH HALF DELAY
C D2=1
C DEL=10
C G=-.05
C GA=-.125
RUN FIG 6.20 IMPULSE RESPONSE WITH PURE INTEGRATION
*
```

B.8 Model for Chapter 8

Basic Version of the Program

This version is based on the influence diagram in Fig. 8.1 but contains a number of parameters to produce Fig. 8.10, loop E from Section 8.5 and Fig. 8.15. There are several other parameters for testing alternatives not specifically mentioned in the text.

```
*   MODEL FOR SYSTEM EXAMPLE IN CHAPTER 8
NOTE
NOTE SALES RATE SECTOR
NOTE
R SR.KL=DS*(100+STEP(40,10))+(1-DS)*(SIN(6.283*TIME.K/PERD)*20
X +100)
C DS=1
C PERD=5
L ASR.K=ASR.J+(DT/TASR)(SR.JK-ASR.J)
CP TASR=4
N ASR=100
NOTE
NOTE INVENTORY AND FACTORY ORDER RATE SECTOR
NOTE
L INV.K=INV.J+DT*(DFF.JK-SR.JK)
N INV=ASR*(WINVD-D2*(1-D1)*TAI)
A DINV.K=WINVD*ASR.K
C WINVD=6
R FOR.KL=D1*ASR.K+D2*((DINV.K-INV.K)/TAI)+D3*((PLD.K-PLA.K)/TAPL)
A DR1.K=(DINV.K-INV.K)/TAI
A DR2.K=(PLD.K-PLA.K)/TAPL
A DR3.K=(OBL.K-RBL.K)/TABL
C D1=1
C D2=1
C D3=0
CP TAPL=4
CP TAI=4
L AOR.K=AOR.J+(DT/TAOR)(FOR.JK-AOR.J)
N AOR=ASR
CP TAOR=4
NOTE
NOTE ORDER BACKLOG SECTOR
NOTE
L OBL.K=OBL.J+DT*(FOR.JK-PSR.JK)
N OBL=RBL+TABL*ASR*(1-DIPL)
C DP=1
A RBL.K=DP*TABHL(TRBL,AOR.K,50,150,25)+(1-DP)*TABHL(TRBL,APL.K,50,150,25
X )
T TRBL=400/525/600/650/675
NOTE
NOTE PRODUCTION PLANNING SECTOR
NOTE
A IPL.K=((OBL.K-RBL.K)/TABL)+DIPL*(D5*AOR.K+(1-D5)*APL.K)
C DIPL=0
C D5=1
CP TABL=4
L APL.K=APL.J+(DT/PAT)(IPL.J-APL.J)
N APL=AOR
CP PAT=3
R PSR.KL=APL.K
R DFF.KL=DELAY3(PSR.JK,PDEL)
C PDEL=6
NOTE
NOTE PRODUCTION PIPELINE AND DELIVERY DELAY SECTOR
NOTE
A PLD.K=PLV.K*(D4*WPLD+(1-D4)*DDR.K)
A PLV.K=DPLV*ASR.K+(1-DPLV)*AOR.K
C DPLV=1
C WPLD=6
C D4=1
L PLA.K=PLA.J+DT*(PSR.JK-DFF.JK)
N PLA=PLV*(D4*WPLD+(1-D4)*DDR)
A DD.K=OBL.K/APL.K
A DDR.K=SMOOTH(DD.K,TSDD)
C TSDD=4
NOTE
```

```
NOTE AMPLITUDE RATIO FOR POWER SPECTRUM
NOTE (ONLY RELEVANT FOR SINE INPUTS)
NOTE
A MAXAMP.K=MAX(PSR.KL,OMAX.K)*CLIP(1,0,TIME.K,40)
L OMAX.K=OMAX.J+DT*PULSE((MAXAMP.J-OMAX.J)/DT,DT,DT)
N OMAX=0
A MINAMP.K=MIN(PSR.KL,OMIN.K)
L OMIN.K=OMIN.J+DT*PULSE((MINAMP.J-OMIN.J)/DT,DT,DT)*CLIP(1,0,TIME.J,40)
N OMIN=10E04
S ARAT.K=(MAXAMP.K-MINAMP.K)/40
PLOT  SR=S,FOR=F,PSR=P,DFF=D(-100,300)/INV=I,OBL=B,PLA=A,DINV=V,PLD=L,
X RBL=R(-500,1500)
PRINT 1)IPL,PSR
PRINT 2)DFF,FOR,AOR
PRINT 3)OBL,RBL,DR1
PRINT 4)DINV,DR2,INV
PRINT 5)SR,ASR,DR3
PRINT 6)APL,ARAT
PRINT 7)PLD,PLA
PRINT 8)DDR,DD
A PRTPER.K=1+STEP(9,11)
C DT=0.125
C PLTPER=1
C LENGTH=120
D AOR=(U/WK) AVERAGE ORDER RATE AT FACTORY
D APL=(U/WK) ACTUAL PRODUCCTION LEVEL
D ARAT=(1) AMPLITUDE RATIO OF PSR AND SR FOR POWER
X          SPECTRUM CALCULATION
D ASR=(U/WK) AVERAGE SALES RATE
D D1=(1)  ZERO-ONE SWITCH TO TEST EFFECTS OF ASR
X         ON FACTORY ORDER RATE
D D2=(1)  ZERO-ONE SWITCH TO TEST EFFECTS OF INVENTORY
X         -CORRECTION ORDERING POLICY
D D3=(1)  ZERO-ONE TO TEST EFFECTS OF PIPELINE ON
X       FACTORY ORDER RATE
D D4=(1)  ZERO-ONE SWITCH TO TEST EFFECTS OF ALTERNATIVE DESIRED
X         PIPELINE FORMULATIONS
D D5=(1)  ZERO-ONE SWITCH TO TEST THE EFFECT OF USING APL
X         VERSUS AOR IN EQUATION FOR IPL
D DD=(WK) ACTUAL DELIVERY DELAY FROM PRODUCTION PIPELINE
D DDR=(WK) DELIVERY DELAY RECOGNISED BY DISTRIBUTION
D DFF=(U/WK) DELIVERY RATE FROM FACTORY
D DINV=(U) DESIRED INVENTORY
D DIPL=(1) ZERO-ONE SWITCH TO TEST EFFECTS OF USING AVERAGE ORDER
X          RATE OR AVERAGE PRODUCTION LEVEL IN EQUATION FOR
X          INDICATED PRODUCTION LEVEL
D DP=(1)  ZERO-ONE SWITCH FOR TESTING EFFECTS OF ORDER RATE VERSUS
X         PRODUCTION LEVEL IN DESIRED BACKLOG
D DPLV=(1) ZERO-ONE SWITCH TO TEST EFFECTS OF USING ASR
X          OR AOR IN EQUATION FOR DESIRED PIPELINE PLD
D DR1=(U/WK) DUMMY VARIABLE FOR CONTRIBUTION OF INVENTORY
X            CORRECTION TO FACTORY ORDER RATE FOR
D DR2=(U/WK) DUMMY VARIABLE FOR CONTRIBUTION OF PIPELINE
X            CORRECTION TO FACTORY ORDER RATE FOR
D DR3=(U/WK) DUMMY VARIABLE FOR CONTRIBUTION OF ORDER BACKLOG
X            CORRECTION TO INDICATED PRODUCTION LEVEL
D DS=(1) ZERO-ONE SWITCH FOR ALTERNATIVE SALES PATTERNS
D DT=(WK) SOLUTION INTERVAL IN SIMULATION
D FOR=(U/WK) FACTORY ORDER RATE
D INV=(U) INVENTORY
D IPL=(U/WK) INDICATED PRODUCTION LEVEL FOR BACKLOG CONTROL
D LENGTH=(WK) LENGTH OF SIMULATED PERIOD
D MAXAMP=(U/WK) MAXIMUM AMPLITUDE OF PSR
D MINAMP=(U/WK) MINIMUM AMPLITUDE OF PSR
D OBL=(U) ACTUAL ORDER BACKLOG
D OMAX=(U/WK) OLD VALUE OF MAXIMUM AMPLITUDE OF PSR
D OMIN=(U/WK) OLD VALUE OF MINIMUM AMPLITUDE OF PSR
D PAT=(WK) TIME TO ADJUST TO PLANNED PRODUCTION LEVEL
D PDEL=(WK) PRODUCTION PROCESS DELAY
D PERD=(WK) PERIODICITY IN SINE INPUT
D PLA=(U) GOODS IN PRODUCTION PIPELINE
D PLD=(U) GOODS DESIRED IN PRODUCTION PIPELINE
D PLTPER=(WK) PLOTTING INTERVAL
D PLV=(U/WK) DUMMY VARIABLE USED IN DESIRED PIPELINE EQUATION
D PRTPER=(WK) PRINTING INTERVAL
D PSR=(U/WK) PRODUCTION START RATE AT FACTORY
D RBL=(U) REQUIRED LEVEL OF BACKLOG
D SR=(U/WK) SALES RATE
D TABL=(WK) TIME TO ADJUST BACKLOG
D TAI=(WK) TIME TO ADJUST INVENTORY
D TASR=(WK) TIME TO AVERAGE SALES RATE
```

```
D TAOR=(WK) TIME TO AVERAGE ORDER RATE
D TIME=(WK) TIME IN SIMULATED PERIOD
D TRBL=(U) TABLE OF REQUIRED BACKLOG
D TSDD=(WK) TIME TO SMOOTH DELIVERY DELAY
D WINVD=(WK) NUMBER OF WEEKS OF AVERAGE SALES DESIRED IN INVENTORY
D TAPL=(WK) TIME TO ADJUST PIPELINE
D WPLD=(WK) WEEKS PIPELINE DESIRED
RUN FIG 8,7 STEP RESPONSE OF INITIAL SYSTEM
C TAOR=12
RUN TAOR=12
T TRBL=200/200/275/300/300
RUN LOWER TRBL
T TRBL=400/525/600/650/675
C TABL=12
RUN TABL=12
C TAI=12
RUN TAI=12,POSITIVE LOOP
T TRBL=675/675/675/675/675
RUN CONSTANT RBL
T TRBL=400/525/600/650/675
C DP=0
RUN FIG 8,10 STEP RESPONSE FOR NEGATIVE SYSTEM
C TAI=12
RUN FIG 8,11 NEGATIVE SYSTEM -LONGER LOOP A DELAY
C TAI=12
C TABL=12
RUN FIG 8,6 CHECK
C TAI=12
C PAT=8
RUN LOW GAIN IN LOOP B
C TABL=2
RUN TABL=12,PAT=3
C DIPL=1
RUN NEW LOOP E
C TAI=12
RUN FIG 8,13 NEGATIVE SYSTEM WITH NEW LOOP
C TAI=12
C TAOR=12
C TABL=12
RUN 8,7 WITH TAI,TAOR,TABL=12
C TAI=12
C D3=1
RUN FIG 8,17 STEP RESPONSE WITH PIPELINE CONTROL
```

Note that the parameter values given above are not necessarily those which will produce Fig. 8.8. This has been done to encourage the reader to study the program carefully before using it.

Non-negativity Constraints

As mentioned in Section 8.3, the basic model produces explosive oscillation with associated negative values of some of the variables. Negative inventory and Factory Order Rate are permissible but negative Backlog and Production are not. Thus FOR could only be negative up to the point where all the Order Backlog had been cancelled.

The approach is to define a Dummy Factory Order Rate, DFOR, using the original equation for FOR. A new equation defines FOR in terms of DFOR and two other dummy variables, DUM1 and DUM2, which respectively verify that OBL is non-negative and that DFOR is positive.

Thus, if

$$\text{OBL} > 0, \quad \text{DUM1} = 1$$

and if

$$\text{DFOR} > 0, \quad \text{DUM2} = 0$$

or

$$(1-DUM2*(1-DUM1)) = 1 \quad \text{and} \quad FOR = DFOR$$

ie. the normal situation. If, however,

$$OBL > 0 \text{ and } DFOR < 0$$

then

$$DUM1 = 1 \text{ and } DUM2 = 1$$

but

$$(1 - DUM2* (1 - DUM1)) = 1 \text{ and } FOR = DFOR$$

In this situation, orders are cancelled because there is a backlog to cancel. This formulation could lead to a slight negative backlog is

$$DFOR*DT > OBL$$

but this is hardly worth bothering about. Finally, if $OBL < 0$ and $DFOR < 0$, $DUM1 = 0$ and $DUM2 = 1$ and $(1-DUM2*(1-DUM1)) = 0$ and $FOR = 0$ so that ordering stops.

Note the use of 0.1 in the CLIP arguments to prevent DUM1 = 1 when OBL = 0 which would make the second case incorrect.

```
A DFOR.K=(D1*ASR.K+D2*((DINV.K-INV.K)/TAI)+D3*((PLD.K-PLA.K)/TAPL))
R FOR.KL=DFOR.K*(1-DUM2.K*(1-DUM1.K))
A DUM1.K=CLIP(1,0,OBL.K,0.1)
A DUM2.K=CLIP(0,1,DFOR.K,0.1)
```

IPL is more easily handled by simply adding MAX(0, . . .) to the existing equation

Version to Give Loop with Three Integrations

Most of the equations are the same as those used in the first program for Chapter 8 but two sections of the program are different:–

```
NOTE
NOTE PRODUCTION PLANNING SECTOR
NOTE
A IPL.K=((OBL.K-RBL.K)/TABL)+DIPL*(D5*AOR.K+(1-D5)*APL.K)
C DIPL=1
C D5=1
CP TABL=4
L APL.K=APL.J+(DT/PAT)(IPL.J-APL.J)
N APL=AOR
CP PAT=3
R PSR.KL=APL.K
R PCR.KL=DELAY3(PSR.JK,PDEL)
L WIFA.K=WIFA.J+DT*(PCR.JK-DFF.JK)
N WIFA=(COVER+(1-D6)*TAWIFA)*AOR
R DFF.KL=(WIFA.K-DWIFA.K)/TAWIFA+D6*AOR.K
C D6=1
C TAWIFA=2
A DWIFA.K=COVER*APL.K
CP COVER=2
C PDEL=4
N PSR=AOR
NOTE
NOTE PRODUCTION PIPELINE AND DELIVERY DELAY SECTOR
NOTE
```

```
A PLD.K=PLV.K*(D4*WPLD+(1-D4)*DDR.K)
A PLV.K=DPLV*ASR.K+(1-DPLV)*AOR.K
C DPLV=1
N WPLD=PDEL+COVER+(1-D6)*TAWIFA
C D4=1
L PLA.K=PLA.J+DT*(PSR.JK-DFF.JK)
N PLA=ASR*(D4*WPLD+(1-D4)*DDR)
A DD.K=OBL.K/APL.K
A DDR.K=SMOOTH(DD.K,TSDD)
C TSDD=4
```

where

```
107 D COVER=(WK) NUMBER OF WEEKS COVER REQUIRED IN FINAL ASSEMBLY
139 D DWIFA=(U) DESIRED WORK IN FINAL ASSEMBLY
150 D PCR=(U/WK) RATE OF COMPLETION OF BASIC MANUFACTURING STAGE
151 X              IE RATE OF ENTRY INTO FINAL ASSEMBLY POOL
167 D TAWIFA=(WK) TIME TO ADJUST WORK IN FINAL ASSEMBLY
174 D WIFA=(U) WORK IN FINAL ASSEMBLY
```

B.9.12 A Model of Corporate Growth

This section shows how a lot of information and understanding can be derived from a model which is really rather simple. The reader should experiment by refining this model to include more detail wherever he feels there is a need for it. In particular he should aim to find the point at which the behaviour of the system becomes radically different. He should then think through the dichotomy between building a simple, clear, understandable, but apparently naïve model, and the alternative approach of adding much more detail in order to boost confidence in it, not least in his own mind.

The following equations ought to be fairly transparent, as the salient ones have already been explained in the text. Some reruns are suggested which are not necessarily those used to produce the diagrams in the text.

```
DOC
*         MODEL FOR CHAPTER 9
NOTE
NOTE  MODEL NAME CHAP9-DYN,USE CHAP9-MAC
NOTE
NOTE ORDER LEVEL EQUATIONS
NOTE
A PDSM.K=TABHL(TPDSE,PPDSE.K/AOL.K,0,2,0.5)/50
T TPDSE=-0.5/-0.2/0/0.5/0.7
R RCOL.KL=PDSM.K*AOL.K
L OL.K=OL.J+DT*RCOL.JK
N OL=AOL
L AOL.K=AOL.J+(DT/TAOL)(OL.J-AOL.J)
C TAOL=4
N AOL=CAPAC
L PPDS.K=PPDS.J+(DT/TPPDS)(PDSR.JK-PPDS.J)
N PPDS=MAX(0,PDSFAC*CFLOW)
C TPPDS=4
NOTE
NOTE PRODUCT DEVELOPMENT SPENDING AND EFFECTS
NOTE
R PDSR.KL=MAX(0,PDSFAC.K*CFLOW.K)
A PPDSE.K=PPDS.K/PDSEC
C PDSEC=1
NOTE
NOTE   CASH FLOW AND ORDER VALUE
NOTE
A CFLOW.K=OV.K*OL.K-CAPAC.K*CMC
A OV.K=TABHL(TDCR,DCR.K,0,2,0.5)
T TDCR=66.7/66.7/66.7/40/30
NOTE
NOTE CAPACITY SPENDING AND EFFECTS
NOTE
```

```
L CAPAC.K=CAPAC.J+DT*(CAR.JK-RLC.JK)
N CAPAC=100
R CAR.KL=DELAY3(CSR.JK,CCDEL)
N CSR=(CAPAC/PLC)
C CCDEL=13
R RLC.KL=CAPAC.K/PLC
C PLC=20
A DCR.K=OL.K/CAPAC.K
L ADCR.K=ADCR.J+(DT/TADCR)(DCR.J-ADCR.J)
N ADCR=1
C TADCR=4
R CSR.KL=MAX(0,((1-PDSFAC.K)*CFLOW.K)/CCOST)
C CCOST=170
C CMC=55
NOTE
NOTE PRODUCT DEVELOPMENT SPENDING CONTROL POLICY
NOTE
A PDSFAC.K=TABHL(TPDSF,ADCR.K,0.8,1.3,0.1)
T TPDSF=1.0/.18/.15/.02/0/0
NOTE
NOTE OUTPUT AND CONTROL EQUATIONS
NOTE
C LENGTH=100
C DT=0.0625
C PLTPER=1
C PRTPER=5
PRINT 1)OL,RCOL
PRINT 2)CSR,CAR,RLC
PRINT 3)CAPAC
PRINT 4)OL,AOL,OV
PRINT 5)PDSFAC,DCR,ADCR
PRINT 6)PPDS,PDSM
PRINT 7)PPDSE
PRINT 8)CFLOW,PDSR
PLOT OL=O,CAPAC=C(50,250)/PDSR=P(0,400)/OV=V(40,80)/
X   CFLOW=£(-2E03,2E03)
RUN TRIAL
T TPDSF=.15/.15/.15/.15/.15/.15
RUN CONSTANT TPDSF FIG 9.14
T TPDSF=1.0/.15/.15/.15/.15/0
RUN ON-OFF TPDSF FIG 9.15
T TPDSF=1.0/.57/.15/.10/.02/0
RUN SMOOTH TPDSF FIG 9.16
T   TDCR=66.7/66.7/66.7/30/20
RUN SMOOTH TPDSF AND SHARP TDCR
T TPDSF=1.0/.18/.15/.02/0/0
T TDCR=66.7/66.7/66.7/40/30
C TADCR=12
RUN SLOWER TADCR,ORIGINAL TPDSF
C TADCR=2
RUN FAST TADCR
C TADCR=4
C PLC=40
RUN LONG-LIFE CAPACITY
C PLC=20
C CCOST=190
RUN DEARER CAPACITY
C CCOST=170
T TPDSF=1/.15/.15/.15/0/0
RUN STEP TPDSF
T TPDSF=1.0/.57/.15/0/0/0
RUN SMOOTH BUT NO OVER TPDSF
C TADCR=2
RUN FAST TADCR AND NO OVER TPDSF
T TPDSF=1.0/.57/.15/.15/.15/.15
RUN FAST TADCR AND FLAT TPDSF
C TADCR=4
RUN ORDINARY TADCR,FLAT TPDSF
T TDCR=66.7/66.7/66.7/30/20
C TADCR=2
RUN SHARP TDCR,FLAT TPDSF,FAST TADCR
T TPDSF=1.0/.57/.15/0/0/0
RUN NO OVER ,FAST TADCR,SHARP TDCR
NOTE
NOTE  DOCUMENTATION SECTOR
NOTE
D ADCR=(1) AVERAGE DEMAND/CAPACITY RATIO
D AOL=(U/WK) AVERAGE ORDER LEVEL
D CAPAC=(U/WK) PRODUCTION CAPACITY
D CAR=(U/WK/WK) CAPACITY ADDITION RATE
D CCDEL=(WK) CAPACITY CONSTRUCTION DELAY
D CCOST=(£/U/WK) CAPITAL COST OF CAPACITY
```

```
D CFLOW=(£/WK) CASH FLOW FROM ORDERS
D CMC=(£/WK/(U/WK)) COST OF MAINTAINING CAPACITY
D CSR=(U/WK/WK) NEW CAPACITY START RATE
D DCR=(1) DEMAND/CAPACITY RATIO
D DT=(WK) SOLUTION INTERVAL
D LENGTH=(WK) SIMULATED PERIOD
D OL=(U/WK) ORDER LEVEL
D OV=(£/U) ORDER VALUE
D PDSEC=(£/WK/(U/WK))  PRODUCT DEVELOPMENT SPENDING COEFFICIENT-
X                      AMOUNT OF AVERAGE PRODUCT SPENDING NEEDED
X                      TO SUSTAIN 1 UNIT PER WEEK OF DEMAND
D PDSFAC=(1) PROPORTION OF CASH FLOW SPENT ON PRODUCT DEVELOPMENT
D PDSM=(1/WK) PRODUCT DEVELOPMENT SPENDING EFFECT MULTIPLIER
D PDSR=(£/WK) PRODUCT DEVELOPMENT SPENDING RATE
D PLC=(WK) PHYSICAL LIFETIME OF CAPACITY
D PLTPER=(WK) PLOTTING INTERVAL
D PPDS=(£/WK) PRODUCT DEVELOPMENT SPENDING LEVEL AS PERCEIVED
X               BY CUSTOMERS
D PPDSE=(U/WK) EFFECTS OF PERCEIVED PRODUCT DEVELOPMENT SPENDING
X                IN TERMS OF MARKET DEMAND
D PRTPER=(WK) PRINTING INTERVAL
D RCOL=(U/WK/WK) RATE OF CHANGE OF ORDER LEVEL
D RLC=(U/WK/WK) RATE OF LOSS OF CAPACITY
D TADCR=(WK) AVERAGING TIME FOR DEMAND/CAPACITY RATIO
D TAOL=(WK) TIME TO AVERAGE ORDER LEVEL
D TDCR=(£/U) TABLE GIVING ORDER VALUE IN TERMS OF DEMAND/CAPACITY
X             RATIO,DCR
D TIME=(WK) TIME INTO SIMULATED PERIOD
D TPDSE=(1) TABLE FOR PRODUCT DEVELOPMENT SPENDING EFFECTS
X           AS A MULTIPLIER OF AVERAGE ORDER LEVEL
D TPDSF=(1) TABLE GIVING POLICY ON PDSFAC
D TPPDS=(WK) TIME FOR CUSTOMERS TO PERCEIVE PRODUCT DEVELOPMENT
X             SPENDING
*
*
```

B.10.7 DMC (Export) Ltd.: Present System

This is the first of the programs written for the case study described in Chapter 10. We shall discuss it in some detail and treat the other programs for the problem rather more cursorily in the ensuing sections.

A

This provides the exogeneous driving force. The first line provides DEMAND either as a step input or as a pattern derived from a table function. The options are switched on or off by the dummy variable DC1.

```
MAP
*   PROGRAM TO INVESTIGATE PRODUCTION-DISTRIBUTION SYSTEM
NOTE
NOTE FOR CHAPTER 10
NOTE
NOTE   PRESENT SYSTEM,PROGRAM NAMED CH101A-DYN,RUN CH101A-MAC
NOTE
A DEMAND.K=DC1*(75+STEP(STH,10))+(1-DC1)*TABHL(TDMD,TIME.K,
X 0,100,10)
T TDMD=75/85/120/95/110/105/100/95/90/82.5/75
C DC1=0
C STH=10
```

B

This section illustrates the way in which DYSMAP allows the modeller to extend his program without his having to worry about exact equation sequence, as he would have to in FORTRAN. The result is a program segment in which the equations may appear to be out of order. We have not re-arranged the equations to show the exact logical flow as we wish to show an actual example of a practical analysis rather than a dressed-up version.

As remarked in the text, actual sales rate ASR may be lower than demand, if the Dealers Stock, DSTOCK, is very low. This is provided for by the equations for DV1 and ASR. If DSTOCK is less than the total demand in the interval DT, recalling that DEMAND is an instantaneous rate, ASR will be reduced to the lesser of DEMAND and the Expected Delivery to Dealers Stock EDDSK.
From the ensuing value of ASR. LSALE is accumulated (at a rate of zero when ASR = DEMAND) and DSTOCK is integrated.

```
L LSALE.K=LSALE.J+DT*(DEMAND.J-ASR.JK)  LOST SALES TOTAL
N LSALE=0
A DV1.K=CLIP(1,0,DSTOCK.K,DEMAND.K*DT)   DUMMY VARIABLE
R ASR.KL=DV1.K*DEMAND.K+(1-DV1.K)*MIN(DEMAND.K,EDDSK.K)
L DSTOCK.K=DSTOCK.J+DT*(DDSK.JK-ASR.JK)   DEALERS STOCK
N DSTOCK=DTSTK*DSSR
C DTSTK=4     DEALER TARGET STOCK WEEKS
```

C

Dealers Sales Rate is smoothed and Company Sales to Dealers, CSD, is calculated as the Dealer Order Rate, DOR, or the lesser of DOR and Deliveries from the Factory, DFF. The dummy variable DV2 fulfils the same kind of purpose as DV1 and is calculated in Section E.

```
L DSSR.K=DSSR.J+(DT/DTC)(ASR.JK-DSSR.J)
C DTC=4
N DSSR=75
R CSD.KL=DV2.K*DOR.K+(1-DV2.K)*MIN(DOR.K,FDR.K)
A EDDSK.K=SMOOTH(CSD.JK,1)
R DDSK.KL=EDDSK.K
N CSD=75
```

D

The Dealer Order Rate, DOR, depends on a number of factors, one of which is that dealers like to group their orders towards the end of the accounting period (APERD, or 1 month of 4.3 weeks). DTIME is the number of weeks into the month. The DOR is in the proportions of 50 in the first 3.3 weeks to 100 in the last week. This is produced by FACTOR, which is the nominal order rate at any point in the month (refer to the example on False Dynamics due to mistakes in timing in Section 4.12). If the nominal ordering pattern was followed exactly a quantity AUX would be ordered during the month.

The Dealers' theoretical average order rate is calculated from a damped sales rate and an inventory correction.

$$\begin{matrix}\text{Basic}\\ \text{Order Rate}\end{matrix} = \alpha \times \begin{matrix}\text{Smoothed}\\ \text{Sales}\end{matrix} + (1-\alpha)\begin{matrix}\text{Dealer}\\ \text{Forecast}\end{matrix} + \frac{\text{Desired Stock} - \text{Actual Stock}}{\text{Adjustment Time}}$$

This basic rate is then distorted to give the actual rate by multiplying by a term which models the proportion of the dealers nominal monthly orders which should be being placed at the appropriate point in the month.

The Dealer Forecast DFCST is based on the demand pattern, shifted backwards by one month and biassed by a simple multiple. The desired stock, DESDSK, is a multiple of the damped sales rate.

```
NOTE
NOTE    DEALERS ORDER RATE EXPRESSED AS MONTHLY PATTERN SCALED UP BY
NOTE   DEALERS SMOOTHED SALES
NOTE
A DOR.K=((APERD*FACTOR.K)/AUX.K)*((ALPHA*DSSR.K+(1-ALPHA)*DFCST.K)+(
X1 DESDSK.K-DSTOCK.K)/TADS.K)
A TADS.K=APERD
A DESDSK.K=DTSTK*(ALPHA*DSSR.K+(1-ALPHA)*DFCST.K)
A DFCST.K=DFAC*TABHL(TDMD,(TIME.K+APERD),0,100,10)
C DFAC=1.0
A AUX.K=50*(APERD-1)+100
C ALPHA=0.7
A FACTOR.K=50+TABHL(TDOP,DTIME.K,APERD-1.1,APERD-1.0,0.1)
C APERD=4.3
T TDOP=0/50
A DTIME.K=TIME.K-OTIME.K
A OTIME.K=SAMPLE(TIME.K,APERD,0)
```

E

This group of equations models the behaviour of the DMC Sales Company. dummy variable, DV2, compares the company stock with the quantity ordered by the dealers during the interval DT, and restricts company sales to dealers, CSD, in the manner shown in the third block of equations in this section.

The inventory at the sales company, IAC, is increased by deliveries from the factory and decreased by CSD, and sampled monthly to give Recorded Company Inventory RCI.

The variable STKOUT compares with LSALE in the second block of equations. It measures the cumulative amount by which the Company was unable meet Dealers Orders at the time they were placed. These are not, in this system, backlogged for later delivery if Company Stocks, IAC, are too low to allow them to be met immediately.

```
L STKOUT.K=STKOUT.J+DT*(DOR.J-CSD.JK)
N STKOUT=0
A DV2.K=CLIP(1,0,IAC.K,DOR.K*DT)
L IAC.K=IAC.J+DT*(DFF.JK-CSD.JK)
N IAC=DCSL*CSSR
C DCSL=8
L CSSR.K=CSSR.J+(DT/TSCSR.J)(CSD.JK-CSSR.J)
N CSSR=75
A TSCSR.K=PERD
R DFF.KL=FDR.K
N FDR=75
A RCI.K=SAMPLE(IAC.K,PERD,CSSR.K*DCSL)
C PERD=4.3      DECISION-MAKING PERIODICITY
```

F

This sector calculates the amount sold by the Sales Company to the dealers in each successive month. The method is very similar to that discussed in Chapter 5.

```
NOTE
NOTE  ACCUMULATION OF ACTUAL SALES IN DECISION PERIODS
NOTE
L NSALES.K=NSALES.J+DT*CSD.JK
N NSALES=0
A DCSALES.K=NSALES.K-OSALES.K
L OSALES.K=OSALES.J+DT*RCOSLES.JK
N OSALES=0
R RCOSLES.KL=PULSE(DCSALES.K/DT,PERD,PERD)
A ACSALES.K=SAMPLE(DCSALES.K,PERD,PERD*75)
```

G

The forecasting sector models the obscure process by which DMC actually makes forecasts. There are two levels of forecast, the actual forecasts and the smoothed values.

The actual forecasts AFCAST1–4 and LTAFCST are for the next four months and a long-term view for an unspecified point in time. These are the demand pattern moved backwards by an appropriate number of PERDs, biassed, and damped by actual sales. This models the actual forecasting process.

The employed forecasts are a first-order smooth of the actual forecasts and this models the way in which the company is reluctant to abandon old forecasts and change immediately to new ones as they are issued.

The difference between ACSALES and ASALES1 is simply in their initial conditions and is intended to take account of the fact the forecasting period will span a rise in demand either from the step or the pattern inputs.

The use of the DEMAND pattern in the forecasting equations means that the results will not be perfectly accurate for the step input. This does not matter as the only purpose of the latter is to give the approximate response time of the system.

```
NOTE
NOTE  FORECASTING SECTOR
NOTE
L FCAST1.K=FCAST1.J+(DT/TFF1.J)(AFCAST1.J-FCAST1.J)
L FCAST2.K=FCAST2.J+(DT/TFF2)(AFCAST2.J-FCAST2.J)
L FCAST3.K=FCAST3.J+(DT/TFF3)(AFCAST3.J-FCAST3.J)
L FCAST4.K=FCAST4.J+(DT/TFF4)(AFCAST4.J-FCAST4.J)
L LTFCST.K=LTFCST.J+(DT/TLTF)(LTAFCST.J-LTFCST.J)
A AFCAST1.K=CFAC*TABHL(TDMD,TIME.K+PERD,0,100,10)*CALPHA*PERD+(1-CAL
X1 PHA)*ACSALES.K
A AFCAST2.K=CFAC*TABHL(TDMD,TIME.K+2*PERD,0,100,10)*CALPHA*PERD+(1-C
X1 ALPHA)*ACSALES.K
A AFCAST3.K=CFAC*TABHL(TDMD,TIME.K+3*PERD,0,100,10)*CALPHA*PERD+(1-C
X1 ALPHA)*ACSALES.K
A AFCAST4.K=CFAC*TABHL(TDMD,TIME.K+4*PERD,0,100,10)*CALPHA*PERD+(1-C
X1 ALPHA)*ASALES1.K
A LTAFCST.K=CFAC*TABHL(TDMD,TIME.K+6*PERD,0,100,10)*CALPHA*PERD+(1-C
X1 ALPHA)*ASALES1.K
C CALPHA=0
A ASALES1.K=SAMPLE(DCSALES.K,PERD,PERD*(75+STH))
N FCAST1=CFAC*PERD*75
N FCAST2=CFAC*PERD*75
N FCAST3=CFAC*PERD*75
N FCAST4=PERD*(75+STH)*CFAC
N LTFCST=PERD*(75+STH)*CFAC
C CFAC=1.0
A TFF1.K=PERD
C TFF2=6.0
C TFF3=7.0
C TFF4=8.0
C TLTF=12.0
S SUPP.K=PERD*DEMAND.K
```

H

This block calculates forecasts of inventory for each of four months. In the last case it is the forecast if nothing is ordered.

```
NOTE
NOTE  INVENTORY FORECASTS
NOTE
A FINV1.K=RCI.K-FCAST1.K+FPT1.K
A FINV2.K=FINV1.K-FCAST2.K+FPT2.K
A FINV3.K=FINV2.K-FCAST3.K+FPT3.K
A FINV4.K=FINV3.K-FCAST4.K
```

J

This is the main equation of the model and gives the quantity to be ordered from the factory by the Sales Company. FCAST4 and LTAFCST are divided by 2*PERD to give the approximate forecast average sales rate for the fourth and subsequent months. Multiplying by DCSL—the weeks of cover required—gives the amount which the Company ought to have in stock. Subtracting the expected amount gives the quantity which would have to be added to stock; and this is the purchase quantity.

```
NOTE
NOTE   PURCHASE QUANTITY FOR MONTH 3
NOTE
A PCHSE.K=SAMPLE(MAX(0,((DCSL/(2*PERD))(FCAST4.K+LTFCST.K)-FINV4.K)
X1 ),PERD,PERD*75)
```

K

This block of equations moves PCHSE forward through three months, transforming it into factory production targets. The same result could be achieved using PULSE functions. The output is the factory despatch rate FDR.

```
NOTE
NOTE   READJUSTMENT OF FACTORY PRODUCTION TARGETS
NOTE
A DCFPT3.K=MAX(0,SAMPLE(PCHSE.K,PERD,PERD*(75+STH)))
R  RCFPT3.KL=DCFPT3.K/PERD
L FPT3.K=FPT3.J+DT*(RCFPT3.JK-RCFPT2.JK)
A DCFPT2.K=MAX(0,SAMPLE(FPT3.K,PERD,PERD*(75+STH)))
R RCFPT2.KL=DCFPT2.K/PERD
N FPT3=PERD*(75+STH)
L FPT2.K=FPT2.J+DT*(RCFPT2.JK-RCFPT1.JK)
N FPT2=PERD*75
A DCFPT1.K=MAX(0,SAMPLE(FPT2.K,PERD,PERD*75))
R RCFPT1.KL=DCFPT1.K/PERD
L FPT1.K=FPT1.J+DT*(RCFPT1.JK-FDR.J)
A DFPT.K=SAMPLE(FPT1.K,PERD,PERD*75)
A FDR.K=DFPT.K/PERD
N FPT1=PERD*75
```

L

The output control sector.

```
PRINT  1)DEMAND,LSALE,ASR
PRINT  2)DSTOCK,DSSR,DDSK,DOR,DFCST
PRINT  3)CSD,STKOUT,IAC,RCI,CSSR
PRINT  4)PCHSE,DFF,FDR
PRINT  5)NSALES,DCSALES,OSALES,ACSALES,ASALES1
PRINT  6)FINV1,FINV2,FINV3,FINV4
PRINT  7)ACSALES,DV1,RCOSLES,DCSALES,ASALES1
PRINT  8)FPT1,FPT2,FPT3,AUX,SUPP
PRINT  9)RCFPT1,RCFPT2,RCFPT3
PRINT  10)OTIME,DTIME
PRINT  11)AFCAST1,AFCAST2,AFCAST3,AFCAST4,LTAFCST
PLOT   DEMAND=D(0,160)/CSD=*/DSTOCK=S,RCI=C,PCHSE=P(0,1200)/LSALE=L,STKOU
X1     T=0(0,1000)
C DT=0.1
C LENGTH=120
A PRTPER.K=1+STEP(9,10)-STEP(9,110)
C  PLTPER=1
RUN  INITIAL
```

M

These cards produce a variety of reruns, testing forecasting error, damping, sharper demand patterns, lower stock levels, etc.

```
C DCSL=4
RUN   LOW
C DCSL=8
C CFAC=1.2
C DFAC=1.2
RUN   OPTIM
C CFAC=0.8
C DFAC=0.8
RUN   PESSI
```

N

The variables are defined as follows.

```
D ACSALES=(U) SALES DURING A PERIOD
D AFCAST1=(U) ACTUAL FORECAST FOR SALES IN NEXT PERIOD
D AFCAST2=(U) ACTUAL FORECAST FOR SALES IN PERIOD 2
D AFCAST3=(U) ACTUAL FORECAST FOR SALES IN PERIOD 3
D AFCAST4=(U) ACTUAL FORECAST FOR SALES IN PERIOD 4
D ALPHA=(1) FACTOR TO REPRESENT DEALER SMOOTHING OF ACTUAL
X           SALES AND FORECASTS IN DERIVING DESIRED STOCK
D APERD=(WK) DEALERS DECISION-MAKING PERIODICITY
D ASALES1=(U) DUMMY FOR ACTUAL SALES USED TO GENERATE HIGHER
X              VALUES FOR INITIAL FORECASTS FOR POINTS BEYOND THE STEP
D ASR=(U/WK) ACTUAL SALES RATE TO PUBLIC
D AUX=(U) DUMMY VARIABLE FOR DEALERS ORDERING PATTERN
D CALPHA=(1) DAMPING CONSTANT TO REFLECT RELATIVE WEIGHTS
X            GIVEN TO TRUE FORECASTS AND AVERAGE SALES
D CFAC=(1) BIAS FACTOR FOR COMPANY FORECAST ERROR
D CSD=(U/WK) COMPANY SALES RATE TO DEALERS
D CSSR=(U/WK) COMPANY'S SMOOTHED SALES RATE
D DC1=(1) ZERO/ONE SWITCH FOR DEMAND PATTERNS
D DCFPT1=(U) DUMMY VARIABLE FOR CHANGE IN FACTORY PRODUCTION
X            TARGET IN PERIOD 1
D DCFPT2=(U) DUMMY VARIABLE FOR CHANGE IN FACTORY PRODUCTION
X            TARGET IN PERIOD 2
D DCFPT3=(U) DUMMY VARIABLE FOR CHANGE IN FACTORY PRODUCTION
X            TARGET IN PERIOD 3
D DCSALES=(U) SALES SO FAR THIS PERIOD
D DCSL=(WK) DESIRED COMPANY STOCK LEVEL AT DISRIBUTION CENTRE
D DDSK=(U/WK) ACTUAL DELIVERIES TO DEALERS STOCKS
D DEMAND=(U/WK) DEMAND FROM PUBLIC FOR PRODUCT
D DESDSK=(U) DESIRED DEALERS STOCKS
D DFAC=(1) DEALERS FORECAST BIAS FACTOR
D DFCST=(U/WK) DEALERS FORECAST ORDER RATE
D DFF=(U/WK) DELIVERIES FROM FACTORY
D DFPT=(U) DUMMY FOR FACTORY PRODUCTION TARGET IN CURRENT PERIOD.
X         USED TO MODEL DESPATCHES
D DOR=(U/WK) DEALER ORDER RATE
D DSSR=(U/WK) DEALERS SMOOTHED SALES RATE
D DSTOCK=(U) DEALERS STOCK
D DT=(WK) SOLUTION INTERVAL
D DTC=(WK) DEALERS TIME CONSTANT FOR SMOOTHING SALES
D DTIME=(WK) TIME SO FAR INTO A DEALERS' PLANNING PERIOD
D DTSTK=(WK) DEALERS TARGET STOCK COVER
D DV1=(1) ZERO-ONE SWITCH FOR OUT OF STOCK POSITION AT DEALERS
D DV2=(1) ZERO-ONE SWITCH TO INDICATE STOCK OUT AT
X         COMPANY DISTRIBUTION CENTRE
D EDDSK=(U/WK) EXPECTED RATE OF DELIVERY TO DEALERS STOCK
D FACTOR=(U/WK) BASE VALUE FOR PROPORTIONS IN DEALERS ORDERING
X               PATTERN
D FCAST1=(U) FORECAST ACTUALLY USED FOR SALES DURING PERIOD 1
D FCAST2=(U) FORECAST ACTUALLY USED FOR SALES DURING PERIOD 2
D FCAST3=(U) FORECAST ACTUALLY USED FOR SALES DURING PERIOD 3
D FCAST4=(U) FORECAST ACTUALLY USED FOR SALES DURING PERIOD 4
D FDR=(U/WK) FACTORY DESPATCH RATE
D FINV1=(U) FORECAST INVENTORY AT COMPANY DISTRIBUTION CENTRE
X           AT END OF PERIOD 1
D FINV2=(U) FORECAST INVENTORY AT COMPANY DISTRIBUTION CENTRE
X           AT END OF PERIOD 2
D FINV3=(U) FORECAST INVENTORY AT COMPANY DISTRIBUTION CENTRE
X           AT END OF PERIOD 3
```

```
D FINV4=(U) FORECAST INVENTORY AT COMPANY DISTRIBUTION CENTRE
X              AT END OF PERIOD 4
D FPT1=(U) FACTORY PRODUCTION TARGET FOR DESPATCH DURING PERIOD 1
D FPT2=(U) FACTORY PRODUCTION TARGET FOR DESPATCH DURING PERIOD 2
D FPT3=(U) FACTORY PRODUCTION TARGET FOR DESPATCH DURING PERIOD 3
D IAC=(U) INVENTORY AT COMPANY DISTRIBUTION CENTRE
D LENGTH=(WK) LENGTH OF SIMULATED PERIOD
D LSALE=(U) CUMULATIVE LOST SALES OVER PRODUCT LIFE CYCLE
D LTAFCST=(U) ACTUAL FORECAST OF LONG-TERM DEMAND.IE FORECAST
X                FOR A TYPICAL PERIOD IN THE LONG-TERM FUTURE
D LTFCST=(U) LONG-TERM FORECAST AS USED,DAMPED BY PREVIOUS
X              FORECASTS
D NSALES=(U) CUMULATIVE SALES TO DEALERS
D OSALES=(U) CUMULATIVE SALES TO DEALERS UP TO END OF LAST
X              COMPLETE PERIOD
D OTIME=(WK) TIME UP TO END OF LAST DEALER'S PLANNING PERIOD
D PCHSE=(U) QUANTITY TO BE ORDERED FROM FACTORY FOR DESPATCH
X             3 PERIODS HENCE
D PERD=(WK) LENGTH OF COMPANY'S DECISION-MAKING PERIOD
D PLTPER=(WK) PLOTTING INTERVAL
D PRTPER=(WK) PRINTING INTERVAL
D RCFPT1=(U/WK) RATE OF CHANGE OF FACTORY PRODUCTION TARGET
X                  IN PERIOD 1.A DUMMY VARIABLE USED TO ENSURE THAT THE
X                  FACTORY PRODUCTION TARGET MOVES PROGRESSIVELY FORWARD
X                  AT THE END OF EACH PLANNING PERIOD SO THAT PCHSE
X                  EVENTUALLY GOVERNS THE FACTORY DESPATCH RATE
D RCFPT2=(U/WK) RATE OF CHANGE OF FACTORY PRODUCTION TARGET
X                  IN PERIOD 2.A DUMMY VARIABLE USED TO ENSURE THAT THE
X                  FACTORY PRODUCTION TARGET MOVES PROGRESSIVELY FORWARD
X                  AT THE END OF EACH PLANNING PERIOD SO THAT PCHSE
X                  EVENTUALLY GOVERNS THE FACTORY DESPATCH RATE
D RCFPT3=(U/WK) RATE OF CHANGE OF FACTORY PRODUCTION TARGET
X                  IN PERIOD 3.A DUMMY VARIABLE USED TO ENSURE THAT THE
X                  FACTORY PRODUCTION TARGET MOVES PROGRESSIVELY FORWARD
X                  AT THE END OF EACH PLANNING PERIOD SO THAT PCHSE
X                  EVENTUALLY GOVERNS THE FACTORY DESPATCH RATE
D RCI=(U) INVENTORY AT FACTORY DISTRIBUTION CENTRE AS RECORDED
X           AT THE END OF THE LAST PLANNING PERIOD
D RCOSLES=(U/WK) RATE OF CHANGE OF OLD SALES'.A DUMMY VARIABLE TO ENSURE
X                   THAT OLD SALES,OSALES,IS UPDATED AT THE END OF EACH
X                   PLANNING PERIOD
D STH=(U/WK) STEP HEIGHT IN DEMAND INPUTS
D STKOUT=(U) CUMULATIVE VALUE OF NET BACKLOG AT COMPANY
X              DISTRIBUTION CENTRE'.IE A MEASURE OF PAST INABILITIES
X              TO MEET DEALER ORDERS AS EXPRESSED BY THE DIFFERENCE
X              BETWEEN DOR AND CSD
D SUPP=(U) SUPPLEMENTARY VARIABLE FOR TOTAL DEMAND IN A PERIOD
D TADS=(WK) TIME TO ADJUST DEALERS STOCKS
D TDMD=(U/WK) TABLE FOR DEMAND PATTERN
D TDOP=(U/WK) TABLE TO ESTABLISH DEALERS NORMAL ORDERING
X                PATTERN OF BUNCHING ORDERS AT THE END OF A PERIOD TO GET
X                LONGER CREDIT
D TFF1=(WK) TIME TO SMOOTH FORECAST FOR PERIOD 1'.A MEASURE OF
X             WILLINGNESS TO RESPOND TO NEW FORECASTS BY ABANDONING
X             THE EXISTING FORECAST FOR THAT PERIOD IN FAVOUR OF THE NEW
X             ONE
D TFF2=(WK) TIME TO SMOOTH FORECAST FOR PERIOD 2'.A MEASURE OF
X             WILLINGNESS TO RESPOND TO NEW FORECASTS BY ABANDONING
X             THE EXISTING FORECAST FOR THAT PERIOD IN FAVOUR OF THE NEW
X             ONE
D TFF3=(WK) TIME TO SMOOTH FORECAST FOR PERIOD 3'.A MEASURE OF
X             WILLINGNESS TO RESPOND TO NEW FORECASTS BY ABANDONING
X             THE EXISTING FORECAST FOR THAT PERIOD IN FAVOUR OF THE NEW
X             ONE
D TFF4=(WK) TIME TO SMOOTH FORECAST FOR PERIOD 4'.A MEASURE OF
X             WILLINGNESS TO RESPOND TO NEW FORECASTS BY ABANDONING
X             THE EXISTING FORECAST FOR THAT PERIOD IN FAVOUR OF THE NEW
X             ONE
D TIME=(WK) SIMULATED TIME IN THE MODEL
D TLTF=(WK) SMOOTHING TIME ANALOGOUS TO TFF1 BUT REFERRING TO THE LONG
X             -TERM FORECAST
D TSCSR=(WK) SMOOTHING TIME FOR COMPANY SMOOTHED SALES RATE
*
```

B.10.8 DMC (Export) Ltd.: Present System without Dealers Stocks

This program is identical with B.10.7 except that the dealers stocks have been eliminated. Sales take place direct from company stocks in the sense that the dealer simply passes DEMAND, which would be met from his own stocks in B.10.7, direct to the company stock where it is either met, or not, depending on company inventory IAC.

Notice that in line 16 we have violated our own rule about not using rate-dependent rates and we have employed DFF.KL in the equation for ASR instead of the formulation involving EDDSK.K which was used in the foregoing program. This is done deliberately in an attempt to provoke the reader to do some simulation experiments to add to his feel for when the strict rules for equation formulation can be broken.

After line 16, B.10.8 is identical with lines 19 to 144 of B.10.7 except that CALPHA, which is 0 in line 84 of B.10.7 is 0.3 in B.10.8. Experiment!

```
 0 DOC
 1 *  PROGRAM TO INVESTIGATE PRODUCTION-DISTRIBUTION SYSTEM
 2 NOTE
 3 NOTE   MODEL NUMBER ONE   PRESENT SYSTEM
 4 NOTE  VERSION WITHOUT DEALERS STOCKS
 5 NOTE
 6 NOTE PROGRAM NAMED CH10IB-DYN
 7 NOTE
 8 A DEMAND.K=DC1*(75+STEP(STH,10))+(1-DC1)*TABHL(TDMD,TIME.K,
 9 X 0,100,10)
10 T TDMD=75/85/120/95/110/105/100/95/90/82.5/75
11 C DC1=1
12 C STH=10
13 L LSALE.K=LSALE.J+(DT)(DEMAND.J-ASR.JK)     LOST
14 N LSALE=0
15 A DV1.K=CLIP(1,0,IAC.K,DEMAND.K*DT)      DUMMY VARIABLE
16 R ASR.KL=DV1.K*DEMAND.K+(1-DV1.K)*MIN(DEMAND.K,DFF.KL)   ACT SLS RTE
```

B.10.11 A Hypothetical System for DMC

The purpose of this model was, it will be recalled, to examine the possibilities of redesigning the logistical system with a view to finding the dynamic characteristics of the new system.

The sectors of the model are listed below.

Note that some of the parameters now have new values which accord with the way DMC said they were going to run this system. The reader may wish to see whether this system would perform better if different parameter values were used.

The variables are documented in B.10.7 or explained in the text.

A

This sector drives the model, calculates actual sales and monitors stock. As described in the text there is no dealer inventory, the relevant physical stock being the Inventory at the Company Distribution Centre, ICDC.

```
0 DOC
1 * PROGRAM TO INVESTIGATE PRODUCTION-DISTRIBUTION SYSTEM
2 NOTE
3 NOTE  MODEL NUMBER TWO - MANUFACTURING FOR COMPANY-HELD STOCKS
4 NOTE   PROPOSED SYSTEM WITH INVENTORY-CONTROLLED ORDERING AT FACTORY
5 NOTE
6 NOTE MODEL CH10IIA-DYN
7 NOTE
8 NOTE  MARKET DEMAND SECTOR
9 NOTE
```

```
A DEMAND.K=DC1*(75+STEP(STH,10))+(1-DC1)*TABHL(TDMD,TIME.K,
X 0,100,10)
C DC1=1.0
T TDMD=75/85/120/95/110/105/100/95/90/82.5/75
C STH=10
L LSALE.K=LSALE.J+(DT)(DEMAND.J-ASR.JK)        LOST SALES TOTAL
N LSALE=0
A DV1.K=CLIP(1,0,ICDC.K,DEMAND.K*DT)            DUMMY VARIABLE
R ASR.KL=DV1.K*DEMAND.K+(1-DV1.K)*MIN(DEMAND.K,DFF.KL)  ACT SLS RTE
L ICDC.K=ICDC.J+(DT)(DFF.JK-ASR.JK)   INV AT COMP DISTRIB CENTRE
N ICDC=SSR*(CDIC-TACDCI)
```

B

The Actual Sales Rate is used to calculate smoothed sales and a delayed version of ASR, OARSC, is used to compute the actual monthly sales total, ACSALES.

```
R OARSC.KL=DELAY3(ASR.JK,DTO)        ORDERS ARRIVING AT SALES CO
C DTO=2                              DELAY IN TRANSMITTING ORDERS.
N ASR=75
C CDIC=8
L SSR.K=SSR.J+(DT/TSSR)(ASR.JK-SSR.J)    SMOOTHED SALES RATE
N SSR=75
C TSSR=4.3                               SMOOTHING CONSTANT
L OASC.K=OASC.J+DT*OARSC.JK
N OASC=0
A DCSALES.K=OASC.K-OSALES.K     CURRENT MONTHS SALES TOTAL
L OSALES.K=OSALES.J+(DT)(RCOSLES.JK)   OLD SALES TOTAL
N OSALES=0
R RCOSLES.KL=PULSE(DCSALES.K/DT,PERD,PERD)
A ACSALES.K=SAMPLE(DCSALES.K,PERD,322.5)  ACT SALES FOR FCST
```

C

The Factory Despatch Rate, FDR, depends on the stocks held at the factory awaiting shipment to the sales company, IASC, and the orders placed by the sales company, DVP. The sales company fixes DPV on the basis of the smoothed sales rate and a stock correction. The desired stock, DIACDC, is modelled as a number of weeks cover of the weighted smoothed sales rate and the forecast.

```
NOTE
NOTE   FACTORY SHIPMENT SECTOR
NOTE
R FDR.KL=MIN(DFDR1.K,DFDR2.K) FACTORY DEL RATE
A DFDR2.K=IASC.K/DT
A DPV.K=((DIACDC.K-ICDC.K)/TACDCI)
A DFDR1.K=MAX(0,DPV.K)
C TACDCI=3
A DIACDC.K=CDIC*(ALPHA*SSR.K+(1-ALPHA)*DEMAND.K) DESD INV AT CO DC
C ALPHA=0.7
R DFF.KL=DELAY3(DDFF.JK,1.5)        DELIVERIES FROM FAC TO DIST CEN
R DDFF.KL=DELAY3(FDR.JK,1.5)        DUMMY FOR 6TH ORDER DELAY
```

D

Production Start Rate Desired, PSRD, is geared to the correction of factory inventory, IASC, plus a term for the Expected Production Rate, EPR.

Production Start Rate, PSR, is a level to represent the smoothing of PSRD. The smoothing time, TAPR, reflects the willingness of the factory to change its activity from that which had been planned. Factory Completion Rate, FCR, is a delayed version of PSR.

```
N FDR=PSR
N DDFF=FDR
C TAPR=4.3                              TIME TO ADJUST PRODN RATE
L PSR.K=PSR.J+(DT/TAPR)(PSRD.J-PSR.J)
N PSR=75
A PSRD.K=((DIASC.K-IASC.K)/TAFI)+EPR.K
A DIASC.K=FIC*SSR.K                    DESIRED INV WAITING SHIPMENT
C TAFI=4
L IASC.K=IASC.J+(DT)(FCR.JK-FDR.JK)  INV WAITING SHIP TO COMP DIST C
N IASC=SSR*FIC
C FIC=6        FACTORY INVENTORY COVER
R FCR.KL=DELAY3(PSR.K,3)               FACTORY COMPLETION RATE
C PERD=4.3 DECISION-MAKING PERIODICITY
```

E

The forecasting sector is the same as before.

```
NOTE
NOTE   FORECASTING AND PRODUCTION PLANNING SECTOR
NOTE
L FCAST1.K=FCAST1.J+(DT/TFF1.J)(AFCAST1.J-FCAST1.J)   PERIOD 1
L FCAST2.K=FCAST2.J+(DT/TFF2)(AFCAST2.J-FCAST2.J)     PERIOD 2
L FCAST3.K=FCAST3.J+(DT/TFF3)(AFCAST3.J-FCAST3.J)     PERIOD 3
L FCAST4.K=FCAST4.J+(DT/TFF4)(AFCAST4.J-FCAST4.J)     PERIOD 4
L LTFCST.K=LTFCST.J+(DT/TLTF)(LTAFCST.J-LTFCST.J)     LONG TERM
A AFCAST1.K=CFAC*TABHL(TDMD,TIME.K+PERD,0,100,10)*CALPHA*PERD+(1-CAL
X1 PHA)*ACSALES.K
A AFCAST2.K=CFAC*TABHL(TDMD,TIME.K+2*PERD,0,100,10)*CALPHA*PERD+(1-C
X1 ALPHA)*ACSALES.K
A AFCAST3.K=CFAC*TABHL(TDMD,TIME.K+3*PERD,0,100,10)*CALPHA*PERD+(1-C
X1 ALPHA)*ACSALES.K
A AFCAST4.K=CFAC*TABHL(TDMD,TIME.K+4*PERD,0,100,10)*CALPHA*PERD+(1-C
X1 ALPHA)*ASALES1.K
A LTAFCST.K=CFAC*TABHL(TDMD,TIME.K+6*PERD,0,100,10)*CALPHA*PERD+(1-C
X1 ALPHA)*ASALES1.K
C CALPHA=0.9
A ASALES1.K=SAMPLE(DCSALES.K,PERD,4.3*(75+STH))
N FCAST1=322.5*CFAC
N FCAST2=322.5*CFAC
N FCAST3=322.5*CFAC
N FCAST4=4.3*(75+STH)*CFAC
N LTFCST=4.3*(75+STH)*CFAC
C CFAC=1.0
A TFF1.K=4.3
C TFF2=6.0
C TFF3=7.0
C TFF4=8.0
C TLTF=12.0
S SUPP.K=PERD*DEMAND.K
```

F

The expected production rate is found by moving FCAST4 progressively forward, in a manner very similar to that used for PCHSE in Model A and B.

```
NOTE
NOTE   ADJUSTMENT OF FACTORY PRODUCTION TARGETS.
NOTE
A DCFPT3.K=MAX(0,SAMPLE(FCAST4.K,PERD,365.5))
R RCFPT3.KL=DCFPT3.K/PERD
L FPT3.K=FPT3.J+(DT)(RCFPT3.JK-RCFPT2.JK)
N FPT3=4.3*(75+STH)
A DCFPT2.K=MAX(0,SAMPLE(FPT3.K,PERD,PERD*(75+STH)))
R RCFPT2.KL=DCFPT2.K/PERD
L FPT2.K=FPT2.J+(DT)(RCFPT2.JK-RCFPT1.JK)
N FPT2=322.5
A DCFPT1.K=SAMPLE(FPT2.K,PERD,322.5)
R RCFPT1.KL=DCFPT1.K/PERD
L FPT1.K=FPT1.J+(DT)(RCFPT1.JK-EPR.J)
N FPT1=322.5
A DFPT1.K=SAMPLE(FPT1.K,PERD,322.5)
A EPR.K=DFPT1.K/PERD                  EXPECTED PRODUCTION RATE.
```

G

The output control sector:

```
PRINT  1)DEMAND,LSALE,ASR,SSR,UASC
PRINT  2)ICDC,DIACDC,IASC,DIASC
PRINT  3)FDR,DFF
PRINT  4)PSR,PSRD,EPR
PRINT  5)FCAST1,FCAST2,FCAST3,FCAST4,LTFCST,DEMAND
PRINT  6)FPT1,FPT2,FPT3
PRINT  7)ACSALES,DV1,RCOSLES,DCSALES,ASALES1
PRINT  8)FPT1,FPT2,FPT3,SUPP
PRINT  9)AFCAST1,AFCAST2,AFCAST3,AFCAST4,LTAFCST
PLOT   DEMAND=D,ASR=A,PSR=P(0,160)/LSALE=L(0,1000)/ICDC=S,IASC=C(0,1200)/
PLOT   FCAST1=1,FCAST2=2,FCAST3=3,FCAST4=4,LTFCST=L,SUPP=D(0,800)/
C LENGTH=100
C DT=0.1
C PLTPER=1
A PRTPER.K=1+STEP(4,21)+STEP(45,31)-STEP(45,81)
```

H

Some suggested reruns:

```
RUN    STEP
C CFAC=1.20
RUN    OPTIM
C DC1=0
C CFAC=1.0
RUN    PTNA
C CFAC=1.20
RUN    PTNO
C CFAC=0.8
RUN PESSI
C CFAC=1.0
C CDIC=4.0
RUN    PTNB
C TACDCI=5
C TAPR=2
RUN    PTND
+
```

B.10.13 Production–Distribution System:

Present System with Smoother Policies

```
MAP
*   PROGRAM TO INVESTIGATE PRODUCTION-DISTRIBUTION SYSTEM
NOTE
NOTE FOR CHAPTER 10
NOTE
NOTE   PRESENT SYSTEM,PROGRAM NAMED CH10IMP-DYN,RUN CH10IMP-MAC
NOTE
NOTE VERSION WITH IMPROVED ORDERING
NOTE
NOTE
A DEMAND.K=DC1*(75+STEP(STH,10))+DC2*TABHL(TDMD,TIME.K,0,100,10)
T TDMD=75/85/120/95/110/105/100/95/90/82.5/75
C DC1=0
C DC2=1
C STH=10
L LSALE.K=LSALE.J+DT*(DEMAND.J-ASR.JK)   LOST SALES TOTAL
N LSALE=0
A DV1.K=CLIP(1,0,DSTOCK.K,DEMAND.K*DT)    DUMMY VARIABLE
R ASR.KL=DV1.K*DEMAND.K+(1-DV1.K)*MIN(DEMAND.K,EDDSK.K)
L DSTOCK.K=DSTOCK.J+DT*(DDSK.JK-ASR.JK)    DEALERS STOCK
N DSTOCK=DTSTK*DSSR
C DTSTK=4     DEALER TARGET STOCK WEEKS
L DSSR.K=DSSR.J+(DT/DTC)(ASR.JK-DSSR.J)
C DTC=4
N DSSR=75
```

```
R CSD.KL=DV2.K*DOR.K+(1-DV2.K)*MIN(DOR.K,FDR.K)
A EDDSK.K=SMOOTH(CSD.JK,1)
R DDSK.KL=EDDSK.K
N CSD=75
NOTE
NOTE   DEALERS ORDER RATE EXPRESSED AS MONTHLY PATTERN SCALED UP BY
NOTE  DEALERS SMOOTHED SALES
NOTE
A DOR.K=((APERD*FACTOR.K)/AUX.K)*((ALPHA*DSSR.K+(1-ALPHA)*DFCST.K)+(
X1 DESDSK.K-DSTOCK.K)/TADS.K)
A TADS.K=APERD
A DESDSK.K=DTSTK*(ALPHA*DSSR.K+(1-ALPHA)*DFCST.K)
A DFCST.K=DFAC*TABHL(TDMD,(TIME.K+APERD),0,100,10)
C DFAC=1.0
A AUX.K=50*(APERD-1)+100
C ALPHA=0.7
A FACTOR.K=50+TABHL(TDOP,DTIME.K,APERD-1.1,APERD-1.0,0.1)
C APERD=4.3
T TDOP=0/50
A DTIME.K=TIME.K-OTIME.K
A OTIME.K=SAMPLE(TIME.K,APERD,0)
L STKOUT.K=STKOUT.J+DT*(DOR.J-CSD.JK)
N STKOUT=0
A DV2.K=CLIP(1,0,IAC.K,DOR.K*DT)
L IAC.K=IAC.J+DT*(DFF.JK-CSD.JK)
N IAC=DCSL*CSSR
C DCSL=8
L CSSR.K=CSSR.J+(DT/TSCSR.J)(CSD.JK-CSSR.J)
N CSSR=75
A TSCSR.K=PERD
R DFF.KL=FDR.K
N FDR=75
A RCI.K=SAMPLE(IAC.K,PERD,CSSR.K*DCSL)
C PERD=4.3 DECISION-MAKING PERIODICITY
```

This is the same as in Section B.10.7.

```
NOTE
NOTE  ACCUMULATION OF ACTUAL SALES IN DECISION PERIODS
NOTE
L NSALES.K=NSALES.J+DT*CSD.JK
N NSALES=0
A DCSALES.K=NSALES.K-OSALES.K
L OSALES.K=OSALES.J+DT*RCOSLES.JK
N OSALES=0
R RCOSLES.KL=PULSE(DCSALES.K/DT,PERD,PERD)
A ACSALES.K=SAMPLE(DCSALES.K,PERD,PERD*75)
NOTE
NOTE  FORECASTING SECTOR
NOTE
L FCAST1.K=FCAST1.J+(DT/TFF1.J)(AFCAST1.J-FCAST1.J)
L FCAST2.K=FCAST2.J+(DT/TFF2)(AFCAST2.J-FCAST2.J)
L FCAST3.K=FCAST3.J+(DT/TFF3)(AFCAST3.J-FCAST3.J)
L FCAST4.K=FCAST4.J+(DT/TFF4)(AFCAST4.J-FCAST4.J)
L LTFCST.K=LTFCST.J+(DT/TLTF)(LTAFCST.J-LTFCST.J)
A AFCAST1.K=CFAC*TABHL(TDMD,TIME.K+PERD,0,100,10)*CALPHA*PERD+(1-CAL
X1 PHA)*ACSALES.K
A AFCAST2.K=CFAC*TABHL(TDMD,TIME.K+2*PERD,0,100,10)*CALPHA*PERD+(1-C
X1 ALPHA)*ACSALES.K
A AFCAST3.K=CFAC*TABHL(TDMD,TIME.K+3*PERD,0,100,10)*CALPHA*PERD+(1-C
X1 ALPHA)*ACSALES.K
A AFCAST4.K=CFAC*TABHL(TDMD,TIME.K+4*PERD,0,100,10)*CALPHA*PERD+(1-C
X1 ALPHA)*ASALES1.K
A LTAFCST.K=CFAC*TABHL(TDMD,TIME.K+6*PERD,0,100,10)*CALPHA*PERD+(1-C
X1 ALPHA)*ASALES1.K
C CALPHA=0.3
A ASALES1.K=SAMPLE(DCSALES.K,PERD,PERD*(75+STH))
N FCAST1=CFAC*PERD*75
N FCAST2=CFAC*PERD*75
N FCAST3=CFAC*PERD*75
N FCAST4=PERD*(75+STH)*CFAC
N LTFCST=PERD*(75+STH)*CFAC
C CFAC=1.0
A TFF1.K=PERD
C TFF2=6.0
C TFF3=7.0
C TFF4=8.0
```

```
C TLTF=12.0
S SUPP.K=PERD*DEMAND.K
NOTE
NOTE   INVENTORY FORECASTS
NOTE
A FINV1.K=RCI.K-FCAST1.K+FPT1.K
A FINV2.K=FINV1.K-FCAST2.K+FPT2.K
A FINV3.K=FINV2.K-FCAST3.K+FPT3.K
A FINV4.K=FINV3.K-FCAST4.K
```

These equations are all the same as those in B.10.7.

```
NOTE
NOTE   PURCHASE QUANTITY FOR MONTH 3
NOTE
A PCHSE.K=SAMPLE(MAX(0,(LTAFCST.K+(((DCSL/PERD)*LTAFCST.K-FINV3.K
X )/TASCI))),PERD,PERD*75)
C TASCI=16
```

This section contains the most salient differences. The equation for PCHSE is as discussed in Chapter 10.

```
NOTE
NOTE   READJUSTMENT OF FACTORY PRODUCTION TARGETS
NOTE
A DCFPT3.K=MAX(0,SAMPLE(PCHSE.K,PERD,PERD*(75+STH)))
R  RCFPT3.KL=DCFPT3.K/PERD
L FPT3.K=FPT3.J+DT*(RCFPT3.JK-RCFPT2.JK)
A DCFPT2.K=MAX(0,SAMPLE(FPT3.K,PERD,PERD*(75+STH)))
R RCFPT2.KL=DCFPT2.K/PERD
N FPT3=PERD*(75+STH)
L FPT2.K=FPT2.J+DT*(RCFPT2.JK-RCFPT1.JK)
N FPT2=PERD*75
A DCFPT1.K=MAX(0,SAMPLE(FPT2.K,PERD,PERD*75))
R RCFPT1.KL=DCFPT1.K/PERD
L FPT1.K=FPT1.J+DT*(RCFPT1.JK-FDR.J)
A DFPT.K=SAMPLE(FPT1.K,PERD,PERD*75)
A FDR.K=DFPT.K/PERD
N FPT1=PERD*75
PRINT 1)DEMAND,LSALE,ASR
PRINT 2)DSTOCK,DSSR,DDSK,DOR,DFCST
PRINT 3)CSD,STKOUT,IAC,RCI,CSSR
PRINT 4)PCHSE,DFF,FDR
PRINT 5)NSALES,DCSALES,OSALES,ACSALES,ASALES1
PRINT 6)FINV1,FINV2,FINV3,FINV4
PRINT 7)ACSALES,DV1,RCOSLES,DCSALES,ASALES1
PRINT 8)FPT1,FPT2,FPT3,AUX,SUPP
PRINT 9)RCFPT1,RCFPT2,RCFPT3
PRINT 10)OTIME,DTIME
PRINT 11)AFCAST1,AFCAST2,AFCAST3,AFCAST4,LTAFCST
PLOT   DEMAND=D(0,160)/CSD=*/DSTOCK=S,RCI=C,PCHSE=P(0,1200)/LSALE=L,STKOU
X1     T=0(0,1000)
C DT=0.1
C LENGTH=120
A PRTPER.K=1+STEP(9,10)-STEP(9,110)
C  PLTPER=1
RUN  INITIAL
```

B.11 Design of an Integrated Oil Supply System

This example has been included to show just how complicated a program can get, and to illustrate a lot of detailed points about DYSMAP equations.

The reader should review the introduction to Chapter 11, and recall that this is a model for a system design study, rather than a disguised description of what happens in a real oil company. There are a few points to note:

1. The program follows DYSMAP conventions but can be run under DYNAMO by

dropping the documentation sector or replacing the D and X cards in the documentation by NOTE cards. A DYSMAP map for the program is an aid to understanding. An ICL magnetic tape for the listing can be obtained from the author at cost.

2. The program went through many months of development and elaboration and the alert reader will be able to see the signs of this.
3. The model described here is very complicated because the design exercise involved making sure that all the relevant aspects had been considered. The model can, of course, be used to test just how significant factors which appear to be important really are when viewed in the context of an interacting system. The student will derive considerable benefit from seeing how far the model can be simplified before appreciable differences are observed in the dynamics of the important variables.

   Despite the suggestion in Chapter 3 that we should start with a simple model and gradually elaborate it, it often happens in practice that the need to satisfy a 'client' that all aspects have been allowed for forces one to build a more complicated than is strictly necessary. It is then a useful discipline to reverse the modelling process and proceed to strip down the model until the really salient features are all that remain.
4. The measure of performance used, CUMEXP, is, as discussed in Chapter 11, very simple.
5. The serious student should realise that what is presented here as a fait accompli is the result of a lot of work, much of which has not been mentioned in the hope that he will think of several times more questions than are answered here or in Chapter 11.
6. In sensitivity testing this model, recall the difference between modes of behaviour and particular values, especially in regard to the smoothest system.

A

The model calls for forecasts of certain variables to be carried forward 12 months for later comparison with actuals. The month does this by using the method of Section 4.8, *calculation of two or more old values*, as mentioned in Section 4.9.

```
* DYNAMICS OF THE SUPPLY SYSTEM AND TANKER OPERATIONS
NOTE
NOTE
NOTE
NOTE  CURRENT VALUE OF CIF FORECAST
NOTE
L CIF11.K=CIF11.J+DT*(PULSE((CIFFC.J-CIF11.J)/DT,PD,PD))
L CIF10.K=CIF10.J+DT*(PULSE((CIF11.J-CIF10.J)/DT,PD,PD))
L CIF9.K=CIF9.J+DT*(PULSE((CIF10.J-CIF9.J)/DT,PD,PD))
L CIF8.K=CIF8.J+DT*(PULSE((CIF9.J-CIF8.J)/DT,PD,PD))
L CIF7.K=CIF7.J+DT*(PULSE((CIF8.J-CIF7.J)/DT,PD,PD))
L CIF6.K=CIF6.J+DT*(PULSE((CIF7.J-CIF6.J)/DT,PD,PD))
L CIF5.K=CIF5.J+DT*(PULSE((CIF6.J-CIF5.J)/DT,PD,PD))
L CIF4.K=CIF4.J+DT*(PULSE((CIF5.J-CIF4.J)/DT,PD,PD))
L CIF3.K=CIF3.J+DT*(PULSE((CIF4.J-CIF3.J)/DT,PD,PD))
L CIF2.K=CIF2.J+DT*(PULSE((CIF3.J-CIF2.J)/DT,PD,PD))
L CIF1.K=CIF1.J+DT*(PULSE((CIF2.J-CIF1.J)/DT,PD,PD))
L CIFVAL.K=CIFVAL.J+DT*(PULSE((CIF1.J-CIFVAL.J)/DT,PD,PD))
N CIF11=CIFIV
N CIF10=CIFIV
N CIF9=CIFIV
```

```
21 N CIF8=CIFIV
22 N CIF7=CIFIV
23 N CIF6=CIFIV
24 N CIF5=CIFIV
25 N CIF4=CIFIV
26 N CIF3=CIFIV
27 N CIF2=CIFIV
28 N CIF1=CIFIV
29 N CIFVAL=CIFIV
30 C PD=1
31 NOTE
32 NOTE  CURRENT VALUE OF FOB FORECAST
33 NOTE
34 L FOB11.K=FOB11.J+DT*(PULSE((FOBFC.J-FOB11.J)/DT,PD,PD))
35 L FOB10.K=FOB10.J+DT*(PULSE((FOB11.J-FOB10.J)/DT,PD,PD))
36 L FOB9.K=FOB9.J+DT*(PULSE((FOB10.J-FOB9.J)/DT,PD,PD))
37 L FOB8.K=FOB8.J+DT*(PULSE((FOB9.J-FOB8.J)/DT,PD,PD))
38 L FOB7.K=FOB7.J+DT*(PULSE((FOB8.J-FOB7.J)/DT,PD,PD))
39 L FOB6.K=FOB6.J+DT*(PULSE((FOB7.J-FOB6.J)/DT,PD,PD))
40 L FOB5.K=FOB5.J+DT*(PULSE((FOB6.J-FOB5.J)/DT,PD,PD))
41 L FOB4.K=FOB4.J+DT*(PULSE((FOB5.J-FOB4.J)/DT,PD,PD))
42 L FOB3.K=FOB3.J+DT*(PULSE((FOB4.J-FOB3.J)/DT,PD,PD))
43 L FOB2.K=FOB2.J+DT*(PULSE((FOB3.J-FOB2.J)/DT,PD,PD))
44 L FOB1.K=FOB1.J+DT*(PULSE((FOB2.J-FOB1.J)/DT,PD,PD))
45 L FOBVAL.K=FOBVAL.J+DT*(PULSE((FOB1.J-FOBVAL.J)/DT,PD,PD))
46 N FOB11=FOBIV
47 N FOB10=FOBIV
48 N FOB9=FOBIV
49 N FOB8=FOBIV
50 N FOB7=FOBIV
51 N FOB6=FOBIV
52 N FOB5=FOBIV
53 N FOB4=FOBIV
54 N FOB3=FOBIV
55 N FOB2=FOBIV
56 N FOB1=FOBIV
57 N FOBVAL=FOBIV
```

B

This sector generates an oil demand in Western Europe by exponential growth. This is based on each 1% increase in GNP having, in the past, produced 2.064% growth in oil consumption (ODGRF). This is then subjected to a variable pattern of GNP growth (TGRAT). RAT1 converts arbitrary GNP units to oil product consumption in tons per years.

The Company's demand comes from a 20.5% market share, MMS, and is seasonal with 11% amplitude (line 68). The seasonality is generated by dummy variable DIME=TIME + 3 to ensure peak seasonality at TIME=0.

Undoubtedly the greatest oversimplification in this case exercise is the assumption of a constant amplitude of seasonality. Thus although the general trend of demand may reasonably be modelled as deriving from the growth of GNP it is a fact of life in all the energy industries that the actual seasonal weather pattern is highly variable and has a very significant effect on actual demand.

The principal exercise for the reader is to adapt the design for a smooth system with constant seasonal amplitude to make it robust with respect to variable seasonality.

```
58 NOTE  DEMAND AND GNP SECTOR
59 NOTE
60 L ODLV.K=ODLV.J+DT*ODGR.JK                         OIL DEMAND LEVEL
61 N ODLV=1
62 R ODGR.KL=(BRAT.K*ODGRF*ODLV.K)/12                 OIL DEMAND GROWTH RATE
63 A MDWE.K=(MMS.K*ODLV.K*RAT1*(1+SEAMP*SIN(6.283*DIME.K/12))/12)
64 C ODGRF=2.064
65 C GRIN=0.048
```

```
A BRAT.K=TABHL(TGRAT,TIME.K,0,120,6)/100
L GNPLEV.K=GNPLEV.J+DT*GNPGR.JK                          LEVEL OF GNP
N GNPLEV=1
R GNPGR.KL=BRAT.K*GNPLEV.K/12                            GNP GROWTH RATE
C SEAMP=0.11
T TGRAT=3.5/4.8/6.1/6.6/5.5/4.7/3.9/2.9/1.9/0.7/1.9/4.0/6.1/6.6/5.5
X1      /4.7/3.9/2.9/1.9/1.3/0.7                         GROWTH RATE TABLE
C RAT1=310E06                                            OIL DEMAND FROM GNP
A MMS.K=MIMS                                             CO MARKET SHARE IN NWE
C MIMS=0.205                                             INITIAL MARKET SHARE
A MRPD.K=RPF*MDWE.K                                      CO REFINED PROD DEMAND
C RPF=0.795                                              REF PRODUCT RATIO
A MCD.K=MDWE.K*(1-RPF)*RRF                               CO CRUDE DEMAND
C RRF=1.10                                               REFINERY RECOVERY
```

C

This section produces some variables based on TIME, DIME having already been mentioned. TDIFF is the time-so-far into each year, MONTH is the number of the current month, M1–M12 are 1 in the corresponding month and zero otherwise, and H1 and H2 are 0–1 variables to indicate the first and second halves of the year.

The factor –DT/2 on the CLIP times is incorporated to ensure that the changeover takes place no later than required. Because of round-off, DYNAMO and DYSMAP both carry out comparisons against TIME to an accuracy of ± DT/2.

```
NOTE
NOTE  DUMMY VARIABLE 'DIME' WAS USED TO ENSURE THAT SINE WAVE FOR
NOTE  SEASONALITY FACTOR IS STARTED IN SIMULATED OCTOBER TO BRING IN MIN
NOTE  OPS AND MAX OPS POLICIES AT THE CORRECT TIMES.THE VARIABLE 'TDIFF'
NOTE  WILL BE USED IN PRODUCT AND CRUDE STOCK-HOLDING SECTORS TO SWITCH
NOTE  ON THE MIN AND MAX OPS POLICIES
NOTE
NOTE  M1 IS JANUARY, M2 IS FEBRUARY, ETC. H1 IS 1ST HALF YEAR, H2 IS 2ND
NOTE
A DIME.K=TIME.K+3
A TDIFF.K=TIME.K-DTIME.K
A DTIME.K=SAMPLE(TIME.K,12,0)
A MONTH.K=1*M1.K+2*M2.K+3*M3.K+4*M4.K+5*M5.K+6*M6.K+7*M7.K+8*M8.K+9*
X1      M9.K+10*M10.K+11*M11.K+12*M12.K
A M1.K=CLIP(1,0,TDIFF.K,0)-CLIP(1,0,TDIFF.K,1-DT/2)
A M2.K=CLIP(1,0,TDIFF.K,1-DT/2)-CLIP(1,0,TDIFF.K,2-DT/2)
A M3.K=CLIP(1,0,TDIFF.K,2-DT/2)-CLIP(1,0,TDIFF.K,3-DT/2)
A M4.K=CLIP(1,0,TDIFF.K,3-DT/2)-CLIP(1,0,TDIFF.K,4-DT/2)
A M5.K=CLIP(1,0,TDIFF.K,4-DT/2)-CLIP(1,0,TDIFF.K,5-DT/2)
A M6.K=CLIP(1,0,TDIFF.K,5-DT/2)-CLIP(1,0,TDIFF.K,6-DT/2)
A M7.K=CLIP(1,0,TDIFF.K,6-DT/2)-CLIP(1,0,TDIFF.K,7-DT/2)
A M8.K=CLIP(1,0,TDIFF.K,7-DT/2)-CLIP(1,0,TDIFF.K,8-DT/2)
A M9.K=CLIP(1,0,TDIFF.K,8-DT/2)-CLIP(1,0,TDIFF.K,9-DT/2)
A M10.K=CLIP(1,0,TDIFF.K,9-DT/2)-CLIP(1,0,TDIFF.K,10-DT/2)
A M11.K=CLIP(1,0,TDIFF.K,10-DT/2)-CLIP(1,0,TDIFF.K,11-DT/2)
A M12.K=CLIP(1,0,TDIFF.K,11-DT/2)-CLIP(1,0,TDIFF.K,12-DT/2)
A H1.K=1-H2.K
A H2.K=CLIP(1,0,TDIFF.K,6-DT/2)
```

D

These equations simulate people forecasting, they DO NOT forecast themselves (review Section 5.9). Future oil demand, FODLY, is generated using TGRAT, and total oil demand FTOD is obtained with the correct seasonality at a forecast horizon of 12 months.

Error and bias ratios are used, as mentioned in Section 11.7 and the total demand MDFC is obtained. The proportion of that demand which will be for crude oil gives the crude demand forecast, MCDFC.

```
NOTE
NOTE   FUTURE DEMAND EQUATIONS. SOME FORECASTS ARE FOR LONG PERIODS
NOTE   AHEAD SO EQUATIONS NOT IN SAME FORM AS IN WORKING MEMORANDUM 3 TO
NOTE   AVOID HAVING TO USE VERY LONG TABLE FUNCTION INPUTS
NOTE
L FODLV.K=FODLV.J+DT*FODGR.JK
N FODLV=1+GRIN*ODGRF
R FODGR.KL=(TABHL(TGRAT,TIME.K+FHOR,0,120,6)*ODGRF*FODLV.K)/1200
C FHOR=12
A FTOD.K=RAT1*FODLV.K*(1+SEAMP*SIN(6.283*(DIME.K+FHOR)/12))
L FGNPLV.K=FGNPLV.J+DT*FGRAT.JK
N FGNPLV=1+GRIN
R FGRAT.KL=TABHL(TGRAT,TIME.K+FHOR,0,120,6)*FGNPLV.K/1200
NOTE
NOTE   FORECAST DEMAND
NOTE
A EBR.K=TABHL(TEBR,TIME.K,0,120,6)                  BIAS FACTOR
NOTE   OPTIMISTIC WHEN GNP ACCELATES
NOTE   PESSIMISTIC WHEN GNP DECELERATES
TP TEBR=1.1/1.1/1.1/1.0/0.9/0.9/0.9/0.9/0.9/1.0/1.1/1.1/1.1/1.0/0.9/
X1     0.9/0.9/0.9/0.9/0.9/0.9
A MDFC.K=(MFMS.K*FTOD.K*EBR.K)/12
A MFMS.K=MMS.K                                      CO F/C MARKET SHARE
A MCDFC.K=MDFC.K*(1-RPF)*RRF                        CRUDE DEMAND FCAST
NOTE
NOTE   FOREGOING EQUATIONS ARE FOR FORECASTS MADE NOW RELATING TO A POINT
NOTE   FHOR MONTHS IN THE FUTURE.ALSO NEED TO HAVE THE CURRENT VALUE OF
NOTE   THE FORECASTS MADE FHOR MONTHS AGO.TO SIMULATE THE SYSTEM THIS IS
NOTE   CALCULATED FROM MC
NOTE   TIME
```

E

Crude oil sales can be CIF, where the seller provides the shipping, or FOB, where the buyer does. In the model the company forecasts CIF and FOB sales up to 18 simulated months ahead. For the first 12 months the forecast is modelled on the current actual sales level, modified for seasonality and subject to error and bias (recall Chapter 5). Beyond 12 months the forecasts are derived from the long-term demand forecasts, as shown in Section K, but modified to get the proper seasonality.

```
NOTE
NOTE   PROFILE OF CIF SALES FORECASTS
NOTE
A CIFFP1.K=CIFAS.K*SRAT1.K*SEBR1.K                  CIF F/C FOR 1 MONTH
A CIFFP2.K=CIFAS.K*SRAT2.K*SEBR2.K                  CIF F/C FOR 2 MONTH
A CIFFP3.K=CIFAS.K*SRAT3.K*SEBR3.K                  CIF F/C FOR 3 MONTH
A CIFFP4.K=CIFAS.K*SRAT4.K*SEBR4.K                  CIF F/C FOR 4 MONTH
A CIFFP5.K=CIFAS.K*SRAT5.K*SEBR5.K                  CIF F/C FOR 5 MONTH
A CIFFP6.K=CIFAS.K*SRAT6.K*SEBR6.K                  CIF F/C FOR 6 MONTH
A CIFFP7.K=CIFAS.K*SRAT7.K*SEBR7.K                  CIF F/C FOR 7 MONTH
A CIFFP8.K=CIFAS.K*SRAT8.K*SEBR8.K                  CIF F/C FOR 8 MONTH
A CIFFP9.K=CIFAS.K*SRAT9.K*SEBR9.K                  CIF F/C FOR 9 MONTH
A CIFFP10.K=CIFAS.K*SRAT10.K*SEBR10.K               CIF F/C FOR 10 MONTHS
A CIFFP11.K=CIFAS.K*SRAT11.K*SEBR11.K               CIF F/C FOR 11 MONTHS
A CIFFP12.K=CIFFC.K                                 CIF F/C FOR 12 MONTHS
A CIFFP13.K=CIFFC.K*SRAT1.K                         CIF F/C FOR 13 MONTHS
A CIFFP14.K=CIFFC.K*SRAT2.K                         CIF F/C FOR 14 MONTHS
A CIFFP15.K=CIFFC.K*SRAT3.K                         CIF F/C FOR 15 MONTHS
A CIFFP16.K=CIFFC.K*SRAT4.K                         CIF F/C FOR 16 MONTHS
A CIFFP17.K=CIFFC.K*SRAT5.K                         CIF F/C FOR 17 MONTHS
A CIFFP18.K=CIFFC.K*SRAT6.K                         CIF F/C FOR    MONTHS
```

F

The company's own fleet may or may not change, so the model assumes changes will be in line with the growth in forecast GNP proportionate to a policy reflected by OTGRF. The term (6 – TDIFF.K)/12 is the proportion of a year remaining before midsummer. There are interesting possibilities for examining how alternative ownership policies would affect the system. (Don't just blunder in–look at the loops!) Check the equations for dimensions.

```
NOTE
NOTE  OWNED TONNAGE FORECASTING EQUATIONS
NOTE
A STFOT.K=(ODWT.K+(((6-TDIFF.K)/12)*(ODWT.K*OTGRF*(FGNPLV.K-GNPLEV.K
X1     ))))*SEBR.K
A LTFOT.K=(ODWT.K+(((18-TDIFF.K)/12)*(ODWT.K*OTGRF*(FGNPLV.K-GNPLEV.
X1     K))))*EBR.K
A FOTNMS.K=(H1.K*STFOT.K)+(H2.K*LTFOT.K)
A FOTDMS.K=(H1.K*LTFOT.K)
A SEBR.K=(M1.K*SEBR6.K)+(M2.K*SEBR5.K)+(M3.K*SEBR4.K)+(M4.K*SEBR3.K)
X1     +(M5.K*SEBR2.K)+(M6.K*SEBR1.K)
```

G

Part of the company's crude sales are CIF (BVALCIF), but this varies with expected spot prices as, to some extent, customers switch their requirements from CIF to FOB. These equations model this and adjust the average CIF fraction which the company is modelled as experiencing.

```
NOTE
NOTE  CIF FRACTION SECTOR
NOTE
L LRCIF.K=LRCIF.J+(DT/ATCIF)(CIFAF.J-LRCIF.J) LONG RUN CIF FRACTION
N LRCIF=BVALCIF                               BLIND GUESS
C ATCIF=6                                     ADJUSTMENT TIME
L CIFAF.K=CIFAF.J+DT*RCCIF.JK                 CIF ACTUAL FRACTION
N CIFAF=LRCIF
R RCCIF.KL=(ICIF.K-CIFAF.K)/RTCIF             CHANGE RATE FOR CIFAF
C RTCIF=6                    BLIND GUESS      RESPONSE TIME
A ICIF.K=TABHL(TCIF,ESCPR.K,0,200,50)*BVALCIF INDICATED CIF FRACTION
T TCIF=0.90/0.95/1.0/1.05/1.10                TABLE OF RESPONSE
C BVALCIF=0.22
```

H

As for the CIF forecast profile, it is necessary to have a product sale (or 'offtake') forecast monthly for the coming 18 months. This is based on the company's overall demand and that fraction which will be for refined product, RPF, biassed and seasonally adjusted. The first 12 months are based on current demand, and thereafter on long-term forecasts.

```
NOTE
NOTE  OFFTAKE FORECAST PROFILE SECTOR   SRAT1-3 ARE CORRECTIONS TO
NOTE  ENSURE THE APPROPRIATE SEASONALITY FACTORS
NOTE
A OFP1.K=MDWE.K*RPF*SRAT1.K*SEBR1.K             OFP 1 MONTH  HENCE
A OFP2.K=MDWE.K*RPF*SRAT2.K*SEBR2.K             OFP 2 MONTHS HENCE
A OFP3.K=MDWE.K*RPF*SRAT3.K*SEBR3.K             OFP 3 MONTHS HENCE
A OFP4.K=MDWE.K*RPF*SRAT4.K*SEBR4.K             OFP 4 MONTHS HENCE
A OFP5.K=MDWE.K*RPF*SRAT5.K*SEBR5.K             OFP 5 MONTHS HENCE
A OFP6.K=MDWE.K*RPF*SRAT6.K*SEBR6.K             OFP 6 MONTHS HENCE
A OFP7.K=MDWE.K*RPF*SRAT7.K*SEBR7.K             OFP 7 MONTHS HENCE
A OFP8.K=MDWE.K*RPF*SRAT8.K*SEBR8.K             OFP 8 MONTHS HENCE
A OFP9.K=MDWE.K*RPF*SRAT9.K*SEBR9.K             OFP 9 MONTHS HENCE
A OFP10.K=MDWE.K*RPF*SRAT10.K*SEBR10.K          OFP 10 MONTHS HENCE
A OFP11.K=MDWE.K*RPF*SRAT11.K*SEBR11.K          OFP 11 MONTHS HENCE
A OFP12.K=MDFC.K*RPF                            OFP 12 MONTHS HENCE
A OFP13.K=MDFC.K*RPF*SRAT1.K                    OFP 13 MONTHS HENCE
A OFP14.K=MDFC.K*PPF*SRAT2.K                    OFP 14 MONTHS HENCE
A OFP15.K=MDFC.K*RPF*SRAT3.K                    OFP 15 MONTHS HENCE
A OFP16.K=MDFC.K*RPF*SRAT4.K                    OFP 16 MONTHS HENCE
A OFP17.K=MDFC.K*RPF*SRAT5.K                    OFP 17 MONTHS HENCE
A OFP18.K=MDFC.K*RPF*SRAT6.K                    OFP 18 MONTHS HENCE
```

I

These seasonality ratios convert current values into seasonal corrections for future months.

```
A SRAT1.K=(1+SEAMP*SIN(6.283*(DIME.K+1)/12))/(1+SEAMP*SIN(6.283*DIME
X1     .K/12))
A SRAT2.K=(1+SEAMP*SIN(6.283*(DIME.K+2)/12))/(1+SEAMP*SIN(6.283*DIME
X1     .K/12))
A SRAT3.K=(1+SEAMP*SIN(6.283*(DIME.K+3)/12))/(1+SEAMP*SIN(6.283*DIME
X1     .K/12))
A SRAT4.K=(1+SEAMP*SIN(6.283*(DIME.K+4)/12))/(1+SEAMP*SIN(6.283*DIME
X1     .K/12))
A SRAT5.K=(1+SEAMP*SIN(6.283*(DIME.K+5)/12))/(1+SEAMP*SIN(6.283*DIME
X1     .K/12))
A SRAT6.K=(1+SEAMP*SIN(6.283*(DIME.K+6)/12))/(1+SEAMP*SIN(6.283*DIME
X1     .K/12))
A SRAT7.K=(1+SEAMP*SIN(6.283*(DIME.K+7)/12))/(1+SEAMP*SIN(6.283*DIME
X1     .K/12))
A SRAT8.K=(1+SEAMP*SIN(6.283*(DIME.K+8)/12))/(1+SEAMP*SIN(6.283*DIME
X1     .K/12))
A SRAT9.K=(1+SEAMP*SIN(6.283*(DIME.K+9)/12))/(1+SEAMP*SIN(6.283*DIME
X1     .K/12))
A SRAT10.K=(1+SEAMP*SIN(6.283*(DIME.K+10)/12))/(1+SEAMP*SIN(6.283*DI
X1     ME.K/12))
A SRAT11.K=(1+SEAMP*SIN(6.283*(DIME.K+11)/12))/(1+SEAMP*SIN(6.283*DI
X1     ME.K/12))
```

J

The short-term error and bias ratios are meant to reflect the company's ability to make accurate three month forecasts (beccause that is all contractual sales), and thereafter to get progressively worse.

```
A SEBR1.K=1.0                                     SHORT
A SEBR2.K=1.00                                     TERM
A SEBR3.K=1.00                               ERROR AND BIAS
A SEBR4.K=1.00+(EBR.K-1.00)*0.25                  RATIOS
A SEBR5.K=1.00+(EBR.K-1.00)*0.50
A SEBR6.K=1.00+(EBR.K-1.00)*0.75
A SEBR7.K=EBR.K
A SEBR8.K=EBR.K
A SEBR9.K=EBR.K
A SEBR10.K=EBR.K
A SEBR11.K=EBR.K
```

K

Crude oil sales, FOB and CIF, are based on the forecast crude demand divided up according to the current average CIF Fraction (Section G). The actual sales FOB and CIF are also found and the gap between FOB forecast and actual calculated, because this is extra shipping which will be needed. The current values of old FOB and CIF forecasts are determined from the macro in Section A.

```
NOTE
NOTE  CRUDE OIL SALES SECTOR
NOTE
A FOBFC.K=MCDFC.K*(1-LRCIF.K)           FOB DEMAND F/CAST 12 MTH HORIZON
A CIFFC.K=MCDFC.K-FOBFC.K               CIF DEMAND F/CAST 12 MTH HORIZON
A FOBAS.K=MCD.K*(1-CIFAF.K)             FOB CURRENT ACTUAL SALES
A FOBGAP.K=MAX(0,CVFOBFC.K-FOBAS.K)
A CIFAS.K=MCD.K*CIFAF.K                 CIF ACTUAL SALES
A CVFOBFC.K=FOBVAL.K
A CVCIFFC.K=CIFVAL.K
N FOBIV=(RAT1*MIMS*FODLV*(1-RPF)*RRF*(1-LRCIF))/12.
N CIFIV=(RAT1*MIMS*FODLV*(1-RPF)*RRF*LRCIF)/12.
```

L

The ideal refinery production for next month but one, TPRTP2, is calculated on the basis of offtake forecast and stock correction. The target production, PRTP2, is then found by applying constraints reflecting the inability to depart too severely from the current month's production (HIFAC and LOFAC), to exceed Refinery Capacity, or to consume more crude oil than is expected to be available as modified by the need to keep a certain proportion of available oil to keep the pipelines full (COSFAC). The calculation is done in terms of equivalent tons of product (hence RRF).

```
250 NOTE
251 NOTE  REFINERY PRODUCTION PLANNING SECTOR
252 NOTE
253 NOTE
254 NOTE  MONTH 2
255 NOTE
256 C HIFAC=1,20                           LIMIT ON  PRODUCTION INCREASE
257 C LOFAC=0,80                           LIMIT ON  PRODUCTION DECREASE
258 C ATPS=3                                         ADJ TIME ON PROD STKS
259 A TPRTP2.K=UFP2.K+(TPSP2.K-APSP1.K)/ATPS         THEORETICAL TARGET
260 A ULIM2.K=MIN((CCP2.K+ACSP1.K-CIFFP2.K)*(COSFAC/RRF),REFCAP.K)
261 A ULIM2A.K=MIN(TPRTP2.K,HIFAC*CRPT.K)            CURRENT PROD CONS
262 A ULIM2B.K=MIN(ULIM2.K,ULIM2A.K)                 NET UPPER LIMIT
263 A PRTP2.K=MAX(ULIM2B.K,LOFAC*CRPT.K)             OVERALL LOWER LIMIT
```

M

Essentially similar in principle to Section L. The dummy variable DC permits rerun testing of alternative refining control equations (note the order of the target and actual terms in the two proportional controllers in line 267). The factor of 0.9 in line 278.3 is a second safety factor against assuming that all the oil expected (NECOA) will actually arrive during the month.

```
264 NOTE
265 NOTE  CALCULATION OF CURRENT MONTH'S ACTUAL PRODUCTION
266 NOTE
267 A ACRPT.K=OFP1.K+((TPSP1.K-APS.K)/ATPS)+DC*((ACS.K-TCS1.K)/ATCS)
268 C DC=0
269 C ATCS=1
270 A ULIMC.K=MIN(REFCAP.K,HIFAC*ORTP1.K)
271 C COSFAC=0,80                          CRUDE OIL STORAGE FACTOR
272 A ULIMCA.K=MIN(ACRPT.K,ULIMC.K)
273 A ULIMCB.K=MIN(ULIMCA.K,COSFAC*(ACS.K+0.9*NECOA.K)/RRF)
274 A CRPT.K=MAX(ULIMCB.K,LOFAC*ORTP1.K)          CURRENT THROUGHPUT
```

N

Strictly speaking, refinery throughput is planned monthly, but in practice there are revisions which justify the continuous approximation in Sections L and M. This section therefore carries forward old Refinery Throughput from IDT previously.

```
275 NOTE
276 NOTE  CALCULATION OF OLD VALUE OF CURRENT REFINERY THROUGHPUT
277 NOTE
278 A PDIFF.K=CRPT.K-ORTP1.K
279 A RCORPT1.K=SAMPLE(PDIFF.K,PERD,0)
280 L ORTP1.K=ORTP1.J+DT*PULSE(RCORPT1.J/DT,PERD,PERD)
281 N PERD=DT
```

O

Refinery Capacity is assumed to grow in line with GNP. What would be the effects of discontinuities in investment?

```
282 NOTE
283 NOTE   REFINERY CAPACITY
284 NOTE
285 L REFCAP.K=REFCAP.J+DT*(GRRCAP.JK)            REFINERY CAPACITY
286 R GRRCAP.KL=(BRAT.K*REFCAP.K*RCGRF)/12        CAPACITY GROWTH RATE
287 N RCGRF=ODGRF
288 N REFCAP=IRC                                  INITIAL VALUE REF CAP
289 N ORTP1=IRC
290 C IRC=4.5E06                                  INITIAL REFINERY CAP.
```

P

A supplementary calculation of aggregate stocks in equivalent product. This could be used to expand the measure of system performance to incorporate investment changes (see previous section).

```
291 NOTE
292 NOTE   TOTAL STOCKS
293 NOTE
294 S ATS.K=(ACS.K/RRF)+APS.K
```

Q

A simple calculation of anticipated and actual product stock positions. The initial equation for APS assumes that actual stocks start at the target position (next section).

```
295 NOTE
296 NOTE   PRODUCT STOCK POSITION SECTOR
297 NOTE
298 S APSP2.K=APSP1.K+PRTP2.K-OFP2.K                    ANTIC PROD STOCKS 2
299 A APSP1.K=APS.K+CRPT.K-OFP1.K                       ANTIC PROD STOCKS 1
300 L APS.K=APS.J+DT*(CRPT.J-PO.JK)                     ACTUAL PRODUCT STOCKS
301 N APS=((MAXPSB+MINPSB)/2)+((MAXPSB-MINPSB)/2)*SIN(6.283*(9-MINPT)/12
302 X1     )
```

R

Target product stocks in the model follow a sine curve, passing through two points MINPS and MAXPS at times MINPT and MINPT–6 (months) respectively. The minimum and maximum points are in proportion to refinery capacity and they define the mean value and the amplitude of the curve (PSLEV and PCAMP). The curve has its mean value at MINPT–9 so the argument in the sine wave of seasonality is TIME–(MINPT–9) as in line 308. From this, the target product stock position at the current time, and for 1 and 2 months ahead is easily found with the aid of the seasonality correctors CRATO–CRAT2 (the first of which is unity).

Because the weather can seriously affect the actual seasonality of demand a certain amount of storage capacity and, of course, refining flexibility has to be provided to cope with such situations as they arise.

```
303 NOTE
304 NOTE   TARGET PRODUCT STOCKS
305 NOTE
306 A PSCURVE.K=PCAMP.K*BSIN.K
307 A BSIN.K=SIN(6.283*(DIME2.K)/12)
308 A DIME2.K=TIME.K+9-MINPT
309 A PCAMP.K=(MAXPS.K-MINPS.K)/2
310 A TPSP0.K=PSCURVE.K*CRAT0.K+PSLEV.K              TARGET PROD STOCKS NOW
311 A TPSP1.K=PSCURVE.K*CRAT1.K+PSLEV.K              TARGET PROD STOCKS MN1
312 A TPSP2.K=PSCURVE.K*CRAT2.K+PSLEV.K              TARGET PROD STOCKS MN2
313 A CRAT0.K=SIN(6.283*(DIME2.K)/12)/BSIN.K
314 A CRAT1.K=SIN(6.283*(DIME2.K+1)/12)/BSIN.K
315 A CRAT2.K=SIN(6.283*(DIME2.K+2)/12)/BSIN.K
316 C MINPT=4                                        MIN PRODUCT STOCK TIME
317 A PSLEV.K=((MINPS.K+MAXPS.K)/2.)
318 A MINPS.K=MINPSB*(REFCAP.K/IRC)                 MINIMUM PRODUCT STOCKS
319 A MAXPS.K=MAXPSB*(REFCAP.K/IRC)                 MAXIMUM PRODUCT STOCKS
320 C MINPSB=1.5E06                                  INITIAL VALUE OF MINPS
321 C MAXPSB=4.2E06                                  INITIAL VALUE OF MAXPS
```

S

Line 322 models an 'out of stock' situation as in Section 5.7. It gives cumulative product offtake and lost sales, and values any lost sales on the basis of a seasonally-variable product profit.

```
322 R PO.KL=MIN(MRPD.K,APS.K/DT)                     ACTUAL PRODUCT OFFTAKE
323 L CUMPO.K=CUMPO.J+DT*PO.JK                       CUMULATIVE PO
324 N CUMPO=0
325 L LSALE.K=LSALE.J+DT*(MRPD.J-PO.JK)              LOST SALES
326 N LSALE=0
327 C SPRPF=2.50
328 C WPRPF=1.50
329 L VALULS.K=VALULS.J+DT*((0.5*(SPRPF-WPRPF)*SIN(6.283*DIME.J/12)+0.5*
330 X1     (WPRPF+SPRPF))*(MRPD.J-PO.JK))            VALUE OF LOST SALES
331 N VALULS=0
```

T

A simple conservation of actual crude stock, with an initial value of the planned level. The Net Expected Crude Oil Arrivals is the amount of oil expected in the coming month from owned and time-chartered shipping (ECCP), recent spot-charters, less the forecast sales of CIF crude.

```
332 NOTE
333 NOTE   CRUDE OIL SECTOR
334 NOTE
335 L ACS.K=ACS.J+DT*(COAR.JK-CRPT.J*RRF)           ACTUAL CRUDE STOCKS
336 N ACS=((MAXCSB+MINCSB)/2)+((MAXCSB-MINCSB)/2)*SIN(6.283*(9-MINCT)/12
337 X1    )
338 A NECOA.K=ECCP.K+MSCT.K*SCEFF-CVCIFFC.K
339 A ACSP1.K=ACS.K-CRPT.K*RRF+NECOA.K              ANTICIPATED CRUDE
```

U

Target crude stocks are modelled on the same basis as product stocks, but not so far ahead.

```
340 NOTE
341 NOTE   TARGET CRUDE STOCKS
342 NOTE
343 A TCS0.K=((MINCS.K+MAXCS.K)/2.)+(CCAMP.K*(SIN(6.283*(TIME.K+9-MINCT)/
344 X1     12)))                                      TARGET CRUDE STOCK NOW
345 A TCS1.K=((MINCS.K+MAXCS.K)/2.)+CCAMP.K*CSINA.K
346 A TCS2.K=((MINCS.K+MAXCS.K)/2.)+CCAMP.K*CSIN.K
347 A CSINA.K=SIN(6.283*(TIME.K+9-MINCT+1)/12)
348 A CSIN.K=SIN(6.283*(TIME.K+9-MINCT+2)/12)
```

```
A CCAMP.K=(MAXCS.K-MINCS.K)/2.
C MINCT=4                                    MIN CRUDE STOCKS TIME
A MINCS.K=MINCSB*(REFCAP.K/IRC)              MIN OPS LEVEL
A MAXCS.K=MAXCSB*(REFCAP.K/IRC)              MAX OPS LEVEL
C MINCSB=1.6E06                          MIN OPS BASE LEVEL  MILLION TONS
C MAXCSB=2.8E06                          MAX OPS BASE LEVEL  MILLION TONS
```

V

The amount of crude required for each of the next 18 months from forecasts of offtake and CIF sales.

```
NOTE
NOTE   CRUDE OIL REQUIREMENTS PROFILE
NOTE
A CRRQP1.K=OFP1.K*RRF+CIFFP1.K           CRUDE REQUIRED, NWE, 1 MTH HENCE
A CRRQP2.K=OFP2.K*RRF+CIFFP2.K           CRUDE REQUIRED, NWE, 2 MTH HENCE
A CRRQP3.K=OFP3.K*RRF+CIFFP3.K           CRUDE REQUIRED, NWE, 3 MTH HENCE
A CRRQP4.K=OFP4.K*RRF+CIFFP4.K           CRUDE REQUIRED, NWE, 4 MTH HENCE
A CRRQP5.K=OFP5.K*RRF+CIFFP5.K           CRUDE REQUIRED, NWE, 5 MTH HENCE
A CRRQP6.K=OFP6.K*RRF+CIFFP6.K           CRUDE REQUIRED, NWE, 6 MTH HENCE
A CRRQP7.K=OFP7.K*RRF+CIFFP7.K           CRUDE REQUIRED, NWE, 7 MTH HENCE
A CRRQP8.K=OFP8.K*RRF+CIFFP8.K           CRUDE REQUIRED, NWE, 8 MTH HENCE
A CRRQP9.K=OFP9.K*RRF+CIFFP9.K           CRUDE REQUIRED, NWE, 9 MTH HENCE
A CRRQP10.K=OFP10.K*RRF+CIFFP10.K        CRUDE REQUIRED, NWE,10 MTH HENCE
A CRRQP11.K=OFP11.K*RRF+CIFFP11.K        CRUDE REQUIRED, NWE,11 MTH HENCE
A CRRQP12.K=OFP12.K*RRF+CIFFP12.K        CRUDE REQUIRED, NWE,12 MTH HENCE
A CRRQP13.K=OFP13.K*RRF+CIFFP13.K        CRUDE REQUIRED, NWE,13 MTH HENCE
A CRRQP14.K=OFP14.K*RRF+CIFFP14.K        CRUDE REQUIRED, NWE,14 MTH HENCE
A CRRQP15.K=OFP15.K*RRF+CIFFP15.K        CRUDE REQUIRED, NWE,15 MTH HENCE
A CRRQP16.K=OFP16.K*RRF+CIFFP16.K        CRUDE REQUIRED, NWE,16 MTH HENCE
A CRRQP17.K=OFP17.K*RRF+CIFFP17.K        CRUDE REQUIRED, NWE,17 MTH HENCE
A CRRQP18.K=OFP18.K*RRF+CIFFP18.K        CRUDE REQUIRED, NWE,18 MTH HENCE
```

W

The crude requirements at the next midsummer are calculated using the zero–one variables M1–11 and on the bais of whether TIME is in the first or second half of the year. The distant midsummer requirement is similarly found. (Review Fig. 11.2, for the principle, but observe that the vertical axis there is in *shipping* whereas here we are working in tons of crude per month. One of the difficulties in this model is the great variety of different units and the presence of monthly and annual values, some of which are in percentages. During debugging of the model several of the variables ran to very high values – in the order of 10^{69} sometimes. This was a sure indication of a failure to divide an annual value by 12, or a percentage by 100.)

```
A CRNMST1.K=M1.K*CRRQP6.K+M2.K*CRRQP5.K+M3.K*CRRQP4.K+M4.K*CRRQP3.K+
X1     M5.K*CRRQP2.K+M6.K*CRRQP1.K
A CRNMST2.K=M7.K*CRRQP12.K+M8.K*CRRQP11.K+M9.K*CRRQP10.K+M10.K*CRRQP
X1     9.K+M11.K*CRRQP8.K+M12.K*CRRQP7.K
A CRNMST.K=H1.K*CRNMST1.K+H2.K*CRNMST2.K  CRUDE REQ FOR NEAR MIDSUM.
A CRDMST.K=M1.K*CRRQP18.K+M2.K*CRRQP17.K+M3.K*CRRQP16.K+M4.K*CRRQP15
X1     .K+M5.K*CRRQP14.K+M6.K*CRRQP13.K     CRUDE REQ FOR DIST MIDSUM.
```

X

What it says, Review Section M.

```
NOTE
NOTE   LOST THROUGHPUT DUE TO LOW STOCKS
NOTE
L LSTPT.K=LSTPT.J+DT*(ULIMCA.J-ULIMCB.J)
N LSTPT=0
```

Y

Parameters for the shipping sector and calculation of owned fleet on the assumption of growth at the same rate as GNP.

```
388 NOTE
389 NOTE   SHIPPING SECTOR
390 NOTE
391 NOTE   SHIPPING PARAMETER VALUES
392 NOTE
393 N CONS=(12*19450)/99580                 CARRYING CAPACITY OF NOTIONAL
394 N OTEFF=330/365                         OWNED TONNAGE EFFECTIVENESS
395 N TCEFF=340/365                         TIME-CHARTERED EFFECTIVENESS
396 N SCEFF=350/365                         SPOT-CHARTERED EFFECTIVENESS
397 L ODWT.K=ODWT.J+DT*(OTNBR.JK)           OWNED DEADWEIGHT TONNAGE
398 N ODWT=4.5E06
399 R OTNBR.KL=(BRAT.K*ODWT.K*OTGF)/12      OWNED TONNAGE NET BUILD RATE
400 N OTGF=ODGRF
401 N OTGRF=ODGRF
```

Z

This is the approach of Section 11.3 for several variants on the short-haul ratio: SHR is the ratio to be operated in the coming month, ESHWF is the existing weighting factor, and SHWFA is the factor which would operate if the Company instantly adopt the short-haul ratio its spot-charter price forecasts indicate as being attractive (if prices are high, it pays to take more short-haul oil).

```
402 NOTE
403 NOTE   SHORT HAUL WEIGHTING FACTOR AND ASSOCIATED VARIABLES
404 NOTE
405 A SHWF.K=SHR.K*(SHVT/LHVT)+(1-SHR.K)          SHORT-HAUL WEIGHTING
406 A SHWFA.K=TSHR1B.K*(SHVT/LHVT)+(1-TSHR1B.K)
407 L ESHWF.K=ESHWF.J+DT*PULSE((SHWF.J-ESHWF.J)/DT,PERD,PERD)   EX SHWF
408 N ESHWF=MASHR*(SHVT/LHVT)+(1-MASHR)
409 C SHVT=25                                 SHORT-HAUL VOYAGE TIME  DAYS
410 C LHVT=67                                 LONG-HAUL VOYAGE TIME DAYS
411 A MPSHWF.K=MPSHR.K*(SHVT/LHVT)+(1-MPSHR.K)    POLICY SH/HAUL FACTOR
```

AA

Average short-haul ratio (MASHR) is calculated and used to modify the price-expectation movement in SHR in line 422–423. This is one of the possible approaches to the problem of controlling a ratio to a long-run value, whilst allowing it to vary under the influence of some other factor. (Compare with Chapter 5.) The target ratio, TSHR1, is constrained by upper and lower bounds meant to reflect contractual relationships with the producing countries. How sensitive is the model to the tightness of these bounds?

```
412 NOTE
413 NOTE   SHORT HAUL RATIO
414 NOTE
415 L MASHR.K=MASHR.J+(DT/SHRAT)(SHR.J-MASHR.J)    CO AVERAGE SHORT-HAUL
416 N MASHR=0.28
417 C SHRAT=4                                 SHORT HAUL RATIO AVERAGING TIME
418 C ATSHR=1                                 ADJUSTMENT TIME FOR SH RATIO
419 A MPSHR.K=0.28                        CO POLICY SHORT-HAUL RATIO
420 A SHRM.K=TABHL(TSHR,ESCPR.K,25,165,35)          SHR MULTIPLIER
421 T TSHR=0.75/0.75/1.0/1.20/1.25                  TABLE OF RESPONSE
422 L TSHR1.K=TSHR1.J+(DT/ATSHR)((SHRM.J*MPSHR.J-
423 X MASHR.J)-(MASHR.J-MPSHR.J))
424 N TSHR1=MASHR
425 A TSHR1A.K=MIN(TSHR1.K,SHRUB.K)
426 A TSHR1B.K=MAX(TSHR1A.K,SHRLB.K)
427 A SHRUB.K=SHRULIM*MPSHR.K                 UPPER BOUND ON SHR
428 C SHRULIM=1.03                            UPPER LIMIT MULTIPLE
429 A SHRLB.K=SHRLLIM*MPSHR.K                 LOWER BOUND ON SHR
430 C SHRLLIM=0.97                            LOWER LIMIT MULTIPLE
```

AB

Section 11.5 described two systems and this is their first reflection in the model. The constrained short-haul ratio, CSHR, is that which fully occupy the existing owned and time-chartered fleets modified by the upper and lower bounds. (Derive lines 432–433!). If there is a surplus of shipping, the shipping deficit ESCDP2A will be negative and SHR is therefore clipped between two alternatives.

```
431 A SHRC.K=CLIP(1.0,0,ESCDP2A.K,0)
432 A CSHR.K=(1-((ODWT.K+TCP1.K)*(1/CONS)*AEFF.K)/(RSCP2.K+FOBGAP.K))/(1
433 X1     -(SHVT/LHVT))
434 A CSHRA.K=MIN(CSHR.K,SHRUB.K)
435 A CSHRB.K=MAX(CSHRA.K,SHRLB.K)
436 A SHR.K=SHRC.K*TSHR1B.K+(1-SHRC.K)*CSHRB.K
```

AC

The required capacity two months hence, RSCP2, (tons of crude per month) is given by usage rates and stock correction. Carrying capacities for the existing fleet in 1 month are calculated for existing SHR, the price-indicated ratio, and the actual ratio for next month. Carrying capacity for the month after is estimated taking account of any presentations from the presentation band or from immediate chartering.

```
437 NOTE
438 NOTE  SHIPPING CAPACITY
439 NOTE
440 A RSCP2.K=(CIFFP2.K+((TCS2.K-ACSP1.K)/TACS)*PRTP2.K*RRF)
441 C TACS=1                                        CRUDE STOCK ADJ TIME
442 A CCP1A.K=(ODWT.K+TCP1.K)(1/CONS)(AEFF.K/SHWFA.K)       AUX CCP1
443 A ECCP.K=(ODWT.K+TCP1.K)(1/CONS)(AEFF.K/ESHWF.K)        EXIST COM CAR
444 S ACCP.K=ECCP.K+(MSCT.K*SCEFF)
445 A CCP1.K=(ODWT.K+TCP1.K)(1/CONS)(AEFF.K/SHWF.K)          NET CCP1
446 A CCP2.K=(0.975*CCP1A.K)+(((ETCP1.K+TCVNMSO.K)*TCEFF)/(CONS*SHWFA.K)
447 X1     )
```

AD

Calculation of average effectiveness and the net shipping deficit, ESCDP2. The shipping deficit is constrained not to be negative because of the presence of the System 1 and System 2 modes of operation described in Section 11.5.

```
448 A AEFF.K=(OTEFF*ODWT.K+TCEFF*TCP1.K)/(ODWT.K+TCP1.K)
449 A ESCDP2A.K=RSCP2.K+FOBGAP.K-CCP2.K
450 A ESCDP2B.K=0
451 A ESCDP2.K=SHRC.K*ESCDP2A.K+(1-SHRC.K)*ESCDP2B.K         SHIP DEFICIT
452 R COAR.KL=CCP1.K+(SCEFF*DELAY3(MSCCR3.JK,MSCOT.K/4))-CIFAS.K
```

AB

Output information for various proportions of ships. HSCF is particularly significant as, from Fig. 11.2 it should peak in winter and trough in summer if the system is doing what it is supposed to do. This is an example of a very useful approach to the assessment of system performance.

```
453 A TDWT.K=ODWT.K+TCP1.K+MSCT.K                TOTAL DWT
454 S COTF.K=ODWT.K/TDWT.K                       CURRENT ODWT FRACTION
455 S CNSCF.K=(ODWT.K+TCP1.K)/TDWT.K             CURR.  NON SC FRACTION
456 S HSCF.K=MSCT.K/TDWT.K                       HISTORICAL SC FRACTION
```

AF

The amounts of shipping from distant chartering which arise in each of the three months of the presentation band. Why should it be three months (Fig. 11.2), why those 3, and why these proportions?

```
457 A ETP10.K=M9.K*PPM1*TCPT.K
458 A ETP11.K=M10.K*PPM2*TCPT.K
459 A ETP12.K=M11.K*PPM3*TCPT.K
460 L ETCP1.K=ETCP1.J+DT*(PULSE(ETP10.J/DT,9,12)-PULSE(TCPT10.J/DT,10,12
461 X1 )+PULSE(ETP11.J/DT,10,12)-PULSE(TCPT11.J/DT,11,12)+PULSE(ETP12.J/DT
462 X2 ,11,12)-PULSE(TCPT12.J/DT,12,12))
463 N ETCP1=0
```

AG

The total oil transported to NWE is another measure of the effectiveness of the design.

```
464 NOTE
465 NOTE  LIFTINGS
466 NOTE
467 L CUMLFT.K=CUMLFT.J+(DT)(CCP1.J+(MSCR.JK*SCEFF)+FUBAS.J)
468 N CUMLFT=0
```

AH

Some simple financial measures of cumulative spending on ships and lost sales. Tanker prices are in World Scales, hence the .04 from Section 11. It will be recalled that the design exercise was conducted with actual prices which are now largely of historical interest.

```
469 NOTE
470 NOTE  SHIPPING EXPENDITURE EQUATIONS
471 NOTE
472 R SCER.KL=PSCD*ESCDP2.K*WS.K*SHWF.K*0.04*SCRM.K      S-C EXP RATE
473 R TCER.KL=DTCER.K*.04*(1/CONS)
474 R OTER.KL=ODWT.K*OTOC                          OWNED TONNAGE EXP RATE
475 L CUMEXP.K=CUMEXP.J+(DT)(SCER.JK+TCER.JK+OTER.JK)         CUM EXPENSE
476 N CUMEXP=0
477 S TOTEXP.K=CUMEXP.K+VALULS.K                   TOTAL EXPENDITURE
478 N OTOC=4/(CONS*OTEFF)    PDS/DWT/MTH OWNED TONNAGE OPERATING COST
479 S TSER.K=SCER.JK+TCER.JK+OTER.JK     TOTAL SHIP. EXPENDITURE RATE
```

AI

Recalling Fig. 11.2, lines 487–488 calculate the two ‘black spot’ positions for the summer troughs. Line 489–491 projects the spot backwards using a linear approximation to the decay profile, the factors of 0.95 being safety margins against over-chartering. During the second half-year a minimization is performed to ensure that the back-projection does not exceed the shipping requirements curve, particularly important for the period just after the half-year before the sine curve has picked up.

```
480 NOTE
481 NOTE  CHARTERING ACTIVITY SECTOR
482 NOTE
483 NOTE  FIRST HALF OF YEAR,H1=1,H2=0
484 NOTE
```

```
485 NOTE  T-C TONNAGES NEEDED AT TROUGHS
486 NOTE
487 A NMSTTC.K=(1-SCF.K)(CRNMST.K*(CONS/AEFF.K)*MPSHWF.K-FOTNMS.K)
488 A DMSTTC.K=(1-SCF.K)(CRDMST.K*(CONS/AEFF.K)*MPSHWF.K-FOTDMS.K)
489 A NMSTCT.K=H1.K*NMSTTC.K*(1+((6-MONTH.K)*0.05)/TCDUR)+H2.K*MIN(NMSTT
490 X1     C.K*(1+((18-MONTH.K)*0.95)/TCDUR),(CRRQP1.K*(CONS/AEFF.K)*MPSHWF.K
491 X2     -ODWT.K))
492 A DMSTCT.K=DMSTTC.K*(1+(6*0.05*H1.K)/TCDUR)
```

AJ

Immediate time-charters are determined using a telescoping adjustment time. The actual immediate T-C rate in lines 498–499 also includes second half-year charters from Section AK, and provision for rerun testing of the policy of preventing any further time-charters when the system is in surplus. Lines 502–505 accumulate shipping chartered but not presented. Distant time-chartering is based on a forecast of the distant midsummer trough and the assumption that the black spot position for near midsummer will be achieved.

```
493 NOTE
494 NOTE  T-C RATES FOR FIRST HALF YEAR
495 NOTE
496 A ITC1.K=H1.K*((NMSTCT.K-TCP1.K-TCVNMSO.K)/ATNTC.K)   IMMEDIATES
497 A ATNTC.K=H1.K*(7-MONTH.K)+H2.K*(13-MONTH.K)   ADJUSTMENT TIME
498 A ITCR.K=MAX(0,(ITC1.K+ITC2.K+((1-PSCD)*(ESCDP2.K*CONS*MPSHWF.K)/TCE
499 X1     FF))*TCRSW.K)
500 A TCRSW.K=SYSVL+(1-SYSVL)*SHRC.K
501 C SYSVL=1
502 L TCVNMSO.K=TCVNMSO.J+(DT)(ITCR.J-ITCPR.JK)
503 N TCVNMSO=0
504 R ITCPR.KL=DELAY3(ITCR.K,1)
505 N ITCR=0
506 A DTCR.K=MAX(0,((DMSTCT.K-NMSTTC.K-TCVDMSO.K)/TATCF.K)*H1.K)
507 A TATCF.K=H1.K*(7-MONTH.K)+(1-H1.K)*1               ADJUSTMENT TIME
508 L TCVDMSO.K=TCVDMSO.J+(DT)(DTCR.J-PULSE(TCVDMSO.J/DT,6+DT,12))
509 N TCVDMSO=0
```

AK

The main charter presentations take place in the second half-year (Fig. 11.2), together with further immediate chartering. TCPT is used to record the amount chartered for the presentation band and the arrival of ships is simulated by taking constant proportions for each of the three months of the presentation band. The actual presentation rate is then given by one of these three proportions, according to the simulated month. Immediate time-chartering takes account of the ships still outstanding from the presentation band.

The section ends with some parameters and a conservation equation for time-charter tonnage.

```
510 NOTE
511 NOTE  SECOND HALF YEAR  MAIN CHARTER PRESENTATION PERIOD
512 NOTE
513 L TCPT.K=TCPT.J+(DT)(PULSE(TCVDMSO.J/DT,6+DT,12)-PULSE(TCPT.J/DT,12+
514 X1     DT/2,12))                                 T-C PRESENTATION TONN.
515 N TCPT=0
516 A TCPT10.K=M10.K*PPM1*TCPT.K
517 A TCPT11.K=M11.K*PPM2*TCPT.K
518 A TCPT12.K=M12.K*PPM3*TCPT.K
519 A PBPR.K=TCPT10.K+TCPT11.K+TCPT12.K               PRESENTATION RATE
520 A ITC2.K=((NMSTCT.K-TCP1.K-TCVNMSO.K-TCPT.K*(M6.K+M7.K+M8.K+M9.K+M10
521 X1     .K+M11.K*(1-PPM1)+M12.K*(1-PPM1-PPM2)))*H2.K)/ATNTC.K
522 R TCAR.KL=DELAY3(ITCR.K,1)+PBPR.K                 T-C ARRIVAL RATE
```

```
523 A TCVOO.K=TCVNMSO.K+TCVDMSO.K
524 C TCDUR=48
525 C PPM1=0.42
526 C PPM2=0.33
527 N PPI'3=1-PPM1-PPM2
528 L TCP1.K=TCP1.J+(DT)(TCAR.JK-TCCR.JK)          T-C TONNAGE NEXT MONTH
529 N TCP1=4.2E06
```

AL

A short section designed to check the accuracy of the linear approximation to the charter decay profile (see Section A1).

```
530 NOTE
531 NOTE   CHECK ON TIME CHARTER PRESENTATION IN FIRST HALF OF YEAR
532 NOTE
533 L DTONS.K=DTONS.J+DT*(H1.J*TCAR.JK-PULSE(DTONS.J/DT,6,12))
534 N DTONS=0
535 A DPROF.K=H1'.K*(TCP1.K-DTONS!.K)
```

AM

The average charter duration is 48 months and 'practically all' are between 3 and 5 years (the really long-term charters for 15 years are treated as though they were owned tonnage). Using the method of the annexe to Chapter 2:

$$4_n = 24 \text{ (a monthly model)}$$

so

$$n = \frac{48^2}{36} = 64$$

therefore a 63rd order delay is approximated by 21 third-order delays.

```
536 NOTE
537 NOTE   EQUATION NECESSARY FOR CALCULATING T-C COMPLETIONS
538 NOTE
539 R TC1.KL=DELAY3(TCAR.JK,TCDUR/21)            )
540 R TC2.KL=DELAY3(TC1.JK,TCDUR/21)             )
541 R TC3.KL=DELAY3(TC2.JK,TCDUR/21)             )
542 R TC4.KL=DELAY3(TC3.JK,TCDUR/21)             )
543 R TC5.KL=DELAY3(TC4.JK,TCDUR/21)             )  DUMMY
544 R TC6.KL=DELAY3(TC5.JK,TCDUR/21)             )
545 R TC7.KL=DELAY3(TC6.JK,TCDUR/21)             )
546 R TC8.KL=DELAY3(TC7.JK,TCDUR/21)             )  RATES
547 R TC9.KL=DELAY3(TC8.JK,TCDUR/21)             )
548 R TC10.KL=DELAY3(TC9.JK,TCDUR/21)            )
549 R TC11.KL=DELAY3(TC10.JK,TCDUR/21)           )  FOR
550 R TC12.KL=DELAY3(TC11.JK,TCDUR/21)           )
551 R TC13.KL=DELAY3(TC12.JK,TCDUR/21)           )
552 R TC14.KL=DELAY3(TC13.JK,TCDUR/21)           )  63RD
553 R TC15.KL=DELAY3(TC14.JK,TCDUR/21)           )
554 R TC16.KL=DELAY3(TC15.JK,TCDUR/21)           )
555 R TC17.KL=DELAY3(TC16.JK,TCDUR/21)           )  ORDER
556 R TC18.KL=DELAY3(TC17.JK,TCDUR/21)           )
557 R TC19.KL=DELAY3(TC18.JK,TCDUR/21)           )
558 R TC20.KL=DELAY3(TC1^.JK,TCDUR/21)           )  DELAY
559 R TCCR.KL=DELAY3(TC20.JK,TCDUR/21)   TC COMPLETION RATE DWT/MONTH
```

AN

Initial conditions are calculated from *very* rough data which indicated roughly the proportions of the existing T-C fleet which had been chartered during each of the past four years (YIFAC–Y4FAC), and an estimate that 75% of the time-charters are placed for the presentation band (CCP). Since TCDUR/21 = 2.286 months, the four years prior to TIME = 0 were divided into 21 periods, from which the presentation bands could be approximately identified. The tonnages could then be allocated to periods and converted to rates by dividing by the appropriate delay, as shown in Section 5.5. An occasional zero entry is needed for rounding error in the time comparison.

```
560 N TCAR=(Y1TCTNS*CCP)/(TCDUR/21)
561 N TC1=((Y1TCTNS*(1-CCP))/3)/(TCDUR/21)
562 N TC2=((Y1TCTNS*(1-CCp))/3)/(TCDUR/21)
563 N TC3=((Y1TCTNS*(1-CCP))/3)/(TCDUR/21)          INITIAL
564 N TC4=0
565 N TC5=(Y2TCTNS*CCP)/(TCDUR/21)
566 N TC6=((Y2TCTNS*(1-CCP))/3)/(TCDUR/21)
567 N TC7=((Y2TCTNS*(1-CCP))/3)/(TCDUR/21)
568 N TC8=((Y2TCTNS*(1-CCP))/3)/(TCDUR/21)         CONDITIONS
569 N TC9=0
570 N TC10=0
571 N TC11=(Y3TCTNS*CCP)/(TCDUR/21)
572 N TC12=((Y3TCTNS*(1-CCp))/3)/(TCDUR/21)
573 N TC13=((y3TCTNS*(1-CCP))/3)/(TCDUR/21)            FOR
574 N TC14=((Y3TCTNS*(1-CCP))/3)/(TCDUR/21)
575 N TC15=0
576 N TC16=(Y4TCTNS/5)/(TCDUR/21)
577 N TC17=(Y4TCTNS/5)/(TCDUR/21)
578 N TC18=(Y4TCTNS/5)/(TCDUR/21)                     RATES
579 N TC19=(Y4TCTNS/5)/(TCDUR/21)
580 N TC20=(Y4TCTNS/5)/(TCDUR/21)
581 N y1TCTNS=TCP1*y1FAC
582 N Y2TCTNS=TCP1*Y2FAC
583 N Y3TCTNS=TCP1*Y3FAC
584 N Y4TCTNS=TCP1*Y4FAC
585 C Y1FAC=0.30
586 C Y2FAC=0.27
587 C Y3FAC=0.24
588 C Y4FAC=0.19
589 C CCP=0.75
```

AO

The spot-charter rate is fixed by the shipping deficit and an implied correction time of 1 month (check dimensions in line 590 to see this). PSCD is the fraction of the deficit spot-chartered (see line 498 in Section AJ for the volume of the deficit being taken on immediate time-charter, and review Section 11.9 The Shipping Controller). Spot-charter rate is multiplied up in the light of price-expectations, but line 594 attempts to damp out any sharp fluctuations, and keep the average multiplier tending to 1.0.

```
590 R MSCR.KL=(ESCDP2.K/SCEFF)*SCRM.K*PSCD
591 C PSCD=1
592 A ISCRM.K=TABHL(TSCM,ESCPR.K-WS.K,-25,25,5)
593 T TSCM=0.8/0.8/0.8/0.8/0.8/1.0/1.2/1.2/1.2/1.2/1.2
594 L SCRM.K=SCRM.J+DT*RCSCRM.JK
595 R RCSCRM.KL=((ISCRM.K-SCRM.K)-(ASCM.K-1))/TAM
596 N SCRM=1
597 C TAM=1
598 L ASCM.K=ASCM.J+(DT/TASCM)(SCRM.J-ASCM.J)
599 N ASCM=1
600 C TASCM=1
```

AP

Conservation equations for tonnage, and calculation of average time the ships are in the system. (The 60 converts days to months, taking account of the fact that the return voyage is empty and the ship has left the S-C fleet by then.)

```
L MSCT.K=MSCT.J+(DT)(MSCR.JK-MSCCR.JK)          S-C TONNAGE DWT
N MSCT=(1/SCEFF)*(MDWE*RRF-FOBAS+RRF*(TPSP1-APS)+0.5*(MAXCSB+MINCSB+
X1      (MAXCSB-MINCSB)*SIN(6.283*((10-MINCT)/12)))-ACS-ECCP)
A MSCOT.K=(SHVT*SHR.K+LHVT*(1-SHR.K))/60      CO S-C OCCUP TIME MTHS
R MSCCR1.KL=DELAY3(MSCR.JK,MSCOT.K/4)          ) DUMMY VARIABLES
R MSCCR2.KL=DELAY3(MSCCR1.JK,MSCOT.K/4)        )   FOR  12TH
R MSCCR3.KL=DELAY3(MSCCR2.JK,MSCOT.K/4)        )     ORDER
R MSCCR.KL=DELAY3(MSCCR3.JK,MSCOT.K/4)         )     DELAY
```

AQ

Initial condition equations similar in principle to Section AN.

```
N MSCR=(MSCT*30)/NSCOT
N MSCCR1=(MSCT*30)/NSCOT
N MSCCR2=(MSCT*30)/NSCOT
N MSCCR3=(MSCT*30)/NSCOT
C NSCOT=27.5
NOTE
```

AR

Tables of 'prices' in World Scale and World Scale Equivalent. As explained in the text, for the purposes of the design study an historical 5-year period was taken and folded over to create a 10-year series.

```
NOTE   TANKER MARKET SECTOR
NOTE
NOTE
NOTE   CHARTER PRICE FORMATION PROCESS
NOTE
A WS.K=TABHL(TSCP,TIME.K,0,120,1.0)                SPOT CHARTER PRICE
A WSE.K=TABHL(TTCP,TIME.K,0,120,1)                 TIME CHARTER PRICE
T TSCP=104/103/92/99/124/129/122/91/93/91/101/99/96/78/84/64/69/69/
X1     79/81/90/102/131/142/116/122/127/133/129/176/213/214/252/289/285/2
X2     18/202/163/147/112/101/77/64/77/62/69/80/94/84/68/56/54/54/62/62/6
X3     2/79/78/90/100/104/103/92/99/124/129/122/91/93/91/101/99/96/78/84/
X4     64/69/69/79/81/90/102/131/142/116/122/127/133/129/176/213/214/252/
X5     289/285/218/202/163/147/112/101/77/64/77/62/69/80/94/84/68/56/54/5
X6     6/69/72/76/97/98/114/128/135
T TTCP=63/63/63/67/74/76/77/76/73/72/72/71/67/64/64/57/57/60/60/60/
X1     60/60/65/66/66/66/72/77/80/99/124/133/139/145/153/152/147/144/133/
X2     122/103/99/86/83/82/79/79/79/79/74/68/65/64/64/64/65/66/71/71/69/6
X3     3/63/63/67/74/76/77/76/73/72/72/71/67/64/64/57/57/60/60/60/60/60/6
X4     5/66/66/66/72/77/80/99/124/133/139/145/153/152/147/144/133/122/103
X5     /99/86/83/82/79/79/79/79/74/68/65/64/64/64/65/66/71/71/69/74
NOTE
```

AS

The model requires spot prices to be predicted. Shipping brokers tend to do this by informal judgement but for the purposes of the design study a technique derived by Zannetos was used to model the forecasting process. Another approach would be to use the table functions of Section AR and the method of Section 5.6. However, the Zannetos method was adopted in an early version of the model which had an explicit model of tanker price formation processes.

```
NOTE  CALCULATION OF EXPECTED CHARTER PRICES,REF ZANNETOS PAGE 219
NOTE
C PERD1=1                                          PERIODICITY
L OSCP1.K=OSCP1.J+DT*PULSE((WS.J-OSCP1.J)/DT,PERD1-DT/2,PERD1)
N OSCP1=IOSCP
L OSCP2.K=OSCP2.J+DT*PULSE((OSCP1.J-OSCP2.J)/DT,PERD1-DT/2,PERD1)
N OSCP2=IOSCP
```

```
643 L OSCP3.K=OSCP3.J+DT*PULSE((OSCP2.J-OSCP3.J)/DT,PERD1-DT/2,PERD1)
644 N OSCP3=IOSCP
645 L OSCP4.K=OSCP4.J+DT*PULSE((OSCP3.J-OSCP4.J)/DT,PERD1-DT/2,PERD1)
646 N OSCP4=IOSCP
647 C IOSCP=100
648 A ETPLUS1.K=0.50*(WS.K/OSCP1.K)+0.27*(OSCP1.K/OSCP2.K)+0.15*(OSCP2.K
649 X1      /OSCP3.K)+0.08*(OSCP3.K/OSCP4.K)
650 A ESCPR.K=ETPLUS1.K*US.K                              EXPECTED S-C PRICE
```

AT

The company's T-C fleet at any one time is made up of ships chartered at various times during the previous 3–5 years and at different price levels. Working out the exact average price of all this could be complicated. We therefore use a method based on calculating the effect on time-charter expenditure rate of increments to the fleet. (DTCER is in WSE and is converted into £/month in line 473, Section AH.) If the time-charter rate is Q [$DWT \times M^{-1}$] and the price is P [$£ \times DWT^{-1} \times M^{-1}$] then the rate of increase in TCER is PQ [$£ \times M^{-2}$]. When Q leaves the fleet, TCER will decrease at the same rate. We therefore simply need to delay the rate PQ by the same amount as Q.

Initial conditions are simplified by assuming that WSE has been 63 for the previous four years. As an exercise, take the first 48 values from TTCP in Section AR, 'fold' them to run backwards, and re-initialize the delay.

```
651 NOTE
652 NOTE  VALUATION OF COMPANYS TIME CHARTER TONNAGE
653 NOTE
654 S ATCP.K=DTCER.K/TCP1.K
655 L DTCER.K=DTCER.J+DT*(RATCE.JK-RRTCE.JK)
656 N DTCER=IWSE*TCP1
657 C IWSE=63
658 R DRATCE.KL=ITCR.K*WSE.K
659 R RATCE.KL=DELAY3(DRATCE.JK,1)+PBPR.K*TABHL(TTCP,TIME.K-6,0,120,1)
660 N DRATCE=IWSE*ITCR
661 R RRTCE.KL=DELAY3(DAR20.JK,TCDUR/21)
662 R DAR20.KL=DELAY3(DAR19.JK,TCDUR/21)
663 R DAR19.KL=DELAY3(DAR18.JK,TCDUR/21)
664 R DAR18.KL=DELAY3(DAR17.JK,TCDUR/21)
665 R DAR17.KL=DELAY3(DAR16.JK,TCDUR/21)
666 R DAR16.KL=DELAY3(DAR15.JK,TCDUR/21)
667 R DAR15.KL=DELAY3(DAR14.JK,TCDUR/21)
668 R DAR14.KL=DELAY3(DAR13.JK,TCDUR/21)
669 R DAR13.KL=DELAY3(DAR12.JK,TCDUR/21)
670 R DAR12.KL=DELAY3(DAR11.JK,TCDUR/21)
671 R DAR11.KL=DELAY3(DAR10.JK,TCDUR/21)
672 R DAR10.KL=DELAY3(DAR9.JK,TCDUR/21)
673 R DAR9.KL=DELAY3(DAR8.JK,TCDUR/21)
674 R DAR8.KL=DELAY3(DAR7.JK,TCDUR/21)
675 R DAR7.KL=DELAY3(DAR6.JK,TCDUR/21)
676 R DAR6.KL=DELAY3(DAR5.JK,TCDUR/21)
677 R DAR5.KL=DELAY3(DAR4.JK,TCDUR/21)
678 R DAR4.KL=DELAY3(DAR3.JK,TCDUR/21)
679 R DAR3.KL=DELAY3(DAR2.JK,TCDUR/21)
680 R DAR2.KL=DELAY3(DAR1.JK,TCDUR/21)
681 R DAR1.KL=DELAY3(RATCE.JK,TCDUR/21)
682 N RATCE=(TCAR*IWSE)
683 N DAR1=TC1*IWSE
684 N DAR2=TC2*IWSE
685 N DAR3=TC3*IWSE
686 N DAR4=TC4*IWSE
687 N DAR5=TC5*IWSE
688 N DAR6=TC6*IWSE
689 N DAR7=TC7*IWSE
690 N DAR8=TC8*IWSE
691 N DAR9=TC9*IWSE
692 N DAR10=TC10*IWSE
693 N DAR11=TC11*IWSE
694 N DAR12=TC12*IWSE
```

```
N DAR13=TC13*IWSE
N DAR14=TC14*IWSE
N DAR15=TC15*IWSE
N DAR16=TC16*IWSE
N DAR17=TC17*IWSE
N DAR18=TC18*IWSE
N DAR19=TC19*IWSE
N DAR20=TC20*IWSE
```

AU

For the purposes of the book, this sector has been reduced to two lines with a constant spot-charter fraction (review Fig. 11.2 *again*).

```
NOTE
NOTE  SPOT CHARTER FRACTION SECTOR
NOTE
A SCF.K=ISCF                                          SPOT CHARTER FRACTION
C ISCF=0.05
```

AV

Fairly straightforward but the serious reader should plot a lot more variables and really get to grips with the problem.

```
NOTE
NOTE   OUTPUT AND CONTROL SECTOR
NOTE
PRINT 1)MDWE,MRPD,MCD,TDIFF,MONTH
PRINT 2)MCDFC,MDFC,CVCIFFC,CVFOBFC,CIFFP2
PRINT 3)CIFAS,FOBAS,CIFFC,FOBFC,LSTPT
PRINT 4)WS,WSE,ESCPR,ATCP,ACSP1
PRINT 5)MINCS,MAXCS,MINPS,MAXPS,COAR
PRINT 6)ACS,TCS2,APS,TPSP1,APSP1
PRINT 7)PRTP2,CRPT,ORTP1,PO,REFCAP
PRINT 8)OFP1,CRNMST,CRDMST,RSCP2,ESCDP2
PRINT 9)CCP1A,CCP1,CCP2,ECCP,ACCP
PRINT 10)ITCR,DTCR,ITC1,ITC2,TCAR
PRINT 11)MSCR,MSCOT,AEFF,CNSCF,HSCF
PRINT 12)TCVOO,TCVNMSO,TCVDMSO,TCPT,CUMPO
PRINT 13)ODWT,TCP1,MSCT,SHWF,SHR
PRINT 14)SCER,TCER,CUMEXP,TOTEXP,LSALE
PLOT   APS=P,TPSPO=X(0,16E06)/TSER=T(0,60E06)/MSCT=M(0,16E06)/COAR=A,PO=O
X1     (0,12E06)
PLOT   ACS=C,TCSO=Y(0,16E06)/CRPT=P,REFCAP=R,PO=O(0,12E06)
C LENGTH=120
A PRTPER.K=1+STEP(7,13)+STEP(32,21)-STEP(20,101)
C PLTPER=1
C DT=0.03125
RUN    EXAGER
```

AW

Definition of Variables   It will be found very useful to have the DYSMAP MAP of the model in understanding it.

```
NOTE
NOTE   PT MEANS TONS OF PRODUCT
NOTE   DWT MEANS DEADWEIGHT TONS OF SHIPPING
NOTE   M   MEANS MONTHS
NOTE   CT MEANS TONS OF CRUDE OIL
NOTE   WS MEANS WORLD SCALE FOR THE SPOT MARKET
NOTE   WSE MEANS WORLD SCALE EQUIVALENT FOR TIME CHARTERS
NOTE
D ACCP=(CT/M) ACTUAL CARRYING CAPACITY
D ACRPT=(PT/M) TARGET REFINERY THROUGHPUT IN COMING MONTH
D ACS=(CT) ACTUAL CRUDE STOCKS
D ACSP1=(CT) ANTICIPATED CRUDE STOCK POSITION AT END OF COMING
X            MONTH
```

```
D AEFF=(1) AVERAGE FLEET EFFECTIVENESS
D APS=(PT) ACTUAL PRODUCT STOCKS
D APSP1=(PT) ANTICIPATED PRODUCT STOCK POSITION AT END OF
X         COMING MONTH
D APSP2=(PT) ANTICIPATED PRODUCT STOCK POSITION AT END OF
X         SECOND MONTH HENCE
D ASCM=(1) AVERAGE SPOT-CHARTER MULTIPLIER
D ATCIF=(M) CIF FRACTION ADJUSTMENT TIME
D ATCP=(WSE) AVERAGE TIME CHARTER PRICE
D ATNTC=(M) ADJUSTMENT TIME FOR NEAR MIDSUMMER TIME-CHARTERS
D ATCS=(M) ADJUSTMENT TIME FOR CRUDE STOCKS IN REFINERY PRODUCTION
X      PLANNING
D ATPS=(M) PRODUCT STOCK ADJUSTMENT TIME
D ATS=(PT) ACTUAL TOTAL STOCKS
D ATSHR=(M) ADJUSTMENT TIME FOR SHORT HAUL RATIO
D BRAT=(%) ANNUAL GROWTH RATE OF NWE GNP
D BSIN=(1) BASE VALUE FOR SEASONAL SINE WAVE USED IN PRODUCT STOCKS
D BVALCIF=(1) BASE VALUE FOR CIF FRACTION
D CCAMP=(CT) AMPLITUDE OF SINE WAVE TO GENERATE TARGET CRUDE
X              STOCKS
D CCP=(1) PROPORTION OF INITIAL TIME-CHARTERED FLEET WHICH PRESENTED
X     IN EARLIER PRESENTATION PERIODS
D CCP1=(CT/M) COMMITTED CARRYING CAPACITY FOR COMING MONTH
D CCP1A=(CT/M) DUMMY FOR CCP1 BASED ON THEORETICAL SHORT-HAUL
X            RATIO
D CCP2=(CT/M) COMMITTED CARRYING CAPACITY FOR SECOND MONTH HENCE
D CIF1=(CT/M) FORECAST MADE 11 MONTHS AGO AND BEING CARRIED
X           FORWARD BY MACRO
D CIF2=(CT/M) FORECAST MADE 10 MONTHS AGO AND BEING CARRIED
X           FORWARD BY MACRO
D CIF3=(CT/M) FORECAST MADE 9 MONTHS AGO AND BEING CARRIED
X           FORWARD BY MACRO
D CIF4=(CT/M) FORECAST MADE 8 MONTHS AGO AND BEING CARRIED
X           FORWARD BY MACRO
D CIF5=(CT/M) FORECAST MADE 7 MONTHS AGO AND BEING CARRIED
X           FORWARD BY MACRO
D CIF6=(CT/M) FORECAST MADE 6 MONTHS AGO AND BEING CARRIED
X           FORWARD BY MACRO
D CIF7=(CT/M) FORECAST MADE 5 MONTHS AGO AND BEING CARRIED
X           FORWARD BY MACRO
D CIF8=(CT/M) FORECAST MADE 4 MONTHS AGO AND BEING CARRIED
X           FORWARD BY MACRO
D CIF9=(CT/M) FORECAST MADF 3 MONTHS AGO AND BEING CARRIED
X           FORWARD BY MACRO
D CIF10=(CT/M) FORECAST MADE 2 MONTHS AGO AND BEING CARRIED
X             FORWARD BY MACRO
D CIF11=(CT/M) FORECAST MADE 1 MONTH AGO AND BEING CARRIED
X             FORWARD BY MACRO
D CIFAF=(1) ACTUAL FRACTION OF CRUDE SALES CIF
D CIFAS=(CT/M) CIF ACTUAL SALES
D CIFFC=(CT/M) CIF DEMAND FORECAST FOR FOUR MONTHS AHEAD
D CIFFP1=(CT/M) CIF FORECAST FOR 1 MONTHS AHEAD
D CIFFP2=(CT/M) CIF FORECAST FOR 2 MONTHS AHEAD
D CIFFP3=(CT/M) CIF FORECAST FOR 3 MONTHS AHEAD
D CIFFP4=(CT/M) CIF FORECAST FOR 4 MONTHS AHEAD
D CIFFP5=(CT/M) CIF FORECAST FOR N MONTHS AHEAD
D CIFFP6=(CT/M) CIF FORECAST FOR 6 MONTHS AHEAD
D CIFFP7=(CT/M) CIF FORECAST FOR 7 MONTHS AHEAD
D CIFFP8=(CT/M) CIF FORECAST FOR 8 MONTHS AHEAD
D CIFFP9=(CT/M) CIF FORECAST FOR 9 MONTHS AHEAD
D CIFFP10=(CT/M) CIF FORECAST FOR 10 MONTHS AHEAD
D CIFFP11=(CT/M) CIF FORECAST FOR 11 MONTHS AHEAD
D CIFFP12=(CT/M) CIF FORECAST FOR 12 MONTHS AHEAD
D CIFFP13=(CT/M) CIF FORECAST FOR 13 MONTHS AHEAD
D CIFFP14=(CT/M) CIF FORECAST FOR 14 MONTHS AHEAD
D CIFFP15=(CT/M) CIF FORECAST FOR 15 MONTHS AHEAD
D CIFFP16=(CT/M) CIF FORECAST FOR 16 MONTHS AHEAD
D CIFFP17=(CT/M) CIF FORECAST FOR 17 MONTHS AHEAD
D CIFFP18=(CT/M) CIF FORECAST FOR 18 MONTHS AHEAD
D CIFIV=(CT/M) INITIAL VALUE FOR CIF FORECAST MADE 12 MONTHS AGO
D CIFVAL=(CT/M) CIF FORECAST CARRIED FORWARD FROM 12 MONTHS AGO
D CNSCF=(1) CURRENT NON-SPOT CHARTER FRACTION
D COTF=(1) CURRENT OWNED TONNAGE FRACTION
D COAR=(CT/M) CRUDE OIL ARRIVAL RATE
D CONS=(DWT/(CT/M)) CARRYING CAPACITY OF NOTIONAL VESSEL
D COSFAC=(1) MAXIMUM PROPORTION OF AVAILABLE CRUDE WHICH CAN BE
X          REFINED
D CRATO=(1) PERIODICITY CORRECTION RATIOS USED IN CALCULATION OF
X        TARGET PRODUCT STOCKS FOR CURRENT MONTH
D CRAT1=(1) PERIODICITY CORRECTION RATIOS USED IN CALCULATION OF
X        TARGET PRODUCT STOCKS FOR 1 MONTH AHEAD
```

```
D CRAT2=(1) PERIODICITY CORRECTION RATIOS USED IN CALCULATION OF
X        TARGET PRODUCT STOCKS FOR 2 MONTHS AHEAD
D CRDMST=(CT) CRUDE REQUIRED FOR DISTANT MIDSUMMER TROUGH
D CRNMST=(CT) CRUDE REQUIRED FOR NEAR MIDSUMMER TROUGH
D CRNMST1=(CT) CRUDE REQUIRED FOR NEAR MIDSUMMER TROUGH WHEN TIME
X            IS IN FIRST HALF OF YEAR
D CRNMST2=(CT) CRUDE REQUIRED FOR NEAR MIDSUMMER TROUGH WHEN TIME
X            IS IN SECOND HALF OF YEAR
D CRPT=(PT/M) REFINERY THROUGHPUT FOR THE COMING MONTH
D CRRQPN=(CT) CRUDE REQUIRED IN NWE N MONTHS HENCE
D CRRQP1=(CT) CRUDE REQUIRED IN NWE 1 MONTHS HENCE
D CRRQP2=(CT) CRUDE REQUIRED IN NWE 2 MONTHS HENCE
D CRRQP3=(CT) CRUDE REQUIRED IN NWE 3 MONTHS HENCE
D CRRQP4=(CT) CRUDE REQUIRED IN NWE 4 MONTHS HENCE
D CRRQP5=(CT) CRUDE REQUIRED IN NWE 5 MONTHS HENCE
D CRRQP6=(CT) CRUDE REQUIRED IN NWE 6 MONTHS HENCE
D CRRQP7=(CT) CRUDE REQUIRED IN NWE 7 MONTHS HENCE
D CRRQP8=(CT) CRUDE REQUIRED IN NWE 8 MONTHS HENCE
D CRRQP9=(CT) CRUDE REQUIRED IN NWE 9 MONTHS HENCE
D CRRQP10=(CT) CRUDE REQUIRED IN NWE 10 MONTHS HENCE
D CRRQP11=(CT) CRUDE REQUIRED IN NWE 11 MONTHS HENCE
D CRRQP12=(CT) CRUDE REQUIRED IN NWE 12 MONTHS HENCE
D CRRQP13=(CT) CRUDE REQUIRED IN NWE 13 MONTHS HENCE
D CRRQP14=(CT) CRUDE REQUIRED IN NWE 14 MONTHS HENCE
D CRRQP15=(CT) CRUDE REQUIRED IN NWE 15 MONTHS HENCE
D CRRQP16=(CT) CRUDE REQUIRED IN NWE 16 MONTHS HENCE
D CRRQP17=(CT) CRUDE REQUIRED IN NWE 17 MONTHS HENCE
D CRRQP18=(CT) CRUDE REQUIRED IN NWE N MONTHS HENCE
D CSHR=(1) SHORT HAUL RATIO NEEDED TO USE UP SHIPPING SURPLUS
D CSHRA=(1) CONSTRAINED SHORT HAUL RATIO AS AFFECTED BY UPPER
X       BOUND
D CSHRB=(1) CONSTRAINED SHORT HAUL RATIO AS AFFECTED BY LOWER
X       BOUND
D CSIN=(1) SINE WAVE FOR PLANNED CRUDE STOCK POSITION
D CSINA=(1) SINE WAVE FOR PLANNED CRUDE STOCK POSITION
D CUMEXP=(£) CUMULATIVE EXPENDITURE ON ALL SHIPPING
D CUMLFT=(CT) CUMULATIVE LIFTINGS
D CUMPO=(PT) CUMULATIVE PRODUCT OFFTAKE
D CVCIFFC=(CT/M) CIF FORECAST MADE 12 MONTHS AGO FOR THE CURRENT
X            MONTH
D CVFOBFC=(CT/M) FOB FORECAST MADE 12 MONTHS AGO FOR THE CURRENT
X            MONTH
D DAR1=(WS/M/M) DUMMY ADDITION RATES TO TIME CHARTER EXPENDITURE
X           RATES USED TO DELAY ADDITION RATE TO MATCH
X           PRESENTATIONS
D DAR2=(WS/M/M) DUMMY ADDITION RATES TO TIME CHARTER EXPENDITURE
X           RATES USED TO DELAY ADDITION RATE TO MATCH
X           PRESENTATIONS
D DAR3=(WS/M/M) DUMMY ADDITION RATES TO TIME CHARTER EXPENDITURE
X           RATES USED TO DELAY ADDITION RATE TO MATCH
X           PRESENTATIONS
D DAR4=(WS/M/M) DUMMY ADDITION RATES TO TIME CHARTER EXPENDITURE
X           RATES USED TO DELAY ADDITION RATE TO MATCH
X           PRESENTATIONS
D DAR5=(WS/M/M) DUMMY ADDITION RATES TO TIME CHARTER EXPENDITURE
X           RATES USED TO DELAY ADDITION RATE TO MATCH
X           PRESENTATIONS
D DAR6=(WS/M/M) DUMMY ADDITION RATES TO TIME CHARTER EXPENDITURE
X           RATES USED TO DELAY ADDITION RATE TO MATCH
X           PRESENTATIONS
D DAR7=(WS/M/M) DUMMY ADDITION RATES TO TIME CHARTER EXPENDITURE
X           RATES USED TO DELAY ADDITION RATE TO MATCH
X           PRESENTATIONS
D DAR8=(WS/M/M) DUMMY ADDITION RATES TO TIME CHARTER EXPENDITURE
X           RATES USED TO DELAY ADDITION RATE TO MATCH
X           PRESENTATIONS
D DAR9=(WS/M/M) DUMMY ADDITION RATES TO TIME CHARTER EXPENDITURE
X           RATES USED TO DELAY ADDITION RATE TO MATCH
X           PRESENTATIONS
D DAR10=(WS/M/M) DUMMY ADDITION RATES TO TIME CHARTER EXPENDITURE
X           RATES USED TO DELAY ADDITION RATE TO MATCH
X           PRESENTATIONS
D DAR11=(WS/M/M) DUMMY ADDITION RATES TO TIME CHARTER EXPENDITURE
X           RATES USED TO DELAY ADDITION RATE TO MATCH
X           PRESENTATIONS
D DAR12=(WS/M/M) DUMMY ADDITION RATES TO TIME CHARTER EXPENDITURE
X           RATES USED TO DELAY ADDITION RATE TO MATCH
X           PRESENTATIONS
D DAR13=(WS/M/M) DUMMY ADDITION RATES TO TIME CHARTER EXPENDITURE
X           RATES USED TO DELAY ADDITION RATE TO MATCH
X           PRESENTATIONS
D DAR14=(WS/M/M) DUMMY ADDITION RATES TO TIME CHARTER EXPENDITURE
```

```
X            RATES USED TO DELAY ADDITION RATE TO MATCH
X            PRESENTATIONS
D DAR15=(WS/M/M) DUMMY ADDITION RATES TO TIME CHARTER EXPENDITURE
X            RATES USED TO DELAY ADDITION RATE TO MATCH
X            PRESENTATIONS
D DAR16=(WS/M/M) DUMMY ADDITION RATES TO TIME CHARTER EXPENDITURE
X            RATES USED TO DELAY ADDITION RATE TO MATCH
X            PRESENTATIONS
D DAR17=(WS/M/M) DUMMY ADDITION RATES TO TIME CHARTER EXPENDITURE
X            RATES USED TO DELAY ADDITION RATE TO MATCH
X            PRESENTATIONS
D DAR18=(WS/M/M) DUMMY ADDITION RATES TO TIME CHARTER EXPENDITURE
X            RATES USED TO DELAY ADDITION RATE TO MATCH
X            PRESENTATIONS
D DAR19=(WS/M/M) DUMMY ADDITION RATES TO TIME CHARTER EXPENDITURE
X            RATES USED TO DELAY ADDITION RATE TO MATCH
X            PRESENTATIONS
D DAR20=(WS/M/M) DUMMY ADDITION RATES TO TIME CHARTER EXPENDITURE
X            RATES USED TO DELAY ADDITION RATE TO MATCH
X            PRESENTATIONS
D DG=(1) RERUN VARIABLE TO TEST ALTERNATIVE REFINING POLICIES
D DIME=(M) DUMMY TIME VARIABLE TO MAKE SINE WAVES PEAK AT
X       TIME=0 SO SEASONALITY PEAKS AT MODEL START
D DIME2=(M) DUMMY TIME VARIABLE IN SEASONALITY CORRECTION
D DMSTCT=(DWT) TARGET T-C TONNAGE FOR DISTANT MIDSUMMER TROUGH
D DMSTTC=(DWT) TONNAGE TO HAVE TIME-CHARTERED FOR DISTANT
X            MIDSUMMER TROUGH
D DPROF=(DWT) AMOUNT OF T-C TONNAGE THAT PRESENTED BEFORE THE
X         START OF THE YEAR
D DRATCE=(WS/M/M) DUMMY RATE OF ADDDING TO TIME CHARTER
X             EXPENDITURE RATE
D DT=(M) SOLUTION TIME INTERVAL
D DTCER=(WSE/M) DUMMY TIME-CHARTER EXPENDITURE RATE
D DTCR=(DWT/M) TIME-CHARTER RATE FOR PRESENTATION IN PRESENTATION
X           BAND
D DTIME=(M) NUMBER OF MONTHS FROM START OF CURRENT YEAR
D DTONS=(DWT) TIME CHARTER TONNAGE PRESENTED DURING FIRST HALF
X         OF YEAR
D EBR=(1) ERROR AND BIAS RATIO IN FORECASTING
D ECCP=(CT/M) EXISTING COMMITTED CARRYING CAPACITY
D ESCDP2=(CT/M) EXPECTED SHIPPING CAPACITY DEFICIT FOR SECOND
X            MONTH HENCE
D ESCDP2A=(CT/M) DUMMY FOR ESCDP2 FOR POSITIVE DEFICIT
D ESCDP2B=(CT/M) DUMMY FOR ESCDP2 FOR ZERO OR NEGATIVE DEFICIT
D ESCPR=(WS) EXPECTED SPOT CHARTER PRICE
D ESHWF=(1) EXISTING SHORT HAUL WEIGHTING FACTOR
D ETCP1=(DWT) EXPECTED PRESENTATION OF T-C SHIPS IN COMING MONTH
D ETPLUS1=(1) RATIO OF EXPECTED SPOT-CHARTER PRICE IN COMING MONTH
X          TO PRESENT PRICE
D ETP10=(DWT) EXPECTED T-C PRESENTATIONS FROM PRESENTATION BAND
X        ONE MONTH IN ADVANCE
D ETP11=(DWT) EXPECTED T-C PRESENTATIONS FROM PRESENTATION BAND
X        ONE MONTH IN ADVANCE
D ETP12=(DWT) EXPECTED T-C PRESENTATIONS FROM PRESENTATION BAND
X        ONE MONTH IN ADVANCE
D FGNPLV=(1) FORECAST GNP LEVEL
D FGRAT=(1/M) FORECAST VALUE OF GNP GROWTH RATE
D FHOR=(M) FORECASTING HORIZON
D FOB1=(CT/M) FORECAST MADE 11 MONTHS AGO OF FOB SALES AND
X         BEING CARRIED FORWARD IN MACRO
D FOB2=(CT/M) FORECAST MADE 10 MONTHS AGO OF FOB SALES AND
X         BEING CARRIED FORWARD IN MACRO
D FOB3=(CT/M) FORECAST MADE 9 MONTHS AGO OF FOB SALES AND
X         BEING CARRIED FORWARD IN MACRO
D FOB4=(CT/M) FORECAST MADE 8 MONTHS AGO OF FOB SALES AND
X         BEING CARRIED FORWARD IN MACRO
D FOB5=(CT/M) FORECAST MADE 7 MONTHS AGO OF FOB SALES AND
X         BEING CARRIED FORWARD IN MACRO
D FOB6=(CT/M) FORECAST MADE 6 MONTHS AGO OF FOB SALES AND
X         BEING CARRIED FORWARD IN MACRO
D FOB7=(CT/M) FORECAST MADE 5 MONTHS AGO OF FOB SALES AND
X         BEING CARRIED FORWARD IN MACRO
D FOB8=(CT/M) FORECAST MADE 4 MONTHS AGO OF FOB SALES AND
X         BEING CARRIED FORWARD IN MACRO
D FOB9=(CT/M) FORECAST MADE 3 MONTHS AGO OF FOB SALES AND
X         BEING CARRIED FORWARD IN MACRO
D FOB10=(CT/M) FORECAST MADE 2 MONTHS AGO OF FOB SALES AND
X          BEING CARRIED FORWARD IN MACRO
D FOB11=(CT/M) FORECAST MADE 1 MONTH AGO OF FOB SALES AND
X          BEING CARRIED FORWARD IN MACRO
D FOBAS=(CT/M) ACTUAL FOB SALES
D FOBFC=(CT/M) FORECAST FOB SALES
```

```
D FOBGAP=(CT/M) GAP BETWEEN ACTUAL AND FORECAST FOB SALES
D FOBIV=(CT/M) INITIAL VALUE FOR FOR FORECASTING MECHANISM
D FOBVAL=(CT/M) FOB FORECAST MADE 12 MONTHS AGO AND CARRIED
X           FORWARD TO PRESENT
D FODGR=(1/M) FUTURE OIL DEMAND LEVEL GROWTH RATE
D FODLV=(1) FUTURE OIL DEMAND LEVEL
D FOTDMS=(DWT) FORECAST OWNED TONNAGE AT DISTANT MIDSUMMER
D FOTNMS=(DWT) FORECAST OWNED TONNAGE AT NEAR MIDSUMMER
D FTOD=(PT/Y) FUTURE OIL DEMAND IN NWE
D GNPGR=(1/M) GROWTH RATE FOR GNP IN NWE
D GNPLEV=(1) GENERAL LEVEL OF GNP IN NWE
D GRIN=(1/Y) FIRST VALUE IN TABLE FOR FUTURE GROWTH RATES
D GRRCAP=(PT/M/M) GROWTH RATE FOR REFINERY CAPACITY
D HIFAC=(1) UPPER LIMIT TO RATIO OF REFINERY THROUGHPUT TO
X        PREVIOUS MONTHS
D H1=(1) DUMMY VARIABLE FOR FIRST HALF OF YEAR
D H2=(1) DUMMY VARIABLE FOR SECOND HALF OF YEAR
D HSCF=(1) HISTORICAL SPOT-CHARTER FRACTION
D ICIF=(1) INDICATED VALUE OF CUSTOMERS CIF FRACTION
D IOSCP=(WS) INITIAL VALUE FOR OLD SPOT-CHARTER PRICES
D IRC=(PT/M) INITIAL REFINERY CAPACITY
D ISCF=(1) INITIAL VALUE FOR SPOT-CHARTER FRACTION
D ISCRM=(1) INDICATED VALUE OF SPOT-CHARTER RATE MULTIPLIER
D ITCPR=(DWT/M) PRESENTATION RATE FOR T-C VESSELS FOR
X           IMMEDIATE PRESENTATION
D ITCR=(DWT/M) IMMEDIATE T-C CHARTER RATE
D ITC1=(DWT/M) VALUE OF ITCR IN FIRST HALF OF YEAR
D LENGTH=(M) SIMULATED PERIOD
D ITC2=(DWT/M) VALUE OF ITCR IN SECOND HALF OF YEAR
D IWSE=(WSE) INITIAL VALUE FOR T-C PRICES
D LHVT=(D) LONG-HAUL VOYAGE TIME
D LOFAC=(1) LOWER LIMIT TO RATIO OF REFINERY THROUGHPUT TO
X        PREVIOUS MONTHS
D LRCIF=(1) LONG RUN AVERAGE CIF FRACTION
D LSALE=(PT) LOST PRODUCT SALES
D LSTPT=(PT) LOST THROUGHPUT DUE TO LOW CRUDE STOCKS
D LTFOT=(DWT) LONG TERM FORECAST OF MIDSUMMER OWNED TONNAGE
D MASHR=(1) COMPANY AVERAGE SHORT HAUL RATIO
D MAXCS=(CT) MAXOPS LEVEL OF CRUDE STOCKS
D MAXCSB=(CT) INITIAL VALUE FOR MAXCS
D MAXPS=(PT) MAXOPS LEVEL FOR PRODUCT STOCKS
D MAXPSB=(PT) INITIAL VALUE FOR MAXPS
D MCD=(CT/M) COMPANY CRUDE SALES DEMAND
D MCDFC=(CT/M) COMPANY CRUDE SALES DEMAND FORECAST
D MDFC=(PT/M) COMPANY DEMAND FORECAST FOR NWE FOR FHOR MONTHS
X         HENCE
D MDWE=(PT/M) COMPANY ACTUAL DEMAND IN NWE
D MFMS=(1) FORECAST COMPANY MARKET SHARE IN NWE FHOR MONTHS
X      HENCE
D MIMS=(1) INITIAL VALUE OF COMPANY MARKET SHARE IN NWE
D MINCS=(CT) MINOPS LEVEL OF CRUDE STOCKS
D MINCSB=(CT) INITIAL VALUE OF MINCS
D MINCT=(M) TIME INTO YEAR WHEN MINIMUM CRUDE STOCKS
X        ARE NEEDED
D MINPS=(PT) MINOPS LEVEL OF PRODUCT STOCKS
D MINPSB=(PT) INITIAL VALUE OF MINPS
D MINPT=(M) TIME INTO YEAR WHEN MINIMUM PRODUCT STOCKS ARE
X        NEEDED
D MMS=(1) COMPANY ACTUAL MARKET SHARE IN NWE
D MONTH=(1) INDEX FOR MONTH IN YEAR
D MPSHR=(1) COMPANY POLICY VALUE FOR SHORT-HAUL RATIO
D MPSHWF=(1) COMPANY POLICY VALUE FOR SHORT-HAUL WEIGHTING FACTOR
D MRPD=(PT/M) COMPANY REFINED PRODUCT DEMAND
D MSCCR=(DWT/M) COMPANY SPOT-CHARTER COMPLETION RATE
D MSCCR1=(DWT/M) DUMMY VARIABLE FOR 12TH ORDER DELAY IN SPOT
X            CHARTER COMPLETIONS
D MSCCR2=(DWT/M) DUMMY VARIABLE FOR 12TH ORDER DELAY IN SPOT
X            CHARTER COMPLETIONS
D MSCCR3=(DWT/M) DUMMY VARIABLE FOR 12TH ORDER DELAY IN SPOT
X            CHARTER COMPLETIONS
D MSCOT=(M) COMPANY SPOT-CHARTER DURATION
D MSCR=(DWT/M) COMPANY SPOT-CHARTER RATE
D MSCT=(DWT) COMPANY SPOT-CHARTER FLEET
D M1=(1) DUMMY VARIABLE TO INDICATE MONTH NUMBER IN YEAR
D M2=(1) DUMMY VARIABLE TO INDICATE MONTH NUMBER IN YEAR
D M3=(1) DUMMY VARIABLE TO INDICATE MONTH NUMBER IN YEAR
D M4=(1) DUMMY VARIABLE TO INDICATE MONTH NUMBER IN YEAR
D M5=(1) DUMMY VARIABLE TO INDICATE MONTH NUMBER IN YEAR
D M6=(1) DUMMY VARIABLE TO INDICATE MONTH NUMBER IN YEAR
D M7=(1) DUMMY VARIABLE TO INDICATE MONTH NUMBER IN YEAR
D M8=(1) DUMMY VARIABLE TO INDICATE MONTH NUMBER IN YEAR
```

```
D M9=(1) DUMMY VARIABLE TO INDICATE MONTH NUMBER IN YEAR
D M10=(1) DUMMY VARIABLE TO INDICATE MONTH NUMBER IN YEAR
D M11=(1) DUMMY VARIABLE TO INDICATE MONTH NUMBER IN YEAR
D M12=(1) DUMMY VARIABLE TO INDICATE MONTH NUMBER IN YEAR
D NECOA=(CT/M) NET EXPECTED CRUDE OIL ARRIVALS
D NMSTCT=(DWT) TARGET T-C TONNAGE FOR NEEDS AT NEAR MIDSUMMER
X        TROUGH
D NMSTTC=(DWT) TONNAGE TO HAVE TIME-CHARTERED FOR NEAR
X        MIDSUMMER TROUGH
D NSCOT=(M) NORMAL SPOT-CHARTER OCCUPATION TIME
D ODGR=(1/M) OIL DEMAND GROWTH RATE
D ODGRF=(1) OIL DEMAND GROWTH RATE FACTOR
D ODLV=(1) OIL DEMAND LEVEL
D ODWT=(DWT) OWNED FLEET SIZE
D OFP1=(PT/M) OFFTAKE FORECAST FOR 1 MONTH AHEAD
D OFP2=(PT/M) OFFTAKE FORECAST FOR 2 MONTHS AHEAD
D OFP3=(PT/M) OFFTAKE FORECAST FOR 3 MONTHS AHEAD
D OFP4=(PT/M) OFFTAKE FORECAST FOR 4 MONTHS AHEAD
D OFP5=(PT/M) OFFTAKE FORECAST FOR 5 MONTHS AHEAD
D OFP6=(PT/M) OFFTAKE FORECAST FOR 6 MONTHS AHEAD
D OFP7=(PT/M) OFFTAKE FORECAST FOR 7 MONTHS AHEAD
D OFP8=(PT/M) OFFTAKE FORECAST FOR 8 MONTHS AHEAD
D OFP9=(PT/M) OFFTAKE FORECAST FOR 9 MONTHS AHEAD
D OFP10=(PT/M) OFFTAKE FORECAST FOR 10 MONTHS AHEAD
D OFP11=(PT/M) OFFTAKE FORECAST FOR 11 MONTHS AHEAD
D OFP12=(PT/M) OFFTAKE FORECAST FOR 12 MONTHS AHEAD
D OFP13=(PT/M) OFFTAKE FORECAST FOR 13 MONTHS AHEAD
D OFP14=(PT/M) OFFTAKE FORECAST FOR 14 MONTHS AHEAD
D OFP15=(PT/M) OFFTAKE FORECAST FOR 15 MONTHS AHEAD
D OFP16=(PT/M) OFFTAKE FORECAST FOR 16 MONTHS AHEAD
D OFP17=(PT/M) OFFTAKE FORECAST FOR 17 MONTHS AHEAD
D OFP18=(PT/M) OFFTAKE FORECAST FOR 18 MONTHS AHEAD
D ORTP1=(PT/M) REFINERY THROUGHPUT IN CURRENT MONTH
D OSCP1=(WS) SPOT CHARTER PRICE 1 MONTH AGO
D OSCP2=(WS) SPOT CHARTER PRICE 2 MONTHS AGO
D OSCP3=(WS) SPOT CHARTER PRICE 3 MONTHS AGO
D OSCP4=(WS) SPOT CHARTER PRICE 4 MONTHS AGO
D OTEFF=(1) OWNED TANKER EFFECTIVENESS
D OTER=(£/M) OWNED TONNAGE EXPENDITURE RATE
D OTGF=(%/Y) ANNUAL INCREASE IN OWNED TONNAGE
D OTGRF=(1) OWNED TONNAGE GROWTH RATE FACTOR
D OTNBR=(DWT/M) OWNED TONNAGE NET BUILDING RATE
D OTOC=(£/DWT/M) OWNED TONNAGE OPERATING COST
D PBPR=(DWT/M) PRESENTATION RATE DURING PRESENTATION BAND
D PD=(M) PULSE INTERVAL IN MACRO TO MOVE FORWARD FOB AND CIF
X   FORECASTS
D PCAMP=(PT) AMPLITUDE OF SINE WAVE FOR TARGET PRODUCT STOCKS
D PDIFF=(PT/M) DIFFERENCE BETWEEN THIS MONTHS AND LAST MONTHS
X        REFINERY THROUGHPUTS
D PERD=(M) DECISION-MAKING AND SAMPLING PERIODICITY
D PLTPER=(M) PLOTTING INTERVAL FOR OUTPUT
D PERD1=(M) RECORDING INTERVAL USED FOR OLD SPOT-CHARTER PRICES
D PO=(PT/M) PRODUCT OFFTAKE
D PPM1=(1) PROPORTION OF TIME-CHARTERED VESSELS ON ORDER IN
X        JUNE WHICH WILL PRESENT DURING THE FIRST MONTH OF THE
X        PRESENTATION BAND
D PPM2=(1) PROPORTION OF TIME-CHARTERED VESSELS ON ORDER IN
X        JUNE WHICH WILL PRESENT DURING THE SECOND MONTH OF THE
X        PRESENTATION BAND
D PPM3=(1) PROPORTION OF TIME-CHARTERED VESSELS ON ORDER IN
X        JUNE WHICH WILL PRESENT DURING THE N MONTH OF THE
X        PRESENTATIOTHIRD BAND
D PRTP2=(PT/M) PLANNED REFINERY THROUGHPUT FOR SECOND MONTH HENCE
D PRTPER=(M) PRINTING INTERVAL IN OUTPUT
D PSCD=(1) PERCENT OF SHIPPING DEFICIT SPOT CHARTERED
D PSCURVE=(PT) SINE WAVE FOR PRODUCT STOCK PLANNING
D PSLEV=(PT) MEAN LEVEL FOR TARGET PRODUCT STOCK CURVE
D RAT1=(PT/1) FACTOR CONVERTING GNP TO OIL DEMAND
D RATCE=(WSE/M/M) RATE OF ADDING TO TIME-CHARTER EXPENDITURE RATE
D RCCIF=(1/M) RATE OF CHANGE OF CIF FRACTION
D RCGRF=(1) REFINERY CAPACITY GROWTH FACTOR
D RCORPT1=(PT/M) RATE OF CHANGE OF OLD REFINERY THROUGHPUT
D RCSCRM=(1/M) RATE OF CHANGE OF SPOT-CHARTER MULTIPLIER
D REFCAP=(PT/M) REFINERY CAPACITY
D RPF=(1) FRACTION OF COMPANY DEMAND WHICH IS FOR REFINED PRODUCT
D RRF=(CT/PT) REFINERY RECOVERY FACTOR
D RRTCE=(WSE/M/M) RATE OF REDUCTION OF T-C EXPENDITURE RATE
D RSCP2=(CT/M) REQUIRED SHIPPING CAPACITY DURING SECOND MONTH
X        HENCE
D RTCIF=(M) RESPONSE TIME FOR CIF FRACTION
D SCEFF=(1) SPOT-CHARTER EFFECTIVENESS
```

```
1153 D SCER=(£/M) SPOT-CHARTER EXPENDITURE RATE
1154 D SCF=(1) SPOT-CHARTER FRACTION - PROPORTION OF MIDSUMMER SHIPPING
1155 X     NEEDS NOT MET BY OWNED OR T-C VESSELS AND LEFT FOR
1156 X     SPOT-CHARTERING
1157 D SCRM=(1) SPOT-CHARTER MULTIPLIER TO REFLECT EFFECT OF PRICE
1158 X     EXPECTATIONS
1159 D SEAMP=(1) SEASONAL AMPLITUDE
1160 D SEBR=(1) SHORT-TERM ERROR AND BIAS RATIO FOR NEAR MIDSUMMER
1161 D SEBR1=(1) SHORT-TERM ERROR AND BIAS RATIO FOR MONTH 1
1162 D SEBR2=(1) SHORT-TERM ERROR AND BIAS RATIO FOR MONTH 2
1163 D SEBR3=(1) SHORT-TERM ERROR AND BIAS RATIO FOR MONTH 3
1164 D SEBR4=(1) SHORT-TERM ERROR AND BIAS RATIO FOR MONTH 4
1165 D SEBR5=(1) SHORT-TERM ERROR AND BIAS RATIO FOR MONTH 5
1166 D SEBR6=(1) SHORT-TERM ERROR AND BIAS RATIO FOR MONTH 6
1167 D SEBR7=(1) SHORT-TERM ERROR AND BIAS RATIO FOR MONTH 7
1168 D SEBR8=(1) SHORT-TERM ERROR AND BIAS RATIO FOR MONTH 8
1169 D SEBR9=(1) SHORT-TERM ERROR AND BIAS RATIO FOR MONTH 9
1170 D SEBR10=(1) SHORT-TERM ERROR AND BIAS RATIO FOR MONTH 10
1171 D SEBR11=(1) SHORT-TERM ERROR AND BIAS RATIO FOR MONTH 11
1172 D SHR=(1) SHORT-HAUL RATIO
1173 D SHRAT=(M) SHORT-HAUL RATIO AVERAGING TIME
1174 D SHRC=(1) SWITCH TO INDICATE PRESENCE OF A SHIPPING SURPLUS
1175 D SHRLB=(1) LOWER BOUND FOR SHORT-HAUL RATIO
1176 D SHRLLIM=(1) MULTIPLIER OF POLICY SHORT-HAUL RATIO TO GIVE LOWER
1177 X        BOUND TO SHORT-HAUL RATIO
1178 D SHRM=(1) MULTIPLIER TO INDICATE CHARTER-PRICE BASED TARGET VALUE
1179 X      FOR SHORT-HAUL RATIO
1180 D SHRUB=(1) UPPER BOUND FOR SHORT-HAUL RATIO
1181 D SHRULIM=(1) MULTIPLIER OF POLICY SHORT-HAUL RATIO TO GIVE UPPER
1182 X        BOUND TO SHORT-HAUL RATIO
1183 D SHVT=(D) SHORT-HAUL VOYAGE TIME
1184 D SHWF=(1) SHORT-HAUL WEIGHTING FACTOR - DEFINED AS THE RECIPROCAL
1185 X      OF THE FACTOR BY WHICH NOTIONAL CARRYING CAPACITY
1186 X      INCREASES AS SHR CHANGES
1187 D SHWFA=(1) SHORT-HAUL WEIGHTING FACTOR AT AVERAGE SHORT-HAUL
1188 X       RATIO
1189 D SPRPF=(£/PT) PRODUCT PROFIT AT MIDSUMMER
1190 D SRAT1=(1) SEASONALITY CORRECTION FACTOR FOR FORECASTING FOR MONTH 1
1191 D SRAT2=(1) SEASONALITY CORRECTION FACTOR FOR FORECASTING FOR MONTH 2
1192 D SRAT3=(1) SEASONALITY CORRECTION FACTOR FOR FORECASTING FOR MONTH 3
1193 D SRAT4=(1) SEASONALITY CORRECTION FACTOR FOR FORECASTING FOR MONTH 4
1194 D SRAT5=(1) SEASONALITY CORRECTION FACTOR FOR FORECASTING FOR MONTH 5
1195 D SRAT6=(1) SEASONALITY CORRECTION FACTOR FOR FORECASTING FOR MONTH 6
1196 D SRAT7=(1) SEASONALITY CORRECTION FACTOR FOR FORECASTING FOR MONTH 7
1197 D SRAT8=(1) SEASONALITY CORRECTION FACTOR FOR FORECASTING FOR MONTH 8
1198 D SRAT9=(1) SEASONALITY CORRECTION FACTOR FOR FORECASTING FOR MONTH 9
1199 D SRAT10=(1) SEASONALITY CORRECTION FACTOR FOR FORECASTING FOR MONTH 10
1200 D SRAT11=(1) SEASONALITY CORRECTION FACTOR FOR FORECASTING FOR MONTH 11
1201 D STFOT=(DWT) SHORT-TERM FORECAST OF MIDSUMMER OWNED TONNAGE
1202 D SYSVL=(1) RE-RUN CONSTANT USED TO TEST T-C POLICIES
1203 D TACS=(M) TIME TO ADJUST CRUDE STOCKS
1204 D TAM=(M) ADJUSTMENT TIME FOR SPOT-CHARTER MULTIPLIER
1205 D TASCM=(M) TIME TO AVERAGE SPOT-CHARTER MULTIPLIER
1206 D TATCF=(M) TIME TO ADJUST T-C DEFICIT
1207 D TCAR=(DWT/M) RATE OF PRESENTATION OF T-C TONNAGE
1208 D TCCR=(DWT/M) TIME-CHARTER COMPLETION RATE
1209 D TCDUR=(M) TIME-CHARTER DURATION
1210 D TCEFF=(1) TIME-CHARTER EFFECTIVENESS
1211 D TCER=(£/M) TIME-CHARTER EXPENDITURE RATE
1212 D TCIF=(1) TABLE FOR PRICE-INDICATED CIF FRACTION MULTIPLIER
1213 D TCP1=(DWT) TIME CHARTER TONNAGE AT CURRENT TIME
1214 D TCPT=(DWT) TONNAGE TO PRESENT DURING REMAINDER OF PRESENTATION
1215 X         PERIOD
1216 D TCPT10=(DWT) TONNAGE TO PRESENT DURING OCTOBER
1217 D TCPT11=(DWT) TONNAGE TO PRESENT DURING NOVEMBER
1218 D TCPT12=(DWT) TONNAGE TO PRESENT DURING DECEMBER
1219 D TCRSW=(1) SWITCH TO STOP IMMEDIATE TIME-CHARTERING WHEN
1220 X       THERE IS A SURPLUS
1221 D TCS0=(CT) TARGET CRUDE STOCKS NOW
1222 D TCS1=(CT) TARGET CRUDE STOCKS FOR 1 MONTH HENCE
1223 D TCS2=(CT) TARGET CRUDE STOCKS FOR 2 MONTHS HENCE
1224 D TCVDMSO=(DWT) T-C VESSELS ORDERED FOR PRESENTATION BEFORE
1225 X          DISTANT MIDSUMMER
1226 D TCVNMSO=(DWT) T-C VESSELS ORDERED FOR NEAR MIDSUMMER TROUGH
1227 D TCVOO=(DWT) T-C VESSELS ON ORDER BUT NOT YET PRESENTED
1228 D TC1=(DWT/M) DUMMY INTERNAL RATE FOR TIME CHARTER
1229 X         COMPLETIONS
1230 D TC2=(DWT/M) DUMMY INTERNAL RATE FOR TIME CHARTER
1231 X         COMPLETIONS
1232 D TC3=(DWT/M) DUMMY INTERNAL RATE FOR TIME CHARTER
1233 X         COMPLETIONS
```

```
D TC4=(DWT/M) DUMMY INTERNAL RATE FOR TIME CHARTER
X          COMPLETIONS
D TC5=(DWT/M) DUMMY INTERNAL RATE FOR TIME CHARTER
X          COMPLETIONS
D TC6=(DWT/M) DUMMY INTERNAL RATE FOR TIME CHARTER
X          COMPLETIONS
D TC7=(DWT/M) DUMMY INTERNAL RATE FOR TIME CHARTER
X          COMPLETIONS
D TC8=(DWT/M) DUMMY INTERNAL RATE FOR TIME CHARTER
X          COMPLETIONS
D TC9=(DWT/M) DUMMY INTERNAL RATE FOR TIME CHARTER
X          COMPLETIONS
D TC10=(DWT/M) DUMMY INTERNAL RATE FOR TIME CHARTER
X           COMPLETIONS
D TC11=(DWT/M) DUMMY INTERNAL RATE FOR TIME CHARTER
X           COMPLETIONS
D TC12=(DWT/M) DUMMY INTERNAL RATE FOR TIME CHARTER
X           COMPLETIONS
D TC13=(DWT/M) DUMMY INTERNAL RATE FOR TIME CHARTER
X           COMPLETIONS
D TC14=(DWT/M) DUMMY INTERNAL RATE FOR TIME CHARTER
X           COMPLETIONS
D TC15=(DWT/M) DUMMY INTERNAL RATE FOR TIME CHARTER
X           COMPLETIONS
D TC16=(DWT/M) DUMMY INTERNAL RATE FOR TIME CHARTER
X           COMPLETIONS
D TC17=(DWT/M) DUMMY INTERNAL RATE FOR TIME CHARTER
X           COMPLETIONS
D TC18=(DWT/M) DUMMY INTERNAL RATE FOR TIME CHARTER
X           COMPLETIONS
D TC19=(DWT/M) DUMMY INTERNAL RATE FOR TIME CHARTER
X           COMPLETIONS
D TC20=(DWT/M) DUMMY INTERNAL RATE FOR TIME CHARTER
X           COMPLETIONS
D TDIFF=(M) TIME INTO SIMULATED YEAR
D TDWT=(DWT) TOTAL SHIPPING IN COMPANY'S SERVICE
D TEBR=(1) TABLE OF ERROR AND BIAS VALUES FOR SUCCESSIVE YEARS
D TGRAT=(%/Y) TABLE OF GNP GROWTH RATES FOR SUCCESSIVE YEARS
D TIME=(M) BUILT-IN VARIABLE FOR PASSAGE OF TIME
D TOTEXP=(£) TOTAL EXPENDITURE
D TPRTP2=(PT/M) TARGET PLANNED REFINERY THROUGHPUT FOR 2 MONTHS
X            HENCE
D TPSP0=(PT) TARGET PRODUCT STOCK POSITION NOW
D TPSP1=(PT) TARGET PRODUCT STOCK POSITION AT END OF NEXT MONTH
D TPSP2=(PT) TARGET PRODUCT STOCK  OSITION AT END OF SECOND
X        MONTH
D TSCM=(1) TABLE OF VALUES FOR INDICATED SPOT-CHARTER RATE
X      MULTIPLIER
D TSCP=(WS) TABLE FOR SPOT-CHARTER PRICE
D TSER=(£/M) TOTAL SHIPPING EXPENDITURE RATE
D TSHR=(1) TABLE FOR SHORT-HAUL RATIO MULTIPLIER
D TSHR1=(1) TARGET SHORT-HAUL RATIO FOR COMING MONTH
D TSHR1A=(1) TARGET SHORT-HAUL RATIO AS CONSTRAINED BY UPPER
X        BOUND
D TSHR1B=(1) TARGET SHORT-HAUL RATIO AS CONSTRAINED BY UPPER
X        AND LOWER BOUNDS
D TTCP=(WSE) TABLE OF T-C PRICES
D ULIM2=(PT/M) UPPER LIMIT ON REFINERY THROUGHPUT FOR SECOND
X            MONTH HENCE
D ULIM2A=(PT/M) SECOND UPPER LIMIT ON REFINERY THROUGHPUT FOR
X            SECOND MONTH HENCE
D ULIM2B=(PT/M) NET UPPER LIMIT ON REFINERY THROUGHPUT FOR
X            SECOND MONTH HENCE
D ULIMC=(PT/M) UPPER LIMIT ON REFINERY THROUGHPUT FOR COMING MONTH
D ULIMCA=(PT/M) SECOND UPPER LIMIT ON REFINERY THROUGHPUT FOR
X            COMING MONTH
D ULIMCB=(PT/M) NET UPPER LIMIT ON REFINERY THROUGHPUT FOR
X           COMING MONTH
D VALULS=(£) VALUE OF LOST SALES
D WPRPF=(£/PT) PRODUCT PROFIT AT MIDWINTER
D WS=(WS) CURRENT SPOT-CHARTER PRICE
D WSE=(WSE) CURRENT T-C PRICE
D Y1FAC=(1) PROPORTION OF INITIAL T-C FLEET WHICH IS 1 YEAR OLD
D Y2FAC=(1) PROPORTION OF INITIAL T-C FLEET WHICH IS 2 YEARS OLD
D Y3FAC=(1) PROPORTION OF INITIAL T-C FLEET WHICH IS 3 YEARS OLD
D Y4FAC=(1) PROPORTION OF INITIAL T-C FLEET WHICH IS 4 YEARS OLD
D Y1TCTNS=(DWT) TONNAGE IN INITIAL T-C FLEET WHICH IS 1 YEAR OLD
D Y2TCTNS=(DWT) TONNAGE IN INITIAL T-C FLEET WHICH IS 2 YEARS OLD
D Y3TCTNS=(DWT) TONNAGE IN INITIAL T-C FLEET WHICH IS 3 YEARS OLD
D Y4TCTNS=(DWT) TONNAGE IN INITIAL T-C FLEET WHICH IS 4 YEARS OLD
```

AX

Rerun cards:

```
T TEBR=0.9/0.9/0.9/1.0/1.1/1.1/1.1/1.1/1.1/1.0/0.9/0.9/0.9/1.0/1.1/
X1     1.1/1.1/1.1/1.1/1.1/1.1
RUN    CONSER
T TEBR=1.0/1.0/1.0/1.0/1.0/1.0/1.0/1.0/1.0/1.0/1.0/1.0/1.0/1.0/1.0/
X1     1.0/1.0/1.0/1.0/1.0/1.0
RUN    PERF
T TEBR=0.9/0.9/0.9/0.9/0.9/0.9/0.9/0.9/0.9/0.9/0.9/0.9/0.9/0.9/0.9/
X 0.9/0.9/0.9/0.9/0.9/0.9
RUN PESSI
T TEBR=1.1/1.1/1.1/1.1/1.1/1.1/1.1/1.1/1.1/1.1/1.1/1.1/1.1/1.1/1.1/
X 1.1/1.1/1.1/1.1/1.1/1.1
RUN OPTIM
C IRC=10E06
RUN 2
C DC=1
C MAXPSB=3.7E06
C MINPSB=2.1E06
C MINCT=2
C MINPT=2
C MAXCSB=2.9E06
C MINCSB=1.9E06
T TSHR=1/1/1/1/1
T TSCM=1/1/1/1/1/1/1/1/1/1/1
C ATPS=1
C ATCS=1
C TACS=1
C PSCD=0
C SYSVL=0
RUN     MXC1
C DC=1
T TSHR=1/1/1/1/1
T TSCM=1/1/1/1/1/1/1/1/1/1/1
C ATPS=1
C MINPT=3
C MAXPSB=3.6E06
C MINPSB=2.1E06
C PSCD=0.5
C MAXCSB=2.5E06
C MINCSB=2.5E06
C ATCS=1
C SYSVL=0
RUN   CRCON2
*
```

B.12 United Metals Ltd.

The salient features of the model have already been examined in Chapter 12. The listing follows, but the reader will find it helpful to obtain a documented DYSMAP run to help in understanding.

The program contains rerun cards in lines 244–279 which will test the situations mentioned in the chapter, and some others.

```
 0 DOC
 1 *   UNITED MODEL
 2 NOTE
 3 NOTE   METAL DEMAND AND COMPANY ORDER RATE
 4 NOTE
 5 A MDMD.K=(400000/12)*(1+STEP(.2,STM))
 6 C STM=60
 7 R UOR.KL=UMS.K*MDMD.K
 8 NOTE
 9 NOTE BACKLOG AND PRODUCTION PLANNING
10 NOTE
11 L OBS.K=OBS.J+DT*(UOR.JK-TPUT.JK)
12 N OBS=COVER*LTPUT
13 C COVER=2
14 A IPUT.K=MAX(0,MIN((OBS.K-TOB.K)/TAOB+LTPUT.K,SFCAP.K))
15 A TOB.K=COVER*LTPUT.K
16 C TAOB=2
```

```
NOTE
NOTE WORKING CAPITAL AND ACTUAL PRODUCTION
NOTE
A WCPUP.K=UPC.K*WCMF+WCSF
N WCMF=WCSF
C WCSF=3018
A WCAPR.K=IPUT.K*WCPUP.K
A WCRAT.K=WCAP.K/WCAPR.K
R TPUT.KL=IPUT.K*MIN(1,WCRAT.K)
N TPUT=IPUT*MIN(1,WCRAT)
R RRWC.KL=((WCAP.K-WCAPR.K)/TRWC)*CLIP(1,0,WCRAT.K,WCRCV)
C TRWC=6
C WCRCV=1.2
NOTE
NOTE AVERAGE THROUGHPUT LEVEL
NOTE
L LTPUT.K=LTPUT.J+(DT/TATP)(TPUT.JK-LTPUT.J)
C TATP=1.0
N LTPUT=UMS*MDMD
NOTE
NOTE CURRENT PRODUCTION CAPACITY
NOTE
L SFCAP.K=SFCAP.J+DT*(SFCCR.JK-SFCDR.JK)
N SFCAP=8330
L MFCAP.K=MFCAP.J+DT*(MFCCR.JK-MFCDR.JK)
N MFCAP=IPCU*SFCAP
R SFCDR.KL=(SFCAP.K*UR)/1200
R MFCDR.KL=(MFCAP.K*UR)/1200
C WR=20
NOTE
NOTE  PROFITS AND CASH FLOW SECTOR
NOTE
R PR.KL=DELAY3(TPUT.JK,SFD)*(1-UPC.K)*GMSF.K
X  +DELAY3(TPUT.JK,MFD)*UPC.K*GMMG.K-MSS.K+(210-UMC.K)*LTPUT.K
X +(INTPC/1200)*(MCNP.K/2+CASH.K-LTDEBT.K)
C INTPC=15
C SFD=1
A GMSF.K=IGMSF*(1+STEP(SFPFAC,PRSTM))
A GMMG.K=IGMMG*(1+STEP(MFPFAC,PRSTM))
C PRSTM=60
C IGMSF=117
C IGMMG=212
C SFPFAC=0
C MFPFAC=0
C MFD=3
L APL.K=APL.J+(DT/TAP)(PR.JK-APL.J)
N APL=SFCAP*GMSF*(1-UPC)+MFCAP*GMMG+(INTPC/1200)(MCNP/2+CASH-LTDEBT)
X -MSS*(210-UMC)*LTPUT
A ATPL.K=((100-TAXDPC)/100)*(APL.K-DCFL.K)
C TAXDPC=50
C TAP=3
NOTE
NOTE FINANCIAL PERFORMANCE INDICATORS
NOTE
A ROI.K=((APL.K*1200)/(VP.K+WCAP.K))
S ROIRAT.K=ROI.K/TROI
C TROI=20
L CUPROF.K=CUPROF.J+DT*PR.JK
N CUPROF=0
NOTE
NOTE CASH GENERATION SECTOR
NOTE
A ACGR.K=ATPL.K+DCFL.K
A DCFL.K=MAX(0,DEPPC*WDV.K)
N DEPPC=BPC/1200
C BPC=30
A ECG.K=ACGR.K*PHOR
C PHOR=12
A MXLTD.K=(CRP.K+SCAP.K+ATPL.K*PHOR)*(GEAR/100)
C GEAR=60
A LTDIFF.K=MXLTD.K-LTDEBT.K
A ARS.K=(MXLTD.K-LTDEBT.K+ECG.K+CASH.K)/PHOR
NOTE
NOTE  FORWARD PLANNING -DEMAND FORECASTS
NOTE
A LTDFC.K=(400000/12)*EBR.K*(1+STEP(.2,STM-PHOR))
A EBR.K=BVAL.K*ERPH.K
A BVAL.K=IBVAL
C IBVAL=1.0
A ERPH.K=TABHL(TERPH,PHOR,12,36,12)
T TERPH=1.0/1.2/1.4
```

```
NOTE
NOTE FORWARD PLANNING - TARGETS
NOTE
A PMS.K=MSGF*UMS.K
C MSGF=1.0
A UPC.K=MFCAP.K/SFCAP.K
N UPC=IPCU
A PPCU.K=(1+GFAC)*IPCU
C GFAC=0
C IPCU=0.30
A ISFCAP.K=LTDFC.K*PMS.K
A IMFCAP.K=PPCU.K*ISFCAP.K
NOTE
NOTE FORWARD PLANNING - CAPACITY REQUIREMENTS
NOTE
A TROSFC.K=MAX(0,(ISFCAP.K-SFCAP.K*(1-DNRSF*(WR/1200)*CDEL)-SFCOR.K)
X /TASFC)
N TASFC=PHOR
A TROMFC.K=MAX(0,(IMFCAP.K-MFCAP.K*(1-DNRMF*(WR/1200)*CDEL)-MFCOR.K)
X /TAMFC)
N TAMFC=PHOR
A IWC.K=ISFCAP.K*(WCSF+PPCU.K*WCMF)
C DNRSF=1
C DNRMF=1
NOTE
NOTE FORWARD PLANNING - SPENDING NEEDS
NOTE
A TIWC.K=MAX(0,(IWC.K-WCAP.K)/TAWC)
N TAWC=PHOR
A RRS.K=TIWC.K+TROSFC.K*CCSFC+TROMFC.K*CCMFC
C CCSFC=5580
N CCMFC=CCSFC*0.25
NOTE
NOTE IMPLEMENTATION OF PLANNING DECISIONS
NOTE
A FRAT.K=ARS.K/(RRS.K+CLIP(0,1,RRS.K,0.1))
A FPMULT.K=TABHL(TFPMULT,FRAT.K,0,2,0.5)
TP TFPMLT=0/0.8/1.0/1.1/1.1
A WCMULT.K=FPMULT.K*WCPF
C WCPF=1.2
A ARCMTNP.K=MAX(0,FPMULT.K*(RRS.K-TIWC.K))
A ARWC.K=WCMULT.K*TIWC.K
A RSAA.K=ARWC.K+ARCMTNP.K
R RALTD.KL=MAX(CDIFF.K,0)
R RRLTD.KL=MIN(CDIFF.K,0)*CLIP(-1,0,LTDEBT.K,0)
R SFCSR.KL=TROSFC.K*FPMULT.K
R MFCSR.KL=TROMFC.K*FPMULT.K
NOTE
NOTE CAPACITY CONSTRUCTION PIPELINE SECTOR
NOTE
L SFCOR.K=SFCOR.J+DT*(SFCSR.JK-SFCCR.JK)
N SFCOR=(SFCAP*WR*(CDEL-TASFC))/1200
R SFCCR.KL=DELAY3(SFCSR.JK,CDEL)
N SFCSR=(SFCAP*WR)/1200
L MFCOR.K=MFCOR.J+DT*(MFCSR.JK-MFCCR.JK)
R MFCCR.KL=DELAY3(MFCSR.JK,CDEL)
N MFCSR=(MFCAP*WR)/1200
N MFCOR=(MFCAP*WR*(CDEL-TAMFC))/1200
C CDEL=24
NOTE
NOTE  VALUATION OF THE FIRM
NOTE
A CAPEMP.K=LTDEBT.K+SCAP.K+CRP.K
L LTDEBT.K=LTDEBT.J+DT*(RALTD.JK-RRLTD.JK)
N LTDEBT=(GEAR/100)*(SCAP+CRP)
A SCAP.K=ISCAP
C ISCAP=31,21E06
L CRP.K=CRP.J+DT*ATPL.J
N CRP=ICRP
CP ICRP=15,01E06
L MCNP.K=MCNP.J+DT*(RCMTNP.JK-RAVP.JK)
R RCMTNP.KL=ARCMTNP.K
N MCNP=SFCOR*CCSFC+MFCOR*CCMFC
L VP.K=VP.J+DT*(RAVP.JK-RLVP.JK)
N VP=(CCMFC*MFCAP+CCSFC*SFCAP)/(1+GEAR/100)
R RLVP.KL=(WR*VP.K)/1200
R RAVP.KL=DELAY3(RCMTNP.JK,CDEL)
N RCMTNP=SFCSR*CCSFC+MFCSR*CCMFC
L WDV.K=WDV.J+DT*(RAVP.JK-DCFL.J)
N WDV=VP
NOTE
NOTE CASH BALANCE SECTOR
```

```
NOTE
L CASH.K=CASH.J+DT*(ACGR.J-RSC.JK-RRLTD.JK+RRWC.JK)
N CASH=CAPEMP-VP-WCAP-MCNP
A FCSR.K=ACGR.K+(CASH.K/PHOR)
R RSC.KL=MIN(RSAA.K,FCSR.K)
A CDIFF.K=RSAA.K-FCSR.K
A CHECK.K=CAPEMP.K-WDV.K-WCAP.K-MCNP.K-CASH.K
NOTE
NOTE WORKING CAPITAL
NOTE
L WCAP.K=WCAP.J+DT*(ARWC.J-RRWC.JK)
N WCAP=LTPUT*WCPUP
NOTE
NOTE  METAL MARKET SECTOR
NOTE
A ID.K=IMSF*MDMD.K
C IMSF=0.35
A ISD.K=MSIM*ID.K
C MSIM=0.45
A UMN.K=LTPUT.K
A TMN.K=UMN.K+ISD.K
A MBOPC.K=(TMN.K-CCAP.K)/TMN.K
L CCAP.K=CCAP.J+DT*PULSE(3000/DT,LENGTH+10,100)
N CCAP=6000
A UMC.K=COM*(1-MBOPC.K)+MBOPC.K*CBM
C COM=150
C CBM=250
NOTE
NOTE  MARKET SHARE SECTOR
NOTE
L UMS.K=UMS.J+DT*(RGMS.JK-RDMS.JK)
N UMS=0.25
L AMSS.K=AMSS.J+(DT/TAMSS)(MSS.J-AMSS.J)
N AMSS=5.E04
C TAMSS=12
R RDMS.KL=UMS.K/ML
C ML=36
R RGMS.KL=TABHL(TMSGR,AMSS.K,0,2E05,5E04)
T TMSGR=0/.00694/.008/.0085/.009
A RMS.K=LTPUT.K/MDMD.K
A MSE.K=PMS.K-RMS.K
A TMSS.K=TABHL(TTMSS,MSE.K,-0.1,0.1,0.05)
T TTMSS=5E04/5E04/5E04/10E04/20E04
A MSS.K=TMSS.K*MIN(1,FPMULT.K*MSSPF)
C MSSPF=1.5
C IMSS=5E04
NOTE
NOTE OUTPUT CONTROL SECTOR
NOTE
A PRTPER.K=1+STEP(9,11)
SPEC DT=0.0625/LENGTH=120/PLTPER=1
PLOT MDMD=M(0,50E03)/UOR=U,SFCAP=S,MFCAP=P(0,20E03)/
X FRAT=F(0,2)/ROI=R(0,20)/CAPEMP=C,LTDEBT=D(0,100E06)/ATPL=A(-100E03,100
X E03)/ARS=£(0.25E06,1.25E06)
X1  /CASH=C(0,1E06)
PRINT 1)MDMD,UOR,OBS,TOB,IPUT,TPUT,LTPUT,UMC
PRINT 2)PR,APL,ATPL,CUPROF,CRP,ROI,ROIRAT
PRINT 3)ARS,RRS,FRAT,FPMULT,WCMULT,RSAA,RCMTNP,CDIFF,FCSR
PRINT 4)ECG,ACGR,MXLTD,DCFL,MCNP,IWC,LTDIFF
PRINT 5)WCAP,WCAPR,WCRAT,,VP,LTDEBT,CAPEMP,WDV,SCAP,RAVP
PRINT 6)CASH,RRLTD,RALTD,RSC,CHECK,RRWC,ARWC
PRINT 7)CCAP,MBOPC,UMN,TMN,PMS,UMS,RMS,MSS,SFCDR
PRINT 8)SFCAP,ISFCAP,TROSFC,SFCSR,SFCCR,SFCOR,MFCAP,UPC,TROMFC
RUN STEP1
C ICRP=14.65E06
RUN NOCASH
T TFPMLT=0/0.6/0.9/1.0/1.0
RUN CONSER
T TFPMLT=0/0.5/1.0/1.5/2.0
RUN PROPOR
C GFAC=.25
RUN DIVERSIFY
C GFAC=-.25
C MSGF=1.25
RUN SEMIFAB
C MSGF=0.7
RUN GETTINGOUT
C DNRSF=0
C DNRMF=0
RUN NO REPLACEMENT
C DNRSF=1
C DNRMF=1
```

```
C MSGF=1.0
C GFAC=0
C SFPFAC=0.2
C MFPFAC=0.2
RUN 20% INCREASE
C MSGF=1.25
C GFAC=0.25
RUN 25PC GROWTH + 20% PRICE RISE
C MSGF=1.0
C GFAC=0
C SFPFAC=0
C MFPFAC=0
C WCPF=1
C MSSPF=1
RUN NO PREFERENCE
C SFPFAC=.1
C PRSTM=0
RUN NOPREF + SF PRICE RISE NOW
D ACGR=(£/M) ACTUAL CASH GENERATION RATE
D AMSS=(£/M) AVERAGE MARKET SHARE SPENDING
D APL=(£/M) AVERAGE PROFIT LEVEL
D ARCMTNP=(£/M) ACTUAL RATE OF COMMITTING MONEY TO NEW PROJECTS
D ARS=(£/M) ACHIEVABLE RATE OF SPENDING ON CAPACITY AND WORKING
X           CAPITAL ADDITIONS
D ARWC=(£/M) ADDITION RATE TO WORKING CAPITAL
D ATPL=(£/M) AFTER TAX PROFIT LEVEL.
D BPC=(%/Y) BASE VALUE FOR CALCULATION OF DEPRECIATION
D BVAL=(1) BASE VALUE FOR RELATING DEMAND FORECAST ERRORS TO LENGTH OF
X          PLANNING HORIZON
D CAPEMP=(£) CAPITAL EMPLOYED IN THE BUSINESS
D CASH=(£) CASH RESERVES IN COMPANY
D CBM=(£/T) COST OF BOUGHT-OUT METAL
D CCAP=(T/M) CONVERTER CAPACITY
D CCMFC=(£/T/M) CAPITAL COST OF MANUFACTURED-METAL CAPACITY
D CCSFC=(£/T/M) CAPITAL COST OF SEMI-FAB CAPACITY
D CDIFF=(£/M) DIFFERENCE BETWEEN SPENDING AND CASH
D CHECK=(£) DUMMY VARIABLE TO CHECK BALANCE SHEET ADDITION
D CDEL=(M) CONSTRUCTION DELAY IN CAPACITY ADDITION
D COM=(£/T) COST OF OWN PRODUCED METAL
D COVER=(M) COVER OF PRODUCTION REQUIRED IN BACKLOG
D CRP=(£) CUMULATIVE RETAINED PROFITS
D CUPROF=(£) CUMULATIVE PROFITS OVER SIMULATED PERIOD
D DCFL=(£/M) DEPRECIATION CASH FLOW
D DEPPC=(1/M) FRACTION OF VALUE OF PLANT ALLOWABLE FOR DEPRECIATION
D DNPMF=(1) ZERO-ONE SWITCH TO TEST EFFECTS OF NOT
X           REPLACING WORN OUT MANUFACTURING CAPACITY
D DNRSF=(1) ZERO-ONE SWITCH TO TEST EFFECTS OF NOT REPLACING
X           WORN-OUT SEMIFAB CAPACITY
D DT=(M) SOLUTION INTERVAL IN SIMULATION CALCULATIONS
D EBR=(1) ERROR AND BIAS FACTOR IN MODELLING OF LONG-TERM DEMAND
X         FORECASTS
D ECG=(£) EXPECTED CASH GENERATION OVER PLANNING HORIZON
D FCSR=($/M) FEASIBLE CASH SPENDING RATE
D ERPH=(1) EFFECTS ON BVAL OF LENGTH OF PLANNING HORIZON
D FPMULT=(1) FRACTION OF PLANNED EXPENDITURE ACTUALLY INCURRED
X            THIS IS THE KEY VARIABLE IN THE FINANCIAL POLICY
X            SECTOR OF THE MODEL
D FRAT=(1) RATIO OF ACHIEVABLE RATE OF SPENDING AND REQUIRED RATE
D GEAR=(%) PROPORTION OF ACTUAL ASSETS WHICH CAN BE TAKEN AS LONG-TERM
X          DEBT
D GFAC=(1) GROWTH FACTOR FOR UPC-THE UNITED PERCENT
D GMMG=(£/T) GROSS MARGIN ON METAL SOLD AS MANUFACTURES (INCLUDES GMSF)
D GMSF=(£/T) GROSS MARGIN ON METAL SOLD AS SEMI-FABRICATED
D IBVAL=(1) FIRST VALUE IN ERROR AND BIAS TABLE
D ICRP=(£) INITIAL VALUE FOR CUMULATIVE RETAINED PROFITS
D ID=(T/M) TOTAL DEMAND FOR INGOT METAL
D IGMSF=(£/T) BASE VALUE FOR GROSS MARGIN IN SEMI-FAB
D IGMMG=(£/T) BASE VALUE FOR GROSS MARGIN IN MANUFACTURING
D IMFCAP=(T/M) INDICATED MANUFACTURING CAPACITY BASED ON INDICATED
X              SEMI-FAB CAPACITY AND PLANNED PERCENT UNITED
D IMSF=(1) PROPORTION OF TOTAL METAL DEMAND WHICH IS FOR
X          INGOT METAL
D IMSS=(£/M) INITIAL VALUE OF MARKET SHARE SPENDING
D INTPC=(%/Y) INTEREST RATE ON CASH RESOURCES
D IPCU=(1) INDICATED VALUE OF UNITED PERCENT
D IPUT=(T/M) INDICATED LEVEL OF THROUGHPUT IN SEMI-FABRICATION
D ISCAP=(T/M) INDICATED VALUE OF SEMI-FAB CAPACITY
D ISD=(T/M) UNITED'S INGOT SALES DEMAND
D ISFCAP=(T/M) INDICATED SEMI-FAB CAPACITY BASED ON DEMAND FORECASTS
D IWC=(£) INDICATED VALUE OF WORKING CAPITAL TO SUPPORT
X       PLANNED GROWTH IN OUTPUT
```

```
D LENGTH=(M) SIMULATED PERIOD
D LTDEBT=(£) LONG-TERM DEBT OF COMPANY
D LTDIFF=(£) AMOUNT BY WHICH LONG TERM DEBT COULD BE EXPANDED
D LTDFC=(T/M) DEMAND FORECAST FOR END OF PLANNING HORIZON
D LTPUT=(T/M) AVERAGE LEVEL OF SEMI-FAB THROUGHPUT
D MBOPC=(1) FRACTION OF TOTAL METAL NEEDS WHICH IS BOUGHT OUT
D MCNP=(£) MONEY COMMITTED TO NEW CAPACITY ADDITION PROJECTS
D MDMD=(T/M) TOTAL DEMAND FOR METAL IN UK
D MFCAP=(T/M) MANUFACTURED-METAL CAPACITY
D MFCCR=(T/M/M) RATE OF COMPLETION OF MANUFACTURED-METAL
X                 CAPACITY
D MFCDR=(T/M/M) RATE OF DECAY OF MANUFACTURED-METAL CAPACITY
D MFCOR=(T/M) MANUFACTURED-METAL CAPACITY ON ORDER
D MFCSR=(T/M/M) MANUFACTURED-METAL CAPACITY START RATE
D MFD=(M) PRODUCTION DELAY IN MANUFACTURING
D MFPFAC=(1) FRACTION CHANGE IN GROSS MARGIN IN
X              MANUFACTURING OPERATIONS
D ML=(M) LIFETIME OF A GIVEN MARKET SHARE POSITION
X          (FIRST-ORDER DELAY)
D MSE=(1) MARKET SHARE ERROR BETWEEN PLANNED AND ACTUAL
D MSGF=(1/) MARKET SHARE GROWTH FACTOR REQUIRED
D MSIM=(1) UNITED'S SHARE OF UK INGOT MARKET.
D MSS=(£/M) ACTUAL MARKET SHARE SPENDING
D MSSPF=(1) MARKET SHARE SPENDING PREFERENCE FACTOR.
X             FACTOR TO REFLECT EMPHASIS GIVEN TO NOT REDUCING
X             MARKET SHARE SPENDING IF AT ALL POSSIBLE
D MXLTD=(£) MAXIMUM PERMISSIBLE LONG-TERM DEBT
D OBS=(T) SIZE OF UNITED'S ORDER BOOK
D PHOR=(M) PLANNING HORIZON IN MODEL
D PLTPER=(M) PLOTTING INTERVAL
D PMS=(1) PLANNED MARKET SHARE OF SEMI-FAB MARKET
D PPCU=(1) PLANNED VALUE OF PERCENT OF SEMI-FAB OUTPUT TO BE
X            RETAINED FOR MANUFACTURE BY UNITED COMPANIES
D PR=(£/M) RATE OF RECEIPT OF PROFIT FROM PRODUCTION
D PRSTM=(WK) TIME AT WHICH PRICE RISES COME INTO EFFECT
D PRTPER=(M) PRINTING INTERVAL FOR OUTPUT
D RALTD=(£/M) RATE OF ADDING TO LONG-TERM DEBT
D RAVP=(£/M) RATE OF ADDING TO VALUE OF PLANT
D RCMTNP=(£/M) RATE OF COMMITTING MONEY TO NEW PROJECTS
D RDMS=(1/M) RATE OF DECLINE OF MARKET SHARE
D RGMS=(1/M) RATE OF GROWTH OF MARKET SHARE
D RLVP=(£/M) RATE OF LOSS OF VALUE OF PLANT
D RMS=(1) REAL MARKET SHARE
D ROI=(%) RETURN ON INVESTMENT
D ROIRAT=(1) RATIO OF ACTUAL AND TARGET RETURN ON INVESTMENT
D RRLTD=(£/M) RATE OF REPAYMENT OF LONG TERM DEBT
D RRS=(£/M) REQUIRED RATE OF SPENDING ON CAPITAL PROJECTS
D RRWC=(£/M) RATE OF REDUCING WORKING CAPITAL IF
X              IT EXCEEDS REQUIREMENTS
D RSAA=(£/M) RATE OF SPENDING ACTUALLY AUTHORISED
D RSC=(£/M) RATE OF SPENDING CASH
D SCAP=(£) SHAREHOLDERS CAPITAL EMPLOYED
D SFCAP=(T/M) SEMI-FINISHED CAPACITY
D SFCCR=(T/M/M) RATE OF COMPLETION OF SEMI-FAB CAPACITY
D SFCDR=(T/M/M) RATE OF DECAY OF SEMI-FAB CAPACITY
D SFCOR=(T/M) SEMI-FAB CAPACITY ON ORDER BUT NOT RECEIVED
D SFCSR=(T/M/M) RATE OF COMMENCEMENT OF SEMI-FAB CAPACITY
D SFD=(M) SEMI-FABRICATION PRODUCTION DELAY
D SFPFAC=(1) FRACTION CHANGE IN GROSS MARGIN IN SEMI-FAB
D STM=(M) STEP TIME
D TAMFC=(M) TIME TO ADJUST MANUFACTURED-METAL CAPACITY
D TAMSS=(M) TIME TO AVERAGE MARKET SHARE SPENDING
D TAOB=(M) TIME TO ADJUST ORDER BACKLOG IN S-F PRODUCTION PLANNING
D TAP=(M) TIME TO AVERAGE PROFITS LEVEL
D TASFC=(M) TIME TO ADJUST SEMI-FAB CAPACITY
D TATP=(M) TIME TO AVERAGE THROUGHPUT
D TAWC=(M) TIME TO ADJUST WORKING CAPITAL POSITION
D TAXDPC=(%) PROPORTION OF PROFIT FLOW LOST IN TAXATION AND DIVIDEND
X              PAYMENTS
D TERPH=(1) TABLE FOR TEST VALUES OF ERPH
D TFPMLT=(1) TABLE VALUE FOR FINANCIAL POLICY MULTIPLIER FPMULT
D TIME=(M) SIMULATED TIME IN MODEL
D TIWC=(£/M) TARGET RATE OF GROWTH IN WORKING CAPITAL
D TMN=(T/M) UNITED'S TOTAL METAL NEEDS
D TMSGR=(1/M) TABLE GIVING MARKET SHARE GROWTH RATE IN TERMS OF
X               AVERAGE MARKET SHARE SPENDING,SCALED TO MATCH
X               NORMAL MARKET SHARE DECAY WHEN MARKET SHARE
X               IS 25% AND SPENDING AVERAGES £50000/M
D TMSS=(£/M) TARGET MARKET SHARE SPENDING
D TOB=(T) TARGET ORDER BACKLOG
D TPUT=(T/M) THROUGHPUT IN MANUFACTURED-METAL SECTOR
```

```
D TROI=(1) TARGET RETURN ON INVESTMENT
D TROMFC=(T/M/M) TARGET RATE OF ORDERING MANUFACTURED-METAL
X                    PRODUCTION CAPACITY
D TROSFC=(T/M/M) TARGET RATE OF ORDERING SEMI-FABRICATED METAL
X                    PRODUCTION CAPACITY
D TRWC=(M) TIME TO REDUCE WORKING CAPITAL WHEN IT EXCEEDS
X            REQUIREMENTS FOR PLANNED THROUGHPUT
D TTMSS=(£/M) TABLE OF TARGET MARKET SHARE SPENDING IN TERMS OF
X               MARKET SHARE ERROR
D UMC=(£/T) AVERAGE COST PER TON OF INGOT METAL
D UMN=(T/M) UNITEDS INGOT NEEDS FOR OWN OPERATIONS
D UMS=(1) UNITED'S MARKET SHARE OF MANUFACTURED-METAL
X          DEMAND
D UOR=(T/M) UNITED'S ORDER RATE AT SEMI-FAB STAGE
D UPC=(1) 'UNITED PERCENT,' - ONE OF THE KEY VARIABLES IN THE MODEL.
X           THE FRACTION OF SEMI-FAB PRODUCTION WHICH IS RETAINED BY
X           UNITED FOR ITS OWN MANUFACTURING ACTIVITIES
D VP=(£) ACTUAL VALUE OF THE PRODUCTION PLANT
D WCAP=(£) WORKING CAPITAL EMPLOYED
D WCAPR=(£) WORKING CAPITAL REQUIRED TO SUPPORT INDICATED PRODUCTION
D WCMF=(£/T/M) WORKING CAPITAL REQUIRED PER UNIT OF PRODUCTION IN
X        MANUFACTURING
D WCMULT=(1) MULTIPLIER TO REGULATE THE GROWTH OF WORKING
X              CAPITAL AS COMPARED TO ITS TARGET GROWTH
D WCPF=(1) MULTIPLIER TO REFLECT RELATIVE PRIORITY GIVEN TO
X            GROWTH IN WORKING CAPITAL AS COMPARED TO GROWTH IN
X            INVESTMENT IN PRODUCTION FACILITIES
D WCPUP=(£/T/M) WORKING CAPITAL REQUIRED PER UNIT OF PRODUCTION
X                 THROUGHPUT,ALLOWING FOR PROPORTION OF PRODUCTION
X                 WHICH GOES ON TO THE MANUFACTURED-METAL STAGE
D WCRAT=(1) RATIO OF ACTUAL WORKING CAPITAL TO THAT REQUIRED TO
X             SUSTAIN THE PLANNED LEVEL OF PRODUCTION IN
X             SEMI-FABRICATION
D WCRCV=(1) CRITICAL VALUE OF WORKING CAPITAL RATIO,WCRAT,
X             ABOVE WHICH REDUCTIONS IN WORKING CAPITAL WILL BE MADE
D WCSF=(£/T/M) WORKG CAPITAL REQUIRED PER UNIT OF THROUGHPUT IN
X                IN SEMI-FABRICATION
D WDV=(£) WRITTEN-DOWN VALUE OF CAPITAL PLANT
D WR=(%/Y) RATE OF PHYSICAL WEAR OF PRODUCTION PLANT
+
```

Appendix C

A Bibliography on System Dynamics

C.1 System Dynamics in General

Books:

Forrester, J. W., *Industrial Dynamics*, MIT Press, 1964
This is the original text in the area, and is very illuminating reading. It is a development from the same author's pioneering paper 'Industrial Dynamics. A major breakthrough for Decision Makers', Harvard Business Review, **36**, No. 4., July–August 1959.

Forrester, J. W., *Principles of Systems*, Wright Allen Press, 1972.
This is an incomplete version of what the author intends to be a text on the properties of feedback loops, written at an elementary level. It will be useful to the new student who has little or no mathematical background.

Ratnatunga A. K., *DYSMAP Users Manual*, preliminary edition, University of Bradford, System Dynamics Research Group, 1975.
An essential item which describes simulation and the use of DYSMAP

Pugh, A. L., *DYNAMO II Users Manual*, MIT Press.
Describes the use of the DYNAMO language.

Coyle, R. G., Sharp, J. A., *System Dynamics Problems and Cases,* University of Bradford, 1977.
A selection of graded problems with solutions, case studies, and research projects.

Jarmain, W. E., *Problems in Industrial Dynamics*, MIT Press, 1963 Contains several graded problems.

Journal:

DYNAMICA, ISSN 0306–7564, published by System Dynamics Research Group, University of Bradford, thrice yearly.

Papers:

Ansoff, M. I. and D. P. Slevin, 'An Appreciation of Industrial Dynamics', *Management Science,* **14**, 383, 1968.

Forrester, J. W., 'A response to Ansoff and Slevin', *Management Science,* **14**, 601, 1968.

These two papers reflect a debate about the value of the System Dynamics approach which was based on the work done up to 1967 in the USA.

Forrester, J. W. 'Industrial Dynamics. After the First Decade', *Management Science,* **14**, 83, 1968.

An interesting discussion of the likely future development of the subject.

Three earlier papers by Professor Forrester had examined some of the implications of treating management problems from a control systems viewpoint.

'The Structure Underlying Management Processes', Proceedings of the 24th Annual Meeting of the Academy of Management, December, 1964.

'A new avenue to management', *Technology Review,* **LXVI**, No. 3, January 1964.

'Common Foundations Underlying Engineering and Management', *JEEE Spectrum*, September 1964.

Other references on System Dynamics in general include:

Coyle, R. G. (1973) 'On the scope and purpose of Industrial Dynamics', *Int. J. Systems Sci.,* 397–406.

Coyle, R. G., *Some Empirical Evidence on the application of Industrial Dynamics*, Unpublished Paper, University of Bradford, System Dynamics Research Group.

C.2 System Dynamics and Corporate Planning

Books:

Nord, O. C., *Growth of a new product. Effects of Capacity–acquisition policies*, MIT Press, 1963.

Packer, D. W., *Resource Acquisition in Corporate Growth,* MIT Press, 1964.

These two early texts were based on generalised models of firms, rather than actual case studies.

General Papers:

Coyle, R. G., 'System Dynamics: An approach to Policy Formulation', *Journal of Business Policy*, Spring 1973.

Swanson, C. V., 'Design of Resource Control and Marketing Policies Using Industrial Dynamics', *Industrial Management Review*, Spring 1969.

Swanson, C. V., 'Information and Control for Corporate Growth', *Sloan Management Review*, Spring 1971.

Coyle, R. G., 'A System view of Forecasting', in H. A. Gordon, (Ed) *Practical Aspects of Forecasting*, Operational Research Society, 1975.

These four papers treat general problems of corporate policy rather than specific case studies (cf. chapter 12).

Industry Models:

Hill, G. W., *A dynamic model of investment by the chemical and petrochemical industries,* M.Sc. Dissertation, University of Bradford, 1972.

Finney, M. J., *The World Petroleum Industry*, M.Sc. Dissertation, University of Bradford, 1972. A general discussion of the industry leading to a very simple model.

Commodity Studies:

Ballmer, R. W., *Copper Market Fluctuations. An Industrial Dynamics Study*, M.S. Thesis, MIT 1960.

Popper, J., *Buffer Stocks for Stabilisation of Commodity Markets*, M.S. Thesis, MIT, 1971.

Meadows, D. L., *Dynamics of Commodity Production Cycles*, Wright–Allen Press, 1970.

Coyle, R. G., *A dynamic model of the Copper Industry. Some Preliminary Results*, Proceedings of 9th International Symposium on the Application of Computers in the Mineral Industry, in Johannesburg.

Shimada, T., 'Industrial Dynamics Model of Weekly Stock Prices. A Case Study', *Bulletin of the Izumi Laboratory of Meiji University*, No. 42, 1968.

Weymar, F. H., *The Dynamics of the World Cocoa Market,* MIT Press, 1968.

C.3 Social and Economic Systems

System Dynamics techniques have been applied to several social and economic problems. This has often resulted in huge controversy followed, when the fuss has died down, by extensive research efforts. There is a very large literature on these subjects which can only be summarized very briefly.

Hamilton, H. R. *System Simulation for Regional Analysis,* MIT Press, 1963.

In many ways this is far and away the best of this literature, mainly because it shows just what can be achieved by a properly integrated approach in which dynamic modelling plays a complementary role to economic and statistical analysis.

The next attempt was to model city problems and the early literature has a very controversial tone. The basic references are

Forrester, J. W., *Urban Dynamics*, MIT Press, 1969.

Mass, N. J. (Ed), *Readings in Urban Dynamics*, Wright–Allen Press, 1974.
This contains a paper which deals with some of the criticisms of *Urban Dynamics.*

Undoubtedly the greatest controversy has arisen over the attempt to model the world social system.

Forrester, J. W., *World Dynamics,* Wright–Allen Press, 1971.
Meadows, D. H. *et al. The Limits to Growth*, Earth Island Ltd., 1972.
Meadows, D. L. (Ed), *Towards Global Equilibrium, Collected Papers*, Wright–Allen Press, 1973.
Meadows, D. L. *et al. Dynamics of Growth in a Finite World*, Wright–Allen Press, 1974.

A critique appears in

Cole, M. S. D. *et al. Thinking about the Future*, Chatto and Windus, 1973.

and in

Nordhaus, W. D., 1973, 'World Dynamics. Measurement without Data', *Economic Journal*, December, 1156–1183.

A very important area which has had far too little attention is the problem of growth in underdeveloped economies. There are, however,

Forrester, N. B., *The Life-Cycle of Economic Development*, Wright–Allen Press, 1972.
Thompson, R., *An Enquiry into the Application of System Dynamics to investment appraisal problems in Developing Economies*, M.Sc. Dissertation, University of Bradford, 1974.

Thompson's dissertion contains a very comprehensively documented analysis of the 'conventional' economic approaches to development and presents a simple dynamic model.

C.4 System Theory

Very little attention has been paid to problems of aggregation and other theoretical matters in System Dynamics methodology. By far the best treatment so far is

Sharp, J. A., *A Study of some problems of System Dynamics Methodology,* Ph.D. Thesis, Bradford University 1974.

Other references are:

Senge, P. M., *The use of additively and Multiplicatively Separable Functions in System Dynamics*, Ph.D Thesis, MIT 1972.
Goodman, M. R., *Elementary System Dynamics Structures,* M.S. Dissertation, MIT 1972.

C.5 Research and Development Management

This, and some related areas have had a fair amount of attention.

Roberts, E. B., *The Dynamics of Research and Development,* Harper and Row, 1964.

Kelly, T. J., *The Dynamics of R and D Project Management*, M.S. Dissertation, MIT 1970.

Nay, J. E., *Choice and Allocation in Multiple Markets. A Research and Development System Analysis*, M.S. Dissertation, MIT 1965.

Pathak, S. U., *Systems Analysis of the Process of Implementing an Innovation*, M.S. Dissertation, MIT 1968.

Perrine, C. H., *The Dynamics of Transition in a large Goverment Research and Development Centre*, M.S. Dissertation, MIT 1968.

Weil, H. B., *Work-load Fluctuations in Management Consulting Firms*, M.S. Dissertation, MIT 1965.

C.6 Functional Problems in Business

Apart from the case studies in this text, and references already quoted, there are comparatively few published reports of the application of System Dynamics to real problems in business and industry. Notable examples are:

Sharp, J. A. and R. G. Coyle, *The Use of a System Dynamics Model for the Redesign of a Production Planning System*, Unpublished paper of the University of Bradford System Dynamics Research Group, 1974.

Roberts, E. B., *Systems Analysis of Apparel Company Problems*, Apparel Research Foundation Conference, November 1967.

Swanson, C. V. and A. C. Thorsten, 'A System Dynamics Design and Implementation of Inventory Policies', *Sloan School of Management, Working Paper,* 539–571.

Wright, R. D., 'An Industrial Dynamics Implementation. Growth strategies for a trucking firm', *Sloan Management Review*, Autumn 1971.

Roberts, E. B. *et al.* 'A systems study of Policy Formulation in a vertically integrated firm', *Management Science*, Aug. 1968.

Barnett, A. B., *A System Dynamics model of an Oilfield's Development*, Ph.D. Thesis, University of Bradford, 1973.

Hughes, K., *Dynamic Model of an Industrial System Subject to Extreme Seasonal Cycles*, M.Sc. Dissertation, University of Bradford, 1971.

For further information on System Dynamics in general see the System Dynamics Newsletter, published annually by the MIT System Dynamics Group.

C.7 Agriculture

Seddon, V., *Simulation Study of a Management of a Small Farm*, M.Sc. Dissertation, University of Bradford, 1975.

C.8 Control Theory Texts

There is a very large literature on the mathematics of control engineering. The reader should shop around until he finds one which suits his mathematical bent. Some suggestions are

Gille, J-C., M. J. Pelegrin and P. DeCaulne, *Feedback Control System*, McGraw-Hill, 1959.

DiStefano *et al.*, *Feedback and Control Systems*, Schaum Publishing Co., 1967.

Elgerd, O. I. *Control Systems Theory,* McGraw-Hill, 1967. A thoroughly excellent book which is highly recommended.

Saucedo, R. and E. E. Schiring, *Introduction to Continuous and Digital Control Systems*, Macmillan, 1968.

Watkins, B. O., *Introduction to Control Systems*, Macmillan, 1969.

The reader will also find considerable benefit from a careful reading of Stafford Beer's various books, notably *Decision and Control*
and *Brain of the Firm.*

Appendix D

Training in System Dynamics

Throughout this book we have laid repeated stress on the importance of training for people involved in SD work. This includes both technical training of analysts and appreciation training for managers. We now discuss some outline training programmes for various situations. These are simply suggestions, based on some experience of teaching SD in a variety of circumstances. The experienced teacher of management science should have no real problem in concocting his own course from these suggested outlines. In all cases, we have found an overhead projector to be a valuable visual aid. (The diagrams in this book can be made into projector overlays by Xerox reproduction, and may be used for educational purposes with due acknowledgement of copyright.)

Programmes D1, D2 and D3 are based on sessions of two hours duration. This allows about half an hour for questions and discussion, spread over the session, or concentrated at its end, according to preference and circumstance. Some of the programmes require supplementing by tutorials, computer programming practice etc, but his is left to the teacher. The problems in *System Dynamics Problems and Cases*, referenced in Appendix C, may be helpful as a teaching supplement, but see also M. R. Goodman *Study Notes in System Dynamics* Wright–Allen Press, 1976.

D.1 Management Appreciation Course (two sessions)

The aim is to present SD as a view of the firm (session 1) and as a practical problem-solving technique (session 2).

Session 1

1. Occurrence of dynamic behaviour (DB) and reasons for concern.
2. Cause of DB as external shocks plus internal actions, which may make DB worse if not designed to fit the system.
3. Conservation equation for inventory to show *choice* of production.
4. Basic feedback loop of Fig. D.1.

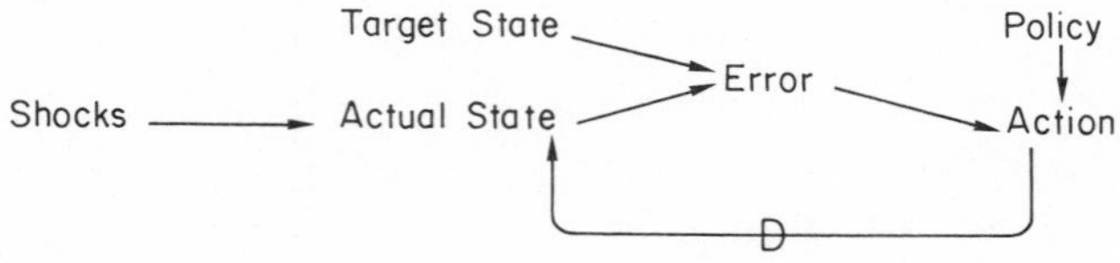

Fig. D.1

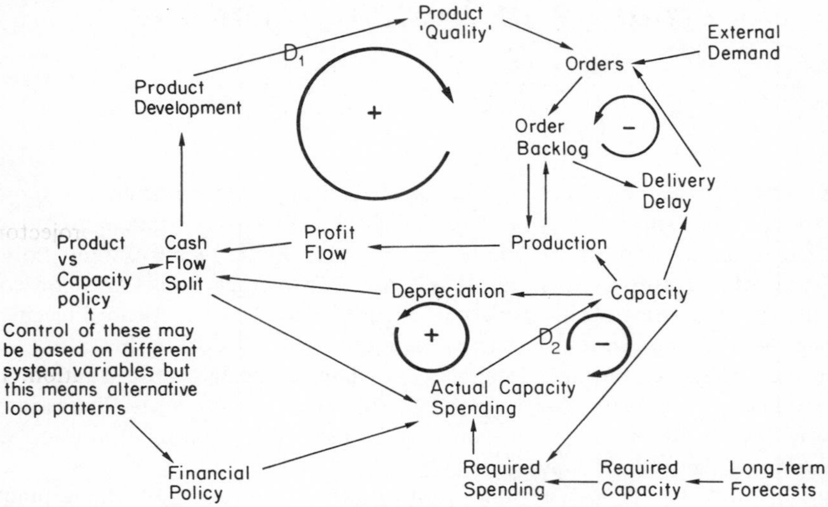

Fig. D.2

5. Emphasize importance of delays and that states may be forecasts.
6. System example from section 2.12.
7. Ideas of negative and positive loops and possibility of conflict.
8. Major types of DB related to + and – loops.
9. Problem of forecasting and system sensitivity (Chapter 9).
10. General model of a firm, as shown in Fig. D.2.
11. Emphasize multiple-loop and possibility of conflict. Need to design policies and idea of robustness. Option of forecasting different variables for capacity requirements. Stress freedom to change system structure and need for subtle, and qualitative, measures of system performance. (Alternatively model in Chapter 9.)

Session

Treatment of a practical case study – Chapter 11 is usually effective.

For in-house training, these two sessions may be followed up by more detailed study of home-grown problems. The case study in session 2 should not be from the firm itself, as two hours is not long enough for all the detailed comment this usually brings out, and one needs to broaden horizons without getting lost in minor details.

D.2 Short Postgraduate Course (five Sessions)

This is an introduction to SD, intended to be taught as an integral part of a course in management science for MBA students.

1. Occurrence of dynamic behaviour. Effects of managerial policy. SD as a policy-design approach. The basic control loop and brief ideas of Levels and Rates Formulation of production system from Section 2.13 to get students involved in simulation from the start.
2. Influence diagramming, list extension and, briefly, coherence.
3. Outline of DYSMAP language. Positive and negative feedback processes. Properties of feedback loops (introduction).
4. A case study.
5. Dynamic processes in the firm, robustness and practical application of SD (Chapters 9 and 13).

D.3 Long Postgraduate Course (20 Sessions)

Intended for students reading for masters degrees in management science, or in an MBA with larger proportion of electives. Two courses are intended – 'Analysis of Corporate Control Systems' and 'Cases in Corporate Control Systems'.

'Analysis' Course

1. Dynamic behaviour–causes and problems. The basic loop.
2. Idea of controller/environment/complement. Levels, Rates, Auxiliaries.
3. Production System from Section 2.12. DYSMAP to start simulation of Section 2.12.
4. Influence diagrams.
5. Introduction of DMC Ltd as a case study and a student project.
6. Equation formulation (Chapter 5).
7. Properties of feedback loops.
8. Analysis of system structure (Chapter 8).
9. Dynamic processes in the firm and practical SD.
10. Review of student project.

'Cases' Course

This has to be done in small groups and very flexibly. The aim is to develop the student's confidence and breadth of understanding. The course can be run either as a classroom study, backed up by simulation experiments by the student, of existing case studies such as Chapters 11 and 12, or as the student's own model of a suitable project. Take care not to try to do too much.

D.4 Practitioners Course

This is based on the programme run at Bradford University for new research students and staff. It lasts two weeks, and is supplemented by tutorials, For practicing management scientists it might require some modifications.

The outline of the programme is shown in the table, to which the following notes refer.

1. The Habitat project is based on a film of a programme shown in the 'This Week' series by Thames Television, on 22nd August, 1974, which comprised interviews with executives of the Habitat chain store company. The students identify the problem and formulate a model to study the reasons for Habitat's problems.
2. 'Foxes and Rabbits' is a simple ecological model which has some striking teaching implications. It illustrates the 'bootstrap method' of building a simple but plausible model without close observation of the system. It shows the general insensitivity of behaviour modes to parameter estimates and it broadens the students horizon away from purely business problems.
3. Stability Analysis is an introduction to state–space mathematics, supported by readings from standard texts.
4. In the case studies the essential factor is to prevent the student from simply battering away at simulation runs. In both cases he has to be led into analysing the structure of the model and the nature of the problem. Once he has grasped these, some simulation experiments can be done. Some of these are implied in the text but many more will suggest themselves.

Day 1	*Morning*		
			DYNAMIC BEHAVIOUR IN SOCIO-ECONOMIC SYSTEMS
			Instances of Dynamic Behaviour Causes of Dynamic Behaviour Business systems as control systems Feedback and Delays The SD approach to systems design.
		Reading	Chapter 1
	Afternoon		
			INFLUENCE DIAGRAMS
			The construction and use of influence diagrams
		Exercise	
			SD Examples 1–10 (The SD Examples are to be selected from 'System Dynamics, Problems and Cases', forthcoming from University of Bradford, System Dynamics Research Group.)
		Reading	Chapter 3

Day 2

PRINCIPLES OF SYSTEM STRUCTURE
The SD approach to modelling
Types of variable: LEVELS, RATES, AUXILIARIES, DELAYS, INITIAL CONDITIONS. Sortability
Production Planning System Model
Special DYNAMO functions TABLE, CLIP, etc.
Fox–Rabbits

Reading Chapters 2 and 4
DYSMAP Users Manual

Day 3 *Morning*

RUNNING DYSMAP PROGRAMS
Facilities of DYSMAP (plus terminal use and program editing as appropriate)

Afternoon

EQUATION FORMULATION
Common Equation Types
Sources of Equation Structure
Sortability
Dimensional Checking
Review Chapter 2 Production Planning Problem

Exercises

SD Examples 19–23
Formulate DYSMAP model for an example

Reading Chapter 5

Day 4 *Morning*

DETERMINATION OF PARAMETERS AND INITIAL CONDITIONS
Role of parameters and initial conditions in SD Models
Determination of initial conditions algebraically and by simulation
Determination of parameters by interview and from data
Simultaneous determination of initial conditions and data

Reading Chapter 6

Exercises

SD Examples 42–46.

Afternoon

Modelling Tutorial
SD Examples 27–41.

Day 5 *Morning*

MATHEMATICAL METHODS I
Formulation of DYSMAP equations as Ordinary Differential Equations
Reduction to State–Space Form
Solution of linear systems by Laplace transforms
Stability

Reading Stroud, *Laplace Transforms Programs and Problems*, Wiley, 1973.

Exercise

SD Examples 53–58.

Afternoon

PROBLEM SELECTION AND FORMULATION
Problem identification and project organization
Useful approaches in model formulation
Robustness

OBTAINING INFORMATION I
Habitat film
Problem formulation
Influence diagram construction
Equation formulation

Reading Chapter 13

Day 6 *Morning*

LOOP ANALYSIS EXERCISE
(An alternative approach is to devote 1½ hours per day for several days so that the student has time to do simulation runs and study the results.)

Reading Chapter 8

Afternoon

Sources of Information
Modelling tutorial.

Day 7 *Morning*

MATHEMATICAL METHODS II
Final Value Theorems
Spectral Analysis
Control Laws

Exercise

SD Examples 62–69.

Afternoon

LOOP ANALYSIS AND CHAPTER 8 SYSTEM

Day 8 *Morning*

CASES IN SYSTEM DYNAMICS
United Metals
Tanker Chartering

Reading Chapters 9, 11 and 12

Afternoon

MATHEMATICAL METHODS III
Linearization
Perturbation Analysis

Exercise

SD Examples 70–73.

Day 9 *Morning*

STAGES OF A PROJECT
WLG Case Studies

Reading WLG Case Series (See *System Dynamics Problems and Cases*)

Afternoon

MODEL TESTING AND METHODS OF SYSTEM REDESIGN
Qualitative and Quantitative comparisons of system and model outputs
Sensitivity Testing
Some mathematical approaches to redesign

Reading Industrial Dynamics, Chapter 13
WLG E and F cases

Day 10 *Morning*

OTHER APPLICATIONS
Oilfield Development
Chemical Plant Investment Cycle
Potential applications of interest to group members

Reading Industrial Dynamics, Chapter 20

Afternoon

DISCUSSION ON SCOPE AND PURPOSE OF SYSTEM DYNAMICS AND THE PRACTICAL PROBLEMS OF CARRYING OUT SD PROJECTS

Index

The reader should study the index so that he learns the general layout and the pattern of cross-indexing. See also the companion Index of Equations.

Page numbers in *italics* refer to particularly important references (e.g. definitions), or the commencement of text sections dealing with the point in more detail.

Index of Equations

This index is to help the reader to find *examples* of equations for particular situations. The reader is again cautioned that these are only examples and they should not be regarded as definitive forms which must *always* be followed. The reader will note that sometimes different equations forms are given for the same item in different parts of the book. This is intentional as different contexts call for different treatment.

The numbers in the index refer to the page in the text when they appear in ordinary type, to a particularly important discussion when they appear in *italics*, and to a line number in the sample programs in Appendix B when they appear in parentheses following a page number. In one or two cases the same line number appears more than once on a given page in Appendix B, but the context makes it clear which is intended.

For levels and rates, the main defining Sections are indexed, together with a general reference to Appendix B.

Note that the following groups of terms are used interchangeably and effectively mean the same thing. They are used in the book for variety of expression and to demonstrate the diversity of industrial practice.

(a) Inventory, stock.
(b) Production, throughput, production order rate, production start rate.
(c) Orders, Sales, consumption.

Depending on the degree of detail, one might have to distinguish between an 'order' when it is placed, and a 'sale' when the goods are delivered. The main thing is not to worry about the terminology but to use the equations intelligently as a guide to the situation being modelled.

Note also that terms such as Production *Rate* and Production *Level* are often used interchangeably in industry without implying the technical System Dynamics meaning of 'Rate' and 'Level'.